9/3
LNTC
33.95

SECOND EDITION

Technical Writing
and
Professional
Communication

For Nonnative
Speakers of English

Thomas N. Huckin
The University of Utah

Leslie A. Olsen
The University of Michigan

D0217419

McGraw-Hill, Inc.

New York   St. Louis   San Francisco   Auckland   Bogotá   Caracas   Hamburg
Lisbon   London   Madrid   Mexico   Milan   Montreal   New Delhi
Paris   San Juan   São Paulo   Singapore   Sydney   Tokyo   Toronto

For our parents, Christiane, Jed, and Neil

**TECHNICAL WRITING AND PROFESSIONAL COMMUNICATION
FOR NONNATIVE SPEAKERS OF ENGLISH**

1 2 3 4 5 6 7 8 9 0   DOC DOC   9 5 4 3 2 1 0

ISBN   0-07-030825-X

This book was set in Century Expanded by Monotype Composition Company.
The editors were Judith R. Cornwell and James R. Belser;
the designer was Wanda Siedlecka;
the production supervisor was Salvador Gonzales.
New drawings were done by Fine Line Illustrations, Inc.
R. R. Donnelley & Sons Company was printer and binder.

The cover illustration is a 3-dimensional graphic of a petro-chemical piping analysis generated
by a customized design/engineering software program.

Photo: Stephen Wilkes/The Image Bank.

Library of Congress Cataloging-in-Publication Data

Huckin, Thomas N.
    Technical writing and professional communication for nonnative
speakers of English / Thomas N. Huckin, Leslie A. Olsen. — 2nd ed.
        p.   cm.
    Rev. ed. of: English for science and technology / Thomas N.
Huckin, Leslie A. Olsen. c1983.
    Includes bibliographical references.
    ISBN 0-07-030825-X
    1. Technical writing.   2. English language—Textbooks for foreign
speakers.   I. Olsen, Leslie A.   II. Huckin, Thomas N.   English for
science and technology.   III. Title.
T11.H823   1991
808'.0666021—dc20   89-14590

Contents

PART FOUR
SPECIFIC APPLICATIONS   185

PART SIX
REVIEW OF GRAMMAR, STYLE, and VOCABULARY BUILDING 501

37   Informal Conversational Expressions   610

38   Pronunciation   627

Preface

While English is the international language of business and science, it is not the native language of many of the students and practitioners of these disciplines.

Technical Writing and Professional Communication for Nonnative Speakers of English, second edition, is the revision of *English for Science and Technology: A Handbook for Nonnative Speakers*. It is intended for those students and professionals in basic science, in the applied sciences such as engineering and medicine, in business and other technically oriented professions, and in technical and professional communications who have the command of English that has enabled them to learn and communicate in that language. Their level of proficiency with English has enabled them to attain their present positions in university or in industry. They can now benefit from focusing on the features of scientific and technical English that are known to be troublesome for them. We wrote this text to offer such a focus. The book emphasizes principles and use rather than usage—it offers functional explanations rather than formal rules.

OUR APPROACH

We have maintained the functional/rhetorical focus of the first edition, and believe it is the text's most important feature. That is, *Technical Writing and Professional Communication for Nonnative Speakers of English* emphasizes the communicative use of language rather than simply its formal aspects. Although we have treated the formal aspects of language where appropriate (in discussing report formats, grammar, punctuation, and visual elements) we have placed more emphasis on the psychological, social, and rhetorical principles underlying effective communication. In doing so, we have focused on language in meaningful contexts, not on sentences or words in isolation.

A second important feature of the book—and one which stems from its rhetorical approach—is its treatment of the early stages of writing. How does a writer think up something to say, find and define a topic, find appropriate things to say about a topic? In our experience, even if students or professional engineers have been assigned to do particular studies and to write reports about them (and thus have found topics), they often do not know how to set up and state a report's problem effectively for a given audience or how to define and apply criteria needed to solve the problem. These skills are critical for scientific and technical professionals, and Chapters 2, 3, 4, and 6 of the book deal explicitly with them in an approach strongly indebted to classical argument and to Richard E. Young, Alton Becker, and Kenneth L. Pike's *Rhetoric: Discovery of Change.*

As the third important feature of the book, we believe we have provided complete and explicit explanations of the main points. We believe that science, engineering, and other professional students prefer systematic, step-by-step instruction and that they like to have reasoned explanations for how things "work." It follows that this approach to how language functions in communication is the most effective for these students.

This point gives rise to yet another feature of the text, namely, its more overt emphasis on the *process* of producing an effective piece of communication for a given audience. Too often, we think, textbooks treat writing as a finished product. Models of good writing for the student to imitate are presented, but adequate guidance by which the student can learn to cope with novel situations on his or her own is not given. We have provided step-by-step procedures that students may follow as guides to the writing process, including a number of flow charts and newly added sections on prewriting, word processing, revising and testing, and footnoting and referencing. At the same time, we continue to emphasize the fact that technical communication is a *social* process in which *text conventions* play a major role.

NEW TO THIS EDITION

In addition to the sections mentioned previously, we have added other new and/or expanded features to reflect the rapidly changing environment in which technical communication occurs.

- Integrated throughout the text is a concern with international dimensions of modern communication that reflects the important fact that science, technology, and business are increasingly becoming global activities.

- Treatment of the ethical dimensions of communication is prompted by the rising interest in ethical concerns as reflected by the many scandals that have plagued government and corporate officials in recent history.

- A concern with multiple authorship and group activities reflects the increased prominence of writing and publication teams in organizational and private settings.

- Many case study exercises provide simulated group activities and organizational settings for students accustomed to working alone in academic environments.

- The treatment of computer-based writing in Chapter 5 provides guidance on how to use the dramatically increased availability of computer resources most effectively.

- The expanded coverage of genres in Part IV satisfies the need for treatment of instructions, procedures, and computer documentation; of oral communications such as those that occur in meetings and negotiations; and of theses and dissertations.

Finally, we hope that *Technical Writing and Professional Communication for Nonnative Speakers of English* will serve not only as a course text, but also as a long-term reference work. As such, it should be especially useful to both university-level students in technical areas who are including English in their studies and who plan to continue into technical careers; and practicing scientists and engineers who need a self-instructional reference book in written and oral English for technical communication.

- This edition contains a partial answer key provided as an aid to learning for both these groups of learners, as well as for their instructors.

- To further increase the book's usefulness, we have included many exercises, lists of supplementary readings, a reference appendix on punctuation, sample reports and letters, and indexes.

- Additionally, an Instructor's Manual is available which contains additional exercise material, suggestions for setting up the curriculum and course syllabus, advice on how to use different chapters, topics for class discussion, and other aids for instructors.

NOTE TO INSTRUCTORS

This text is divided into two basic parts. The first twenty-eight chapters (which are identical to those in its companion text, *Technical Writing and Professional Communication*) cover the basic principles of technical communication. These chapters can be used by both native and nonnative speakers of English. The remaining ten chapters and appendixes in this version for nonnative English speakers focus on the special problems of grammar, style, vocabulary, and pronunciation that nonnative speakers are known to have when using English. The two versions are supplied so that one or the other can be used when addressing both native and nonnative speakers in the same classroom.

THOMAS N. HUCKIN
LESLIE A. OLSEN

Acknowledgments

The preparation of this book has been a long but exciting project and one which would not have been possible without the help and encouragement of our families, friends, and colleagues, many professional acquaintances, and many students. We would like to thank all those who have contributed in one way or another to the completion of this book, but in particular we would like to thank the following people for their special contributions: Christiane Huckin for her patience and support, as well as for other invaluable behind-the-scenes assistance; our parents, grandparents, and siblings for their unflagging moral support; and our teachers and later colleagues—Heles Contreras, J. C. Mathes, Dwight W. Stevenson, Louis Trimble, Virginia J. Tufte, W. Ross Winterowd, Richard E. Young—for our early training in the field and their later influence and inspiration.

In addition, we would like to thank many others who have provided special research materials, intellectual stimulation, and other forms of assistance: Richard C. Anderson, Charles Bazerman, Robert Caddell, Edward P. J. Corbett, Barbara Couture, Mary Dieli, Richard Easton, Richard Enos, Hansford W. Farris, Linda Flower, Mary Sue Garay, Jone Rymer Goldstein, M. A. K. Halliday, Blanchard Hiatt, Margaret Kantz, David Kieras, Walter Kintsch, Moira McFadden, Carolyn Miller, Chaim Perelman, Marcia Petty, Gerri Power, Larry Selinker, Jack Selzer, Henrietta Nickels Shirk, David Steinberg, Alfred Sussman, John M. Swales, Julie Wei, and the many other authors we have cited in the references and other readings (a list of these authors appears at the end of the book). We would like to thank those who have provided special research funds: The University of Michigan—particularly, the College of Engineering, the Horace H. Rackham School of Graduate Studies, the Center for Research on Learning and Teaching, and the Division of Research Development and Administration—and the National Endowment for the Humanities; our colleagues Thomas M. Sawyer, Lisa Barton, Peter Klaver, and James Zappen for their ready counsel and support; and our office staff, Linda Bardeleben and Connie Christman for cheerfully putting up with us even in times of stress.

We would also like to thank our reviewers, Patricia Sullivan, Karl Drobnic, Quentin Johnson, Louis Holschuh, John Lackstrom, Carol Romett, Christopher Gould, William Woods, Douglas Wixson, Ruth Falor, O. Allen Gianniny, Thomas E. Gaston, Douglas L. Ewing, Jay Webster, Carolyn R. Miller, Deborah B. Normand, Dean Huber, Shelia K. Webster-Jain, Robert Kribbs, Terence Odlin, James M. Cunningham, Margaret Kantz, Michael G. Moran, C. William Brewer, and Jack Selzer for conscientiously and expertly examining various drafts of our manuscript and giving us numerous helpful suggestions. We wish to thank Phillip Butcher, Judy Cornwell, James Belser, and Claudia Tantillo of the McGraw-Hill staff for their help in producing the book.

Finally, we would like to acknowledge the contribution of many students in our classes at the University of Michigan and Carnegie Mellon University. They have been patient and helpful experimental subjects and sources of examples, suggestions, and new ideas.

THOMAS N. HUCKIN
LESLIE A. OLSEN

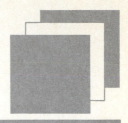

PART ONE
INTRODUCTION

Why Study Technical and Professional Communication?

Why should engineers, scientists, and other technical professionals study technical writing or speaking? Their primary training and interests lie in *technical* areas; most science and engineering students successfully pursue their *technical* subjects without extra writing courses; and practicing engineers and scientists in government and industry work on *technical* projects. It might seem, then, that writing and communication are superfluous to a technical education.

In fact, this is not the case. Scientists and engineers may be technically brilliant and creative, but unless they can convince coworkers, clients, and supervisors of their worth, their technical skills will be unnoticed, unappreciated, and unused. In a word, if technical people cannot communicate to others what they are doing and why it is important, it is they and their excellent *technical* skills that will be superfluous. From this perspective, communication skills are not just handy; they are critical tools for success, even survival, in "real world" environments.

Every technical person stands to gain from improving his or her communication skills. Most scientists and engineers work in organizational settings where teamwork is essential. Good teamwork is impossible without good communication. And those scientists and engineers who work independently have to communicate with clients, sponsors, or other interested parties. For many technical professionals, the ultimate product of their work is a written document. If that document is badly written, it reflects badly not only on the individual involved but on the entire organization. Organizations know this, of course, and sometimes base hiring and promotion decisions on writing ability. To maximize your chances for landing a good job and doing well at it, you should work on your communication skills—even if you think they're already pretty good. Communication will become even more important as you progress through your career; the better you are at it, the more likely you are to be promoted.

1.1 THE IMPORTANCE OF GOOD COMMUNICATION SKILLS

The importance of good communication skills has been emphasized in survey after survey. For instance, the American Society for Engineering Education conducted a survey[1] to determine which academic subjects are most needed for engineering careers in industry. Responses were received from 4057 engineers, all having at least several years of on-the-job experience behind them. The results, given in Table 1-1, show that communication skills rank above any other type of skill, capturing five of the nine "most needed" categories. These include technical writing (2), public speaking (4), working with individuals (6), working with groups (7), and talking with people (9). In contrast, technical skills rank toward the bottom of the list. Of course, for any individual a particular technical subject might rank higher than it does for the group; but for the group as a whole, the more general "people-oriented skills" are the ones that are seen as having across-the-board importance.

In more focused surveys, 245 distinguished engineers[2] and 837 experienced businesspeople[3] were asked, "How important is the writing that you do?" Their responses, as can be seen in Table 1-2, indicated that for most of them, writing was very important to their work.

Similar findings emerge from other surveys. For example, when 595 engineering alumni of the University of California, Berkeley, were asked if writing skills had aided their advancement, 73% answered yes. And almost all (95%) said that they would consider writing ability in making hiring or promotion decisions.[4]

Writing is not the only communication skill that's important to success on the job. Oral communication skills—conversing, interviewing, listening, giving oral presentations, etc.—are equally important. In the Berkeley survey, for example, 60% of the respondents said that speaking skills had aided their career advancement. In another survey, 305 engineering technologists were asked, "How important to your job is the ability to speak clearly and effectively?" Almost all the respondents (97%) answered either "very important" or "important."[5]

The need for good communication skills in industry is nicely articulated by Russel R. Raney, former Chief Engineer at International Harvester. Raney points out that most industries are composed of internal units that include (at least) a marketing unit, a manufacturing unit, an engineering or design unit, and a management unit. These units, or internal organizations,

Table 1-1 ❑ SUBJECTS MOST NEEDED FOR ENGINEERING
CAREERS IN INDUSTRY

Rank	Subject
1	Management practices
2	Technical writing
3	Probability and statistics
4	Public speaking
5	Creative thinking
6	Working with individuals
7	Working with groups
8	Speed reading
9	Talking with people
10	Business practices (marketing, finance, economics)
11	Survey of computer uses
12	Heat transfer
13	Instrumentation and measurements
14	Data processing
15	Systems programming
16	Economics
17	Ordinary differential equations
18	Logic
19	Economic analysis
20	Applications programming
21	Psychology
22	Reliability
23	Vector analysis
24	Electronic systems engineering (circuit design)
25	Laplace transforms
26	Solid-state physics
27	Electromechanical energy transformation
28	Matrix algebra
29	Computer systems engineering
30	Operations research
31	Law practice (patents, contracts)
32	Information and control systems
33	Numerical analysis
34	Physics of fluids
35	Thermodynamics
36	Electromagnetics
37	Human engineering
38	Materials engineering

SOURCE: Middendorf, Reference 1, © 1980, American Society for Engineering
Education.

Table 1-2 ❑ RESPONSES OF DISTINGUISHED ENGINEERS AND EXPERIENCED BUSINESSPEOPLE
(To the question "How important is the writing that you do in your present position?")

	Distinguished Engineers	Experienced Businesspeople
Critically important	45%	30%
Very important	51%	44%
Somewhat important	4%	23%
Minimally important	0%	3%

SOURCES: Davis, Reference 2, © 1977, American Society for Engineering Education. Storms, Reference 3, © 1983, The ABCA Bulletin.

are quite similar to factions in a political party in that, while they all strive toward the same end, they do not necessarily agree among themselves as to the means to be employed in achieving that end. So it is to be expected that the normal operation of the business will be marked by a certain amount of internal conflict as each organization seeks to promote its own particular interest.[6]

This natural conflict, the many details and changes to be managed, and the incredible complexity of a large organization all combine to create a demand for communication skills unforeseen by most science and engineering students:

In school you can attain your goal of a degree on your own individual efforts with practically no dependence on anyone else. On the other hand, it should be evident from the fore-going description of industrial operation that you will be able to accomplish very little by yourself in an organization. . . .

[This] means that for the results of your efforts to be incorporated in any end result you must present them effectively in competition with many conflicting considerations. It will, therefore, inevitably be a part of your responsibility to transmit understandably and persuasively the results of your work to other people, usually both verbally and in writing. Since success in engineering school does not depend much on communication abilities, you may not have developed your skills and may mistakenly resent the absolute dependence you will have on them for accomplishment in an organization.[7]

The Social and Ethical Aspects

We live in a world that is increasingly diversified and interactive. There is more commerce among more people than ever before, and it is cutting across geographical, occupational, cultural, and linguistic boundaries everywhere. Computer programmers talk to business managers, pediatricians confer with psychologists, molecular biologists exchange ideas with chemists, patent lawyers meet with mechanical engineers. Even though much international communication today is carried out in one language, English, there are still many possibilities for miscommunication. The English of any one occupation is often quite different from that of another, and the English of each country is somewhat different, too. In such a diverse world, it is important to recognize, as Gregory Clark has pointed out, that technical communication is not a simple process of conveying information but rather is a process of *social interaction.* "Instead of a commodity to be transported, [technical] information is an understanding to be shared with particular people for particular purposes. . . . In this view, ethical technical communication mediates conflicts between the incomplete understandings of those who have information and those who need it, helping people develop the shared knowledge that will meet their common needs."[8]

Legal Considerations

Failure to appreciate the ethical dimensions of technical communication can have serious human and legal consequences. In the last few years alone, the news media have reported on numerous accidents in the home and in the workplace resulting from poor technical communication. One need only think of the nuclear meltdown at Chernobyl, the scare at Three Mile Island, the Bhopal chemical plant disaster, and the tragedy of the *Challenger* space shuttle. Many people lost their lives and many more were injured. Yet follow-up investigations showed that these accidents could have been averted by better technical communication. In some of these cases, companies have had to face litigation as well. Indeed, product liability laws are getting tougher in many countries. If consumers misunderstand assembly instructions or operating manuals and injure themselves, they are more likely today than ever before to file lawsuits against manufacturers, and win. The best way for a company to protect itself against such lawsuits, obviously, is to have the product documentation written in such a way that the consumer is extremely unlikely to misunderstand it.

1.2 THE FREQUENCY OF TECHNICAL AND PROFESSIONAL COMMUNICATION

The actual amount of time spent on communication tasks varies from job to job, of course, but the overall average for engineers, scientists, and other technical professionals is quite consistent across the board. Davis's 245 distinguished engineers spent 24.4% of their time writing, Spretnak's 595 engineers and managers from Berkeley spent 25.3% of theirs, and Storms' 837 businesspeople spent 24.9% of theirs. If we add in the time spent on other communication activities, the total rises dramatically. For example, Spretnak found that her Berkeley alumni spent 23% of their time reading other people's writing, 11% supervising other people's writing, and another 7% giving oral presentations. In addition, they presumably spent considerable time attending meetings and conversing. A veteran industrial consultant, Ed Gilbert, has estimated that in most organizations one can expect to spend more and more time communicating as one rises in rank (see Table 1-3).[9]

Table 1-3 ❏ TYPICAL COMMUNICATION TIME WITHIN ORGANIZATIONS

Hierarchy	Portion of Total Work Time Spent on Communication, %		
	Verbal Media	Informal Writing	Formal Writing
Chief executive	65	10	2
Division manager	60	15	2
Department manager	45	20	5
Section head	40	20	5
Project head	30	20	10
Engineer or scientist	25	15	10
Junior engineer	20	10	10
Technician	15	10	2

SOURCE: Gilbert, Reference 9.

Obviously, if you can make such time-consuming activities more efficient, both you and your company will be well served. One easy way to improve efficiency is to develop effective strategies for analyzing communication tasks and audiences as well as strategies for producing reports and oral presentations. This text, and any technical or professional communication course based on it, should help you develop such strategies.

1.3 THE COLLABORATIVE NATURE OF TECHNICAL AND PROFESSIONAL COMMUNICATION

Scientific, technical, and managerial professionals frequently collaborate with other people, especially in activities which are large or important or inherently interdisciplinary. The collaboration may involve many individuals at geographically separated sites and routinely involves planning, writing, meeting, negotiating, designing, and decision making.

The extent of this collaboration is illustrated in a survey of 125 authors of National Science Foundation proposals from a variety of disciplines in the sciences, engineering, and social sciences.[10] The number of proposals with collaborative activity is relatively high in the planning stage (42%) and reviewing stage (67%) and surprisingly high in the actual writing stage (30%). Considering the creation process as a whole, the amount of collaboration is quite high. As shown in Table 1-4, 70% of the proposals involved collaboration at some stage in their creation, and 26% involved collaborators who were geographically dispersed—that is, not in the same building. About half the dispersed collaborators were off campus, and most of these were out of state or country.

Table 1-4 ❑ COLLABORATION IN PROPOSAL CREATION TEAMS
(Totals for planning, writing, and reviewing)

30%	1 person
31%	2 persons
15%	3–4 persons
24%	more than 4 persons
26%	of readers/writers/planners are geographically dispersed. Of these
	48% were on campus, in separate buildings
	52% were off campus, and of these
	20% were out of state, in the USA
	24% were outside the USA

SOURCE: Olsen, Reference 10, © 1989, Baywood Publishing Co., Inc., p. 100.

Collaboration can be important even if there's only one writer. In a series of case studies at seven large American companies, Rachel Spilka studied technical professionals who were in charge of major writing projects.[11] Five of these projects resulted in successful, well-written documents; the other two did not. What distinguished the successful cases, according to Spilka, was that the writer interacted with his or her target readers *early on*, even before writing the first draft. By bringing those readers into the communication process at an early stage, the writer was able to create a collaborative environment in which the readers themselves were able to suggest ideas, provide feedback, and otherwise "guide" the development of the writing. Thus, when the finished document finally appeared, the readers were primed to approve it. In the two unsuccessful cases, by contrast, the writers did none of these things.

1.4 THE COMPUTERIZATION OF TECHNICAL AND PROFESSIONAL COMMUNICATION

It is widely recognized that many scientists, engineers, professional technical communicators, and other professionals have some access to computer equipment. However, access to computing power is increasing at dramatic rates, as illustrated by the National Science Foundation study of collaboration cited above. In 1986, 58% of the prominent scientific and technical professionals in the study had ready access to personal computers and only 15% had access to high-powered workstations (significantly more powerful computers). However, by 1987 (just one year later) all the scientists had access to at least one personal computer (many had access to more) and 51% had access to high-powered workstations.[12] The technology for workstations now allows the production of very complex documents with easily embedded and editable pictures, graphs, spreadsheets (computerized budget programs), mathematical expressions, charts, graphs, and automatically numbered footnotes and bib-

Table 1-5 ❑ DISTRIBUTION OF SPECIFIC MEDIA TYPES IN NATIONAL SCIENCE FOUNDATION PROPOSALS

78%	had a bibliography (average length = 56 items)
49%	had endnotes and/or footnotes
	42% had endnotes
	10% had footnotes
	51% had no endnotes or footnotes
	(Note: most text-processing programs used did not support footnotes, endnotes, or bibliographies)
43%	had math
38%	had figures (line drawings, cross-sections, etc.)
34%	had tables
26%	had graphs
9%	had photos
7%	had other media types (detailed artists' illustrations, questionnaires, depictions of computer screens, maps, ads, boat schedules, etc.)

SOURCE: Olsen, Reference 10, © 1989, Baywood Publishing Co., Inc., p. 104.

liographies. This rapidly increasing access to workstations and high-powered personal computers is already changing the look of technical documents.

For instance, the capabilities of the 1986 systems were usually poor for anything but prose text (words), and as a result, the documents produced on them were largely prose text: one-third of the documents used only text, one-third used text plus one other medium (photograph, line drawing, mathematics, chart or graph, etc.), and one-third used 2–5 additional media types. This limited use of media types other than text, outlined in more detail in Table 1-5, may well reflect the last-minute handwork required to paste these media types into the document. If writers have access to a computer system that makes it very easy to include nonprose elements, those elements should appear more frequently and carry more of the important information of a text. This conclusion is supported by anecdotal evidence and the responses of the authors using such systems in 1987.

1.5 THE INTERNATIONAL NATURE OF TECHNICAL COMMUNICATION

We are moving more and more toward a single global economy, an economy based on a flow of information as much as it is on material products. We have indeed entered the "information age." Since much of this information is technical in nature, it is no surprise that technical communication itself has become highly internationalized. There are now more than 40,000 science journals worldwide, publishing more than one million articles a year. And only about half are in English. The United States has more than 350,000 foreign students, and other countries have many thousands more; most of them are studying technical subjects. Multinational corporations are commonplace, and all of them depend on technical communication.

If you are a technical professional or planning to become one, you should prepare yourself for work in this increasingly internationalized world. You should seek out opportunities for studying foreign cultures and languages. You should learn how to deal with people who may not share your cultural norms.

1.6 THE SPECIFIC SKILLS NEEDED

So far, we have argued that scientific and technical writing is important, time-consuming, and often collaborative, is computerized and becoming more so, will be relying even more on nonprose elements, and has important international analogues and legal implications. Let us now consider what communication skills are needed to support the needs of technical professionals in the near future.

What are the specific communication skills most needed by technical professionals? In terms of particular types of communication, the most common written forms are memos, letters, and short reports, followed by step-by-step procedures and proposals to clients. The most common oral forms are one-to-one talks, telephone calls, and small group or committee meetings. Frequency, however, is not necessarily the best measure of importance. Even infrequent forms of communication can be extremely important. For example, at one research laboratory we know of, each staff scientist is required to give a 10-minute oral presentation to upper management once a year summarizing the work he or she has done during that year. It may be an infrequent form of communication, but you can imagine how important it is to these scientists! (In fact, some of them practice it for weeks ahead of time.)

Typically, technical professionals need to master many different types of communication, both oral and written. This variety can be seen in Table 1-6, which gives the results of a survey conducted by Peter Schiff of more than 800 engineers with 5–10 years on the job.[13] On a scale of 1 to 7, with 7 being "most important," these practicing engineers rated almost four-fifths of the 30 items listed as "important."

Furthermore, technical professionals often command a variety of composing processes. For example, in a 1981 survey of 160 technical professionals and managers, Lester Faigley and Thomas Miller[14] found that 73.5% occasionally engaged in *team-writing*, 26% regularly *dictated* letters or reports, and 25.5% used *word processors*. It seems likely that the percentage of those using word processors would be even higher today.

As for the basic language skills needed for good technical communication, both the Davis and Storms surveys, cited earlier, found that "writing clearly," "clearly stating one's purposes to readers," "knowing how to organize a communication," and "writing concisely" were the top choices of experienced

Table 1-6 ❑ TYPES OF COMMUNICATION RANKED BY IMPORTANCE TO ENGINEERING PRACTICE

Rank	Type of Communication	Mean*	Std. Dev.
1	One-to-one talks with technically sophisticated personnel	5.9391	1.3109
2	Writing using graphs, charts, and/or other illustrative aids	5.3955	1.7141
3	Project proposals (written)	5.3324	1.8167
4	Participation in a small group or committee made up of only technically sophisticated members (oral)	5.2694	1.7733
5	Instructions for completing a technical process	5.2444	1.8332
6	One-to-one talks with nontechnical personnel	5.0718	1.9065
7	Project progress reports (written)	5.0609	1.6301
8	Project proposal presentations (oral)	5.0587	1.9787
9	Writing technical information for technical audiences	4.9157	1.9396
10	Oral presentations using graphs, charts, and/or other aids	4.9139	1.9623
11	Technical description of a piece of hardware (written)	4.8687	1.8793
12	Memoranda	4.8169	1.7175
13	Short reports (less than 10 pages typewritten, double-spaced)	4.7967	1.8507
14	Writing technical information for nontechnical audiences	4.7458	2.1377
15	Project feasibility studies (written)	4.7124	1.9305
16	Project progress report presentations (oral)	4.6667	1.9277
17	Participation in a small group or committee including nontechnical members (oral)	4.6639	1.9608
18	Business letters	4.6257	1.8856
19	Telephone reports	4.5838	1.9682
20	Project feasibility study presentations (oral)	4.4258	2.0780
21	Form completion	4.2842	1.8847
22	Formal speeches to technically sophisticated audiences	4.0452	2.2108
23	Writing in collaboration with one or more colleagues	4.0313	2.0175
24	Laboratory reports (written)	4.0184	2.1800
25†	Long reports (10 or more pages typewritten, double-spaced)	3.8249	2.2120
26	Formal speeches to nontechnical audiences	3.5869	2.2020
27	Writing requiring library research	3.2062	2.0723
28	Abstracts/summaries of others' writing (written)	2.8864	1.9910
29	Articles submitted to professional journals	2.8704	2.1499
30	Reports submitted to professional societies	2.7224	2.0827

*Rating scale: 1 = least important; 7 = most important
†Skills 25 through 30 were ranked "below average to unimportant."
SOURCE: Schiff, Reference 13, © 1980, American Society for Engineering Education.

Table 1-7 ❑ COMPLAINTS ABOUT TECHNICAL WRITING
FROM 182 SENIOR OFFICIALS IN SCIENCE AND INDUSTRY

Complaint	Number Responding
Generally foggy language	182
Inadequate general vocabulary	173
Failure to connect information to point at issue	169
Wordiness	164
Lack of stressing important points	163
All sorts of illogical reasoning	163
Too much "engineering gobbledygook"	160
Poor overall organization	153
No clear overview	143
No clear continuity	142
Very little concept of writing for anyone but fellow specialists	136
Poor grammar	133
General lack of flexibility to suit circumstances	127
Deliberate obscurity	92
Poor punctuation	81
No sense of proper tone for circumstances	72
Poor adaptation of form to purpose	64
Almost meaningless introduction	54

SOURCE: MacIntosh, Reference 15.

engineers and businesspeople. These same findings are mirrored in a survey by Fred MacIntosh,[15] who asked 182 senior scientists, government agency heads, corporate executives, and other senior officials in science and industry to list their complaints about the technical writing they saw. The results are given in Table 1-7.

1.7 THE COMPLEXITY OF TECHNICAL AND PROFESSIONAL COMMUNICATION

Why do technical professionals often communicate so badly? One reason is poor training. In trying to keep up with rapid advances in technical knowledge, technical schools are often forced to make sacrifices and sometimes it is the teaching of communication skills that gets shortchanged—despite repeated complaints from industry. Not surprisingly, this neglect of communication skills creates a series of problems. Technical students are trained to follow certain standardized procedures in their work, to use correct technical language, to try to be as objective as possible—in short, to "let the facts speak for themselves." This, of course, is as it should be. But communication is different. Technical students need to understand that, when they go to work in the real world, they will sometimes "have to speak a different language," especially if they want to be understood by nonspecialists. They will sometimes have to avoid following the standard narrative style of lab reports and scientific papers. They will sometimes have to cut down on their use of technical terminology (or at least explain what it means). They will sometimes have to inject their personal opinion into their writing or speaking, emphasize important points, make arguments and recommendations. If communicators do not understand how to do these things, they will communicate badly.

But the main reasons people communicate badly, we think, have to do with the sheer cognitive *complexity* of technical communication and the physical complexity of the process by which it occurs. What makes technical communication cognitively complex? In the technical, scientific, and business worlds, the most attention-consuming activities usually address some kind of problem. Indeed, if you are an engineer, scientist, or other technical professional, you probably see yourself primarily as a problem solver. Problems come in different sizes and shapes: some are narrowly technical and can often be solved by a single person; others are organization-wide and can only be "solved," if at all, by reaching some kind of consensus among many people. Typically, it is this latter kind of problem that requires the most skilled communication. The proposal you are trying to get funded, the project that's falling behind schedule, a personnel evaluation—these are the kinds of things that are essentially communication problems, not purely technical ones. These things are also problems which require a sophisticated level of intellectual activity—sophisticated thinking, analysis, evaluation, and design skills—to deal with the content of your problem and the audience needs driving your communication. If you are writing about unfamiliar content, you have the added complexity of mastering difficult new content.

What Is a "Problem"?

A problem is some kind of incongruity, a discrepancy between what we expect or want to find and what we actually do find. It is a conflict between what *should be* and what *is*. Problems are not inherent in nature; they are a product of our expectations and desires. As the philosopher of science Stephen Toulmin has said, "The problems of science have never been determined by the nature of the world alone, but have arisen always from the fact that, in the field concerned, our ideas about the world are at variance either with Nature or with one another."[16]

Technical communication can be seen as a way of mediating such incongruities. As shown in Figure 1-1, the technical communicator is usually faced with some kind of organizational problem, a discrepancy between the desired state of affairs and the actual state of affairs.[17] The desired state of affairs is determined by certain goals, constraints, and criteria, which are

Figure 1-1 ☐ TECHNICAL COMMUNICATION AS PROBLEM SOLVING
(Courtesy J. C. Mathes © IEEE and ASEE, 1979; Adapted by J. C. Mathes from Reference 17, p. 111.)

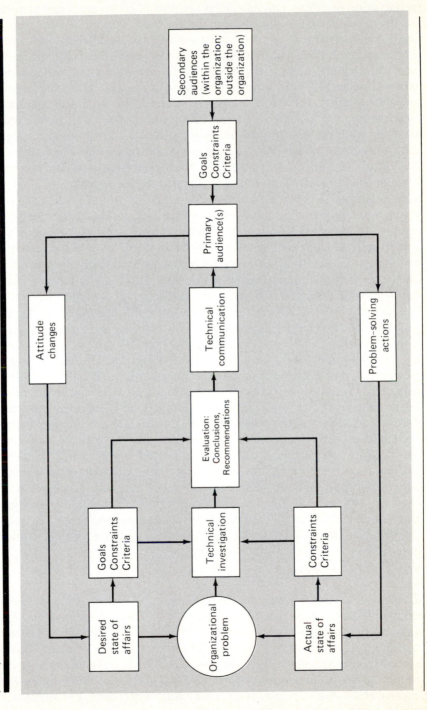

themselves determined by various primary and secondary audiences. The actual state of affairs may also be a product of constraints and criteria, but if so, they would only be constraints and criteria from an earlier time, not the same ones that the current audiences have established. To resolve the organizational problem, the technical communicator conducts a technical investigation and arrives at certain conclusions and recommendations, all the while keeping in mind the various goals, constraints, and criteria determined by the desired and actual states of affairs. The technical communicator's conclusions and recommendations are presented to a primary audience of decision makers, who in turn are responsive to various secondary audiences (those who are affected by the primary audience's decisions). The primary audience's decisions should produce attitude changes or problem-solving actions which serve to reduce or eliminate the original incongruity.

We wish to emphasize two features of this way of looking at problems. First, most technical problems are embedded in some kind of organizational context and are therefore *socially-conditioned*. They are created by the wants and needs of communities of people. Likewise, they are most effectively resolved when the technical investigator keeps those wants and needs in mind. This means that communication with one's audiences is a major part of the problem-solving process. If you are a technical professional and you want to produce the best solutions to the problems you are working on, you must maintain good communication with your audiences throughout the lifetime of a project.

Second, most technical problems are *dynamic*. They can change even as people are working on them. Indeed, they may change precisely *because* people are working on them. In a problem-solving environment where good communication is taking place, there is likely to be a continuous process of attitude changes and/or problem-solving actions which change the nature of the problem. Figure 1-1 depicts not so much a linear process as a *recursive* one. Notice the feedback loops. The attitude changes and problem-solving actions represented by these loops do not have to be one-time events. There can be—indeed, there should be—frequent interaction between technical investigators and their audiences. For example, investigators can make sure they are pursuing viable solutions by using progress reports, conferences, phone calls, preliminary drafts, etc. to stay in touch with their audiences.

Types of Problems

There are many different kinds of problems. As Walter Reitman and Herbert Simon[18,19] have pointed out, some problems are well defined, others are ill defined, and many are somewhere in between. A well-defined problem has a small and relatively simple "problem space" in which to work out the problem. It usually has a single correct solution. Technical calculations, even those that require a computer, are usually well-defined problems. An ill-defined problem, on the other hand, has a much larger and more complex problem space in which many different *kinds* of variables can play a role. It typically does *not* have a single correct solution. For example, if you were put in charge of a work group and told to improve its performance, you would have an ill-defined problem on your hands. There would be many different ways you could go about finding a solution, and there would be many possible improvements.

In school, most of the problems you deal with are well-defined ones. But in the real world, many of the most important problems you have to confront are ill-defined ones. And often these are precisely the problems that require good communication skills. If you are asked simply to do some mathematical calculations and report on them in a memo, you are probably dealing with a well-defined problem. You will probably have no more trouble writing the memo than you will doing the calculations. But if you are asked to write a report about the feasibility of buying some new equipment, you will be facing an ill-defined problem and will probably have to put a lot of thought and effort into writing that report.

Intellectual Skills in Problem Solving and Communication

It is tempting to think of communication as a simple matter of conveying information: Person A knows something that Person B doesn't, so all A needs to do is pass that information on to B—like pouring information into one end of a pipe and having it come out the other end, unchanged. But although this conception applies nicely to electronic data flow and other forms of nonhuman information processing, it does not work at all for human communication. Human communication is simply not that simple. Human beings are always *interpreting* information, and when they do, they draw on their personal experience to try to gain better understanding. To the extent that people vary in their experience—and of course everyone has a unique background of individual experience in this world—they will naturally vary in how they interpret new information.

For example, let's say you're a staff physician in a hospital and have written a memo asking the director to install computer terminals with "Grateful Med" interfaces on every floor so that doctors can get information directly from international, online databases. Although the hospital library has such a terminal, you believe that putting one on every floor is an idea whose time has come. For one thing, it would spare you all those time-consuming trips to the library you've been taking. The director, however, may remember a time when the introduction of a new recordkeeping system at another hospital led to major changes in the overall operation of the hospital. Thus, she may interpret your idea differently, seeing your memo as a first step toward general decentralization. And the librarian, who has been treated by certain doctors with condescension, may interpret this memo as yet another sign of disrespect: instead of working with him to access information, "now they want to go around me." The different personal backgrounds of these people will lead to different interpretations of the "same" information. So it is likely that your memo will not be well received.

Situations like these are commonplace in professional life. They illustrate the point that *good communication often requires socially negotiated problem solving.* If you acknowledge the fact that other people have a different body of experience and perhaps a different perspective than you do, there are two ways in which you can benefit from negotiating a solution to problems. First, you can use their experience to generate possible solutions that might not have occurred to you otherwise. Second, the solution you end up with is more likely to be acceptable to everyone involved, and therefore more likely to be effective.

Project Management Skills

It should be clear from the preceding discussion that you need intellectual skills to handle the content of the communication and the needs of the audience. However, you need another set of skills—project management skills—related to the physical process by which you produce a communication product. Project management skills include being able to estimate the time and money and resources it will take to complete a project or one of its parts, being able to meet the deadlines set for a stage in the production process, and being able to deal effectively with coworkers with different backgrounds. The importance of these skills becomes evident when you see the production process as it is presented in Figure 1-2. This view stresses the complexity of the production process, especially for projects with multiple authors, illustrators, editors, and production specialists.[20] However, even simple projects have many of the stages represented in Figure 1-2, and each stage involves a deadline which must be met for the project to continue successfully. In many contexts, timing is almost everything. If you miss a deadine, you disrupt the tight coordination of many people's schedules and cause the communication product to be late. This in turn may cause the late release or failure of a technical product, or a poor decision by managers missing vital information. Thus, even if you have very highly developed skills in identifying and communicating about problems, your skills will be only minimally useful or even useless in an organizational context if you cannot combine them with project management skills and meet deadlines.

Conclusion

In sum, technical communication is very complex. The problems it attempts to resolve are typically ill-defined ones that sometimes change character even as you are working on them. It requires sophisticated techniques of cognitive and social problem solving. It requires a good command of both technical and nontechnical language and a sense of how to tailor language to the needs and capabilities of specific audiences. It requires a familiarity with many different genres of communication. As if this weren't enough, technical communication must also frequently consider the ethical dimensions of a problem.

When something is as complex as this, learning it is not easy. We hope this book will help.

Figure 1-2 ☐ TYPICAL TECHNICAL INFORMATION DEVELOPMENT PROCESS
(From Deardorff et al., Reference 20.)

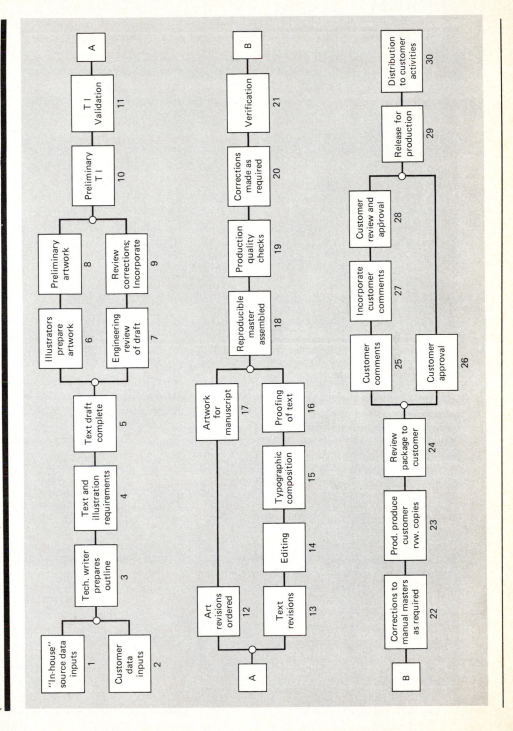

REFERENCES

1   William H. Middendorf, "Academic Programs and Industrial Needs," *Engineering Education*, May 1980, pp. 835–837.

2   Richard M. Davis, "Technical Writing: Who Needs It?" *Engineering Education*, November 1977, pp. 209–211.

3   C. Gilbert Storms, "What Business School Graduates Say about the Writing They Do at Work: Implications for the Business Communication Course," *The ABCA Bulletin*, December 1983, pp. 13–18.

4   Charlene M. Spretnak, "A Survey of the Frequency and Importance of Technical Communication in an Engineering Career," *The Technical Writing Teacher*, Spring 1982, pp. 133–136.

5   Carol Barnum and Robert Fischer, "Engineering Technologists as Writers: Results of a Survey," *Technical Communication*, Second Quarter 1984, pp. 9–11.

6   Russel R. Raney, "Why Cultural Education for the Engineer?" Speech to the Iowa-Illinois Section of the American Society of Agricultural Engineers, April 29, 1953. Quoted in John G. Young, "Employment Negotiations," *Placement Manual: Fall 1981* (Ann Arbor: The University of Michigan, College of Engineering, 1981), p. 49.

7   Ibid.

8   Gregory Clark, "Ethics in Technical Communication: A Rhetorical Perspective," *IEEE Transactions on Professional Communication 30*(3):190–195 (September 1987).

9   Ed Gilbert, "Technical Communication and Writing: Management's Perspective." Speech to the Conference on Teaching Technical and Professional Communication, co-chaired by Dwight W. Stevenson and J. C. Mathes, Department of Humanities, College of Engineering, The University of Michigan, Ann Arbor, 1975. Used with the permission of J. C. Mathes.

10   Leslie A. Olsen, "Computer-Based Writing and Communication: Some Implications for Technical Communication Activities," *Journal of Technical Writing and Communication 19*(2):97–118 (June 1989).

11   Rachel Spilka, *Adapting Discourse to Multiple Audiences: Invention Strategies of Seven Corporate Engineers*, Unpublished doctoral dissertation, Carnegie Mellon University, Pittsburgh, PA, 1988.

12   Olsen, p. 102.

13   Peter M. Schiff, "Speech: Another Facet of Technical Communication," *Engineering Education*, November 1980, p. 181.

14   Lester Faigley and Thomas P. Miller, "What We Learn from Writing on the Job," *College English 44*(6):557–569 (October 1982).

15   Fred H. MacIntosh, "How Good Is Our Product?" Speech delivered at the annual meeting of the Conference on College Composition and Communication, Louisville, KY, April 8, 1967.

16   Stephen Toulmin, *Human Understanding* (Princeton, NJ: Princeton University Press, 1972), p. 150.

17   J. C. Mathes, "Problem Solving: The Engineering in Technical Communication," in *Proceedings of the Ninth Annual Conference on Frontiers in Education*, edited by Lawrence P. Grayson and Joseph M. Biedenbach (Niagara Falls, Canada:

American Society for Engineering Education and The Institute of Electrical and Electronics Engineers, Inc., October 1979), pp. 109–112.

18 Walter R. Reitman, "Heuristic Decision Procedures, Open Constraints, and the Structure of Ill-Defined Problems," in *Human Judgments and Optimality*, edited by Maynard W. Shelly and Glenn L. Bryan (New York: Wiley, 1964), pp. 282–315.

19 Herbert Simon, *Models of Discovery* (Dordrecht, Holland: D. Reidel, 1977).

20 D. D. Deardorff, K. C. Hageman, W. Hehs, and J. M. Norton, *Development of a Quality Assurance Methodology for the Technical Information Generating Subsystem of NTIPS* (NAVAIRSYSCOM 78-C-1075-0003). Naval Air Systems Command, Washington, DC, June 1979.

ADDITIONAL READING

Carol J. German and William R. Rath, "Making Technical Communication a Real-World Exercise: A Report of Classroom and Industry-Based Research," *Journal of Technical Writing and Communication* 17(4):335–346 (1987).

H. K. Jenny, "Heavy Readers Are Heavy Hitters," *IEEE Spectrum* 15(9):66–68 (1978).

Patrick Kelly, M. Kranzberg, S. R. Carpenter, and F. A. Rossini, *The Flow of Scientific and Technical Information in the Innovation Process: An Analytic Study* (Atlanta: Georgia Institute of Technology, 1977).

William Kimel and Melford E. Monsees, "Engineering Graduates: How Good Are They?" *Engineering Education*, November 1979, pp. 210–212.

Diana Railton, "Communication Skills Training for Engineering Students in British Universities," *IEEE Transactions on Professional Communication* 29(2):7–13 (June 1986).

Hedvah L. Shuchman, *Information Transfer in Engineering* (Glastonbury, CT: The New Futures Group, 1981).

Nicholas D. Sylvester, "Engineering Education Must Improve the Communication Skills of Its Graduates," *Engineering Education*, April 1980, pp. 739–740.

PART TWO

GENERAL STRATEGIES FOR THE WRITING PROCESS

2

Generating Ideas

Before you begin to draft a report or speech about a topic, you first need to figure out what you know and want to say; you need to generate some ideas about the topic. Only later can you can shape your ideas into an argument and present them to a particular audience. The first stage of generating ideas about the topic is the subject of this chapter. Shaping or organizing the ideas into an argument is treated in Chapter 4; presenting the ideas to various audiences is treated in Chapter 3 and later chapters.

The stage of generating ideas about a topic has traditionally been called *invention*, or prewriting. In the study of communication, *to invent* means "to come upon" or "to find" an argument or idea, often one which the writer has already been exposed to but which does not necessarily "spring to mind." In everyday speech, we also use *invention* to mean the act of creating something new. Both of these senses are appropriate for the discussion of invention or idea generation here, since we are going to discuss ways of finding or retrieving already known information as well as information which is not already known but which is relevant to the topic. The latter type of information is perhaps the most troublesome: How do you find information which is relevant when you don't even know that it exists?

There are two basic visions of invention or idea generation or "thinking about" a topic. In the first, thinking is seen as a "black box" which is mysterious and unknown, a creative process which is beyond control.[1] In this vision, the best way to help the invention process is to provide a comfortable and secure environment in which the writer can think and then—once some thoughts are produced—to help the writer organize and edit those thoughts. In the second vision of invention, thinking is seen as a somewhat mysterious and not totally controllable process, but a creative process that can be helped along by

25

procedures for thinking systematically and inclusively. Even though many writers like a comfortable and secure environment, we feel that most writers can be helped by systematic and inclusive thinking. Thus, the rest of this chapter will outline some approaches to invention or idea generation which have proved helpful to writers for up to 2000 years. These approaches include brainstorming and the use of systematic thinking.

It should be noted that both of these approaches work better if you use them just to generate your ideas; you should save the evaluation and organization of the ideas for a later stage in the writing process.

Sometimes it is very easy to think of the ideas you need to cover in a document or speech. For instance, in some technical environments, the ideas for a final report or journal article are often developed long before the actual writing stage. Before you can write up an experiment, you have to come up with a potentially interesting hypothesis, survey the literature, refine the hypothesis, run the experiment, and analyze the results—all the while thinking about how the experiment fits in with other work in the field.

All these activities occur before the actual process of writing begins—that is, they are prewriting activities. For a journal article or a final report, the prewriting stage may take months, and it will probably involve a lot of interaction with other people. You will be generating ideas and refining them as you go along, so that when the time comes to write up your results in the form of a journal article, say, you will probably have already thought of most of the information you will need to cover. There is not much we can tell you about this kind of idea generation, since it is highly domain-specific. Every technical field has its own special methods for generating information. You have learned—and perhaps are still learning—some of these special methods through your technical training. To write your article or report, you will need to gather up the ideas you have been thinking about for the length of your project and then check to see that you have covered everything relevant. Your gathering and checking should be aided by the suggestions in the rest of this chapter, since these suggestions form a kind of checklist that you can use to see that you have gathered in everything relevant.

Sometimes it will be much harder to think of the ideas you need to cover in a document or speech. For instance, if you are asked to write a proposal to a funding agency on a new project, you will not have months of technical work and thinking on which to rely. Thus, your idea-generating tasks will involve coming up with a good and complete set of ideas from scratch. Again, the suggestions in the rest of this chapter should form a checklist that you can use to help you generate your ideas.

2.1 DEFINING THE PROBLEM AND YOUR TASK

Before you begin to generate ideas, you need to have something to generate ideas about. This usually means (1) identifying the problem you need to address, and (2) determining exactly how you are supposed to address it.

As we noted in Chapter 1, most professional communication is problem-oriented. Therefore, when you begin to do a writing project or prepare an oral presentation, you should start by asking yourself, "What is the main problem here?" In an organizational setting, the main problem is usually some *organizational problem*, something that is affecting the organization as a whole and is causing you to respond. Unlike some other problems, an organizational problem is one that is overtly identified as such and one that is recognized as needing a solution. Do not confuse your personal goals or problems with the organizational problem. For example, if you are writing a job letter, you may think that the problem you're addressing is one of "finding a job." On a personal level, of course, it is. But the real problem—the organizational problem—is the fact that the company needs someone to fill a position. By keeping the organizational problem in mind, you are more likely to generate appropriate ideas and write a good letter than if you think only about your own needs.

Although you can sometimes easily identify an overriding organizational problem, at other times you may be faced with a situation where there seems to be not one but a number of problems. In these cases, it is important to sort through the problems and try to find the most important one. You may find it useful to list out the problems and then ask yourself how they relate to each other. By embedding some of them as *sub-problems* under others, you can create a hierarchy or tree-diagram of problems, with the most important one on top. This way, when you try to generate ideas, you can focus your attention on the most important problem first, on the next most important problem second, and so on.

In addition to defining the problem, you need to have a clear idea of what your *task* is. What are you expected to do about the problem? Are you supposed to describe it? To propose an approach to solving it? To recommend a solution? To do something else? Are you supposed to deal with the entire problem—or only part of it? To generate good ideas, you have to know your assigned task. Otherwise, you may end up spending a lot of time thinking up ideas that are thoroughly inappropriate to your situation and task.

To a great extent, your task depends on what your *role* is. Are you an "insider" or an "outsider"? Where are your loyalties? Are you a specialist in this area? Do you have the power to make decisions? If not, do you have the power to influence the decision makers? How do other people see you? You'll want to generate ideas that are appropriate not only to the organizational problem and the task but also to your role in this situation.

2.2 BRAINSTORMING

One way of generating ideas is just to generate them—that is, to talk out or write down whatever comes into your head about the topic you want to write about. Don't worry about whether the ideas are good or adequate or complete; just get them down. The purpose of brainstorming is to provide yourself with some notes that you can use to further stimulate your thinking and organizing. If you feel intimidated by a blank page, you might try talking into a tape recorder about your topic and then listening to what you've said and writing down the good ideas that are there.

Brainstorming is a very intuitive process. You just follow your ideas wherever they lead and take notes to record your mental journey. It's important that you not censor an idea before it has a chance to flower, but just generate it. Evaluation and organization of the ideas will come later, but you can't evaluate or organize something that isn't there.

Brainstorming is often best done in collaboration with other people. By "bouncing ideas" off of others, you can stimulate their thinking and get them to produce ideas of their own. As with any kind of brainstorming, the important thing is not to evaluate or criticize ideas but just to generate them.

If you are collaborating with others who are not always in your room or building, you might see if you and your collaborators can collaborate via computer. To do this, you will all need access to computers on a network which connects each computer. This network may be a Local Area Network (LAN) or one of the more extensive networks serving your institution or industry.

❏ EXERCISE 2-1

To see how brainstorming works, we suggest that you take a minute to imagine the following scenarios.

A You work for a newly formed company that manufactures avant-garde office furniture. The company has plans to manufacture a special desk chair that can be adjusted to fit any user's anatomy perfectly and to reduce the stiffness and pain that can result from sitting all day, especially in front of a computer terminal. You are part of a marketing team that has been told to come up with a novel sales strategy for this ergonomic chair and then to write up your strategy in a report to the company President and the Director of Marketing. The team includes a design engineer, a mechanical engineer, an advertising specialist, a graphics designer, a marketing specialist, and a professional writer. Taking one of these roles, try brainstorming about *ergonomic chairs* to see what "leaps to mind."

B You are working in the Marketing Department of a consulting firm and your department has decided that your company should get into the business of providing security to its industrial customers. Your boss has decided that your company could design an electronic security system for industrial use and has asked you to survey the market for such a system, to see what other electronic security systems are available, and then to come up with a list of features that would make such a system attractive to industrial customers. Try brainstorming about the desired features of such an electronic security system.

C Brainstorm about a topic on which you have to write a paper, report, or article.

In the next two sections, we will help you apply some systematic thinking to your topic and illustrate how such thinking can help you generate even more ideas. However, for now it's important that you brainstorm about the topic so that later you'll see what systematic thinking can do for you.

2.3 USING SYSTEMATIC QUESTIONS AS PROMPTS

Once you've brainstormed about your subject, you might try a more systematic way of thinking about or exploring it to make sure that you have thought about all the relevant material and not overlooked some critical piece of information. Relevant material might arise from exploring

1   The subject matter you want to talk about (ergonomic chairs or electronic locks)

2   The requirements of the subject field (industrial design, marketing, electrical engineering) and its demands on argumentation within that field

3   The requirements of the genre or type of writing (marketing report, technical report, journal article, proposal, letter)

4   Your audience and its impact on your communication

Another area has been suggested by Carolyn R. Miller and Jack Selzer that is potentially quite important and easy to overlook:

5   Institutional concerns—that is, political issues or public relations issues or technical issues that your company or institution wants to stress in its communication[2]

Each of these considerations will present perspectives from which you can view your task and in so doing will help you see relevant things about your subject. For instance, in the ergonomic chair situation in Exercise 2-1, considering the requirements of the field (industrial design and marketing) and genre (the technical report) will generate some ideas for the report. Considering the subject (the ergonomic chair) will produce some other ideas for the report, and considering the audience (the President, the Director of Marketing, and the ultimate buyers who will read the sales literature) will generate still more ideas.

In the rest of this chapter, we will outline ways of exploring each of these in turn. In particular, we will outline the perspectives and the kinds of questions you can ask about your subject. We won't always answer them, but we will suggest the questions, since they are the critical items. Once you have generated a question, you can answer it, either from your experience or through research.

Thus, we'll ask lots of questions and see where they lead and what ideas they generate. The goal now is for you or your group to get a lot of ideas "on the table" and then later decide what they add up to, how they can be used, which ones will be used, and how they'll fit together. We should note that in any writing project, you'll probably use some of the ideas you generate in the invention phase, but not all of them.

Exploring Field and Genre Impacts Using Special Topics

Over 2000 years ago, Aristotle outlined his *topics:* a powerful set of tools for thinking about subjects.[3] The word *topic* comes from the Greek word *topos* (place) and means an intellectual place in which relevant information about a subject may be found. Aristotle divided his topics into the *common topics*, which are relevant to thinking about all subjects, and the *special topics*, which are relevant only to thinking about a special or particular subject. The special topics will be discussed first, since they are probably the most important type of prompt for most technical communication.

Special topics are questions or perspectives which need to be considered for a particular subject or type of discourse (type of writing or speaking). In Aristotle's terms, there are three general types of discourse: deliberative discourse, judicial or legal discourse, and ceremonial discourse. There are special topics associated with each of these kinds of discourse, and we will briefly review these and the types of special topics associated with each.

Deliberative Discourse

Deliberative discourse often operates in the world of politics and business and tries to convince its audience by appeals to fact or to practical advantage or inherent goodness. The generation of ideas in deliberative discourse revolves around questions of existence, definition, quality, practical advantage, and inherent worth. For instance, if you wanted to explore the need for the ergonomic chair (stiffness and pain from sitting all day in a conventional chair in front of a computer terminal), you might generate the following questions:

Existence:	Is there evidence of stiffness and pain from sitting all day in a conventional chair in front of a computer terminal?
Definition:	Is it the kind of stiffness and pain that is regulated by law, health guidelines, etc.?
Quality:	Is the amount of stiffness and pain illegal or unsafe?
Practical advantage:	Is there some practical benefit to reducing or eliminating the stiffness and pain?
Worth or goodness:	Is it inherently good to remove or reduce the stiffness and pain?

Some situations already have well-defined categories and questions for exploring and generating ideas. For instance, suppose you have been asked to do a technical evaluation of somebody's research. The specific questions you would want to ask depend on the methodology used (e.g., experimental, analytical, simulation, or theoretical). Here are some examples:

Experimental methods (which collect data to verify theory)

1   Is this an idealized or a "real" situation? What are the characteristics that make it idealized or real? To what degree is it idealized?

2   Is the experiment repeatable?

3   Does it use standard or invented or adapted procedures?

4   What kind of experimental apparatus is used?

5   Was the experiment looking at the "right" parameters (what is the adequacy of the experimental design)?

6   Was it using a good measurement scheme (was it looking at the "right grain" or level of detail)?

7   Was it using appropriate criteria? Did these change?

8   How did you treat calibration problems and experimental error?

9   Was there a sufficiently large body of data?

10 What is the validity of the data?

11 Do the results support or contradict existing theory?

12 Is this the best method?

Analytical methods (which use existing theory to analyze a body of data)

1   What is the nature of the data? Is it the same as the available experimental data?

2   How were the data collected?

3   How should the data be classified?

4   Is there a solution to the problem being analyzed? Is an exact solution known? If so, how do the data used for the analysis compare with the known solution(s)?

5   Can existing theory explain the results of the analysis? If not, is there a new or alternative explanation of the result?

6   What is the analytical approach? If this is a new approach, how does it compare with the old approach?

7   What are the assumptions behind the analytical method?

8   What is the model behind the analytical method?

9   Why are the model and assumptions applicable to the data?

10 What are the initial and boundary conditions?

Simulation methods, both computerized and noncomputerized (which present alternative ways to generate data for complicated situations in which it would be too expensive or too hard to use normal experimental techniques)

1   Can the reality being studied be simulated?

2   Where does the simulation fall on a scale of representation from reality to idealization?

3   How does the model compare with theoretical and experimental results?

4   What is the method of simulation?

5   How many simulations were run and what was the size of the runs?

6   What are the inputs to the simulation?

7   What is the purpose of the simulation?

8   What are the underlying assumptions/simplifications?

9   What are the validity measures?

10   What were the calibration measures and efforts?

11   What were the error criteria for an acceptable simulation?

12   What were the boundary conditions?

13   What are the yields of the simulation? Do they make sense? How do they compare with historically accepted data? How do they compare with the results of analytical methods?

14   Was there an evaluation of the first set(s) of results and then feedback to subsequent runs of the simulation?

Theoretical methods (which develop new theory to explain existing data or predict new data)

1   Why is the new theory needed? What are the problems with the existing theory?

2   Do you have enough proof to argue that the new theory is adequate?

3   How does the new theory compare with the old theory? Is is a generalization of the old theory? An extension or different point of view? A contradiction? A replacement?

4   What are the new theory's assumptions? Are they valid?

5   What are the limitations and boundary conditions?

6   Does the new theory lead to a more accurate explanation of existing data or theories?

7   Does it predict any new data or theory?

8   What are the implications of the new theory? Its applications?

9   What kinds of data can the theory handle?

10   How do the theoretical perspective and models and expectations affect the "reality" that we see?

11   How can you prove or defend your theory?

12   Is the theory provable?

13   What does the field accept as an adequate standard of proof for the theory?

14   How does the theoretical result perform in practice?

   If you need to write about some technical subject for which there are no special technical topics, you can probably prepare a list of such topics and questions by thinking strategically about your subject for a while. What you are really trying to define are those issues which might normally be considered in dealing with your subject or that general type of subject. Subject matter constraints are considered for communication about economics in McCloskey[4] and for the experimental article in science in Bazerman.[5]

   However, to give you a start on defining your own "special topics" for technical subjects, later chapters of this book discuss such technical writing forms as reports, journal articles, proposals, and letters, along with the requirements for each type of communication (its parts and what those parts must cover). For instance, if you are writing a report, it must have an Introduction or Foreword which defines the problem to be addressed, tells what you did in addressing the problem, and provides an overview of the contents of your report. Thus, you can turn this discussion of form into a list of special topics to generate information:

   What should the Foreword or Introduction cover?

   What is the problem being addressed?

   What did I do in addressing the problem?

   What are the main subject divisions that I need to cover in my report?

A number of the discussions in the chapters on specific forms of communication also provide lists of questions which should be considered (special topics for the particular situation); other discussions will allow you to generate additional questions easily. In this sense, as Winkler (Mikelonis) and Miller and Selzer have argued, the conventional features of a particular type of writing (genres such as lab reports, technical reports, proposals, and job letters) serve as prompts for ideas that need to be included.[6]

   In addition, you can always look at previous instances of discourse similar to the one you are trying to produce. In this sense, the previous instances can serve as models for you. You should, of course, be sure to look at good and successful instances, since it will not be too helpful if you are trying to imitate a poor model. We have provided a number of examples in this book which can serve as models for you.

Judicial or Legal Discourse

Judicial or legal discourse is discourse that relies heavily on a dispassionate analysis of evidence and often operates in the courts. The special topics for legal discourse have received much attention over the years and include *evidence, definition of crime,* and *motives* or *causes for crime.*[7] Evidence will be explored here as our sample special legal topic. Notice the kinds of questions which are raised in Edward P. J. Corbett's treatment of this topic in his discussion of classical Aristotelian and Ciceronian rhetoric. These questions are relevant to any treatment of evidence, whether in a legal context or in a marketing report evaluating the credibility of a marketing analysis.

The major subtopic here is *evidence.* In developing this topic, we pursue questions like these:

1 What is the evidence?

2 How, when, where, and by whom was the evidence gathered?

3 What is the reliability of the evidence?
   a. Is it accurate?
   b. Is is relevant?
   c. Is it consequential?
   d. Is it merely circumstantial?

4 What about the credibility of the witnesses adducing the evidence?
   a. Are they prejudiced?
   b. Are they reliable?
   c. Are they competent?
   d. Are the consistent?

5 What about the conflicting evidence?[8]

In answering each of these questions, a lawyer would systematically explore the evidence against and for the client and use the questions to generate ideas about the evidence. A technical communicator could use the same list of questions to explore the evidence about a technical topic.

Ceremonial Discourse

The third type of discourse described by Aristotle is ceremonial discourse. It is designed mainly to praise or censure someone, and includes introductions of speakers, job recommendations, and nominating speeches for offices in business, professional organizations, and politics. (In the nontechnical world, ceremonial discourse also includes obituaries or funeral eulogies, holiday or graduation speeches, and some kinds of sermons.) If you were going to prepare a ceremonial speech (such as the introduction of a speaker) or a written document such as a letter of recommendation, you should consider at least the following special topics originally suggested by Aristotle and evaluate the presence or absence of the characteristics they cover:

1 Moral virtues, including honesty, fairness, temperance, courage, generosity, prudence, kindness, and diligence

2 Philosophy or approach

3 Activities and accomplishments

4 Personal characteristics such as intelligence, education, diligence, organization, and communication skills

5 Testimonials of a subject's status from respected colleagues or peers

6 Other special qualities

❏ EXERCISE 2-2

Choose a subject you have to write about, or the ergonomic chair or electronic security system situation in Exercise 2-1, or the situation outlined at the end of this chapter—"The Clinic Case"—and then explore it using special topics. Define the field and genre or type of communication for your subject, and then decide what special topics would be appropriate for it.

Exploring the Subject Using Aristotle's Common Topics

The common topics presented here can be grouped into five categories: *definition, comparison, relationship, circumstance,* and *testimony*.[9] The common topics are outlined below, along with the kinds of questions they might raise about the ergonomic chair from Exercise 2-1. Although some of the questions seem to be pointed directly at the ergonomic chair for our illustration, they are really general questions which can be applied to any concept or thing. They will look more general for future uses if you mentally replace each instance of "ergonomic chair" with "it" in the following questions.

Definition
   A Genus: What kind of thing is it?
   B Division: What are its constituent parts?
Comparison
   A Similarity: What is it similar to?
   B Difference: How is it different from similar things?
   C Degree: What things does it differ from only in degree?
Relationship
   A Cause/Effect: What effect does it have? What causes that effect?

    B   Antecedent/Consequence: What did we have before the advent of the ergonomic chair? What are we likely to have in the future?

    C   Contraries: What is its likely opposition (other possible but opposing solutions)?

    D   Contradictions: Is there anything that argues against the use of the ergonomic chair?

Circumstance

    A   Possible/Impossible: What can it do? What can't it do?

    B   Past Fact/Future Fact: How has the ergonomic chair performed in the past? How will it probably do in the future?

Testimony

    A   Authority: What do experts say about it?

    B   Testimonial: What do users say about it?

    C   Statistics: What quantitative data can we use to talk about it?

    D   Maxims: Do we know of any platitudes about the ergonomic chair?

    E   Law: Do any laws pertain to it?

    F   Precedents/Examples: Have there been previous attempts to make or market it? If so, how successful were they?

Not all of these questions will be equally useful, but it is important to ask them anyway. Remember, the purpose of asking such questions at this stage is not to come up with definitive answers but simply to generate ideas. By going systematically through an inventory of questions, you will have more sources of information to draw from than if you just brainstorm randomly.

❏ EXERCISE 2-3

Choose a subject you have to write about, or the ergonomic chair in Exercise 2-1, or the case outlined at the end of this chapter—"The Clinic Case"—and then explore it using Aristotle's common topics.

2.4 USING SOCIAL AND ETHICAL CONSIDERATIONS TO EXPLORE AUDIENCE IMPACTS

The Impacts of Cooperative Idea Generation

There have been a series of impressive considerations and studies of "real world" communication in organizational or professional settings. Taken as a group, these studies outline the importance for a communicator of involving others in the planning and design of communication. Lester Faigley,[10] Lee Odell,[11] Charles Bazerman,[12] Karen LeFevre,[13] and Steven Doheny-Farina[14] have argued that to be effective in the workplace, technical and professional communicators must understand the social contexts in which they are working. They need to understand what kinds of questions their readers are likely to ask about their topic, what concerns these readers will want to see addressed. They need to know which members of their organizations they should contact for information and how to do this politely. In addition, in a case study of seven on-the-job writers, Rachel Spilka found that communicators who planned in isolation "made composing decisions that their readers considered to be socially insensitive and rhetorically inappropriate."[15]

Further, she argued that "writers need to interact with readers and other project participants early in the composing process to refine, revise, or expand their initial impressions of their audience and the unique social features" of their situation. When this occurred in her case studies, there were many benefits to the writers: readers tended to trust the writers' judgments, readers admired the writers' adherence to social norms and recommended writers for upcoming special projects, and writers found more ways to appeal to a multiple audience. Susan Kleimann[16] and Spilka also found that the most successful communicators interacted with their audiences and other project participants through a series of discussions and meetings and through circulating drafts of the communication for reader response. Kleimann found that many cycles of drafts plus feedback were needed to produce successful documents.

All these studies suggest that successful documents are often "negotiated" or cooperative products in that they involve the planning and editing inputs from many people. The planning inputs may occur (1) through formal idea generation or planning sessions; (2) through informal conversations or inquiries about institutional values, goals, concerns, or constraints; political considerations; technical considerations; or special reader considerations; or (3) through formal or informal reviews of written products. These results lead to the first set of questions for exploring social and ethical considerations:

1   Have you consulted with a range of readers and other project participants in determining the ideas which should appear in your communication and the form they should take?

2   Have you consulted with these readers in a variety of ways, including formal and informal conversations and reviews of drafts of your document once a draft is available?

The Ethical Impacts of Distortion, Incompleteness, Inaccuracy, and Bias

When people consider ethics in communication, the first thing which usually leaps to mind is the unethical use of information to portray a false or misleading impression. However, this leads to the not very helpful guideline "Avoid unethical uses of information." The converse of this is more interesting and more useful to us and provides our first ethical goal: Present information so that it is complete and accurate and clear. This goal will be relatively easy to achieve if we have information from the invention process which is accurate and complete and if we consciously desire to present the information as accurately and completely as possible. However, we will have problems meeting this goal if either our information or our intention is faulty. The issue of faulty intention is one which we will not address here, since it is not really an invention problem; it is treated under the topic of argumentation in Chapter 4. The issue of faulty information—inaccurate or incomplete information—is an invention problem, since the information is developed during the invention state.

Inaccurate or incomplete information can arise from several sources: poor research, incomplete thinking or biased analysis during the invention stage, unavailable information, or unconscious bias during analysis of available information. The first two problems—poor research and incomplete or biased thinking—are problems of conscious behavior and can be corrected consciously by changing your behavior and doing things properly. The third problem, unavailable data, is not correctable unless data suddenly become available. However, the last—unconscious bias during invention—is both hard to notice and hard to correct.

What kind of problem is bias in the real world? Bias could possibly arise from your desire or your company's desire to promote some point of view. For example, in our discussion of the ergonomic chair, you may be under some pressure to produce a marketing report which says that the chair is needed so that your department can have the business of working on the chair and employing you and your colleagues. If subtle pressure to keep your department employed is operating, you may have answered some of the questions in the previous sections in a more favorable way than you might have in other situations. If this is the case, a subtle bias has arisen in your analysis which may show up in the generation of inaccurate or incomplete information. Even if you are trying to do a fair and honest job on the marketing report, you may find yourself working with poor but apparently adequate information, since you generated it by carefully following the methods outlined above. Thus, the next set of questions for exploring social and ethical considerations is as follows:

3 Has the assessment of the prior topics and perspectives been conducted in an open, honest, and unbiased manner?

4 Have you tried to provide complete and accurate information?

The Ethical Impacts of Communication as a Cooperative Activity

Communication has often been depicted as a one-directional activity: writers and speakers "send" messages to readers and listeners who "receive" them. But this implies that readers and listeners are nothing more than passive recipients of information, which we know to be false. Extensive research in cognitive psychology during the past 15 years has shown that readers and listeners do not just receive information but actively interpret it. Indeed, each reader or listener uses his or her unique personal experience to help *construct* an interpretation. Thus, meaning is not the exclusive property of the writer (or speaker); meaning is *negotiated* between the writer (or speaker) and the reader (or listener). Communication is successful when the interpretation the writer intends is similar to the interpretation the reader constructs.

In other words, communication is inherently a *cooperative, social activity*. As with any cooperative activity, there must be a basic level of trust and goodwill among the participants. This does not mean that each participant always has to be completely honest with the others (there is sometimes a need to tell "white lies"); it does mean that each participant should not try to deceive the others in a harmful way. Otherwise, communication is not true cooperation. Plato and Aristotle argued that the ethical communicator is one who tries to promote the best interests of everyone involved in the communication; and that he or she does so not unilaterally but communally, that is, by providing readers or listeners with the information they need to make judgments that will serve their interests. As Dorothy Winsor[17] and Gregory Clark note, this often entails more than simply writing a report and sending it on.

> The engineers who monitored the performance of the *Challenger* booster rockets submitted reports documenting problems with the O-ring seals to the officials who made launch decisions. Apparently, however, those reports did not result in the officials being able to fully share the engineers' understanding of how the O-rings were functioning.[18]

Clark observes that what was needed in this situation was an active *exchange* of information,

> a continuing interaction in which the officials who had received the reports might have responded to the engineers with particular questions and offered them specific judgments, in response to which the engineers might have articulated further concerns and provided further interpretations that they had not thought to include at first.[19]

Thus, ethical communication means more than just providing complete and accurate information; it also means trying to *present* that information in a way that allows the reader to fully understand it.

How do such ethical considerations contribute to idea generation? First, they encourage communicators to explore the subject from the reader's perspective. If we know that readers draw on their own experience in constructing meaning, and if we see communication as a cooperative activity, we should try to put ourselves in the reader's place and generate ideas from his or her point of view. For example, if we're going to target our ergonomic chair advertisements at a certain segment of the market, we might ask ourselves questions like these: What kinds of prior notions might readers have about ergonomic chairs? What use can they get out of such a chair? What information do they need in order to decide whether buying such a chair would be in their best interest? Clearly, these considerations overlap strongly with audience analysis, which will be discussed further in Chapter 3. They lead to two additional prompts:

5   What is the reader's perspective?

6   How can information be presented to allow the reader to fully understand it?

Second, in many communication situations, the reader represents not only herself but other people as well. If the situation calls for the reader to make some kind of decision, she will want information that allows her to take the interests of these other people into consideration. After all, she is *answerable* to them; her own security depends on their support. Therefore, in generating ideas, we should broaden our perspective to include these other people, this "secondary" audience.[20] For example, suppose our marketing plan for the ergonomic chair has to be approved by the company president. If we know that the president will be concerned about what the company stockholders think, we would do well to generate ideas from *their* perspective. What are they likely to think of ergonomic chairs? How does the ergonomic chair fit in with the rest of the company's product line? What image will it give the company in the eyes of the larger society? And so on. Again, these kinds of considerations involve audience analysis; but it is a larger audience than just the intended reader. By recognizing the existence of a larger audience and by including this audience in the process of negotiating meaning, we implicitly embrace a more societal view of ethics, one that is more likely to keep the larger community's collective interests in mind. This discussion leads to our next social and ethical guidelines for invention:

7   What is the secondary audience for this communication?

8   Has that audience been considered in the invention stage?

Finally, ethical considerations contribute to idea generation when they cause us to consider the wider ethical implications of our arguments. What impacts will our arguments have on affected people and situations and are these impacts desirable? Such considerations can radically modify the argument and point which might otherwise be made. For instance, Steven Katz[21] has analyzed some of the memos written in Nazi Germany about the construction of the gas chambers and found that they are concerned with strictly technical issues—how to build a better and more efficient gas chamber. There is no sense that the writer considered the ethics of the entire situation—should one be building gas chambers in the first place? Or making them more efficient? If the writer had considered such issues and applied an ethical standard, the memos on increasing the efficiency of the gas chambers would not have been written at all. This leads to our final social and ethical guideline for invention:

9   What impacts will our arguments have on affected people and situations and are these impacts desirable?

2.5 FINDING INFORMATION IN LIBRARIES OR DATABASES

Once you've discovered the issues that you need to consider and generated the information you can by yourself, you may want to check the library to find books or articles on your topic. To do this, you will look up relevant authors, topics, and books in your library's card catalog and in indexes of journals in your field. You may also browse through relevant sections of the library. Depending on the capabilities at your institution, you may be able to search for items through a computerized version of your card catalog, and you may be able to do this yourself from your computer or from one in the library. (Some libraries require the librarian to conduct such a search for you.) If your institution does not have a computerized card catalog, you may have to physically go to the library's card catalog and indexes and to look for things by hand. While you're checking for information at your library, you might also check with your librarian about doing searches of national and international databases for information on your topic. There may be a charge for this service, but a good search of these databases can save you a lot of time. Once you have found information from the library or databases, you need to add this information to the ideas you have already generated.

2.6 FORMING A POTENTIAL THESIS OR POINT OF VIEW

Once you have explored your topic, found the answers to your questions, and generated a pile of ideas, you need to gather these together and see what they add up to. You need to identify the main point you think you want to make at this stage, realizing that it may change if you change your assessment of your audience (Chapter 3) and may change when you start organizing and structuring your ideas into a formal argument (Chapter 4). However, you need to begin forming an argument somewhere, and trying to pull out a possible thesis or point of view from the ideas you have generated is a good place to start.

Once you have identified a main point, you need to organize the rest of your ideas in some way to support that point. There are several ways this can be done, but two common ones are the outline and the idea (or node or tree) diagram.

Organizing Your Ideas in an Outline

If you decide to set up an outline, you should first decide what your main point is, then what your main lines of supporting argument will be (what the Roman numerals will be in the outline), and then what your subarguments will be. At each stage, you should think about only one level of your outline at a time. Later on, when you are specifically considering the nature of your argument (Chapter 4), you will probably rearrange the outline and fill it out, but outlining at this stage provides a way of beginning to organize your ideas.

Stage 1   State your main point.

Stage 2   State the main lines of argument to support your main point. At this stage your outline will be very simple, and it will be easy to reorder any of the main lines of argument if you need to. If you expect to be writing to busy readers who are likely to quit reading before reaching the end of the document, you should order your arguments from most important to least important. If your readers quit early, you want them to have heard your best arguments. Other issues of ordering are discussed in Chapter 4. At the end of Stage 2, your outline will contain only the main point and the main lines of argument, as indicated below. The outline on the right is a summary of the one with explanations on the left.

Main Point	Main Point
I first main line of argument	I
II second main line of argument	II
III third main line of argument	III

Stage 3   Add in the first level of subarguments to support your main lines of argument (the Roman numeral level of argument). At this stage your outline will be a bit more complicated, but you will be able to see where you are trying to go. Again, the outline on the right is a summary of the one with explanations on the left.

Main Point		Main Point	
I first main line of argument		I	
	A first subargument for I		A
	B second subargument for I		B
II second main line of argument		II	
	A first subargument for II		A
	B second subargument for II		B
III third main line of argument		III	
	A first subargument for III		A
	B second subargument for III		B

Stage 4   Add in the next levels of argument, one level at a time. If the arguments are not all equally important, consider devoting more space and proof to the more important ones. For instance, in the following outline, the arguments are presented in descending order of importance. Argument I is more important than argument II, which is more important than argument III. This level of importance is reflected in the amount of space and proof devoted to each argument: argument I has more proof than II, which has more proof than III.

Main Point			Main Point	
I	first main line of argument		I	
	A	first subargument for I	A	
		1 first proof for IA		1
		2 second proof for IA		2
		3 third proof for IA		3
	B	second subargument for I	B	
		1 first proof for IB		1
		2 second proof for IB		2
		3 third proof for IB		3
II	second main line of argument		II	
	A	first subargument for II	A	
		1 first proof for IIA		1
		2 second proof for IIA		2
	B	second subargument for II	B	
		1 first proof for IIB		1
		2 second proof for IIB		2
III	third main line of argument		III	
	A	first subargument for III	A	
	B	second subargument for III	B	

Organizing Your Ideas in an Idea Diagram

Sometimes it is helpful to get a visual idea of the relationships among your ideas by using an idea diagram (or tree or node diagram). An idea diagram is simply a visual outline of your main topics, with lines and spacing to help indicate the relationships among the ideas. Two sample idea diagrams appear in Figures 2-1 and 2-2.

❏ EXERCISE 2-4

Return to the subjects you chose in Exercises 2-2 and 2-3 and explore them once more using social and ethical considerations. Once you have finished exploring the topic, answer any unanswered questions, and then try to define a main point or point of view.

44

Figure 2-1 □ A SAMPLE IDEA DIAGRAM

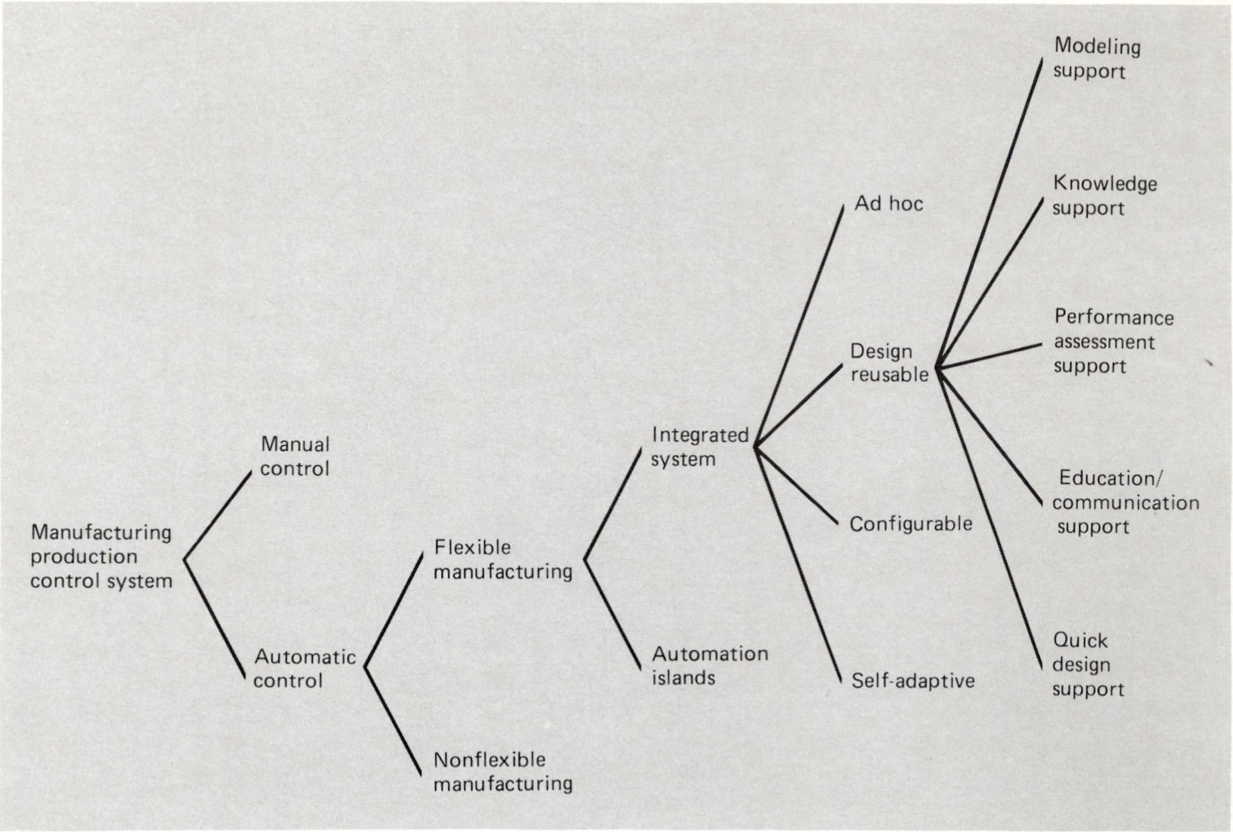

Figure 2-2 ❑ A SAMPLE IDEA DIAGRAM

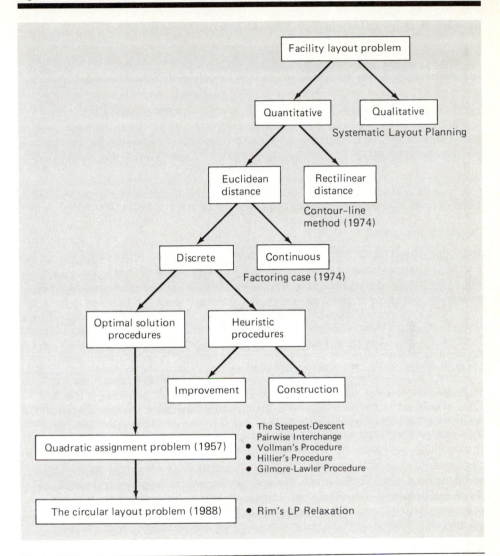

2.7 SUMMARY

In considering each of the ideas or approaches identified in the discussions above, you may have discovered that some were interesting lines of exploration while others were not. However, many of the most relevant ideas probably did not turn up in your original brainstorming session about chairs, but arose only through the process of systematic thinking. The point to note here is that if you can apply such systematic thinking to the subjects you need to write or talk about, you will probably generate more relevant ideas than if you only brainstorm or randomly think about your subject.

Situation 2-1 ❑ THE CLINIC CASE (by Mary Sue Garay[22])

Monday, June 5, 8:05 A.M.

You pull onto the Interstate and head south to Houston. You've accepted the 45–minute commute as the price you pay for living in the serene Woodlands subdivision north of Houston's sprawling and unplanned metropolis. The management consulting firm you work for, Molly Buchanan Associates (MBA), is conveniently located close to some of its major clients: the many doctors, clinics, and support services spawned by the five major hospitals in Houston's vast medical center. As you cruise along, you think about your main project for the day, a report on patient services for the Westheimer Clinic for Women.

The Westheimer Clinic

The work for Westheimer is involved but interesting. MBA is conducting a survey to assess patient satisfaction with the clinic's services. Dr. Larry Miller, who was on Houston General's board of directors when MBA helped turn things around there, is responsible for MBA's contract: Miller is now Director of the Westheimer Clinic.

Miller's clinic is in a precarious position. Although it is one of the oldest and most prestigious medical clinics in Houston, the Westheimer recently went through a much publicized malpractice suit. The suit was expensive in terms of increased insurance premiums and a tarnished reputation. Moreover, over the last 2 years the Westheimer has lost many patients because of one doctor's bout with cancer. Although other Westheimer doctors have covered ably for Dr. Johnson, the clinic has been unsuccessful in switching 40% of his patients to other clinic doctors.

In addition to these developments, the Westheimer is having to adjust to social and economic changes that have occurred during the 25 years it has been in business. First, there are more female doctors than ever before, and patients are increasingly preferring them to male doctors. Second, there are more hospitals and clinics in the Houston area. Finally, preferred provider organizations (PPOs) have come onto the scene. In a PPO agreement, a doctor, clinic,or hospital can contract with another party, usually an insurance company, to provide services to that company's patients. Westheimer has declined to be a PPO because the clinic fears that it would lose its independence. However, PPOs have proved popular in Texas, and some Westheimer patients have left the clinic to go with the PPOs their insurors selected.

The result of all these changes is a buyer's market for medical services in the Houston area which has adversely affected Westheimer. The clinic has lost 2000 patients in 5 years!

Despite this serious situation, some of Dr. Miller's colleagues are reluctant to have an "outsider"—even you and MBA—watching them. Consequently, the doctors have specified just what MBA can and cannot do. The survey is not to be "scientific." The doctors have stipulated that they each have anonymity as far as that is possible. They aren't interested in "witch hunts" or "personality contests." They have also specified that *the MBA report present only the findings of the survey, no recommendations.* The doctors will make their own decisions.

The questionnaire you devised was completed once by each patient who attended the clinic during the past month. The clinic has 19,000 patients and estimates that it sees 50 nonrepeat patients out of the 130 patients seen daily. That should have produced 1050 questionnaires over the 21 business days in the month, but owing to procedural irregularities only 1000 were actually completed and turned in. Julie, your assistant, collected the question-naires from the clinic each afternoon.

Your Task: Define the organizational problem and determine your task. Then use the survey results given below to identify any subproblems. Try brainstorming and using systematic prompts to generate ideas for possible solutions to the organizational problem and the subproblems. Then sift through these ideas to form a potential thesis or point of view. If you have a computer, you might try using a spreadsheet program or database program to sort the data and to check for patterns which emerge from the data.

WESTHEIMER CLINIC QUESTIONNAIRE

Instructions: Please help us evaluate our clinic by answering the following questions. These questionnaires will not be seen by any of our staff. They will be analyzed by the medical research/public relations firm of Molly Buchanan Associates. MBA will collect these questionnaires daily during the month of May, and then report to us. Please complete only one questionnaire during this time. We intend to use the results of the survey to improve our service to you.*

Thanks for your cooperation.

1. How long have you been a patient at this clinic?
   (20) 0–1 yr.   (390) 1–3 yrs.   (100) 3–6 yrs.   (490) 6 + yrs.

2. How old are you?
   (50) Under 20   (530) 20–35   (330) 36–50   (90) 51+

3. How did you choose the Westheimer Clinic for Women?
   (590) friend   (360) another doctor   (50) Yellow Pages

   If another doctor, please name. _____

      Roper (70)

      Amiss (60)

      Sears (50)

      Johnson (50)

      Mallard (50)

      S. Smith (30)

      A. Smith (30)

      Hardison (10)

      Fagliaux (10)

      Jemison (10)

4. How many appointments have you had in the past year?
   (60) 0–1   (20) 2–5   (20) 6–8   (170) 9–12   (730) more

5. How would you rate the overall performance of the clinic?
   (560) excellent   (260) good   (10) fair   (170) poor

*All numbers have been rounded to the nearest tenth. The number 00 indicates 4 or fewer responses.

6. I am satisfied with the way the clinic meets my emergency needs.
   (<u>610</u>) very satisfied   (<u>270</u>) somewhat   (<u>120</u>) not satisfied

7. What time of day is most convenient for an appoinment?
   (<u>770</u>) 9:00–11:00 A.M.   (<u>70</u>) 11:00 A.M.–2:00 P.M.   (<u>160</u>) 2:00–5:00 P.M.

8. What day is most convenient for an appointment?
   (<u>340</u>) Mon.   (<u>340</u>) Tues.   (<u>120</u>) Wed.   (<u>160</u>) Thurs.   (<u>40</u>) Fri.

9. Would you like the option of Saturday morning appointments?
   (<u>920</u>) yes   (<u>80</u>) no

10. When you make an appointment, what is the average length of time between your call and the appointment you receive?
    (<u>480</u>) 1 week   (<u>110</u>) 2 weeks   (<u>70</u>) 3 weeks   (<u>340</u>) 1 month

11. Receptionists are friendly and helpful when you call.
    (<u>10</u>) always   (<u>00</u>) usually   (<u>160</u>) rarely   (<u>830</u>) never

12. Is the waiting room large enough?
    (<u>70</u>) yes   (<u>930</u>) no

13. Are there enough reading mateials in the waiting room?
    (<u>270</u>) yes   (<u>730</u>) no

14. Check the magazines/newspapers you would like to see in the waiting room.

Houston Post (110)	*Houston Chronicle* (220)
Sports Illustrated (10)	*Reader's Digest* (160)
National Geographic (20)	*Ebony* (50)
People (100)	*Newsweek* (60)
Ladies Home Journal (90)	*Vogue* (80)
Redbook (110)	*McCall's* (70)
Prevention (150)	*Golden Years* (60)
Time (40)	*Wall Street Journal* (20)
USA Today (100)	*Women's Wear Daily* (30)

    Name any others: *Family Weekly* (120), *Crewel World* (10), *Perspectives on Infertility* (00), *Oui* (00), *Playboy* (00), *Playgirl* (00)

15. In what ways would you change the waiting room?

    Make the waiting room larger (340)

    Provide access to telephone (110)

    Buy more comfortable chairs (40)

    Add more windows (10)

    Put a TV in the waiting room (00)

    Provide coffee and doughnuts (00)

16. Nurses are helpful when I come in for an appointment.
    (<u>380</u>) always   (<u>330</u>) usually   (<u>100</u>) rarely (<u>190</u>) never

17. Nurses are helpful when I talk to them over the phone.

    (<u>610</u>) always   (<u>20</u>) usually   (<u>100</u>) rarely   (<u>270</u>) never

18. The examination area is clean.

    (<u>770</u>) always   (<u>140</u>) usually   (<u>70</u>) rarely   (<u>20</u>) never

19. How satisfied are you with the help you're given in filling out Medicare/Medicaid and insurance forms?

    (<u>00</u>) very satisfied   (<u>220</u>) somewhat   (<u>780</u>) not satisfied

20. I feel that my doctor cares about me as a person.

    (<u>280</u>) yes   (<u>410</u>) somewhat   (<u>310</u>) no

21. I am satisfied with my doctor.

    (<u>790</u>) always   (<u>70</u>) usually   (<u>130</u>) rarely   (<u>10</u>) never

22. My doctor explains my problems to me.

    (<u>240</u>) always   (<u>490</u>) usually   (<u>220</u>) rarely   (<u>10</u>) never

23. My doctor outlines treatment options for my condition.

    (<u>680</u>) always   (<u>190</u>) usually   (<u>60</u>) rarely   (<u>70</u>) never

24. I feel free to ask my doctor questions.

    (<u>40</u>) always   (<u>200</u>) usually   (<u>260</u>) rarely   (<u>500</u>) never

25. My doctor answers my questions to my satisfaction.

    (<u>10</u>) always   (<u>70</u>) usually   (<u>890</u>) rarely   (<u>30</u>) never

26. My doctor rushes me during conferences

    (<u>150</u>) always   (<u>230</u>) usually   (<u>190</u>) rarely   (<u>430</u>) never

27. I understand my doctor's answers and explanations.

    (<u>170</u>) always   (<u>220</u>) usually   (<u>160</u>) rarely   (<u>450</u>) never

28. In what way could your doctor improve the service he gives you? Please explain.

    - Be on time for appointment. (170)

    - Spend more time with patient. (120)

    - Be friendlier. (90)

    - Provide more information about treatment. (80)

      (<u>Example</u> [extreme]: "He could have given me more information on how devastating chemotherapy is, perhaps tested me to see if I could take it. As it was, I just did what he told me when I got breast cancer at age 53. Dr. Lewis said that I should have a lumpectomy and chemotherapy instead of a radical mastectomy. It's been 2 years since the chemotherapy began, and I am in constant pain. I may sue, especially since the last malpractice suit against Westheimer was so lucrative. My expenses are, and, according to my new doctor (from a different clinic), will continue to be exhorbitant for the rest of my life—however long that is. I'm just here today to sign the form that allows my medical records to be sent to my new doctor.")

    - Stop being so cheerful. (40) (<u>Example</u>: "If I felt good, I wouldn't be here.")

    - Remember patient's individual needs before examination/conference. (10)

      (<u>Example</u>: "I think that my doctor forgets who I am because I only come in once a year if I stay healthy. I hear his nurse whispering the things that I've told her about myself outside the door of the examination room. That's unnerving. It makes me feel like I have no history and that I'm getting ad hoc treatment. I would like to think when I come that I come with a history, that the doctor remembers my case and has reflected on it before he examines me. He could spend 30 minutes or so in the morning and afternoon prior to seeing us just going over our charts to jog his memory of who we all are and what we <u>as individuals need</u>.")

- Prescribe generic drugs. (10)

- Haven't been coming long (1–4 appts.) but doctor seems uninterested in them. (10)

- Make better recommendations. With clinic from 2½ years to 7 years. (10) (<u>Example:</u> "I have done everything that my doctor recommended, but I'm still not pregnant, and we're out a lot of money. We've probably averaged $1500 a year for the last 5 years. We're thinking about switching clinics.")

- Overly moralistic attitude. (10) (<u>Example:</u> "He could stop looking down at me. So what if I'm not a virgin! Lots of girls aren't virgins at 29. I'm fed up with his supercilious, moralistic attitude. I just want to get rid of this damn infection. My love life's shot to hell. I'll probably leave this place and try another doctor. Maybe a woman.")

29. Are there any services you would like to see added? Please explain.

- Take more time explaining procedures, drugs, options, how body works, by doctors, nurses, etc. (<u>Example:</u> "Somebody needs to answer general health questions. If the doctors don't have time—how about their nurses or some paramedical person?") (270)

- Take Mastercard (160)

- Provide on-site drugstore (70)

30. What do you like most about the clinic? Please explain.

- Doctor. (270) In order of descending mention: Miller, Lewis, Brady, Jones, Johnson, Simpson. (<u>Example:</u> "I love Dr. Miller. He's the only doctor who ever took the time to read my medical history. I spent a lot of time writing that out because you always hear about how heredity affects a person's response to disease. Still, Dr. Miller is the only one of my three doctors who ever read it. He's the best doctor in the world—he really understands me.")

- Doctors' availability for emergency needs. (110)

- Received successful infertility treatment. (40) (<u>Example:</u> "Finally, after 8 years and treatment from three other Houston clinics, Dr. Miller helped me have my first child last year. Now I'm pregnant again, thanks to him.")

- Location. (40) (<u>Example:</u> "I can get here on my lunch hour. I like that because then I don't have to take off from work.")

- Doctors available and solicitous throughout troubled times. (20)
(<u>Example 1:</u> "I had miscarried three times before my first baby was born, so I was pretty frantic during the pregnancy and all the hospitalization that went with it. But Dr. Johnson always called about me, even when he was sick himself. I don't know many people who would do that, much less doctors." <u>Example 2:</u> "When I had to have a hysterectomy 3 years ago, when I was 27, I nearly went crazy because I wanted to have children. However, Dr. Lewis was just wonderful. He put me in touch with a number of adoption agencies, and we're supposed to get a baby next month.")

- Clinic flexible about bills. (10) (<u>Example:</u> "Hospital bills for my premature baby were very high. The clinic allowed me to postpone paying my bills with them so I could pay the hospital first. I don't know what I would have done if the clinic hadn't allowed me to wait. I'm a working mother, and my husband left me a year ago.")

- Nurses. (10)

31. Which of the clinic's services would you most like to see changed? Please explain.

- Chart number system. (240) (<u>Example 1:</u> "I don't like always being asked to give my chart number or my husband's name. My chart number doesn't correspond with any other number I have to remember, so I always either have to look it up— a hassle—or give my husband's name. Why give <u>my husband's</u> name? He's not the patient and he's not paying my bills." <u>Example 2:</u> "Why can't they use Social Security numbers instead of the number they make up for the charts?")

- Have doctors remember drugs patients are allergic to. (200)

- Provide duplicates for bills to make filing insurance claims easier. (30)

- Have doctors offer classes for clinic patients every so often on items of general interest, e.g., breast cancer. (30)

Situation 2-2 ❑ PLURIBUS: A QUESTION OF INVESTOR INFORMATION (By Scott A. Goodhue. Used with permission.)

You work as an analyst at Pluribus, a mutual fund house which has operated for 5 years. Pluribus is a relative newcomer to the investment industry, and you landed a trainee position there 2 years ago. Because of the company's recent entry into the field, Pluribus managers are intent on providing steady returns and maintaining fine relations with their customers. Your responsibilities include writing the quarterly *Arrow Fund Newsletter*, which updates customers on the performance of the Arrow Fund, Pluribus's aggressive 3-year-old growth fund. You realize that the *Arrow Fund Newsletter* should inspire and secure investments in the Arrow Fund, and such a premise makes it a sort of marketing tool. Materials on other funds in the Pluribus family also accompany the newsletter as riders to prompt further investments.

Ian Monroe manages the Pluribus Arrow Fund, which is failing to perform as well as anticipated. Monroe's concern centers on improving the fund's performance, and he has been investigating potential vehicles. The Arrow Fund has returned no more than 16% for the past 2 years, against an expected 20–30% return. All hope that the Arrow Fund will rebound to its previous first-year high of 23% or beyond. This is Monroe's first position as a portfolio manager; but his experience includes several years as an assistant manager of a successful and similar fund at a reputable company.

In accordance with the Federal Securities Act of 1933, the Arrow Fund issued a prospectus to investors explaining the policies, objectives, and investment activities of the fund. Investors are expected to read and understand the prospectus prior to investing in the fund. The Pluribus Arrow Fund prospectus states that 80% of monies will be invested in common stocks of small, fast-growing companies and 20% in cash reserves. The investor should thus be well aware that the Arrow Fund is a high-risk investment, as it "aims at capital appreciation" rather than the preservation of income.

Market conditions of the late 1980s did not help the fledgling Arrow Fund. October 19, 1987—Black Monday—shook the entire investment industry, and shares of all Pluribus funds slid greatly. However, most Pluribus funds slowly rebounded from their lows, and some of those investors who stayed in began to approach or surpass the amount of their principal investments. But the Arrow Fund continued to suffer despite market revivals. One week ago, the Dow Jones

Industrial Average closed up 3.5 points, but the Arrow Fund dropped from $14.70 to $12.55 the same day, incurring a $2.15 loss per share. Because the Dow was up this day, Arrow Fund investors were startled at the drop in share price. An overwhelming number of phone calls from irate investors poured into Pluribus. The general response to the drop is one of confusion both inside and outside the company.

You realize that you need to explain the drop in the upcoming issue of the newsletter. Moreover, the Arrow Fund pays dividends next month, which will produce another drop in share value. It is obvious to you that the small investor worries about the entire market's climate, and you know that Monroe worries about a migration from the Arrow Fund toward more conservative investments. Being an Arrow Fund shareholder yourself, you sense that something peculiar happened on April 27. You overhear conversations which imply that Monroe took an unusually large gamble and lost. Your manager then asks you into his office to review the outlines for the *Arrow Fund Newsletter*. This is standard procedure.

Your manager tells you that Monroe did, in fact, take a great risk and the fund suffers from it. Because your manager is close to Monroe, and Monroe survived higher-level meetings regarding this matter, he wants to find ways to keep investors' confidence in the fund. Your manager asks you to "get your juices flowing" to help write a report that prevents investors from abandoning the fund. You then learn confidentially that Monroe took some insider advice that turned into a "bum steer." Instead of switching monies into areas with proven growth records, Monroe received a tip which seemed too good to ignore. In hopes of recouping and gaining on the Arrow Fund's returns, Monroe believed that some gold options were going to "certainly roll over, giving an opportunity to double an investment within a week's time." Monroe transferred 15% of the entire cash reserves into the options and lost a substantial amount. Luckily, "more stable investments" kept the loss from being devastating.

Gold investments are the highest risk of all, but Monroe thought gold was a "sure thing," and went ahead with it. Because gold options do not constitute possible investments for the Arrow Fund as laid out in the prospectus, you point out that Monroe put his investments in an unspecified category. Your manager tells you that the intentions were good and that a new strategy has been worked out that will remain within the fund's guidelines. You respond that you are in disagreement with Monroe's action. Your manager then speculates that if the options turned in Monroe's favor, there would be no reason to worry.

You say, "This whole situation makes me feel uneasy."

"How do you think I feel?" your manager says. "Because of the mammoth fluctuations since October 1987, no one can be sure how extreme the market may turn. Monroe realizes the fund should follow less-risky investments, and the fund will surely perform well with upcoming projections. We've been in meetings all week, and now it looks like we finally have the Arrow Fund on target."

Your manager then asks you to point to interest rates, computerized training, unemployment figures, and/or the national debt as the major overall problems with the Arrow Fund. Because these areas are improving, you then are to say that the Arrow Fund will likewise improve. Although these factors affect the fund's performance, they obviously are not the sole reasons. The fact that some of the issues held are in small high-tech companies with negligible growth is to be pointed to, together with the culmination of outside factors, for the April 27 drop. Your manager then tells you that at this point investors would benefit only by riding out the fund until it goes past its previous highs. Your manager hands you financial data to create the report.

Within the next few months, you plan to leave Pluribus because of a potential job offer from Lane and Company, another mutual fund house. The offer would lead to a promotion and a salary increase. Lane and Company wants to make an appointment in the near future to meet with you. You are waiting for an appropriate time to ask your manager for a letter of recommendation and know that for now the timing is poor.

Working for Pluribus has become disappointing, because lately you find that your job revolves around restoring investor support for a fund on the wane. At Lane and Company, you would be working with a successful fund with moderate investment risk. The position at Lane and Company is open, but not a definite offer. You would like to make the change, but want to remain on good terms with your manager so you can get a letter of recommendation from him.

Your Task:

1 You, as the writer, must make a decision to write up the report with the slanted information, or refuse to do so. If you refuse to write up the report, you are not sure whether Lane and Company will laud you for ethical behavior or label you a "turncoat." What do you do?

2 If the Federal Securities and Exchange Commission decides to investigate the matter, will this influence your decision?

REFERENCES

1 "Vitalist assumptions, which have dominated our thinking about the composing process since Coleridge, appear to be inconsistent with the rational processes and formal procedures required by an art of invention. Vitalism leads to a view of writing ability as a knack and to a repudiation of the possibility of teaching the composing process; composition tends to dwindle to an art of editing." Richard E. Young, "Invention: A Topographical Survey," in *Teaching Composition: 10 Bibliographical Essays*, edited by Gary Tate (Fort Worth, TX: Texas Christian University Press, 1976), pp. 20–21.

2 Carolyn R. Miller and Jack Selzer, "Special Topics of Argument in Engineering Reports," in *Writing in Non-Academic Settings*, edited by Lee Odell and Dixie Goswami (New York: Guilford Press, 1985), pp. 309–341.

3 *The Rhetoric of Aristotle*, translated by Lane Cooper (Englewood Cliffs, NJ: Prentice-Hall, 1932, 1960), pp. 55–89.

4 Donald N. McCloskey, *The Rhetoric of Economics* (Madison: University of Wisconsin Press, 1985).

5 Charles Bazerman, *Shaping Written Knowledge: The Genre and Activity of the Experimental Article in Science* (Madison: University of Wisconsin Press, 1988).

6 Victoria M. Winkler (Mikelonis), "The Role of Models in Technical and Scientific Writing," in *New Essays in Technical and Scientific Communication: Research, Theory, Practice*, edited by Paul V. Anderson, R. John Brockmann, and Carolyn R. Miller (Farmingdale, NY: Baywood, 1983), pp. 111–122. See also Miller and Selzer, op cit.

7 Explorations of legal systems can be conducted through a variety of well-defined approaches, including Aristotle's special judicial topics and Hohfeld's taxonomy of rights, responsibilities, and powers. See Wesley N. Hohfeld, "Fundamental Legal Conceptions as Applied in Judicial Reasoning," *23 Yale Law Journal* 16, 29 (1913); Layman E. Allen and Charles S. Saxon, "Analysis of the Logical Structures of Legal Rules by a Modernized and Formalized Version of Hohfeld's Fundamental Legal Conceptions," *Proceedings of Consiglio Nazionale delle Ricerche, IInd International Congress, Logica Informatica Diritto: Automated Analysis of Legal Texts*, Florence, Italy (September 3–6, 1985).

8 Edward P. J. Corbett, *Classical Rhetoric for the Modern Student*, 2d ed. (New York: Oxford University Press, 1971), p. 150.

9 These topics are discussed at some length in Corbett, pp. 107–155.

10 Lester Faigley, "Non-Academic Writing: The Social Perspective," in *Writing in Non-Academic Settings*, edited by Lee Odell and Dixie Goswami (New York: Guilford Press, 1985), pp. 231–248.

11 Lee Odell, "Beyond the Text: Relations between Writing and Social Context," in *Writing in Non-Academic Settings*, edited by Lee Odell and Dixie Goswami (New York: Guilford Press, 1985), pp. 249–280.

12 Charles Bazerman, "Scientific Writing as a Social Act: A Review of the Literature of the Sociology of Science," in *New Essays in Technical and Scientific Communication*, edited by Paul V. Anderson, R. John Brockmann, and Carolyn R. Miller (Farmingdale, NY: Baywood, 1983), pp. 111–122.

13 Karen Burke LeFevre, *Invention as a Social Act* (Carbondale: Southern Illinois University Press, 1987).

14 Steven Doheny-Farina, "The Individual and the Organization: The Relations of the Part to the Whole in Nonacademic Discourse." Paper delivered at the Conference on College Composition and Communication, Seattle, March 1989.

15 Rachel Spilka, "Negotiating with Readers from Multiple Corporate Cultures: An Exploratory Study of Writing in the Workplace." Paper delivered at the Conference on College Composition and Communication, Seattle, March 1989.

16 Susan Kleimann, "Negotiating to a New Text: Document Cycling in One Government Agency." Paper delivered at the Conference on College Composition and Communication, Seattle, March 1989.

17 Dorothy Winsor, "Communication Failures Contributing to the *Challenger* Accident: An Example for Technical Communicators," *IEEE Transactions on Professional Communication 31*(3):101–107 (September 1988).

18 Gregory Clark, "Ethics in Technical Communication: A Rhetorical Perspective," *IEEE Transactions on Professional Communication PC 30*(3):194 (September 1987). © 1987, IEEE.

19 Clark, p. 194.

20 The concept of secondary audience has been developed by J. C. Mathes and Dwight W. Stevenson, *Designing Technical Reports* (Indianapolis: Bobbs-Merrill, 1976), Chapter 2.

21 Steven Katz, "The Ethics of Expediency: Classical Rhetoric, Technology, and the Holocaust," *Journal of Technical Writing and Communication*, forthcoming.

22 From Mary Sue Garay, "Clinic Case: A Case to Elicit Point-Making during Revision of Survey-Based Reports," *The Technical Writing Teacher*, *16*(2):155–168 (Spring 1989), with slight modifications. Used with permission.

ADDITIONAL READING

James Adams, *Conceptual Blockbusting: A Guide to Better Ideas* (San Francisco: Freeman, 1974).

Thomas W. Benson and Michael H. Prosser (eds.), *Readings in Classical Rhetoric* (Davis, CA: Hermagoras Press, 1988).

Cicero, *De Oratore*, Loeb Classical Library, Books 1 and 2 (1959); Book 3 (1960) (Cambridge, MA: Harvard University Press; London: William Heinemann Ltd.).

Cicero, *Ad C. Herennium*, Loeb Classic Library (Cambridge, MA: Harvard University Press; London: William Heinemann, 1977).

Chaim Perelman and L. Olbrechts-Tyteca, *The New Rhetoric: A Treatise on Argumentation*, translated by John Wilkinson and Purcell Weaver (Notre Dame, IN: Notre Dame Press, 1969).

D. N. Perkins, *The Mind's Best Work* (Cambridge, MA: Harvard University Press, 1981).

Quintilian, *Institutia Oratoria*, Loeb Classical Library, 4 vols. (Cambridge, MA: Harvard University Press; London: William Heinemann, Ltd.).

Richard M .Weaver, *The Ethics of Rhetoric* (Davis, CA: Hermagoras Press, 1985).

James P. Zappen, "A Rhetoric for Research in Sciences and Technologies," in *New Essays in Technical and Scientific Communication*, edited by Paul V. Anderson, R. John Brockmann, and Carolyn R. Miller (Farmingdale, NY: Baywood, 1983), pp. 123–138.

3

Identifying Audiences and Purposes

Early in the process of thinking about a topic, you should identify your audiences and your purposes. Who are your readers or listeners? Why are you communicating with them? What do you expect or hope them to do? Unless it's clear in your mind what the answers are, you won't be able to construct a clear and effective piece of communication.

3.1 AUDIENCES

Before we proceed any further, we would like you to read the technical reports excerpted in Figures 3-1 and 3-2. Each was written by an engineer. Try very hard to spend *no more than 1 minute on each report.* When you have finished reading them, take another minute or two to decide which one you feel is better and why.

Have you now read the two reports? Which one do you think is better? Most people at the beginning of a technical writing course prefer the report in Figure 3-1; but at the end of the course, most prefer the report in Figure 3-2. Why does this happen? The report in Figure 3-1 looks very businesslike: it has a nicely indented list and lots of impressive data. And at first glance it appears to provide just what the senator's public relations officer has asked for—data on the Y-Ships.

However, if we take a second look at this report and consider its audience and use, we get a somewhat different picture. The public relations officer, or someone else on the senator's staff, will be using this report to write a speech for the Y-Ship launching. Since one usually expects the most important

Figure 3-1 ❑ BEGINNING OF AN INFORMAL REPORT BY A NAVAL ARCHITECT
(Used with the permission of J. C. Mathes.)

To: XXXX, Public Relations   December 6, 19xx

From: YYYY, Naval Architect

        Y-SHIPS: VESSEL CHARACTERISTICS

I understand Senator Q's Office has requested data on the Y-Ships for use in connection
with his participation in launching the PRESIDENT Y-SHIP at Pascagoula, Mississippi.

Following are physical characteristics of these ships:

Length overall	572'-0"
Beam	82'-0"
Depth to Main Deck	45'-6"
Designed Draft	28'-4"
Maximum Draft	30'-7"
Displacement at 30'-7" Draft	21,230 Tons
Cargo Deadweight	10,000 Tons
General Cargo Capacity	770,000 Cubic Feet
Refrigerated Cargo Capacity	48,000 Cubic Feet
Liquid Cargo	40,000 Cubic Feet or 1,000 Tons
Container Capacity (8' x 8' x 20')	144 containers
Passengers Carried	12 in Eight Staterooms
Crew	45

Propulsion: Steam turbine developing 24,000 Horsepower driving a single 22'-6"-diameter
five-bladed propeller.

Cruising Speed: 23.0 Knots

Cruising Radius: 11,600 Nautical Miles

PRESIDENT Y-SHIP is the first of a new class of vessels known as the Y-Ships (Design
C4–S-69a). The design was developed by ABC Company, Naval Architects, New York, to
meet design and performance characteristics and service requirements established by ST
Lines for vessels operating on Trade Route Number 17. This route links both the east and
west coasts of the United States with the Orient and involves a round trip in excess of
30,000 miles. The 23–knot continuous sea speed capability is optimum for the long ocean
legs in this service.

The Y-Ships are notable in several respects. They are the first merchant vessels in the
world to be built almost entirely of low-alloy, high-strength steel. The result is a weight
saving of approximately 330 long tons, as compared with a similar ship built of conventional
shipbuilding steel. This is reflected in an equivalent increase in cargo capacity.

information to come first, if the speechwriter is not paying careful attention, he or she may produce an unusual speech:

> Good afternoon, ladies and gentlemen. We are gathered today to launch a Y-Ship of overall length 572', beam 82'-0", depth to main deck 45'-6", designed draft 28'-4" . . .

Obviously, such a speech would not fit the occasion. The information needed for an appropriate speech is at the end of the report segment: that this Y-Ship is the first of a new class, that it will link the east and west coasts of the United States with the Orient, that these ships are the first merchant vessels in the world to be built almost entirely of low-alloy, high-strength steel, and that they are unusually light and have greatly increased cargo space.

Given the purpose of this report, why do you suppose the writer put the technical information first and the more general information second? Perhaps because he was much more interested in the technical information. This naval architect had just spent months designing the Y-Ships, and each of the numbers in that list represents many decisions he took pride in. The problem is that he was writing for the writer, not for the reader.

By contrast, if we look more carefully at the report to Lt. Cousins in Figure 3-2, we notice that it is designed for the reader. It puts the information that Lt. Cousins needs at the beginning and saves the details for later; it tells where to find the details but does not clutter up the beginning of the report with them. Notice, however, what *is* at the beginning: a polite thank you for a recent visit to the ship, a statement about the report's topic, the specifications that need to be met, the acceptability of the proposed clutch (it will meet the specifications), and the costs of the clutch. These are the kinds of information Lt. Cousins needs to do his job: he must be able to see that the proposed clutch will do the job, be installable, and be affordable. Thus, Figure 3-2 presents a report written for the audience. It orders information in a helpful way, pushing generalizations to the top of the report and details to the bottom. It doesn't waste the time of a busy reader.

How do you learn to write for your audience? You must first learn the characteristics of the audience and then figure out what the audience needs to know. Let us consider each of these in turn.

Characteristics of Real-World Audiences

If you think about the audiences you have written for in school, you'll notice they have the following characteristics. They usually consist of one reader (the teacher) whom you know personally. This reader usually knows more about the subject than you do. You can expect this reader to read your entire paper. (In fact, you would have a legitimate complaint if the reader didn't!) Since the purpose of school writing is to "show what you know," the teacher often puts a great deal of emphasis on the details that show in-depth knowledge. Thus, the teacher reads carefully, trying hard to figure out what you have to say (usually the teacher is trying to decide how much you've learned in order

Figure 3-2 ❑ BEGINNING OF AN INFORMAL REPORT BY A SALES ENGINEER
(Used with the permission of J. C. Mathes.)

January 29, 19xx

CGC <u>Boutwell</u>
c/o U.S. Coast Guard Base
427 Commercial Street
Boston, Massachusetts

Attention: Lt. (j.g.) G. L. Cousins

Subject: Clutch to disengage turbine starting pump

Gentlemen:

Thank you for the courtesy extended our representative, Ed Driscoll, on his recent visit aboard the <u>Boutwell</u>, at which time he discussed with you your requirement of the clutch to be used to disengage the turbine starting pump from the main generator engine.

We understand you wish to mount the clutch on a shaft which would be turning 720 RPM, and that the duty of the turbine pump is 117 HP. Based on this, the torque requirement is 853 pound feet, and a clutch having 16.3 HP per 100 PRM is indicated. Twin Disk Model #CL-310 is rated 873 pound feet, 16.6 HP per 100 PRM, and 135 HP normal duty. The #CL-310 therefore would seem to fill the bill quite nicely.

We refer you to Twin Disk Bulletin #326–B, enclosed. On page 10 you will find the description, capacities, etc., of this clutch. You will also note, it is available in 2¼″ and 2⁷⁄₁₆″ bore. Accessories are described on page 11. Two possible spider drive arrangements are suggested, as described in Figures 2 and 3 on the back cover of the bulletin. The spiders and their dimensions are indicated on pages 18 and 19. We are pleased to quote as follows:

1	XA5752 Model # CL-310 Twin Disk Clutch in standard bore of 2¼ or 2⁷⁄₁₆	$131.20
2	Part #3507 Throwout Yoke	2.25
3	Part #3039 Hand Lever	1.49
4	Part #1144–E Operating Shaft	2.18

---

to give you a fair grade). If something isn't clear, the teacher can often supply missing information or see where an unclear argument is going.

In contrast to this ideal situation, the audiences you will normally have to deal with in a real-world setting are much more difficult. First of all, these audiences are likely to consist of *a variety of readers*; instead of a single audience for a single communication, you may well have *multiple audiences* for that one communication. These readers will probably know less about the subject than you do, which means you'll have to *explain* things to them. They may differ in background knowledge, in needs and purposes, and in reading conditions. Thus, they may differ in their *reading strategies*, some reading only one part of the document, others skipping from section to section, still others studying every word. If something isn't clear to them, they may make no effort to figure it out.

In a real-world setting, it's no longer easy to "psych out the professor"—that is, to put into a paper just what you know a single reader wants. Nonetheless, it's just as important to try to understand and satisfy all of your various readers' needs. In the next few pages, we discuss several of the largest and most important types of audiences you are likely to encounter: managerial audiences, nonspecialist audiences, peer audiences, international audiences, and mixed audiences. This is not an exhaustive list, but it serves to point out some major variables you should be aware of.

Managerial Audiences

Managers are often the most important audiences for technical communicators because they make decisions that affect projects and careers. If scientists, engineers, and other technical professionals hope to influence these decision makers favorably, they had better understand how managers work and what will catch their attention.

The nature of managerial work has been studied by a number of researchers, and one of them, Henry Mintzberg, has made some particularly telling observations. Mintzberg notes that managers fill many roles at one time. These roles largely determine how a manager communicates. According to Mintzberg, the fundamental managerial roles are *interpersonal roles:* figurehead, leader of the unit, liaison to external units. However, since managers have many contacts from their interpersonal roles and their power, they also have important *informational roles:* they are the "nerve centers of information." As such, they monitor and disseminate information and serve as the spokesperson for their units. Finally, managers have *decisional roles:* they are initiators of change—supervising up to 50 new projects at a time—disturbance handlers, resource allocators, and negotiators.[1] In short, as Leonard R. Sayles has put it, a manager

> is like a symphony orchestra conductor, endeavouring to maintain a melodious performance in which the contributions of the various instruments are coordinated and sequenced, patterned and paced, while the orchestra members are having various personal difficulties, stagehands are moving music stands, alternating excessive heat and cold are creating audience and instrument problems, and the sponsor of the concert is insisting on irrational changes in the program.[2]

The effects of this environment on communication are severe; managers obviously have little time or attention to spare for careful reading or listening. For instance, according to Mintzberg:

1 Half the activities engaged in by the five chief executives of his study lasted less than nine minutes, and only 10% exceeded one hour.[3]

2 These five chief executives treated mail processing as a burden to be dispensed with. One came in Saturday morning to process 142 pieces of mail in just over three hours, to "get rid of all the stuff." This same manager looked at the first piece of "hard" mail he had received all week, a standard cost report, and put it aside with the comment, "I never look at this."[4]

Rather than being reflective, systematic planners, as is commonly believed, managers actually "work at an unrelenting pace, . . . are strongly oriented to action and dislike reflective activities."[5]

Given this almost impossible set of handicaps, what can you do to make a memo or report more easily read and more fully understood by managers? One thing is to make key information maximally accessible by *foregrounding* it—that is, by putting it up front where the manager can easily find it. A study of Westinghouse executives by Professor James Souther found that all read the Abstract or Executive Summary and most read the Introduction, Background, and Conclusions sections. By contrast, only 15% read the body of a report. In general, these managerial readers looked for important generalizations and tended to ignore details.[6]

Nonspecialist Audiences

Nonspecialist audiences are often the most difficult audiences to write for. These are the readers who know little about a subject but will be reading your writing in detail to find out more. (Managers are also nonspecialists, but they typically ignore most details.) If you are writing a proposal to a potential client, you are probably writing for a nonspecialist audience. If you are responding to a letter of complaint from an angry customer, you are probably writing for a nonspecialist audience. If you are writing a set of operating instructions, you are probably writing for a nonspecialist audience. There are many other situations like these. You yourself become a member of the nonspecialist audience whenever you read something out of your field.

The difficulty of writing for nonspecialist readers is this: As a specialist, you become used to thinking about certain topics in certain specialized ways. Your knowledge of these topics is so great that you've organized it in your mind into a network of manageable chunks, each consisting of many smaller chunks. And you've given special labels, or "technical terms," to many of these chunks. By simply using these terms, you've found it easier to think about these topics. In fact, it may be somewhat difficult for you to think about such topics *without* using your technical terms. The problem, of course, is that nonspecialists do not know those terms and do not have all of those chunks of knowledge organized into a nice, coherent network. If you insist on using only your technical language, you fail to create a bridge of common knowledge between you and your nonspecialist reader. And this makes communication difficult, perhaps even impossible.

There are a number of things you can do to make life easier for the nonspecialist reader. They boil down to one basic principle: Refer to "common knowledge" as much as you can without distorting the technical content of your message. You can begin doing this at the very outset by using a *conventional mode of presentation*. Standard genres like the proposal, the short report, and the operating manual are familiar to most people with whom you're likely to be communicating. By sticking to the conventions of a well-known genre, you enable the reader to have a better sense of the overall flow of your logic. Providing an *overview* at the beginning of your document also helps.

Another way of making the nonspecialist reader feel more at home is to provide some *background information*. You do not necessarily have to put it all up front (where it might get in the way of more knowledgeable readers). Rather, you can work it into the document here and there, much as journalists do in their news reports. But it should be true background information in the sense that it is explanatory and free of jargon.

In general, writing for nonspecialists means using lots of *explanations*. For relatively simple concepts, this can usually be done by using embedded *definitions* . Earlier in this chapter, for example, when we used the technical linguistic term *foregrounding*, we made a point of immediately defining it for you. You can do the same with many of your own technical terms. The least intrusive way of defining a term is by using a short paraphrase enclosed in parentheses: "A common health problem in many countries is hypertension (high blood pressure)."

For more complex concepts that are crucial for understanding the text as a whole, you may want to use *examples*. By taking something abstract and making it more concrete, examples are extremely powerful aids to understanding. However, they should be chosen carefully. They should highlight important features in a very direct and obvious way. Nonessential features should be avoided. And everything about the example should already be familiar to the reader.

Illustrations (photographs, drawings, graphs, etc.) are also helpful in making things clear to a nonspecialist reader. Like examples, illustrations are usually powerful and vivid. They attract attention, and they are memorable. But as with examples, they must be carefully chosen or their power will work against you. Make sure that your illustrations are not so specialized that a nonspecialist audience will not know how to interpret them. And make sure they focus on the concept you're trying to illustrate, not something else. (See Chapters 8 and 9 for more advice on using illustrations.)

Finally, for special occasions, you may want to try *analogies* (or "verbal illustrations"). These often require quite a lot of imagination, and they can be misleading if they are not designed carefully. Therefore, you should use analogies only as a last resort, when none of the techniques mentioned above will work.

More extensive discussion of how to make your writing accessible to nonspecialists can be found in Chapter 5.

Peer Audiences

Sometimes you have the luxury of writing for peers, people who know as much about the subject as you do. For example, if you're a condensed-matter physicist and are writing an article on condensed matter for *Physical Review*, you're probably writing primarily for other condensed-matter specialists. If you're part of a design team that's been working on a project for several months, you probably write memos and notes to other team members who are as familiar with the project as you are.

In cases like these, you effectively "speak the same language" as the people you are writing to. Therefore, you don't need to do many of the things you would do for a nonspecialist audience, such as giving lengthy explanations, defining terms, and using many examples. In fact, if you used such devices with your peers, they might accuse you of being patronizing. With peers, you should use language the way most other people in your field use it. You should

1 Use standard technical terms.

2 Use a conventional format.

3 Emphasize data and display it in standard ways, using graphs, tables, equations, or other appropriate forms.

4 Use standard forms of reasoning and argumentation.

5 Make your main points clear and accessible.

6 Be careful not to overstate your claims.

Given the basic facts of a situation, experts can usually fill in whatever gaps there might be. They can make *inferences*. This means that you do not need to spell things out for them. And this allows you to be concise in your writing. Experts are somewhat like managers in that they like to scan a piece of writing for its main points. A concise, well-structured piece of writing makes this easier for them. But experts differ from managers in that they often want to examine certain minute pieces of data. As an expert yourself, you should anticipate this and decide what data you think your peers will want to look at. Then you should present it in a clear, accessible form.

Chapter 18, "Theses and Journal Articles," and Chapter 21, "Making Your Writing Readable: Information Selection," give further advice on how to write for your fellow experts.

International Audiences

The world is getting smaller. National economies are becoming more and more interdependent. International scientific conferences have become common-place, multinational corporations abound, and foreign trade is at an all-time high. Technical professionals have been caught up in all this activity as much as anybody. More and more scientists, engineers, administrators, and businesspeople are traveling abroad, interacting with foreign colleagues, and negotiating with foreign companies and clients.

Increasingly, the language being used for these kinds of activities is English. Indeed, English has become the international language of science, technology, and commerce. If you happen to overhear a conversation between a Singaporean businesswoman and a Brazilian businesswoman, it's likely to be in English. If you glance at a list of biomedical articles in the *Index Medicus*, you'll probably find that most of them are written in English. Today, there are more nonnative speakers of English in the world than there are native speakers.

We are fortunate to have a language like English for international communication. But this does not mean that all language problems are eliminated. Most nonnative speakers of English do not have full command of the language. You should keep this in mind and try to use a "controlled" form of English. Specifically, you should:

■ *Avoid long or complicated sentences.* Other languages often have different ways of forming sentences, and forcing nonnative speakers to figure out English syntax will only make things difficult for them. This is especially important to remember when writing step-by-step instructions.

■ *Avoid idiomatic vocabulary.* Although many nonnative speakers have large vocabularies, many others do not. They may depend on a few thousand basic words, the kind that are found in small pocket dictionaries. They may easily be confused by idiomatic language, slang, or multiword phrases. Instead of saying, "I'm feeling a little *under the weather*," say, "I'm feeling a little sick." Avoid sports-oriented and other culture-specific slang like *out in left field* and *do an end-run*. Be aware that English combines common verbs (*make, turn, put*, etc.) and particles (*up, out, over*, etc.) to create very different meanings; just think of the differences between *turn down, turn into, turn up, turn over*, and *turn out*—each of which has at least two different meanings of its own. Use such expressions only if you cannot think of a simple one-word equivalent.

If you are using examples or analogies, try not to make them too culture-specific. Look for opportunities to use visual aids instead of verbal explanations.

Mixed Audiences

Probably the most difficult audience for a writer to deal with is an audience consisting of managers, nonspecialists, experts, and nonnative speakers, or some smaller combination of these. Unfortunately, such mixed audiences are commonplace. For example, proposals to do funded research are read by both generalists and technical specialists. The generalists usually make the final decision, but their judgment is deeply influenced by that of the specialists. If you are writing such a proposal, you'd better make sure it's comprehensible and persuasive to both groups. Or consider the many situations where you write a letter or memo to a single person but list other people to receive copies. Those people on the copy list constitute an audience whose knowledge and interests may differ significantly from those of the primary addressee. In some cases, the copied audience may actually be more important than the addressee.

International communication, of course, is likely to involve both native speakers and nonnative speakers, as well as managers and experts. We could easily give many other examples. How can you best meet the needs of such mixed audiences? There are basically two ways:

1 "Layer" the document so that different sections are aimed at different audiences.

2 "Democratize" your writing so that all audiences can understand all parts of it.

A common example of the layered approach is the intermediate-length or long technical report: the first page contains background information and recommendations for the managerial reader, while the body of the report and the appendixes contain details that are usually of more interest to the specialist. Long proposals are usually layered, as are many user manuals. The layered strategy works well in many circumstances, and is generally easy to follow: you can "switch hats" from one section to another or use a mixed team of writers.

There are many situations, though, where a mixed audience will want to read an entire document rather than just selected parts of it. This is where the "democratic" strategy works better. You aim each part of the document at its most important audience, but you add little touches that make the parts accessible to other audiences as well. For example, let's say you've written an in-house proposal to modify the electronic mail system in your company. You figure that the proposal will be distributed to upper management, to department heads, to computer systems technicians, and, indirectly, to other users of the system (engineers, clerical staff, etc.). You expect that upper management will be interested primarily in the main points of the proposal, so you gear the Executive Summary to that audience. You expect that department heads and other users will want a detailed account of how your modification will work, and so you tailor the Narrative of the proposal to them. You assume that the technicians will be most interested in the technical details, so you provide a section for them discussing implementation procedures, security problems, systems reliability, etc.

What you've created so far is a traditional layered document. But you know that some of the upper-level managers who do not use the current mail system would be interested in using your modified one. They might well want to read the Narrative of your proposal. Most of the technicians will also want to know how the new system is supposed to work—but in their language. Some of the engineers in the user group, meanwhile, will be interested in, though not extremely knowledgeable about, some of the technical details. To satisfy the diverse needs of all these readers, you need to do more than create a layered document. You need to look closely at each part of the proposal and add special information for each of your secondary audiences. For example, the Narrative may contain terms that are unfamiliar to nonusers, such as those in the upper-management audience; you should include brief definitions or even short explanations for those readers. At the same time, you might want to insert technical terms at appropriate places for the benefit of interested technicians. The engineers, meanwhile, might appreciate some brief definitions or explanations of technical concepts in the Technical Details section.

Writing a "democratized" document for a mixed audience is not easy. But in many situations it is more effective than the more traditional layered document. You can find more detailed advice on how to write definitions, introductions, and other explanatory material in Chapters 4, 6, 12, 13, 16, and 18.

❑ EXERCISE 3-1

Rewrite the informal report in Figure 3-1 so that it will be more immediately useful for its intended audience.

❑ EXERCISE 3-2

(For native speakers only.)
Revise the following electronic message so that it will be easier for an international reader to understand:

> Dear Ms. Poirot:
>
> Just thought I'd touch base after our recent get-together in Lyons to let you know that I'm interested in hearing more about your new MAX synthesizer. Considering my company's needs, it may be just what the doctor ordered. First, though, I need to know more about the actual nuts and bolts. For example, how does it cope with in-between pitch levels? Or off-key problems? Sometimes we get our hands full of those, with the F_1 signal fluttering like a knuckleball and the F_2 going absolutely bananas. If MAX can cut the mustard with these kinds of problems, we're definitely talking business!
>
> Sincerely,

A Procedure for Audience Analysis

Identifying audiences, their characteristics, and their needs is one of the most important jobs communicators have; it determines what kind of information must be provided and how it should be presented. However, identifying audiences can be difficult, since audiences vary with the situation and with the type of communication involved. For instance, technical reports and memos written within an organization usually have a wide range of audiences, perhaps including laboratory technicians, technical specialists, managers, lawyers, and specialists in marketing and finance. In contrast, technical articles written for specialized journals have a narrow range of audiences, usually just the specialized readers of the journal.

The five-step procedure outlined below can help you identify your audience. If you consciously follow these five steps, you should be able to produce a good assessment of your audience(s) and a more effective communication.

Identify the Communication's Uses and Routes

To make sure you identify all of your audiences, you should think carefully about the uses your communication will have and the routes it may travel, since these will often be more diverse than expected. Here is an example: A computer engineer was working for a company that makes small computers to control industrial machines. He had been asked by his immediate supervisor to improve the program by which a user tells a computer, the Model ABX, what to do. The engineer wrote a new program for the ABX and sent it off to his supervisor with the note, "Here's the program you asked me to write last week to improve the control of the ABX." The supervisor forwarded the engineer's note and program to the head of the Programming Department, who then forwarded it for review to a second computer engineer, to the head of the Hardware Design Department, and to the head of the Marketing Department. The second computer engineer knew a bit about the project because she had talked to the first engineer about it briefly over lunch one day, but the two department heads knew nothing about it. Since they could not review the new program themselves, they forwarded it to people in their departments who they hoped could do the review: a junior engineer in Hardware Design and a project engineer in Marketing.

All of these reviewers were faced with the task of evaluating the first computer engineer's new program. The second computer engineer had to check the program for programming adequacy. The hardware engineer needed to see how difficult it would be to implement the new program, since his department would have to redesign some circuitry to accommodate it if the company adopted the program for the ABX. The marketing engineer had to evaluate the sales potential of the new program. How much would it increase the cost of the ABX? Would it be something customers would want and be willing to retrain their employees to use? In such a situation, this engineer normally surveyed customers, calling them up, explaining the features and costs of a new program, and getting their response to it.

Unfortunately, the first engineer had not anticipated these reviews, so his communication did not include an overview or adequate documentation for the new program or a discussion of its improvements on the old program. Thus, the reviewers could not review the new program as it stood. To do their jobs, they had to go to the writer and interview him to get the information that should have been provided in a short report or memo attached to the new program.

The communication situation at first seemed straightforward for the first engineer. He assumed that he had to report only to his immediate supervisor, who had handed him the assignment. In fact, he communicated with—in addition to his supervisor—his department head, the heads of two other departments, two engineers in different departments, and one in his own. In addition, if the report had been adequate, the marketing engineer would probably have used it as the basis of a communication to a number of customers. Thus, as illustrated by this example, *you should identify the uses your communication will have and the routes it will travel.*

Identify All Possible Audiences

Because technical people work in organizations, their work is often picked up and extended by other people in other places. For instance, they may work on a project that goes on for years and passes through many stages, such as those as illustrated below.

Problem Perceived →	Problem Evaluated →	Product Designed →	Product Evaluated →	Product Constructed →
Product Tested →	Product Refined →	Product Retested →	Product Produced →	Product Improved → etc.

During each of these stages, someone will provide information about the project which may be used immediately by someone else in the communicator's organization, or by someone outside the organization, or later on by someone connected with the project. Thus, as a second step in audience analysis, *you should try to identify all the possible audiences, current or future, for a given communication.*

Identify the Concerns, Goals, Values, and Needs of Each Audience

Technical people operate in organizations composed of units with different concerns, goals, values, and needs. Engineers in industrial settings, for instance, must cope with, at least, a marketing unit, a manufacturing unit, an engineering or design unit, and a management unit. That these units have different perspectives is illustrated by the following observation:

> Sales-minded persons tend to recommend the development of machines which resemble those being marketed by successful rival firms—partly because the competitive models are known to be in good demand and partly because a program of imitation takes a minimum of time to execute. The design engineering department, on the other hand, tends to propose the development of machines which are, at the very least, improvements over the competitor's design, or, in many instances, machines which are totally new by comparison to the offerings of the current market. Although the end result is usually more desirable, the development time required is always much greater and less ponderable because of the unknown factors which may be involved. These viewpoints may be summed up by saying that the sales department takes a short-run view while the design engineering department takes the long-run view. Before we yield to our natural bias as engineers and exponents of progress (whatever that is) and condemn the sales department for its short-sightedness, let us bear in mind that the money is made in the short run, and in the long run we are all dead.[7]

Thus, to ensure that they communicate effectively to their audiences, technical communicators should be aware of the differing perspectives of their audiences. In other words, *you should identify the concerns, goals, values, and needs of your audiences.*

Make Communication Appropriate for Managers

Because technical communicators work in organizations, any project in which they are involved requires the integration of many perspectives and goals. This integration must be achieved at the technical level but is directed at the managerial level. Thus, to make appropriate integration possible and to gain the needed support of management for your own projects, *you should make your communication accessible and useful to busy managers.*

Identify Each Audience's Preferences for and Objections to the Arguments

The integration of perspectives and goals mentioned above will necessarily involve compromises or even occasional setbacks when conflicts between units occur. Thus, because there may be others within your organization who are competitors for scarce resources and are potential critics of your work, *you should try to identify those arguments and approaches that will be most effective with your audiences and to anticipate any objections that might be raised.*

3.2 PURPOSES

At this point in the writing process, you've identified the problem you need to address, determined what your task is, generated a lot of ideas, formed a potential thesis or point, and analyzed your audience. This activity has probably enabled you to formulate a fairly clear sense of purpose. Now you should pause and make sure that, indeed, you do know exactly what your purpose is (or, as is often the case, what your purposes are).

All communication is purpose-driven. Writers and readers, speakers and listeners—everyone who engages in the process of communication has one or more purposes in mind. But these purposes may differ from one person to another and even from moment to moment. For example, let's say you're asked to take minutes at a committee meeting. You take notes and then later write them up. Your main purpose in doing this is simply to fulfill an obligation and to do so in a professional manner. But the committee members who *read* your minutes may do so for other purposes: to review what was said and agreed upon, to prepare for the next meeting, etc. The department head (who doesn't attend these meetings but wants to be kept informed of them) may first read the minutes just for the purpose of seeing who was there and what was talked about. If something catches his attention, though, he may change his purpose in midstream and start to read more closely.

As a writer or speaker, you should try to anticipate the different purposes that readers and listeners may bring to an act of communication. Having your own purpose clearly in mind helps you organize your thinking in a coherent way. And if you make that purpose clear to your audience, they will find it easier to understand your message. If you fail to do this, your audience may well try to "create" a purpose for you. This can result in a breakdown of communication: what *your readers* see as your purpose may not be what *you* see.

It is important to distinguish between your *ostensible purpose* and other, *unstated purposes* you might have. The ostensible purpose is usually dictated by the situation and is usually quite conventional. For example, if you were writing up a trip report, your ostensible purpose would be to tell your supervisor about the trip and convince him or her that it was worth the company's expenditure. But you might have unstated purposes as well. For instance, you might want to use the occasion to make a subtle plug for some change of company policy. Or maybe you ran into somebody you've long thought your company should try to recruit, and you see this as an opportunity to mention her name. Or maybe you want to show your boss that you have excellent writing skills. Or perhaps you want to do all three. You could have several unstated purposes or even conflicting unstated purposes.

Whatever the situation, it's a good idea to keep your ostensible purpose clear in your mind and separate from whatever unstated purposes you might have. It is the ostensible purpose that makes communication coherent, not the unstated ones. It's like the punchline in a joke. When you tell a joke, you should concentrate on the punchline and mention only those details that contribute to its effect. It's what "makes" the joke. And listeners have the same expectations. They are quick to spot an irrelevant detail or a mangled punchline. Regardless of how many unstated purposes you might have, focus on your ostensible purpose and use it to guide the reader. In some cases, e.g., long formal reports, you may even want to use an explicit purpose statement: "The purpose of this report is to . . . ." In most communications, though, it is not necessary to do this, as long as your ostensible purpose is made clear in other ways.

❑ EXERCISE 3-3

Re-read Situation 2-1, "The Clinic Case." Using the Procedure for Audience Analysis given on pages 66–69, analyze the audience(s) for the report you are supposed to write. Then determine what your purposes will be.

❑ EXERCISE 3-4

Re-read Situation 2-2, "Pluribus: A Question of Investor Information." Analyze your audience(s) and purpose(s) for this case.

REFERENCES

1 Reprinted by permission of *Harvard Business Review*. Excerpted from Henry Mintzberg, "The Manager's Job: Folklore and Fact," *Harvard Business Review*, July–August 1975, pp. 54–59. Copyright © 1975 by the President and Fellows of

Harvard College: all rights reserved. Mintzberg has also written *The Nature of Managerial Work* (New York: Harper & Row, 1973).

2 Leonard R. Sayles, *Managerial Behavior* (New York: McGraw-Hill, 1964), p. 162.

3 Mintzberg, p. 50.

4 Mintzberg, p. 52.

5 Mintzberg, pp. 50–55.

6 James W. Souther, "What to Report," *IEEE Transactions on Professional Communication 28*(3):6 (1985). © 1985, IEEE.

7 Russell R. Raney, "Why Cultural Education for the Engineer?" quoted in John G. Young, "Employment Negotiations," *Placement Manual: Fall 1981* (Ann Arbor: The University of Michigan College of Engineering: 1981), p. 49.

ADDITIONAL READING

Thomas J. Allen and S. E. Cohen, "Information Flow in Research and Development Laboratories," *Administrative Science Quarterly 14*(1):12–19 (1969).

Aristotle, *The Rhetoric of Aristotle*, translated by Lane Cooper (Englewood Cliffs, NJ: Prentice-Hall, 1932, 1960).

Susan Artendi, "Man, Information, and Society: New Patterns of Interaction," *Journal of the American Society for Information Science 30*(1):15–18 (1979).

C. West Churchman, *The Design of Inquiring Systems: Basic Concepts of Systems and Organization* (New York: Basic Books, 1971).

Mary J. Culnan, "An Analysis of the Information Usage Patterns of Academics and Practitioners in the Computer Field," *Information Processing and Management 14*(6):395–404 (1978).

Linda Flower, "The Construction of Purpose in Writing and Reading," *College English 50*(5):528–550 (1988).

William J. J. Gordon, *Synectics: The Development of Creative Capacity* (New York: Harper & Row, 1961; Macmillan, 1968).

R. Johnson and M. Gibbons, "Characteristics of Information Usage in Technological Innovation," *IEEE Transactions in Engineering Management EM 22*:27–34 (1975).

J. C. Mathes and Dwight W. Stevenson, *Designing Technical Reports* (Indianapolis: Bobbs-Merrill, 1976).

Lee Odell, "Beyond the Text: Relations between Writing and Social Context," in *Writing in Non-Academic Settings*, edited by Lee Odell and Dixie Goswami (New York: Guilford Press, 1985), pp. 249–280.

M. A. Overington, "The Scientific Community as Audience: Toward a Rhetorical Analysis of Science," *Philosophy and Rhetoric 10*(1):1–29 (1977).

Gene Piche and Duane Roen, "Social Cognition and Writing," *Written Communication 4*(1):68–89 (1987).

James W. Souther, "What Management Wants in the Technical Report," *Journal of Engineering Education 52*(8):498–503 (1962).

Richard E. Young, Alton Becker, and Kenneth Pike, *Rhetoric: Discovery and Change* (New York: Harcourt Brace & World, 1970).

4

Constructing Arguments

As a technical professional, you are expected to do more than simply report information. You are expected to exercise your judgment about that information, to make recommendations, to propose solutions to problems. In many such cases, whether you are making a feasibility report, writing a proposal, or presiding over a teleconference, you cannot expect the information to simply "speak for itself." The interests of your audience may be too diverse, or the situation too complicated, for that. Instead, you have to present information in such a way that your audience will see how it applies to the problem at hand. In short, you have to be *persuasive*. This is where argumentation skills are important.

When we talk about arguments in this book, we do not mean quarrels or debates. Indeed, that kind of argumentation is usually counterproductive in professional settings and should generally be avoided. Rather, an argument is simply a claim that something should be believed or done, plus proof or good reasons for believing or doing it. If you have done a survey of pine bark beetle damage to the lodgepole pine forests of northwestern Wyoming and are recommending that natural predation be used against them instead of chemical sprays, you are *arguing* for that particular point of view. If you are writing a cover letter and résumé for a software engineering position, you are *arguing* that you are better qualified than others.

To make effective arguments, you must first describe a problem that the reader wants to have solved (as discussed in the preceding chapter). Then you must:

1   Make appropriate claims (e.g., solutions) about it.

2 Find and recognize proof or good reasons for these claims.

3 Know when you have enough proof.

4 Link your arguments together to build a case.

This chapter gives you some advice on how to build effective arguments.

4.1 EXPECTATIONS ABOUT CLAIMS AND PROOF

When they first begin writing on the job, many technical professionals fail to provide enough support, or "backup data," for the claims they make in their reports. This often makes the reports ineffective or even useless when audiences disagree with the writer's conclusion, see it as threatening, or need a solid foundation of support on which to base their own decisions. For instance, consider the following first page of a report dashed off by a junior biologist:

Subject
Establishing a computer file for our library: Cost estimate

Foreword
Frequent reference to the books in our library is essential for the research workers in the office. However, because of imperfections in the card cataloging system, precious time is wasted in locating books. To solve this problem, we have decided to use the computers for book searching; this will be faster and far more convenient. I have been asked to study the feasibility of such a project under our current budget. The purpose of this report is to estimate the cost of establishing a computer file for our library.

Summary
For my test run on the computer, I experimented with 50 books from the library and estimated the cost for establishing a computer file for these 50 books. Since there are approximately 1500 books in our library, that cost figure was then multiplied by 30 to give the total cost figure for the whole library. It is estimated that $2000 will be needed to establish the computer file for the library.

Assume that you are the department manager to whom this memo was sent, that your department has limited funds, and that you can spend absolutely no more than $2000 on this project, if you can spend even that. Further, you can't waste your money: if you invest in the project, you must get a completed library system in return. A partially completed project will be useless and divert money from other needed projects.

Under these very realistic conditions, what would you want to see as proof for the claims made in the summary? Wouldn't you want to know that the sample of 50 books was a representative sample? How could the writer convince you? Wouldn't you like some proof that the estimated cost for the project was accurate? How would you expect to find that out? Perhaps by seeing a breakdown of the total cost into its component parts—$1500 for labor, $400 for computer time, $100 for miscellaneous expenses. Even if the writer provided those figures, how could you be convinced that the $1500 charge for labor would be sufficient, that you wouldn't be asked to provide more money later on?

Similarly, consider the next beginning of a report, written by a test engineer to report the mileage for a weight-reduced T-car.

Foreword
Increasing the fuel economy of our automobiles has been the top priority of our company for the past 2 years. In your memo dated October 6, 1988, you requested that I perform a simulated road test for our T-car with a weight reduction of 1000 lb and compare its highway mileage rating (MPG) to our present model. I have completed the test and have placed the results in our file. The purpose of this report is to present the test results and my recommendation.

Summary
Upon completion of five simulated road tests, I have determined that the weight-reduced T-car will only give an average 2.5-MPG increase in highway mileage, while its lightness creates handling and safety problems. Therefore, for safety reasons, I recommend that sources other than solely reducing the weight of the car be investigated.

If you had to make decisions about fuel economy on the basis of this summary, wouldn't you want proof that the five road tests were sufficient, that the mileage increased only 2.5 miles per gallon, and that there were handling and safety problems with the car? If you were a hostile audience— perhaps you believed in the weight-reduced car, say, or had even helped design it—wouldn't you need to see the same kinds of proof before you believed the claims made in the Summary?

The following exercises provide practice in identifying claims which need to be proved. The rest of this chapter considers ways of proving such claims.

❏ EXERCISE 4-1

This exercise presents excerpts from three different reports. Predict what claims or numbers need to be proved to make a convincing argument.

A   ### Foreword
The wind turbine project is a new undertaking of Energy Systems. The goal of the project is to design a marketable system which makes wind energy accessible to private residents. To aid in the selection and development of a preliminary design, Mr. Zondervan asked that I research the practicality of using a flywheel for energy storage. He also requested that I identify the organizations doing development work with flywheel rotors so that further information could be obtained if a flywheel is incorporated in the design. This report addresses these requests by identifying characteristics of flywheel energy storage which affect its practicality and by listing those institutions involved in flywheel research.

Summary
New flywheel designs show much promise as energy storage devices, making possible energy densities up to 40 Wh/bl and reducing costs to as low as $50 per kWh. These figures compare favorably with other storage devices, such as batteries. In addition, the new designs have solved major problems with bearings and energy conversion, making flywheel energy storage quite practical.

Flywheel development projects are being carried on by the following people at the listed institutions.

1   David Rabenhorst, The Applied Physics Laboratory of Johns Hopkins University

2   J. A. Rinde, Lawrence Livermore Laboratory, Livermore, CA

3   A. E. Raynard, AiResearch Division of Garret Corporation, Torrance, CA

4   Frances Younger, Wm. Brobeck and Associates

5   R. P. Nimmer, General Electric Company, Corporate Research and Development, Schenectady, NY

B   ## Foreword
In the past 6 months, the Production Department has received complaints from workers on the main floor that their working environment has become too noisy. They complain that intense noise from machines around them is nerve-wracking and that they can communicate with one another only by shouting. Furthermore, some workers have even said that their hearing is deteriorating. Consequently, I measured the sound level in various spots on the main floor to locate the source of the problem. The purpose of this report is to recommend methods for reducing the noise level on the main floor.

Results
I collected data and compared my results with those obtained in December 1989; the total noise level produced by the machinery has increased by 30%. I then discovered that the noise is produced by the vibration of large housings and coverplates and by friction between loose and worn-out parts of machinery. In addition, the noisy condition on the main floor is worsened by reverberations from bare concrete ceilings and metal walls, which reflect a high percentage of incident sound waves.

Recommendations
My recommendations for reducing the noisy condition on the main floor are as follows:

1   Add acoustic material on the concrete ceiling and metal walls to reduce reverberation.

2   Add vibration-damping material beneath large housings and coverplates to dissipate vibration energy in the form of heat.

3   Replace worn-out parts of machines and fix loose components.

4   Provide earmuffs to prevent damage of the workers' ears caused by prolonged working in noisy condition.

5 Consider noise specifications when buying new equipment to reduce noise production in the future.

C Foreword

The present economic trends have caused smaller fraternities to have problems competing financially with larger fraternities, dormitories, and cooperatives. Specifically, Alpha Epsilon Zeta Fraternity must address the possibility of expansion into nearby apartments not only to remain competitive but also to gain the many social advantages of a larger membership. The purpose of this report is to demonstrate the economic feasibility of such an annex and to recommend its optimal size and location.

Summary

Rent prices vary only slightly within walking distance. Therefore, the best location is the closest one, that being Plaza Apartments at 605 E. Monroe St. Rent per member living in the annex will cost the fraternity substantially more than for those living in the lodge. However, excess revenue generated in the board account will counter that expense. The annex must hold at least six members to be successful, and anyone in addition could even cause a slight reduction in bills.

4.2 THREE BASIC STRATEGIES OF ARGUMENT

Over 2000 years ago, the Greek philosopher and rhetorician Aristotle identified three main strategies or bases for argument:

1 Logic and reason

2 The character and credentials of the communicator

3 Emotion

If you look through many types of technical writing, you will notice that most use (or try to use) arguments based overtly on *logic and reason*. For instance,

Pierlest Inc. should adopt a proposed pollution control device because (1) it will cut pollution to an acceptable level, (2) it will be easy to install, (3) it has a reputation for being very reliable, and (4) it will be cost effective.

In contrast, relatively few arguments are based overtly on *emotion*, though some are. For instance, the argument "Pierlest Inc. is violating federal pollution standards and will be closed down unless it greatly reduces pollution" appeals to the emotion of fear, the fear of being closed down.

You will note that most arguments are combinations of two or all three strategies, with one predominating. For instance, when you write a report recommending a particular solution to a problem, you probably give a series of arguments based on logic and reason to support your recommendation. However, you also base the argument at least partly on your character and credentials. Since you were hired because of your qualifications, your recommendation is often accepted at least partly, and sometimes fully, because your readers believe that you're competent to speak on your topic. (It is, of course, important to *look* competent and to reinforce your credentials.)

You will also note that in different situations, different strategies of argument will be more important. As indicated above, most technical communication appeals heavily to logic and reason, because this strategy of argument is usually appropriate and convincing. However, sometimes technical communicators need to base arguments on emotion. For instance, when scientists and engineers confront deeply emotional issues, such as the fear that a community might have over the location of a nuclear power plant, overreliance on logical arguments can be ineffective. Scientists have often argued very logically and "objectively" about a nuclear plant's backup systems for controlling dangerous situations and the very slight chance of an accident, but they rarely convince listeners that a plant should be located in their community. For many, the "fact" that a plant has only a one-in-a-million chance of blowing up does not outweigh the fear-producing knowledge that the reactor at Three Mile Island almost broke through several security systems to become a radioactive hazard to the surrounding communities.

❏ EXERCISE 4-2

Read the following passages and identify the strategies being used in each.

A   Social Control of Research
When I promised to speak here on the topic of social control of research, I thought it would be a much easier job than it has turned out to be. I have some practical experience in the management of research programs in the federal bureaucracy, but more relevant was seven years helping the Congress deal with research and development (R&D) matters. There, I learned the intricacies of grooming and passing authorization bills for large research programs, and negotiating complex conference agreements. On a number of occasions I had translated what society—represented by its elected officials—wanted in legislation that established new programs of research and development, or institutions to control such programs and their results, such as the Office of Technology Assessment. Another important part of that congressional experience was informing and cajoling those with responsibility for the appropriations process, in order to assure funding for those authorized programs. I had even received the satisfaction which came from coordinating the override of a presidential veto, only the 89th such veto override in the history of our Republic.

My confidence that social control of research would be an easy topic was further increased by my knowledge of and involvement with a number of AAAS programs on such topics as scientific freedom and responsibility, the scientific and legal interface, regulation of recombinant DNA research, analysis of the federal research and development budget, and other policy issues.[1]

B   Recombinant DNA Research
James D. Watson of Harvard, Nobel Laureate for his discovery of the molecular structure of DNA, [said] . . . that theoretically all forms of higher animal life

may be capable of clonal reproduction, and that the details of the research related to that, as well as to its implications, had not so far been communicated to the public to any substantial degree. He went on to ask,

> Does this effective silence imply a conspiracy to keep the public unaware of a potential threat to their basic ways of life? Could it be motivated by fear that the general reaction would be a further damning of all science and thereby decreasing even more the limited money available for pure research? Or does it merely tell us that most scientists do live such an ivory tower existence that they are capable of rationally thinking only about pure science, dismissing more practical matters as subjects for the lawyers, students, clergy, and politicians to face up to in a real way.[2]

C   Hard Times for Basic Research

Despite the general understanding that the federal government should carry the major responsibility for the funding of basic research, and, indeed, does provide about 70 percent of the monies available for this purpose, unyielding preoccupation with quick practical payoff on the part of both the Congress and the executive leadership has meant year-to-year uncertainty for an intellectual activity that should be built on long-range continuity. At least one fundamental reason for this seems to lie in our inability to make clear to the larger public the place of serendipity in science. The layman seems not to fully appreciate the fact that one cannot provide on cue the antecedent information needed for innovative technology, and therefore that knowledge must be stockpiled like scarce materials against the day when it per chance may be needed. I am confident that this lacuna in the public's understanding of science can be filled in the long run with the right kind of precollegiate science education. But, meanwhile, we must give serious attention to creating a more vigorous, more comprehensive, and more carefully calculated program of education for members of the Congress and their legislative aides than we have been willing to give time to in the past.[3]

4.3 BASIC TYPES OF ARGUMENT

There are two main types of argument: the argument of fact and the argument of policy. An argument of fact is an argument that something *is* or *exists* (or *is not* or *does not exist*); it can also be an argument that something *is* or *is not true, necessary, justified,* etc. The key word for the argument of fact is *is* or *are:* this company *is* (or *is not*) discharging pollutants into public waters; increased domestic production of oil and natural gas *is necessary* if the United States is to maintain its standard of living without being too dependent on Middle East oil; the costs of buying a new house *are* too high for the average consumer.

In contrast, an argument of policy is an argument that something *should* or *should not* be done. The key word for this argument is *should*: this company *should* be stopped from discharging pollutants into public waters; increased domestic production of oil and natural gas *should* be encouraged in the United States so that people can maintain their standard of living (or increased production of gas and oil *should not* be encouraged in the United States because it would be too harmful to the environment); the costs of buying a new house *should* be reduced for the average consumer.

Notice that your support for an argument of policy would differ from your support for the corresponding argument of fact; in some cases you would need to make an argument of fact as a subargument in an argument of policy. For instance, if you were to argue that the costs of buying a new house should be reduced for the average consumer (an argument of policy), you would need first to argue that the current costs are too high (an argument of fact).

The Argument of Fact

Arguments of fact can be derived from three sources:

1   Questions or subarguments of existence

2   Questions or subarguments of definition

3   Questions or subarguments of quality

These are outlined and illustrated in Table 4-1. It can be seen from the examples that, as you move from the question or subargument of existence to the subargument of quality, you logically build an argument. For instance, if you want to argue that a certain manufacturer should be stopped from discharging pollutants into public waters, you first need to address the question of existence (Is this company discharging material into public waters?). You then need to address the question of definition (Are the discharged materials regulated by law? dangerous? etc.). Finally, you must address the question of quality (Are the discharged materials present in illegal or unsafe amounts?). Only when all these questions have been addressed can you argue convincingly that the company should be stopped from discharging into public waters. You make this main argument by stringing together a chain of subarguments of existence, definition, and quality:

> Subargument of existence:
>   The company is discharging material into public waters.
>
>   ↓
>
> Subargument of definition:
>   The materials being discharged are regulated by law and are dangerous.
>
>   ↓
>
> Subargument of quality:
>   The materials being discharged are present in public waters in illegal and
>   unsafe amounts.

Note that your main argument could fail on any of these subarguments. Your argument might fail to establish the subargument of existence. It could happen, for instance, that although there are pollutants in the public waters, another company put them there, not Pierlest Inc. In such a case, placing restraints

Table 4-1 ❏ OUTLINE OF THE ARGUMENT OF FACT

Argument of fact: argument that something *is* or *exists* (or *is not* or *does not exist*)

Question or subargument of existence: Does the thing actually exist? Has something actually happened?

Example A   Are there Soviet troops in Cuba?

Example B   Is Pierlest Inc. discharging material from its manufacturing processes into public waters?

Question or subargument of definition: If it exists or has happened, what kind of thing or event is it?

Example A   If it is granted that there are Soviet troops in Cuba, are the troops educators and advisers or attack troops or something else?

Example B   If it is granted (or proved) that Pierlest Inc. is discharging material from its manufacturing processes into public waters, are the discharged materials regulated by law, considered dangerous, considered nontoxic even in large amounts, etc.?

Question or subargument of quality: If it exists and has been defined, how is it to be judged?

Example A   If Soviet troops are in Cuba and are educators and advisers, are they justified, necessary, present in appropriate numbers, desirable, etc.?

Example B   If Pierlest Inc. is discharging material into public waters, and if those materials are regulated by law (or dangerous), are the materials present in legal (or safe) amounts, illegal (or unsafe) amounts, desirable (or undesirable) amounts, avoidable (or unavoidable) amounts, etc.?

on Pierlest would be both unfair and ineffective. Similarly, your argument could fail on the subargument of definition. It could be that Pierlest is discharging material into public waters but that these materials are not regulated by law or are not harmful. In such a case, it would probably be wasteful to spend time, effort, and money trying to regulate the company. Finally, your argument could fail on the subargument of quality. Pierlest might well be discharging dangerous materials regulated by law, but only those amounts allowed under the law. In such a case, if you still wanted to stop those discharges, you would have to argue that the law should be changed; you could not effectively argue that the company was breaking the law.

The Argument of Policy

The argument of policy, the second basic type of argument, can be derived from two sources:

1   Questions or subarguments of worth or goodness

2   Questions or subarguments of expediency, advantage, or use

These are outlined and illustrated in Table 4–2.

Table 4-2 ❑ OUTLINE FOR THE ARGUMENT OF POLICY

Argument of policy: argument that something *should* (or *should not* be done)

Subargument of worth or goodness: Is a proposed activity or course of action worthy or good in itself?

Example C   The United States should not protest the presence of Russian troops in Cuba because it is right that they are there (because they have the right to be there).

Example D   Pierlest Inc. should stop polluting public waters because it is right (or Pierlest has a social duty) to protect the environment.

Example E   Companies should make honest claims about the merits of their products because it is good and worthy to be honest.

Subargument of expediency, advantage, or use: Is the proposed activity or course of action good for the audience in that it is expedient, advantageous, or useful?

Example C   The United States should agree to the presence of Russian troops in Cuba because such agreement would strengthen the presence of U.S. troops on Russian borders.

Example D   Pierlest Inc. should stop polluting public waters because doing so would improve its public image and thus its sales.

Example E   Companies should make honest claims about the merits of their products because it is advantageous in dealing with customers to have a reputation for honesty and because such honesty will protect the company from charges of fraud, expensive lawsuits, and costly penalties imposed by both the government and the courts (because it will be advantageous for them to do so).

You should note that arguments based on expediency, advantage, or use are much more frequent in most types of technical writing—indeed, in most areas of our lives—than are arguments based on intrinsic worth, goodness, or merit. People are more likely to believe or do something if it is to their own advantage. Let us consider, for example, the arguments in Table 4-2 and the following situation. Company administrators intend to make false claims about the merits of their product in order to increase sales and profits. You want to argue instead that they should submit honest claims about the product. If you appeal only to the issue of worth (it is good and worthy to be honest), you will probably be unsuccessful. In the value scheme of these administrators, the desire for profit is probably stronger than the desire for honesty. Instead, you are more likely to succeed if you appeal to expediency and advantage: the administrators should make honest claims about the merits of their product because

1   It is advantageous in dealing with customers to have a reputation for honesty.

2   Such honesty will protect the company from charges of fraud, expensive lawsuits, and costly penalties imposed by the government and the courts.

Figure 4-1 ❏ FLOW CHART FOR CONSTRUCTING EFFECTIVE ARGUMENTS

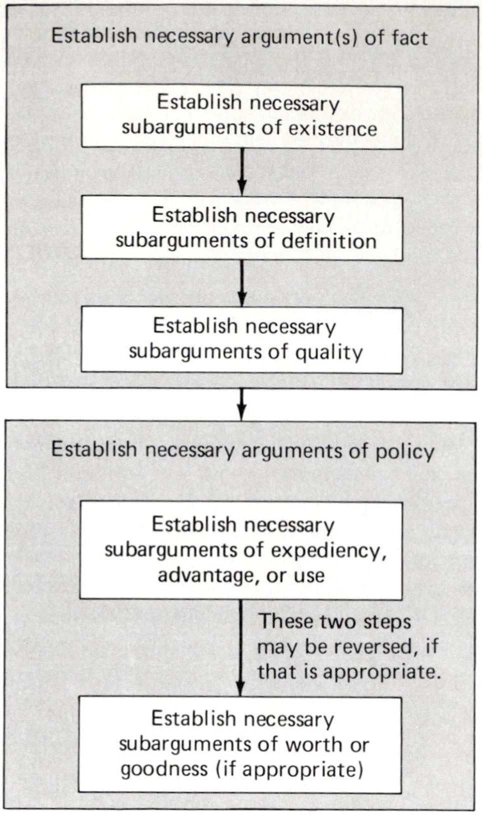

Establish necessary argument(s) of fact

> Establish necessary
> subarguments of existence

> Establish necessary
> subarguments of definition

> Establish necessary
> subarguments of quality

Establish necessary arguments of policy

> Establish necessary
> subarguments of expediency,
> advantage, or use

These two steps
may be reversed, if
that is appropriate.

> Establish necessary
> subarguments of worth or
> goodness (if appropriate)

This kind of argument better "fits" the supposed value scheme of the administrators. Can you think of a situation in which the appeal to worth or goodness would be as strong as or stronger than the appeal to advantage?

The Relationship between Arguments of Fact and Arguments of Policy

You have seen that arguments of fact should be made in a logical sequence and that arguments of policy often depend upon prior arguments of fact. You can combine these insights to produce an overall plan or flowchart for ordering arguments of fact and policy, as demonstrated in Figure 4-1.

Notice that if you are only trying to make an argument of fact, you will never "get to" an argument of policy. In contrast, if you are making an argument of policy, you will need to establish all the necessary parts of the

Figure 4-2 ❑ INTRODUCTION TO AN ARGUMENT OF POLICY FOR AN AUDIENCE ALREADY ACCEPTING THE PRIOR ARGUMENT OF FACT

We all know that Pierlest Inc. has been illegally dumping polyvinyl chloride into the Huron River [*argument of fact, already accepted*]. We are here to decide what to do about that problem. I would like to argue that the company should be required to follow the cleanup plan proposed by the State Department of Health [*argument of policy, to be proved*]. Pierlest has a responsibility to the people of the state to clean up the river [*subargument of worth, perhaps needing to be proved*], and this plan will allow the company to do so most quickly and thoroughly [*subargument, advantage to the public, to be proved*] and at a minimum cost [*subargument, advantage to the company, to be proved*].

argument of fact: the subargument of existence, the subargument of definition, and the subargument of quality. You do not, however, need to spend a lot of time establishing these subarguments if your audience already agrees to them.

Suppose that, in a speech or report, you were trying to present an argument of policy about Pierlest Inc. and its pollution of public waters: "Pierlest Inc. should be required to stop polluting public waters and to follow a cleanup plan advocated by the State Department of Health." If you were addressing an audience that knew nothing about the pollution problem, you would have to carefully prove each part of the argument of fact before you could argue for your policy. However, if everyone already knew that the company had been polluting the waters with a dangerous and illegal substance, you might summarize all the subarguments of fact in a quick opening statement:

> We all know that Pierlest Inc. has been illegally dumping polyvinyl chloride into the Huron River. We are here to decide what to do about the problem. I would like to argue that Pierlest should be required to follow the cleanup plan proposed by the State Department of Health. The company has a responsibility to the people of the state to clean up the river, and this plan will allow Pierlest to do so most quickly and thoroughly and at a minimum cost.

You would then continue (1) to outline the proposed plan, (2) to demonstrate that the plan will allow quick and thorough cleanup of the river, and (3) to demonstrate that the cost will be minimal. The various subarguments presented are identified in Figure 4-2.

4.4 BUILDING A CASE

Once you have decided on the basic arguments you are going to make and on their logical order, you need to decide *how* to present the entire speech or report. What goes where? How should things be organized? These are especially difficult questions if you have limited amounts of time and a large amount of information to organize.

Table 4-3 ❑ OUTLINE FOR BUILDING A CASE

I   Problem/Introduction/Foreword

    A   Direct the audience's attention toward the problem.

II Credentials*

    A   If it is useful, give your credentials, i.e., explain why you can speak with authority on the subject, and establish common ground by pointing out shared beliefs, attitudes, and experiences.

III Position/Solution/Summary*

    A   Briefly state your position or your proposed solution.

    B   Briefly state the major reasons for advocating your position or solution.

IV Background of the Problem*

    A   Point out the nature of the problem: (1) its historical background and (2) its causes.

    B   Explain how it concerns the audience.

V   Argument for Position or Solution

    A   State the criteria for judgment, i.e., the standards or characteristics any acceptable solution or position must meet. Include explanation where necessary.

    B   State your position or solution to the problem, along with any necessary clarification.

    C   Demonstrate the soundness of your position or solution by showing how it meets the criteria established in III-A. This step should be accompanied by ample evidence: facts (illustrations, statistics, examples of successful application of the solution) and statements of authority. Be sure to identify the authorities if they aren't widely known.

    D   If there are competing positions or solutions, demonstrate the superiority of yours by showing how the others fail to meet the criteria as completely as yours do.

VI Conclusion

    A   Explain briefly the benefits to be gained by accepting your position or solution or the dangers of rejecting it.

    B   Summarize your argument: (1) restate your position or solution (III-B); (2) restate the reasons your position or solution should be accepted (III-C).

* Items II, III, and IV may be deleted depending on the situation, the needs of the audience, and the accepted formats for a particular type of technical communication.
SOURCE: Adapted from Richard E. Young, Alton Becker, and Kenneth Pike, *Rhetoric: Discovery and Change* (New York: Harcourt, Brace & World, 1970), p. 234.

      One psychologically satisfying and widely useful organization is presented in Table 4-3, which gives an outline for building a case. This organization is so effective that it is still being used over 2500 years after it was first developed in Greece and Rome for speaking in public forums and courts of law. By learning this one outline, you can have at your fingertips a general organization for essentially any speech or report.

      This outline is a tool for *organizing* and *planning* your case; it is not a pat formula for the final draft. You should be prepared to adjust it according to the audience's values and knowledge, your goals, the discourse conventions, and so on. Although some kind of problem statement or other form of orientation is always needed at the beginning (item **I** in the table), the other components are flexible.

For example, if your credentials (**II**) are well known by the audience, you do not need to state them. Indeed, stating your credentials when the audience already knows them could actually work against you, because your audience might think you are showing off or trying a power play. Sometimes, however, the format requirements take this decision out of your hands. For instance, proposals have a standard and very important section explicitly presenting the credentials of the proposers. You must fill this out even if you think the proposal evaluators already know your credentials. On the other hand, most technical reports do not have a place for full credentials, only for your name and organizational role. Even if the audience does not know your qualifications, you cannot state them.

Likewise, the amount of background information you should provide (**IV**) depends on how much the audience already knows about the problem. For good communication to take place, the audience needs to share with you certain background knowledge and assumptions; it needs to have enough "old information" to make sense of the "new information." But if you give them too much old information, the audience will see it as a waste of time and will be irritated with you. Generally, it's better to err on the side of giving too much information, but you shouldn't overdo it.

It's always necessary to state your position or proposed solution (**III**), but *where* you state it depends in good measure on your audience and your goals. In most professional communication, it's a good idea to state your position or solution early. This facilitates selective reading and also makes it easier for readers or listeners to follow the discussion. But there are situations where you might want to delay taking a position until the end. For example, if you are addressing a hostile audience, it might be a good idea to construct your argument step by step, gradually breaking down your audience's resistance, *before* you tell them where you stand.

The heart of persuasive communication is the way you link arguments together to support your position or proposed solution (**V**). Every argument you make, every case you build, depends on the criteria you are using (**V-A**). But the criteria you choose to emphasize, and the degree to which you emphasize them, will depend greatly on the situation and on certain characteristics of your readers and listeners, such as their background knowledge, their interests, their values, and their expectations. The kind of evidence you use (**V-C**) depends on these same characteristics. For a technical audience, you can call on your training in school and on the job to bring forth the right kinds of criteria and evidence. For a more general audience, you should consider at least the following general criteria.

1 *Effectiveness:* Is the solution effective? Will it solve the problem posed? Why? How do you know?

2 *Technical Feasibility:* Can the solution be implemented? Does it require technology or resources that are unavailable? How do you know?

3 *Desirability:* Would one want to implement the proposed solution? Does it have any undesirable effects? Does it have desirable effects? Why? What are they?

4 *Affordability:* What will the solution cost to implement? To maintain? Is this cost reasonable? Is it affordable? Will it reduce costs in the future? Why?

5 *Preferability:* Is the solution better than or preferred over any other possible solution? Why?

There are at least six major strategies for linking arguments together and building a case that are commonly found in technical and scientific communication: (a) simple problem-solution, (b) criteria, (c) chain of reasoning, (d) process of elimination, (e) experimental research, and (f) improving the system. Each of these has its own strengths and weaknesses and is appropriate only for certain situations. Various combinations of these case types are possible.[4]

Simple Problem-Solution

In a simple problem-solution case, you describe a problem and propose a solution without explicitly stating the criteria for judgment or evaluating any competing solutions. In short, you omit items **V-A** and **V-D** of the outline in Table 4-3. This is an easy type of case to build, and it is easy for an audience to understand. For that reason, it is especially appropriate with an audience that is uninformed about a subject. On the other hand, it can easily be attacked by well-informed opponents.

Criteria

A case built on criteria closely follows the outline in Table 4-3 and *explicitly emphasizes* the criteria. Its success depends crucially on the acceptance of the criteria by the audience as being both (1) correct and (2) complete. If your criteria are found acceptable, then you have only to show that your proposed solution meets those criteria in order to win your case. But if your criteria are seen as being either incorrect or incomplete, the evidence you provide as support will be less relevant and therefore less persuasive. Thus, this approach is most effective in situations where you are confident that your criteria are complete and that your audience subscribes to them. If these requirements are satisfied, the case can be effective even with audiences that might initially disagree with your proposed solution. Indeed, international negotiators have found that in cases where two sides have taken strongly opposing positions, they can make progress toward a compromise only if they first concentrate on establishing mutually agreed-upon criteria.

The sample case given in Table 4-4, which closely follows the outline in Table 4-3, can easily be turned into a criteria case.

Chain of Reasoning

In a chain-of-reasoning case, you try to link arguments together in a sequence of logical steps. It is generally a very deliberate and analytical form of argumentation, and for that reason you need a patient, dedicated audience. Usually, a chain-of-reasoning case works inductively, proceeding from more acceptable claims and details to more controversial ones. Since it "breaks the news" slowly, it is especially effective with a hostile or skeptical audience. But, like the criteria case, it is very committing: a single misstep in the chain of reasoning can cause it to fail. The chain-of-reasoning approach is often used in support of other cases, such as the experimental research type of case discussed below.

Process of Elimination

In a process-of-elimination case, you present a number of possible solutions and gradually eliminate all but one. In effect, you invert steps **V-C** and **V-D**. It is a very effective form of argumentation in situations where the possible solutions are limited and well known. It combines naturally with the criteria case: you first establish the relevant criteria for selection, and then list the alternative solutions; then you test each of the solutions against the criteria, eliminating all but one. The process-of-elimination case works well when your audience concurs with your choice of possible solutions. On the other hand, if you fail to consider all possible solutions, the argument can quickly fall apart. Certain kinds of readers—scientists, for example—are particularly adept at coming up with "rival hypotheses" and undermining a process-of-elimination case.

Experimental Research

Yet another variant of the outline in Table 4-3 is that used to present experimental research. It has four parts: problem, method, results, and discussion. The problem statement corresponds to items **I**, **II**, and **IV** of the outline. The method section is equivalent to item **V-A**. The results section corresponds to items **V-C** and **V-D**. And the discussion corresponds to items **V-B**, **VI-A**, and **VI-B**. The power of this case resides in its apparent commitment to "objectivity": a hypothesis emerges from the problem statement and is tested with methods that have been independently validated by the scientific community. Thus, whatever the results may be, they are supposed to constitute accurate and useful information about nature. This type of case is required in certain settings (e.g., journal articles), and so you have little choice about whether to use it. Its success depends on how well you satisfy all of the conditions that the scientific community has placed upon it (testable hypothesis, valid and appropriate methods, etc.).

Table 4-4 ❏ SAMPLE ARGUMENT OF FACT:
"ELECTRONIC LOCK DESIGN PROPOSAL"

I Introduction

   A   The need for effective yet flexible security systems is more critical today than ever. In many applications, the traditional mechanical key or combination lock is no longer adequate. Electronic systems using electronic locks are proving to be the optimal solution, providing advantages unavailable in mechanical systems.

   B   Wiley Electronics is about to begin development of an electronic security system of its own. Mr. Silvers has asked me to design a lock to be used as part of such a system.

II Credentials: I have studied various systems.

III Summary

An electronic lock with an 8–digit combination is the optimal lock for Wiley's electronic security system. The proposed lock is effective, simple, and versatile, and it costs only $10.35.

IV Background of Problem

   A   Better and more efficient security systems are needed by both industry and individuals.

      1   Most industrial complexes maintain large areas in which stringent security is essential, yet through which large numbers of employees must pass unhindered.

      2   Many individuals also find themselves in situations in which they require a level of security beyond what is normally adequate.

   B   Strong interest in better security systems has encouraged the manufacture of many different security systems for the potential customer.

      1   Basically, the customer must choose between ordinary key or combination locks and an electronic lock.

         a   A clear trend toward electronic security systems has prevailed over the last few years.

   C   Wiley Electronics is about to begin developing a security system to compete in this market.

V Argument

   A   The criteria for choosing a system: What does a consumer expect from a security system?

      1   The lock must keep out unauthorized persons and allow those authorized to enter freely.

      2   The lock must be tamperproof.

      3   The lock must have a combination that is easy to alter.

      4   The lock must be easy to interface with a larger controlling system.

      5   The lock must be capable of independent operation.

      6   The lock must have an acceptable cost and market appeal.

   B   Solution: the electronic lock meets all of the criteria as well as or better than the mechanical key or combination lock.

   C   Argument

      1   The electronic lock is as effective as the other two locks in screening authorized from unauthorized persons.

      2   The electronic lock is more tamperproof.

      3   Its combination is easier to alter.

Table 4-4 ❑ SAMPLE ARGUMENT OF FACT:
"ELECTRONIC LOCK DESIGN PROPOSAL" (*Continued*)

    4   It is easier to interface with a larger controlling system.

    5   It is as capable of independent operation as the other two locks.

    6   It has an acceptable cost of only $10.35 wholesale.

    7   It will appeal to a wider range of the market than the other two types.

D   Description of the proposed general-purpose electronic lock

    1   Security features of the lock

       a   The lock uses an 8-digit combination providing 65,536 different possible combinations for a sufficient level of security for most applications.

       b   To provide absolute security, a minor modification will allow the lock to operate with a 16-digit combination, providing over 4 billion unique combinations.

    2   Operation of the lock

       a   General operation of the lock

       b   Procedure for opening the lock

       c   Procedure for setting a new combination

    3   Internal construction and operation of the lock

       a   The push buttons

       b   Counter #1

       c   The memory

       d   The comparator

       e   Counter #2

E   Possible alternative: the "key card" lock

    1   Description: a "key card" lock is the type of lock that accepts a small wallet-size card which is inserted by the operator into a reader. The card is usually impregnated with a magnetic code that the card reader can detect and transmit as a digital signal.

    2   Evaluation: there are two possible means of processing the key card's digital signal, local and remote, both of which can be rejected as inadequate solutions.

VI Conclusion

SOURCE: Adapted from "Electronic Lock Design," *Student Report Handbook*, edited by Leslie A. Olsen, Lisa Barton, and Peter Klaver (Ann Arbor, MI: Professional Communications Press, 1979).

Improving the System

Sometimes you may want to argue for some change from the status quo, not because there is a particular problem but because you think a good situation can be made even better. For example, let's say your company already has pretty good public relations with the surrounding community, but you think they could be even better if the company would sponsor a local road race. In such a situation, you would be smart to place your idea within the context of the company's larger public relations effort. Instead of treating it as a "problem" (item **I** in Table 4-3), you would treat it as an *opportunity*. The criteria for

Figure 4-3 ❑ TECHNICAL ARGUMENT: SAMPLE 1
[From Charles Maurer, in "Tomorrow's Train Is Running Today in Canada," *Popular Science* 217(6):75–76 (1980).]

Tomorrow's Train Is Running Today in Canada

We are standing by the 1.4-mile test track of Ontario's Urban Transportation Development Corporation. Around us are empty fields that the UTDC hopes will become the foremost center of urban mass transit in North America. The elevated train I've come to hear and see—the Intermediate Capacity Transit System (ICTS)—is the vehicle that could bring that plan to fruition. It's tomorrow's means of mass transportation today. Subways, of course, are ideal—fast, unobtrusive, impervious to bad weather. But at $80 million per double-track mile, precious few cities can afford to build them. Few can even afford to maintain the ones they have.

Yet streetcars and buses are limited to 10,000 passengers per hour, and to carry that many they would have to be so close together you could almost climb out on the vehicles' roofs and walk—and arrive at your destination faster.

Lightweight, relatively inexpensive rail systems can fill the gap, but if they're to run at grade level they need a right of way. And with sidewalk-to-sidewalk skyscrapers, there's no right of way to be had—except over the streets. That's why our largest cities long ago installed elevated street railways. Sensible monstrosities, but to judge from their noise and looks, the same firm engineered them that handled hell.

To make a quiet, unobtrusive yet affordable elevated railway was the goal of UTDC. Visionary affairs like monorails were out of the question. "They're fine for fairgrounds," explained Dick Giles, a mechanical engineer who serves as UTDC's sales manager, "but for a raft of reasons they're impractical." With monorails, performing maintenance, doing switching, designing stations, and providing emergency evacuations are all far more difficult. And monorail trains require a higher, more obtrusive track.

Magnetic levitation was also out—"too complicated, too many developmental risks." So the train had to run on wheels. To make it unobtrusive, stations had to be short, so trains had to be short, too—120 feet maximum (four 30-foot cars). Short trains meant fast trains—45 MPH—spaced at intervals of less than 1 minute apart.

Spacing trains so closely at those speeds meant computers had to be in control. It also meant unusually quick accelerating and braking. This, plus the ability to take steep grades in stride—essential if the railway is to be shoehorned into existing space—ruled out conventional motors driving steel wheels: too much slippage. And though conventional motors can drive rubber wheels quietly and surely, the line must be kept free of ice and snow, impossible in Canada.

---

judgment (item **V-A**) would be essentially the same as those that were used in the company's other public relations campaigns. A case based on improving the system is most effective with an audience that supports the current system and does not want to change it in any major way.

❑ EXERCISE 4-3

Figures 4-3 and 4-4 present two samples of technical argumentation. Analyze each and determine what kind of case the author is presenting and what criteria are used.

4.5 WHISTLEBLOWING

Ideally, the goals of any private enterprise harmonize with those of the larger society: what's good for the company, we would like to hope, is good for everyone. But of course we do not live in such an ideal world. Instead, we live in a world where companies sometimes try to maximize their profits *at the expense of* society. Thus, it sometimes happens that technical professionals are asked by their companies to write or say something that they feel not only works against the public good but also prevents those affected from understanding the true situation. In short, they are asked to violate the kinds of ethical standards we discussed in Chapter 2. In extreme cases, this sort of situation can pose a grave danger to public safety.

When faced with ethical dilemmas of this type, what should the technical professional do? Of course, each case is different and there are no simple answers to cover them all. In general, it is a good idea to first try to work out the problem within the normal chain of command. Talk to your boss and explain your concerns; it could be that there is more to the story than you know, or that there is an easy solution available. If that doesn't work, you might be able to find a way around your boss and talk to *his* or *her* boss. Or, in some cases, you might be able to find a technical solution to the problem.

If none of these remedies works and if the situation is serious enough, you may feel that you have no recourse but to "go public." Such a move can have a serious effect both on the company and, of course, on your career, so it is not a step to be taken lightly. Before you decide to "blow the whistle," we recommend that you read a thoughtful article by Richard T. DeGeorge titled "Ethical Responsibilities of Engineers in Large Corporations: The Pinto Case."[5] DeGeorge discusses the famous 1978 accident in which three young women died of burns when their 1973 Ford Pinto was rammed from behind by a van. The Indiana state prosecutor charged the Ford Motor Company with three counts of reckless homicide, claiming that the company had placed the gas tank in a dangerous position. As it turns out, Ford was acquitted of all charges. However, as DeGeorge notes, this was not the only case of fuel tank problems with 1970-era Pintos. Between 1971 and 1978 some fifty suits were filed against the Ford Motor Company by citizens claiming damages arising from rear-end accidents in Pintos, and eventually Ford recalled the car for alterations. So it appears that there may well have been some design flaw in those early model Pintos.

Should Ford's engineers have gone public? It appears that they knew of the design problem and had even informed higher management that a part costing only $6.65 could alleviate it. But DeGeorge argues that, just like the Indiana state prosecutor, the Ford engineers who designed the Pinto did not have enough documented evidence to make a solid case and thus were right not to blow the whistle. He goes on to offer the following guidelines:

> I would suggest as a rule of thumb that engineers and other workers in a large corporation are morally *permitted* to go public with information about the safety of a product if the following conditions are met:

Figure 4-4 ❑ TECHNICAL ARGUMENT: SAMPLE 2

Fasano Manufacturing Co.
Interdepartmental Memo

Date: June 24, 1989

To: V. Voros
        Supervisor—Computer Operations

From: R. Sholander
        Analyst—Computer Operations

Subject: Production and Information Control System:
             Preliminary Report on Features and Implementation

Foreword

Mr. Bauer, Vice President of Production, wants to centralize control of production and purchasing to optimize both inventory levels and assembly shop loads. New information about purchasing and production is becoming dispersed throughout the system. This makes it difficult for managers and sales representatives to know the status of a job quickly, resulting in delays in dealing with customers. Further, inventory is often at unnecessarily high levels as stock waits to be used in production that has been delayed, and the machine shops are sometimes idle while waiting for parts because purchase orders were delayed or production is ahead of schedule. The Computer Operations Department was given the task of researching and designing solutions to the problem. This report will present my solution and its implementation by computer.

Summary

The solution I propose is an integrated system combining sales forecasting, engineering, inventory control, purchasing, and job shop scheduling. The Production and Information Control System, called PICS, will provide a central database in which all pertinent information will be collected. It will also provide the following:

- Access only to authorized users

- A materials requirements planning system

- Optimal shop capacity through simulation procedures

- A status report on any job at any time in the system

- Faster response to market demands

This system is well modularized and can be implemented gradually by our departments. A rough time estimate for completion would be about 7 months. The cost would be minimal if PICS were operated as a manual system, but this would still require an overwhelming amount of paperwork. A gradual change to a computer system would cost approximately $150,000, but our paperwork could be handled much more efficiently if it were entered

Figure 4-4 ❏ TECHNICAL ARGUMENT: SAMPLE 2 (*Continued*)

directly on a computer disk. Therefore, I suggest a rapid conversion to a computerized approach.

Why a New System Is Needed

Information concerning the inventory system is becoming increasingly more difficult to interpret and understand. Problems arise because each assembly shop keeps its own records of transactions. One problem occurs in Purchasing, where orders from shops must be combined to save money in bulk ordering. This combining is often impossible if orders come in separately. A second problem is that it is hard to find the exact status of a job in production if you are not directly involved in the daily operation of that job.

The Computer Operations Department was given the assignment of researching all possible solutions to these problems with particular emphasis on:

- A mechanized system that would facilitate a change to a computerized approach

- A central databank where all departments contribute and receive information

- A low-cost system that could be phased in gradually

This report presents the alternative I was assigned to investigate and is divided into four sections:

- System overview

- Database requirements

- Necessary changes

- Cost of new system

---

1   if the harm that will be done by the product to the public is serious and considerable;

2   if they make their concerns known to their superiors; and

3   if, getting no satisfaction from their immediate superiors, they exhaust the channels available within the corporation, including going to the board of directors.

To have a moral *obligation* to go public, two other conditions have to be fulfilled in addition to the first three:

4   The employee must have documented evidence that would convince a reasonable impartial observer that the employee's view of the situation is correct and the company policy wrong; and

5   there must be strong evidence that making the information public will in fact prevent the threatened serious harm.[6]

Notice how the ability to maintain your ethical standards in a situation like this may depend on your ability to assemble evidence, to build a case—in short, to *argue*. This is something that the classical thinkers pointed out more than 2000 years ago, and it is as true today as it was then.

❏ EXERCISE 4-4

Decide what type of case you should build for the Westheimer report ("The Clinic Case" presented in Chapter 2). Then, following the outline for building a case in Table 4-3, write a complete outline for the report.

REFERENCES

1   J. Thomas Ratchford, in *Proceedings of the 31st National Conference on the Advancement of Research* (Denver: Denver Research Institute, 1978), pp. 59–60.

2   Ratchford, p. 61.

3   William Bevan, in *Proceedings of the 31st National Conference on the Advancement of Research* (Denver: Denver Research Institute, 1978), p. 32.

4   This section is inspired by Richard D. Rieke and Malcolm O. Sillars, *Argumentation and the Decision-Making Process*, 2d ed. (Glenview, IL: Scott, Foresman, 1984), Chapter 8.

5   Richard T. DeGeorge, "Ethical Responsibilities of Engineers in Large Organizations: The Pinto Case," *Business and Professional Ethics Journal* 1(1) (Fall 1981).

6   DeGeorge, p. 6.

ADDITIONAL READING

Aristotle, *The Rhetoric of Aristotle*, translated by Lane Cooper (Englewood Cliffs, NJ: Prentice-Hall, 1932, 1960). Another good edition of this work is *The "Art" of Rhetoric*, translated by J. H. Freese (Cambridge, MA: Harvard University Press, 1926, 1967).

Wayne Booth, *Modern Dogma and the Rhetoric of Assent* (Chicago: University of Chicago Press, 1974).

Jeanne Fahnestock and Marie Secor, *A Rhetoric of Argument* (New York: Random House, 1982).

S. Michael Halloran, "Technical Writing and the Rhetoric of Science," *Journal of Technical Writing and Communication* 8(2):77–88 (1978).

James A. Kelso, "Science and the Rhetoric of Reality," *Central States Speech Journal* 31:17–29 (1980).

Carolyn R. Miller, "A Humanistic Rationale for Technical Writing," *College English* 40(6):610–617 (1979).

Chaim Perelman and L. Olbrechts-Tyteca, *The New Rhetoric: A Treatise on Argumentation* (Notre Dame, IN: University of Notre Dame Press, 1969).

Richard D. Rieke and Malcolm O. Sillars, *Argumentation and the Decision-Making Process* , 2d ed. (Glenview, IL: Scott, Foresman, 1984).

Stephen Toulmin, *The Uses of Argument* (London: Cambridge University Press, 1958).

Richard E. Young, Alton Becker, and Kenneth Pike, *Rhetoric: Discovery and Change* (New York: Harcourt, Brace & World, 1970).

5

Stating Problems

Once you have worked out what you want to say, you need to figure out how to capture your audience's attention. One of the best ways to do this is to start your communication with a clear and interesting statement of the problem. Psychologist Leon Festinger[1] explains how this works, in his theory of "cognitive dissonance." Festinger claims that the mind sometimes finds itself with a sense of unease about some conflict or inconsistency between two things it believes to be true. When such a conflict or inconsistency occurs, the mind wants it eliminated to avoid chaos, worry, and psychological discomfort.

5.1 INTRODUCING A PROBLEM

Richard Young, Alton Becker, and Kenneth Pike[2] note that you can make use of cognitive dissonance by stating a problem in the form of two directly conflicting terms joined by a word signaling the conflict. The conflicting terms may be perceptions, facts, values, expectations, or beliefs, as indicated in Figure 5-1. Frequently, a value, expectation, or belief in our mind conflicts with an external perception or fact. This pattern is illustrated at the bottom of Figure 5-1, where a belief (the A term) conflicts with a perception (the B term):

> *Our new stereo system must appeal to the teenage market* [belief, A term]. *However, it apparently does not appeal to this important market* [perception, B term].

AUTHORS' NOTE: We would like to acknowledge the influence of Richard E. Young, Dwight W. Stevenson, and J. C. Mathes on the evolution of this chapter and the thinking behind it.

Figure 5-1 ❑ OUTLINE OF A FOREWORD OR INTRODUCTION BY PROBLEM DEFINITION, STEP 1

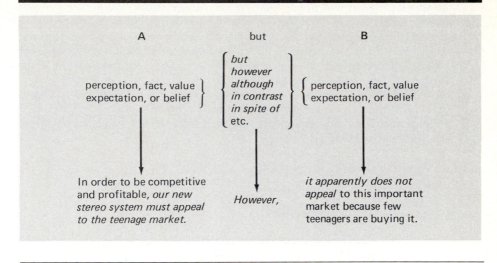

This same pattern appears in the following Introduction from an argument that doctors "are slowly but surely extending cancer patients' lives." (Italics added.)

IS THE GREAT CANCER CURE ALREADY WITH US?

Since the National Cancer Act of 1971, there has been a continual hubbub about conquering cancer. The public has been led to believe that some morning it will wake up and newspaper headlines will glare: Cancer Cure Found! *The cure, of course, will have come about because some smart or lucky scientist anguished away days and nights until he found the key.*

But science rarely works that way. Breakthroughs come in bits and pieces. And it's exactly in this slow, undramatic way that cancer scientists have greatly increased the number of long-term or even permanent cancer remissions since cancer therapy was first tried during the 19th century. During the 1930s only one out of five cancer patients could expect a remission for 5 years or perhaps longer. Then one out of three could expect such a remission. Now the rate is approaching one out of two patients. And some of these patients will go on to live out their natural lifespan.[3]

This conflict between a value, expectation, or belief and a perception or fact appears again as the writers of the following Foreword set up a problem for the managers of the Belle River Power Plant. Note that the authors indicate the magnitude of the problem here—replacing coal mills is serious. (Italics added.)

The Belle River Power Plant Unit 2 has had over 40 years of on-line operation. *Although Unit 2 is still functioning, its increasing demand for both major and minor repairs indicates that its two existing coal mills will need to be replaced soon.* Thus, you requested that we solicit and evaluate bids for the design and fabrication of the two new coal mills. We understand that if you concur, our recommendation will be used as a basis for awarding a purchase order for the design and fabrication of the two new mills. Since the guidelines published by the Public Service Commission require that at least three bidders be considered for such a project, we performed a Dun and Bradstreet analysis of assets and project capabilities on every bidder and then solicited and evaluated bids from the best-qualified bidders:

1   American Coal Mill Corporation

2   Babcock and Willis Incorporated

3   Western Electric Incorporated

The following report summarizes our recommendation and supplies the supporting material generated in developing our recommendations.

These authors have definitely chosen to phrase the problem statement to catch the audience's attention: managers are concerned that their plants run smoothly and not break down.

Often you only need to state the A-but-B situation and then move on, as illustrated in the two examples above. However, with a less knowledgeable audience you may have to not only state the A-but-B situation but go on to give details about the A or the B or both. This is illustrated in the following example. (Italics added.)

A NEW PUZZLE IN PHYSICS: THE "COSMION"

The latest cherished principle of subatomic physics to be threatened by experiment is the charge independence of the strong interaction. Observations that violate it are reported in the December 30 *Physical Review Letters* as a result of work done partly at the Nuclear Research Center at Demokritos near Athens.

The strong interaction is the force that holds atomic nuclei together. It acts equally upon the electrically charged proton and the neutral neutron. *This independence of charge was crucial in convincing physicists that in the binding of the nucleus they were dealing with a new kind of force* and not a manifestation of electromagnetism as early thoughts on the subject had tended to suppose. *Now the charge independence doesn't look as complete as it once did* thanks to the present work. The work was done by T. E. Kalogeropoulos of Syracuse University and the Demokritos center and nine colleagues from the two institutions.

In the experiments, hydrogen and deuterium nuclei were bombarded with antiprotons to see what happened as the antiprotons met the protons in the nuclei and annihilated each other. Out of such annihilations certain debris is expected, including pi mesons and photons. The basic experimental finding is that there are too many photons, surprisingly too many, almost a whole extra photon per annihilation. The interpretation of this seemingly simple fact leads down a number of important paths in subatomic physics.[4]

At other times a good problem statement needs a line of reasoning leading up to the A or B statement. This is illustrated in the second paragraph of the example immediately above and in the following example from a student paper. (Italics added.)

THE EMERGENCE OF GOD FIGURES IN SCIENCE FICTION

During the 20th century, humanity has enchained itself with a world of science, technology, machines, and inventions. Arising with this philosophy is the feeling that the universe is a machine. Human beings, who portray themselves as chief engineers, should have every right to probe, study, tinker with, and eventually control the mechanism of the world. As scientific humanity learns more tricks and sheds its old "religious crutch," it is better able to rationally exercise control over nature. *Hence, one would expect science fiction, in keeping with this trend, to depict humanity as conqueror, leader, and master of its future, free of the influence of "gods." But this often is not the case.* Much of science fiction is devoted to the inferiority of humanity or its inability to execute its free will. Instead of futuristic beings finding total fulfillment in expressions of individuality and personal creativity, science fiction often depicts humanity as passively subservient to a supreme organization, as in Orwell's *1984* or Zamiatin's *We*.

    The organization with its godhead is idolized and obeyed more fervently than any god ever known to religion. At other times, the controlling figure is a less tangible supernatural entity, as in Lindsay's *A Voyage to Arcturus*. No matter what the form, even in science fiction humanity finds itself at the control of the gods.

If you use the A-but-B form to identify a problem or conflict for the audience, you must be careful that the A and B terms really conflict or else the whole introduction will not make sense.

❑ EXERCISE 5-1

The following passages were chosen from a variety of scientific and technical sources. See if you can identify the problem stated or implied in each passage, and then try to strengthen or clarify the problem statement. (Although all these passages are understandable, many would benefit from a sharper problem statement.) Be sure that the magnitude of the problem is clear. Note that several of the passages will be outside your area of specialization; they were selected to help develop your "feel" for the nonspecialist audiences with whom you will often need to communicate.

A   Subject: Aluminum Pitting Corrosion: A Progress Report
    Foreword: Aluminum is a desirable material for automobile cooling systems because its use provides less corrosion and up to a one-third weight saving and a one-third cost saving over conventional copper-brass cooling systems. However,

aluminum exposed to chloride ion solutions undergoes localized corrosion, or pitting.

The reaction of chloride ion with aluminum metal is a well-documented phenomenon, but the interaction between the chloride ion and the passive oxide or hydroxide film which always exists on aluminum surfaces is yet unclear. This interaction is currently being studied by potentiostatic testing to determine the effect of chloride ions on the initiation of pitting corrosion. The purpose of this report is to present the results of these potentiostatic tests.

B   Supergravity and the Unification of the Laws of Physics
A catalog of the most basic constituents of the universe would have to include dozens or even hundreds of particles of matter, which interact with one another through the agency of four kinds of force: strong, electromagnetic, weak, and gravitational. There is no obvious reason why nature should be so complicated, and perhaps the most ambitious goal of modern physics is to discover in the diversity of particles and forces a simpler underlying order. In particular, a more satisfying understanding of nature could be achieved if the four forces could somehow be unified. Ideally they would all be shown to have a common origin; they would be viewed as different manifestations of a single, more fundamental force.

In the past 50 years remarkable progress has been made in identifying the elementary particles of matter and in understanding the interactions between them. Of course, many problems remain to be solved; two of the most fundamental ones concern gravitation. First, it is not understood how gravitation is related to the other fundamental forces. Second, there is no workable theory of gravitation that is consistent with the principles of quantum mechanics. Recently a new theory of gravitation called supergravity has led to new ideas on both these problems. It may represent a step toward solving them.[5]

D   Life May Be Hazardous to Your Health
The message we seem to be getting from the media is that our health, if not our very lives, is endangered by the very same technology that supports our existence. The series of warnings is ceaseless: DDT, PCB, polyvinyl chloride, nitrous oxides in the air, unsafe automobiles, radiation, food poisoning, sweeteners, contraceptives, fallout, fluorocarbons, noise pollution, x-rays, cosmetics, . . ., and on and on. One might begin to wonder how we ever managed to survive past the age of 25. Controversies begin and crises of this type often arise because of the uncertainties and complexities of the concept of safety.[6]

C   Science Fiction: A Warning to Humanity
From the beginning of the human race, through prehistoric times to the present, humanity has always sought power. Human beings have wanted to control themselves and their environment. They have wanted to be able to predict the future. In short, they have wanted to be omnipotent. The examples of this desire, this greed, to become an ideal, perfectly powerful being are many. We have, in the tale of Adam and Eve, a desire to become one's own master. More recently, we have leaders such as Caesar and Hitler, who wanted to control the world. Even today we see this quest for power, not only in political leaders but in everyday, common people. People are always trying to make more money, to learn more, in order that they may improve their status.

But this quest for power is not without effect. Things go wrong, people are hurt, and the human condition deteriorates. You may ask, "What can be done about this situation?" My answer is that we must be more careful about what we do; we must more closely analyze our situation, what we are planning on doing, and what its effects will be. In this paper, I will present the role that science fiction plays in showing us why we should be more careful about what we do, and what the consequences are if we are not careful.

❑ EXERCISE 5-2

Survey your experience for a hobby or some other subject you know something about. If you have worked at a job or done an experiment in a laboratory course, then pick a particular problem or experiment you could talk about. For your topic, write an Introduction by problem statement for someone who knows nothing about it. Be sure to define your audience and to keep its interests in mind in setting up the Introduction.

5.2 IDENTIFYING YOUR STRATEGY AND PURPOSE

Once you have established an important problem for the audience, you need to show that your communication addresses this problem. This is done by

1 Indicating the missing information you're providing and the strategy you're taking on the problem (you do this by stating the question or task you're addressing)

2 Announcing the purpose of the communication

For example, let us reconsider the example in Figure 5-1:

> Our new stereo system must appeal to the teenage market.
>
> However, it apparently does not appeal to this market.

After having stated the problem directly, we could have tried to solve it by asking questions such as the following:

> What is wrong with teenagers?
>
> What is wrong with the stereo system?
>
> Why aren't teenagers buying the stereo system?
>
> Where is it being sold?
>
> Who is buying the stereo system?

If we had asked the first question, we would have assumed that something is wrong with teenagers and have committed ourselves to this point of view. It may be that something is wrong with them, but it may be that something else is wrong, something we would never see given this perspective. The same complaint could be advanced about the second question. However, the third, fourth, and fifth questions would not have committed us to a limited point of view and thus would probably have produced better information. If, for instance, the cause of poor sales is poor advertising, the last three questions would be much more likely to guide us to this cause.

This example illustrates two important properties of such questions. First, since at the investigation stage we don't want to close off fruitful lines of inquiry, the questions re-creating this stage should be open-ended, that is, they should not limit us unnecessarily and should usually begin with such words as *who, what, where, when, why, how,* and *to what degree.* Second, the questions we ask affect the solutions we reach; in fact, the questions define our *strategy* for solving a problem. For instance, all the questions above point to potential causes; they indicate that our strategy for finding a solution is to find the cause of the problem. A different set of questions would have defined a different strategy. If we had asked the following questions, for instance, our strategy would have been to improve the product:

How can we make the stereo more cheaply?

How can we improve the performance of the stereo?

How can we reduce the size and weight of the stereo to make it more portable?

Can we make the stereo look "richer" and more elegant?

Note that if the real problem had been poor advertising, the strategy of improving the product would probably have been ineffective.

How do we define our strategy and our purpose in an Introduction? Let's say that we have explored the question "Why aren't teenagers buying the stereo system?" and decided that the reason was poor advertising, appearing at the wrong places and at the wrong times. If we were to report on this situation to a vice president and a manager at the stereo company, we might produce an Introduction similar to one of the following:

Version 1
In order to be competitive and profitable, our new stereo system must appeal to the teenage market. However, it apparently does not appeal to this important market, since few teenagers are buying it. What caused this situation? It appears that a poor sales campaign is the only major cause.

Version 2
In order to be competitive and profitable, our new stereo system must appeal to the teenage market. However, it apparently does not appeal to this important market, since few teenagers are buying it. I was asked to investigate this situation and to identify the cause. This report will argue that a poor sales campaign is the only major cause.

Figure 5-2 ❏ OUTLINE OF A FOREWORD OR INTRODUCTION, STEPS 2 AND 3

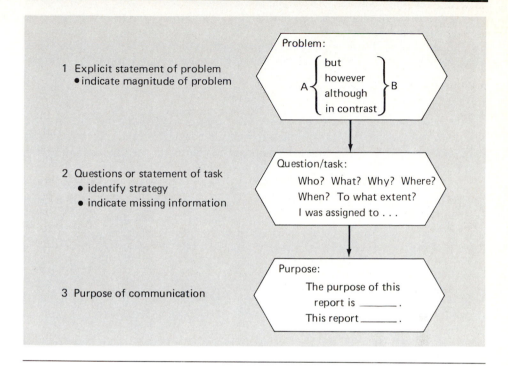

Notice that each of these has a clear problem statement followed by a statement of question or task and a statement of purpose. Notice also that in version 2 the statement of task is merely an unstated question. "I was asked to investigate this situation and to identify the cause" means "I was asked to find out what caused this situation." The process of writing an Introduction is summarized in Figure 5-2.

Another example of an Introduction by problem statement follows, written by a student to administrators at his university. The student hoped that the report, from which this Introduction was taken, would help decrease or eliminate the use of the test mentioned.

In 1962, the Opinion Attitude and Interest Survey (OAIS) Test, more commonly known as the "raw carrot" test, was first introduced to the University of Michigan as part of the testing program administered during an entering student's orientation. It is a psychological test, testing for motivation, vocational aptitude, vocational interest, and certain factors relating to personality. This exam is supposed to be used as a placement device by the counselors in chemistry and math courses. However, if one surveys the counselors on campus, the test scores are not used in this manner; in fact, they are not used at all. Thus, the largest school at the University—Letters, Sciences, and Arts—stopped "coercing" its students to take this examination. Some schools have followed, but others have not—notably, Nursing, Education, and Engineering.

If the results are not used as the test examiners have projected, then why should the exam be forced on the non-LSA students in Nursing, Education, and Engineering? There is no compelling reason. In fact, the test should not even be encouraged by these colleges for large groups of students; it has internal problems and can be misused and misinterpreted, especially in combination with other, less volatile sources of information at the counselors' disposal.

Notice that this author has considered his audience carefully in phrasing the problem statement. He has emphasized the fact that the test scores are not used and implied that the university is wasting its resources by continuing to administer the test.

The author of the above report on the OAIS also had three other complaints about the test, which he developed in his report: (1) that it was an invasion of privacy, (2) that it was poorly designed and thus gave poor results, and (3) that it wasted valuable student time during orientation week.

❑ EXERCISE 5-3

Define some audiences for which each of these three complaints could be used in an A-but-B problem statement. You will thus have at least three audiences and three problem statements.

❑ EXERCISE 5-4

Go back to the passages quoted in Exercise 5-1. See if you can identify the question or task and the statement of purpose in each one. Then see if you can add a question or task and purpose for the introduction you wrote in Exercise 5-2.

5.3 A SHORT FORM FOR STATING PROBLEMS

Section 5.1 presented a general form for starting letters, reports, or speeches based on an A-but-B problem statement. This format is commonly used and important. However, if you look at much communication in science and technology, you will find examples which seem to violate the A-but-B format because they seem to be missing either the A term or the B term. Consider the following example from a report by a research forester to the Superintendent of a state forest nursery (notice how the problem statement addresses the concerns of the Superintendent and indicates the magnitude of the problem):

Foreword
During my visit to your nursery on September 15, 1989, you and I observed the high mortality of the 2-year-old red pine seedlings in bed 19. As you stated, this would lead to a production shortage in 1990. You requested my help in finding

the cause of this problem so that it could be corrected in the near future. The purposes of this report are (1) to explain that parasitic nematodes are the probable cause of the high seedling mortality and (2) to recommend a solution to the problem.

What is happening in this example and others like it is that one term in the A-but-B format is unstated, but assumed. It's assumed because it's believed by all members of the audience (the Superintendent and other employees at the nursery) and so obvious that saying it would seem ridiculous. For instance, in the example above, the author merely states the B term: "you and I observed the high mortality of the 2-year-old red pine seedlings . . . this would lead to a production shortage in 1990." The author assumes any reader of the report would know the A term: one does not want a high mortality rate of red pine seedlings or a production shortage in 1990. Just stating that they might occur puts the ideas of high mortality and production shortage in clear conflict with a reader's deeply held values and assumptions that they should not occur.

The "short form" problem statement just described is represented in Figure 5-3, where brackets have been placed around unstated but assumed material. The short form appears frequently in scientific and technical communication, since many of the problems scientists and technical people write about arise from commonly held assumptions. Almost everyone wants and is interested in

Efficiency

Low cost

Competitiveness in cost and performance

Products of good and even quality

Freedom from pollution violations, labor unrest, and legal action by consumers or the government

Elimination of wasted time, energy, or money

Simplicity rather than complexity in organizational structures and product design

When such values are used in a problem statement, they are usually not stated directly. However, they are there and are often the values needed to catch the interest of managers and others not directly involved with a project. Interestingly, these values also provide the context needed by someone outside a project to understand its real importance and purpose. Look back, for instance, at the example on mortality in red pine seedlings. Could you really understand why someone was worried about parasitic nematodes (the probable cause of the high seedling mortality) without calling on the values and assumptions implied in the Foreword?

Figure 5-3 ❏ OUTLINE FOR A SHORT-FORM PROBLEM STATEMENT

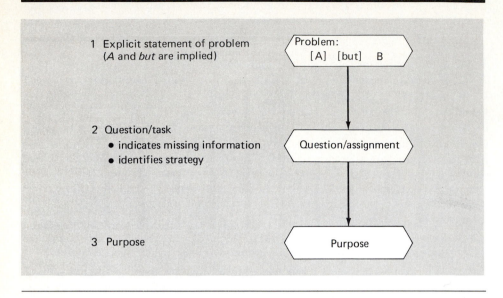

We might consider another example of a short-form problem statement from a report written by a civil engineer to the Chief Engineer of a manufacturing company. The readers of this report might well include managers, architects, financial analysts, constructors, and even a vice president. The author rightly assumes that any reader would know that the manufacturing company does not want to damage its building foundations. Again, this is so obvious that an author would not state it, but it is unquestionably there in the mind of the reader.

On January 22, 1990, you requested the analysis of two samples of sand to see if they could be used as a foundation for several high-speed presses. You expressed concern that this sand, which is below the punch press foundations, might settle due to the vibration of the presses, thus damaging your building foundations. Along with the two samples, the following data were forwarded:

Void ratio of white sand (Ottawa Sand)   .65

Dry unit weight of black sand (Turkish Emery)   118 lb/ft^3

The purpose of this letter is to recommend that neither sand is sufficient for use as a foundation for high-speed punch presses. The vibratory loadings transferred from the machines will cause the foundations to settle.

❏ EXERCISE 5-5

Examine the following Forewords and Introductions from technical articles and reports. Each is missing an overt statement of one of the conflicting terms of its problem statement. See if you can identify or easily reconstruct both the A and B terms of each problem statement.

A   Subject: Vacuum Filter Sludge Cake:
Accuracy of Present Production Data
Foreword: Slowdowns and necessary repairs have occurred in the incineration area of the Sewage Treatment Plant, costing $2 million last year alone. These problems may have occurred because of inaccurate data on the amount of sludge cake being produced and delivered to individual furnaces. Thus, Dennis Moore, the Chief Project Engineer, requested that I investigate our data on sludge cake production and suggest improvements in monitoring it if those are necessary. The purpose of this report is to document the unreliability of our present monitoring technique and to recommend a better one.

B   The Search for Life on Mars
Is there life on Mars? The question is an interesting and legitimate scientific one, quite unrelated to the fact that generations of science fiction writers have populated Mars with creatures of their imagination. Of all the extraterrestrial bodies in the solar system, Mars is the one most like the earth, and it is by far the most plausible habitat for extraterrestrial life in the solar system. For that reason a major objective of the *Viking* mission to Mars was to search for evidences of life.[7]

5.4 GUIDELINES FOR CHOOSING BETWEEN FULL-FORM AND SHORT-FORM PROBLEM STATEMENTS

Sections 5.1 and 5.3 describe two related forms for problem statements, the full form and the short form. The purpose of this section is to summarize guidelines for choosing one form or the other.

As already suggested, you should use the full form of the problem statement when your audience or part of your audience might not know clearly and obviously what each term is. You may choose the short form when one term is so clearly known and believed by all members of your audience that it does not need to be stated to make the problem clear. The possible choices are summarized in Figure 5-4.

For instance, in the competitive business system of the United States, shared values include those presented in Section 5.3: the desire for low cost; efficiency; good and uniform product quality; freedom from legal prosecution, labor disputes, and pollution violations; simplicity; and elimination of wasted time, energy, or money.

In a scientific or technical field, shared knowledge, beliefs, and expectations are based on the experiences common to members of the field. It would be impossible to give an inclusive list of all the shared values that could be unstated in some particular situation. Thus, a communicator must consider various audiences for a communication and produce an Introduction aimed at the least knowledgeable member of the audience.

Figure 5-4 ❑ GUIDELINE FOR CHOOSING BETWEEN FULL- AND SHORT-FORM PROBLEM STATEMENTS

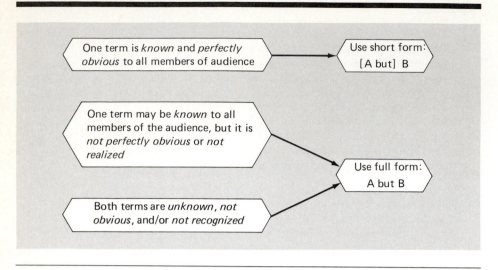

❑ EXERCISE 5-6

Look at the outline you've written for the Westheimer Clinic case in Exercise 4-4 and add an appropriate problem statement to it.

REFERENCES

1   Leon Festinger, *A Theory of Cognitive Dissonance* (Palo Alto, CA: Stanford University Press, 1957).

2   Richard E. Young, Alton Becker, and Kenneth Pike, *Rhetoric: Discovery and Change* (New York: Harcourt, Brace & World, 1970).

3   Joan Arehart-Treichel, *Science News* 107(2):26 (1975).

4   *Science News* 107(2):20 (1975).

5   Daniel Z. Freedman and Peter van Nieuwenhuizen, "Supergravity and the Unification of the Laws of Physics," *Scientific American* 238(2):126 (1978). Copyright © 1978 by Scientific American, Inc. All rights reserved.

6   Ray Berry, "Caution: Life May Be Hazardous to Your Health," *Michigan Technic* 96(4):6 (1978).

7   Norman H. Horowitz, "The Search for Life on Mars," *Scientific American* 237(5):52 (1977). Copyright © 1977 by Scientific American, Inc. All rights reserved.

ADDITIONAL READING

Elliot Aronson, "The Rationalizing Animal," *Psychology Today*, May 1973, pp. 46–52, 119.

Scott Consigny, "Rhetoric and Its Situations," *Philosophy and Rhetoric*, Summer 1974, pp. 175–186.

Otto Dieter, "Stasis," *Speech Monographs 17*(4):345–369 (1950).

J. C. Mathes and Dwight Stevenson, *Designing Technical Reports* (Indianapolis: Bobbs-Merrill, 1976).

Thomas Nickles, "What Is a Problem That We May Solve It?" *Synthese 47*:85–118 (1981).

6

Drafting and Word Processing

So far, we have talked about a number of basic topics related to technical documents—identifying audiences and purposes, generating ideas, and constructing arguments. These topics have been covered because they deal with generic problems that arise in any kind of technical writing—whether it is done on a computer or with a pencil. Solving these problems largely involves developing a method of composition which explicitly addresses them, independent of the physical mode you use to record your thoughts.

As more and more technical communicators come to rely on computers to produce and/or edit drafts of their work, they also need to be aware of the potential effects of computer use on the process of writing. Thus, in this chapter we will first consider some generic methods for preparing a draft, whether it is written by hand or by computer, and then focus on some of the additional difficulties and opportunities offered by computer-based writing.

6.1 WRITING A FIRST DRAFT

Your goal in writing a draft is to turn the purposes, ideas, and arguments you have generated into a text. You should not expect to produce a finished text in one sitting, not if it's an important text. Instead, you should plan on writing at least several drafts, with the first draft being only a rough approximation of what you want to say. If you've put a lot of thought into your subject and have analyzed your audience, purposes, and arguments (Chapters 2–4), you will probably find it relatively easy to sit down and knock out a first draft. Conversely, if you haven't done the appropriate prewriting steps, you may well find it difficult to write a draft. Even with the best preparation, however, writers sometimes find it hard to "get going." Here are some suggestions for overcoming "writer's block" and getting that first draft written:

1 Write an outline (Chapters 2 and 4) and then flesh it out in whatever way suits you best. Some writers like to work top-down, others like to work on the easiest parts first, still others like to skip around and do a little bit here, a little bit there. At this stage, just do what feels most comfortable for *you*.

2 Sit down and write whatever comes into your mind about your topic. Don't worry about its organization, or style, or even its sense; just get something down. This technique for getting started is called *focused free writing*[1]— focused because your writing is focused on a particular topic, and free because you are free from all the normal constraints of organization, style, grammar, spelling, audience, etc. Once you get some version written, you will have something that you can later evaluate, organize, rewrite, and edit.

3 If you use a computer, try "invisible writing." Turn the illumination on your computer screen down so that you can't see anything on it. Then write. When you've done as much as you can, turn the illumination back up and see what you've got. This is a form of free writing that's particularly unconstrained.

4 Sit down and write a version of what you want to say and then throw it away. You don't need to worry about how it turns out because no one will see it; you won't even read it over. Write another version and then read it to see if it is good enough. If not, throw it away and write another version. Do this a few times until you like something enough to work on improving it.

5 Talk about your subject to a friend, coworker, or family member. Don't worry about organizing what you say; just have a conversation in which you tell someone friendly about your project and why it is important. Then talk about it again to a second person and then to a third person. At this point, try to write a version, assuming that you have now honed your points and approach in your oral "trial runs."

6 Talk about your subject to a friend, coworker, family member, or some imaginary person and tape-record what you say. Explain what you have done or want to do and why it is important. Then listen to the tape and write down your main points, or literally transcribe what you say into a draft.

Each of these six approaches will produce something about your subject, although that something may be very rough. At this stage, that's all you need.

Combating the Psychological Need for a Perfect First Draft

Many inexperienced writers try to write perfect sentences in their first drafts, whether they are writing by hand or on a computer. This strategy often "freezes" up the writing process and creates a kind of writer's block, since it puts too many demands on the writer at one time. A first draft will do its job if it just gets the ideas down in some form. Trying to get the ideas down in perfect form is like trying to memorize a speech in one pass: it can't be done.

Once you get a rough draft, you will need to evaluate it. Do this by first considering its content. Does it make sense? Is it what you want to say? Have you covered all your important points? If the answer to all these questions is yes, you're ready to move on to editing, testing, and revising. If not, you probably should take another look at your outline and try to figure out what's wrong.

6.2 USING THE COMPUTER

Whether you write your rough draft by hand or on a computer, you will find it helpful to use a computer the rest of the way. A computerized word processor allows you to do a variety of things that you cannot do with any other technology. You can write faster; you can make more changes faster; you can use outliners, spelling checkers, and other tools; you can print out hard copy in a variety of fonts and type sizes; you can integrate visual aids and special formatting features; and you can send the result electronically to distant recipients. Whether you use a personal computer or a mainframe, you can find a word-processing program that will suit your tastes and your bank account.

But research has shown that not everyone benefits equally from using a computer. For example, many students apparently lack an understanding of the computer as a writer's machine.[2] If you are careful about your planning, and if you can type well, you will probably get maximal benefit out of writing on a computer. If not, you probably won't. The rest of this chapter discusses some of the pros and cons of using a word processor. Although the former, in our view, far outweigh the latter, everyone should be aware of possible difficulties in using the computer and how to overcome them. These difficulties include a range of writing process problems, computer problems, and visual design problems.

6.3 MINIMIZING DISTRACTIONS FROM TYPING OR RUNNING THE COMPUTER

The first set of problems to confront a computer-based writer are the many things to learn and control when first using a new word processor or computer system. Many people find that when they start with a new system, they seem to focus all their attention on their typing skills or on the commands needed to run the computer. In such cases, they may have little or no attention "left over" to devote to the cognitive acts of composition, to developing an appropriate point and structure and focus for the audience to which they are writing. Thus, if you are a new or moderately new user of a computer system, you may want to minimize the attention you need to spend on running the computer by trying the following strategies.

Define a Minimal Instruction Set

Instead of trying to learn everything about your new system right away, make a list of the smallest number of tasks you need to learn to get started. Then, make a reference sheet of the commands for these tasks and tape it to your computer or set it up close by while you work. The list will probably include:

- Turning on your computer
- Starting up your text-processing system
- Opening a new document
- Calling up an existing document
- Entering new material
- Deleting material
- Adding new material to existing material
- Saving your document
- Printing your document
- Turning off your text-processing system and computer

Once you have these actions under control, you might add the following to your list:

- Changing left margins
- Setting up lists
- Copying and moving text

If you can master this subset of text-processing tasks, you will be able to get started on your writing with a minimum of things to remember and thus a minimum of distractions to your writing activity. Once you are comfortable with these tasks, you can learn other things as you need to use them.

Get Formal Training

Take a course on your text-processing system or teach yourself to use it by going through all the tasks listed in the documentation and then practicing on "junk" documents. This approach has the advantage of letting you know how to run the computer before you start to do real work, but the serious disadvantage of forcing you to spend a lot of time practicing instead of doing real work while you learn.

Improve Your Typing

If you are a poor typist, try to improve your typing by taking a typing course or practicing with a "typing tutor" on your computer. While you're learning, try not to worry about your typing as you write. One great advantage of using a computer is that it is very easy to correct mistakes when you are editing. If you can shift your attention from your typing to your thinking and writing, you will be way ahead.

6.4 SPECIAL HIGHLIGHTING FEATURES

Many computers give writers the capability of using different fonts—that is, different sizes, shapes, thicknesses, and styles of letters. Since these are interesting and new, writers often get carried away and use most of them in the same document, giving their work a hodgepodge look. If your system has different fonts, try to design a use of the fonts which is functional. Choose one major font as the base for the text and then use its different sizes and styles to highlight important things. For instance, you might use bold and large sizes for headings, and use bold or italics alone for particular kinds of emphasis. If you use too many fonts in a text, you'll detract from its ultimate readability.

6.5 COMPUTER FAILURES

One of the easiest ways to save a lot of time is to *save your document frequently* when writing on a computer and to *make at least one backup copy* of the document. You should update a backup copy of your text frequently as you work and keep it in a separate, safe place so that if your computer has a failure, you won't lose hours of work. This is especially critical if you're working during an electrical storm (which is not a good idea) or working on a system that experiences occasional power surges or losses. Although you may find it annoying to save your text and make backup copies so frequently, you won't find it nearly as annoying as losing all you've written in a power failure.

6.6 PROBLEMS IN PLANNING WHEN WRITING ON A COMPUTER

Planning involves the writer's selecting and organizing material and considering how the audience and purpose affect what is being said. It is essential to good writing, and therefore should be done not just before writing but throughout the writing process, even during final editing and formating. Although planning presents problems for writers working by hand, there is some evidence that planning is even more of a challenge for writers working on a computer. When writers use a computer, they tend to do less planning overall, less planning before writing, and less conceptual or high-level planning than when they write by hand.[3] Though the evidence could be stronger, it suggests certain "downside" effects of the computer that correspond to many people's observations and comments:

1 The empty, blinking computer screen "pressures" some people to start writing before they are really ready,[4] makes them feel that "I just can't sit and think with my fingers on the keyboard—I feel I have to start typing."[5]

2 The small screen limits the amount of text the writer can see and thus focuses attention on local issues rather than on larger issues of overall organization, audience, and purpose.

3 The ease of making small editing changes biases the writer toward stylistic editing rather than basic organizational changes.[6]

If you have trouble with planning or feel especially pressed to write in front of a computer screen, there are several strategies you can use to increase the amount and effect of your planning activities.

Plan before Writing

Do as much explicit planning before writing as you can. Make sure you have followed the recommendations in Chapters 2–4 and have identified the problem, your audience, your purposes, your main point(s), and your arguments. Keep a hard-copy (printed) outline handy so that you can keep a sense of the whole document in mind as you write.

Use the computer in your idea generation stage—that is, to take and organize notes for writing. (You might want to try one of the "idea generating" programs which ask a series of questions about a topic to help you think of relevant things to say.) This approach allows you to plan and also to relieve the sense of "pressure" from the blank screen by doing some typing. It also saves you from having to write notes once by hand on note cards or paper and then to retype needed parts into the computer. Use the computer to set up your bibliography and footnotes.

Plan during Organizing and Writing

Use an outlining program. A good outliner allows you (1) to create and view an outline of a document, (2) to write text under each entry in the outline, and (3) to use the entries as headings and subheadings for the actual document and thus as writing guides to topic structure. Such programs typically allow you to view the outline alone, even if there is a large amount of text under some or all of the entries, and to easily add to or rearrange the outline and associated text as reorganization is needed. Outlining programs thus enable you to gain high-level views of the organization of the document, even within the limitations of a small computer screen.

If you don't have access to an outlining program with the capabilities described, make an outline and keep it handy to guide your writing and to remind you of the overall structure of the document as you work on its parts.

Scroll back and forth in the document frequently. Scrolling allows you to review what comes before and after in the document and to better connect the text being written to its context. There is some evidence that writers who do less planning when they write on the computer make up some of their planning time in rereading the text after it is written.[7]

Plan throughout the Writing Process

Print out copies of the document or parts of the document at critical stages in the writing process, and use these copies for serious editing and revision. A hard (printed) copy eliminates the viewing limitations imposed by a small computer screen, allows you to see the overall structure and relationships among parts, and—if it is a good copy—may be easier and faster to read.

Reevaluate the document after any changes. It is very easy to move a paragraph from one place to another without any typographic indication that a move has occurred, though there may be a conceptual "hole" where the paragraph came from and a lack of integration in its new location. Be particularly careful to check the organizational implications of a move. Also, make sure you still have good transitions from one paragraph to another.

6.7 COMPUTER-AIDED EDITING

Once you start editing, proceed in stages to make the editing easier and more effective. Do one pass for organizational issues, another for paragraph structure, another for grammatical or stylistic issues, another for spelling, punctuation, etc. (see Chapters 21–28 and Appendix A). Make a list of your special problems and add a pass for each problem.

Feel free to use the copying and moving capabilities of your computer. They allow you to reorganize your text and create multiple versions of it so you can compare two or more different ways of doing something. Use the search and replace capability to find multiple forms of reference for the same item and to pinpoint your typical wordy expressions and words that may need to be defined for your various audiences.

Use the spelling checker and other editing tools available on your system. If you don't have a spelling checker, try to get one, and add your special "problem" words to it if they're not already there. Other computer-based editing tools which might be useful include a thesaurus program for finding synonyms and style programs for isolating stylistic problems such as overly long sentences, overly abstract or "fancy" words, and sexist terms.

If you do use a spelling checker or style program, be sure to critically evaluate the advice it gives you, since much of the advice will be inappropriate for given situations. For instance, the spelling checker can tell if *there* or *their* is misspelled as *theer*, but not if one is used incorrectly for the other. Thus, it will not catch *their* used for *there* as a misspelling. Also, some style programs single out passive verbs as weak writing, but they cannot identify particular sentences in which a passive verb may be better than an active one.

Finally, set up your writing area with the tools you will need so that you can work without the interruptions of finding a dictionary and other missing items.

6.8 ELECTRONIC MAIL

One form of writing which is becoming increasingly popular is electronic mail and its variants of interactive electronic messages and electronic conferences. In each case, the computer is used to send notes, comments, questions, or full documents to a variety of readers through a computer network. As many of you already know, an electronic mail system puts a piece of communication in a "mailbox" owned by the recipient of the communication and then allows the recipient to read it whenever he or she wants to. Interactive messages are messages exchanged between two computer users who have signed on at the same time, and conferences (or forums) are electronic meeting places where people interested in a particular topic can exchange information. If you're not using electronic mail now, you probably soon will be. In one survey of university researchers, the use of electronic mail doubled from 1986 to 1987.[8] In another study in 1986 at IBM, 86% of the participants spent more than 10% of their working hours using computer-mediated communication, while 28% spent more than 30% of their working hours doing so.[9]

Although electronic mail is becoming very popular, certain conventions still need to be worked out so that the audiences who receive it will feel that it is appropriate in a variety of ways. Today, many electronic mail systems allow junior employees to send messages directly to the highest-ranking officers in the institution. In this sense, electronic mail seems to be a somewhat "democratizing" influence on a hierarchical organization and a way of providing access to the highest officers.

However, the style of much electronic mail is not that of formal communication aimed at busy and preoccupied audiences. It is more like that of conversation but without all the politeness conventions you normally find in conversation. For instance, electronic mail is often used to dash off short notes or responses to readers. These pieces of communication are often "shot from the hip," in that they are not very carefully reasoned or composed or edited for style and image. They may assume the context instead of defining it for the audience, or they may talk in what seems to be a conversational mode to the writer but a curt or even impolite mode to the reader.

The effects of electronic mail have not yet been studied in depth, but four studies suggest some interesting conclusions. One found that groups communicating electronically had more antisocial behavior and made more extreme decisions than groups communicating in traditional ways.[10] Two others found that responses to an electronic questionnaire were more extreme, more revealing, and less socially acceptable than were responses to a paper version of the questionnaire.[11] A fourth study found that the use of electronic mail within an organization tends to "break down" organizational status and structure—that is, the normal lines of communication with the organization.[12]

Thus, in dealing with this ultimately computerized form of communication, a writer must shift out of the conversational mode of writing, treat the messages as other written documents, and evaluate them in terms of the criteria outlined in other chapters of this book.

6.9 HYPERTEXT AND HYPERMEDIA

Hypertext and hypermedia are a system of organizing a document that allows a reader to read hierarchically—that is, in a nonlinear fashion. Normally, when a document is read, the words appear on the page in sequence: heading, first sentence, second sentence, etc. In this mode of presentation, readers must read all the words in their normal order, or if readers want to skim the document, they must skip from title to title, heading to heading, topic sentence to topic sentence, graphic to graphic, etc. But they must confront all the words and other material in the text and explicitly deal with them by skipping or reading. This can be annoying to readers who are not interested in all the details, but just want a quick overview of the document and its arguments.

By contrast, hypertext and hypermedia allow a document to be organized so readers can self-define the level of detail that is presented. If they want to see only an overview of the document, they can see just that; if they want to see all the details, they can see that; and if they want first an overview and then details about some of the parts, they can see just that. The writer makes this possible by writing each heading and unit and then specifying paths which readers can use to read through the headings and units. In hypertext, a unit may be a section or paragraph or sentence. In hypermedia, a unit may be a unit of text or some other kind of material, such as a graphic or budget.

Since hypertext and hypermedia are still very new, there is little research on the best way to use them to organize and write documents. Some early discussions of hypertext and hypermedia include those by Barrett,[13] Conklin,[14]

Figure 6-1 ❏ POTENTIAL MOVEMENT THROUGH AN ARGUMENT STRUCTURE IN A HYPERTEXT DOCUMENT

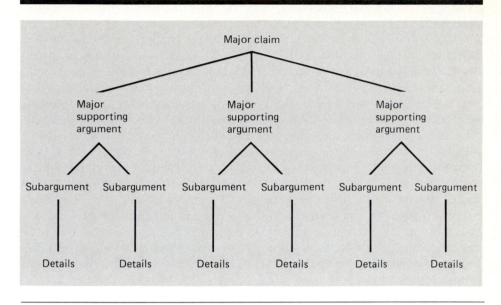

Kaehler,[15] Marchionini and Shneiderman,[16] Tessler,[17] Yankelovich et al.,[18] and articles on HyperCard in *Macworld, IEEE Computer*, and other computer publications. If you are using one of these systems, you might organize your document as shown in Figure 6-1. Have the text move clearly from major claim to major supporting argument to subargument to details, as discussed in Chapter 7, or from general to particular, as discussed in Chapter 12. Then provide paths which allow a reader to read across at each level of argument tree and down each line of argument to the details, or across each level of generalization and down each level of specificity, as discussed in Chapter 12. This will allow a reader to read in all the ways described above: to see an overview of the document; to see all the details; or to get an overview at some stages and then details at other stages.

6.10 DESKTOP PUBLISHING

Desktop publishing gives a writer the ability—without the help of a designer or typesetter—to produce camera-ready copy of a document that has complex formats and text interspersed with graphics, mathematics, pictures, charts, spreadsheets, and other visual aids. Desktop publishing is possible because of the availability of inexpensive hardware (computer systems and laser printers) and of software to do complex layouts and page formatting. Although a variety of software has been available to do computer-based layout and page formatting on large, time-shared computers, only recently has this capability become available on personal computers.

The first full desktop publishing systems were Pagemaker for the Macintosh line and Ventura for IBM personal computers and clones. Although these systems were very advanced for their time, now many text-processing programs have some of the formatting features (such as multiple-column pages) associated with these systems. As time goes on, more text-processing systems are getting more of the capabilities associated with desktop publishing.

Desktop publishing is spreading rapidly in the technical writing world because it provides a number of significant advantages over older text-processing systems. It provides a high-quality, interesting-looking document, much "flashier" than straight wall-to-wall, single-column text. It gives the writer control of the final appearance of the document in a way that is not possible if a designer or someone else does the formatting. It reduces the cost of producing the "master copy" of a document, since it eliminates designing and typesetting costs. It allows the writer to easily experiment with alternative ways of formatting the document, again without the costs of a designer and typesetter or the need to spend a large amount of time "pasting up" the document by hand. It reduces the amount of time it takes for the document to cycle through the layout and design stages, since it takes the writer only a few minutes to run a document through desktop publishing software, in contrast to the days or weeks it can take to get the document to and from a designer or typesetter.

Although all these benefits are very appealing to a harried writer with a limited budget, doing your own desktop publishing has some significant costs which must be paid. First and most important, you have to invest a significant amount of time to learn to use the desktop publishing software itself. The programs are sophisticated and much more complicated to learn than the typical word-processing programs. Second, to use the software well, you have to become a good page designer, and that means learning something about design (unless you use only the few formats already provided with the software). Learning about page design is an interesting activity, but it does take time, and—even with some effort—the result may still look like the work of an amateur rather than a professional. Some of the technical and visual difficulties to be mastered appear in Table 6-1; many of these were encountered by the three students and two professionals whose desktop publishing experience is summarized in the table.[19] Finally, even though the final product of a desktop publishing system looks quite good, it is not typeset quality; desktop documents now have about half the sharpness of typeset material. (The average laser printer prints about 300 dots per inch, whereas typeset material is usually printed at between 590 and 750 dots per inch.)

In spite of these costs, desktop publishing is becoming more popular. It is changing the look of documents and our expectations about the standards for document appearance. This revolution in quality of appearance is similar

Table 6-1 ❏ SUMMARY OF THE PROBLEMS OF FIVE DESKTOP PUBLISHING USERS

Problem Category	Total	Problem Category	Total
Computer Technology Problems		***Visual Document Problems***	
Computer knowledge	4	Obeying conventions of document	5
Using operating system	4	Misapplying newspaper or magazine layout	2
Manipulating files	4	Logic of headings	2
Revising scanned images	3	Labeling of images	3
Fixing minor problems	3	Making important information prominent	4
Creating sample graphics in Macdraw or Macpaint	4	Knowledge of two-dimensional layout	5
Entering/saving text	2		
Backing up text	4	Knowledge of typefaces	4
Printing	2	Overusing	
		Lines and rules	5
Process Problems		Boxes	4
Planning project	3	Typefaces	4
Identifying subtasks	3	Images	5
Starting subtasks on time	3	Effects (e.g., shadow)	4
Finishing subtasks on time	5		
Keeping alternative versions	5		
Trying more than one idea	3		
Doing math for layout specs	4		
Using a grid or mock-up	3		

SOURCE: Adapted from Patricia Sullivan, "Writers as Total Desktop Publishers: Developing a Conceptual Approach to Training," in *Text, Context, and Hypertext*, edited by Edward Barrett (Cambridge, MA: MIT Press, 1988), p. 273.

to that which occurred with the advent of computer programs that facilitated editing and produced clean final drafts. Even in the recent past, typed copies had visible corrections (in some cases, handwritten corrections), but today our eye has grown accustomed to the clean, uncorrected computer look. Our eye of the future will be even more demanding.

❏ EXERCISE 6-1

Review Situation 2-1, "The Clinic Case" by Mary Sue Garay, on pages 46–51. The results from the survey in Situation 2-1 on pages 47–51 have to be drafted into a report *today*, Monday, and a final report is to be completely finished by Wednesday. You determine a set of interim deadlines over the 3 days preceding the Wednesday deadline. You have to produce a really fine report in a minimum of time:

- Julie will read your draft tonight (Monday) and

- tell you her reaction to the draft first thing tomorrow morning so that

- you can revise the draft Tuesday afternoon in time for

- one of the word processor operators to have the final report finished by 1:30 Wednesday afternoon, as promised.

That means that your task for this afternoon is twofold: (1) to compile the results of the survey and then (2) to write up those results in report form. After asking the receptionist not to interrupt you, you begin compiling the data.

The survey results are given in Situation 2–1 on pages 47–51. Use the advice given in Chapters 2–6 to write a draft report of this survey for the administration and staff of the Westheimer Clinic. Be sure to (1) look for patterns in the survey data and comments to see what's causing patient dissatisfaction, or review the results of the brainstorming and systematic thinking you did about the problems in that situation; (2) analyze the members of your audience or review your answer to Exercise 3-3 and then ask yourself what they want to know and need to know; (3) synthesize steps 1 and 2 so as to identify and define the major problem and most important secondary problems facing the clinic; and (4) construct arguments (in your mind) for ways to solve these problems or review your results for Exercise 4-4. Although the doctors have specifically said they don't want recommendations from you, by having certain arguments in mind you can create a more focused presentation of your findings.

Your report should be in letter form, about two pages long. (See Chapters 12 and 13 for examples.) It should be addressed to Dr. Larry Miller (who is now the Director of the Westheimer Clinic), with copies distributed to all other administrators and staff physicians.

REFERENCES

1 Ken Macrorie, *Telling Writing* (New York: Hayden, 1970), pp. 12–17.

2 Patricia Sullivan, "What Computer Experience to Expect of Technical Writing Students Entering a Computer Classroom: The Case of Purdue Students," *Journal of Technical Writing and Communication* 19(1):53–70 (1989).

3 Christina Haas, "Planning in Writing: The Influence of Writing Tools." Paper presented at the Conference on College Composition and Communication, St. Louis, March 1988.

4 Lillian Bridwell-Cowles, P. Johnson, and S. Brehe, "Composing and Computers: Case Studies of Experienced Writers," in *Writing in Real Time: Modeling Production Processes*, edited by A. Matsuhashi (Norwood, NJ: Ablex, 1987), pp. 81–107.

5 Christina Haas, *Computers and the Composing Process: A Comparative Protocol*

Study, Technical Report 34 (Pittsburgh, PA: Carnegie-Mellon University Communications Design Center, 1987).

6 Lillian Bridwell, G. Sirc, and R. Brook, "Revising and Computing: Case Studies of Student Writers," in *The Acquisition of Written Language: Revision and Response*, edited by S. Freedman (Norwood, NJ: Ablex, 1985), pp. 172–194. See also D. Case, "Processing Professional Words: Personal Computers and the Writing Habits of University Professors," *College Composition and Communication* *36*(3):317–322 (1985).

7 Haas (1988).

8 Leslie A. Olsen, "Computer-Based Writing and Communication: Some Implications for Technical Communication Activities," *Journal of Technical Writing and Communication* *19*(2) (June 1989).

9 D. E. Murray, *A Survey of Computer-Mediated Communication*. IBM Corporation Scientific Center Report, 1986.

10 Jane Siegel, Vitaly Dubrovsky, Sara Kiesler, and Timothy W. McGuire, "Group Processes in Computer-Mediated Communication," *Organization Behavior and Human Decision Processes* *37*(2):157–187 (1985).

11 Sara Kiesler and Lee Sproull, "Response Effects in the Electronic Survey," *Public Opinion Quarterly* *5*(3):402–413 (1986). See also Lee Sproull, "Using Electronic Mail for Data Collection in Organizational Research," *Academy of Management Journal* *19*:159–169 (1986).

12 Lee Sproull and Sarah Kiesler, "Reducing Social Context Cues: Electronic Mail in Organizational Communication," *Management Science* *32*:1492–1512.

13 Edward Barrett (ed.), *Text, Context, and Hypertext* (Cambridge, MA: MIT Press, 1988), p. 273.

14 Jeff Conklin, "Hypertext: An Introduction and Survey," *IEEE Computer*, September 1987.

15 Carol Kaehler, *HyperCard Power: Techniques and Scripts* (Reading, MA: Addison-Wesley, 1988).

16 Gary Marchionini and Ben Shneiderman, "Finding Facts vs. Browsing: Knowledge in Hypertext Systems," *IEEE Computer*, January 1988.

17 Franklin Tessler, "Build Your First HyperCard Application," *Macworld*, May 1988.

18 Nicole Yankelovich, Norman Meyrowitz, and Andries van Dam, "Reading and Writing the Electronic Book," *Computer*, October 1985.

19 Patricia Sullivan, "Writers as Desktop Publishers: Developing a Conceptual Approach to Training," in *Text, Context, and Hypertext*, edited by Edward Barrett (Cambridge, MA: MIT Press, 1988), p. 273.

ADDITIONAL READING

Mike Rose (ed.), *When a Writer Can't Write* (New York: Guilford, 1985).

7

Testing and Revising

Testing and revising are essential parts of the writing process. They are not "things you do if you have the time"—they are *essential*. Even the best, most experienced writers are seldom satisfied with a single draft. Ernest Hemingway, for example, sometimes wrote more than 20 drafts of a story before being satisfied with it. Once you have done your prewriting, analyzed your audience(s), built an argument (if you need to have one), and written a rough draft, you need to take time to test your writing and, if necessary, revise it. As with steel ingots or electronic watches or any other manufactured product, a piece of writing needs to undergo quality control. If it looks like it's not going to do what it's supposed to do, you want to find out and make corrections before it's too late.

Documents differ in the degree to which they need to be tested and revised. Documents that are likely to be closely scrutinized (proposals, resumes, and manuals, for example) may need to undergo multiple cycles of careful testing and revision. By contrast, documents that will be read more casually (such as informal notes, letters, and memos) may need only a quick checking-over. The amount of time and effort you should expend on testing and revising depends on how important the document is. This is a judgment you have to make based on your perception of the organizational problem, your purpose(s) in writing, and your audiences (see Chapters 2 and 3). The Procedure for Audience Analysis given in Chapter 3—in which you are advised to identify the document's intended uses and routes; its likely audiences; and the concerns, goals, values, and needs of each audience—should help you determine the importance of your document and the kind of testing and revising it warrants.

Collaborative projects carried out within large organizations routinely require extensive reviews of final reports. The review process, as Susan Kleimann has shown in several studies of writing in bureaucratic organizations,[1] gives other project members as well as managers a chance not only to evaluate what the writers have written but to provide additional ideas. Even simple questions from a reviewer can cause the project writers to do some valuable rethinking. In most large organizations, the review process goes through many cycles at various levels and thus takes a long time to complete. But it can be of great value in improving both the quality and the acceptability of a document.

7.1 TESTING

The basic principle of good testing is to try to see the document as your readers will see it—to put yourself in their place. Generally, the best way to do this is by having some of your intended readers look at the document, use it, and give you their reactions. This is called *field testing*. To get maximum benefit from field testing, (1) make sure you identify your target audience(s) accurately and select representative members of each audience, and (2) be receptive to whatever comments and criticisms they make about your document. Do not argue with your readers or take their criticisms personally; if you do, you will only inhibit them from giving you honest and useful feedback.

In many cases, of course, you may not be able to test a document on your target audience. For example, if you are applying for a job, you cannot very easily send your potential employers a draft version of a resume and ask them to critique it for you. A better strategy might be to use *role playing*: You recruit some friends and ask them to play the role of your intended readers, giving you appropriate feedback. Of course, for role playing to be effective, you need to analyze your target audience very carefully and make sure that your friends know their roles. You can always try *self-evaluation*, in which you do the role playing yourself, but you may find that you're too close to the document to get the necessary detachment and objectivity. This can be especially problematic if you're trying to evaluate instructions or other forms of procedural writing. If you do want to test your writing yourself, at least try to distance yourself from it by letting some time pass. Do not try to evaluate it immediately after you have written it. Instead, wait a while, sleep on it, turn your attention to other things. Then, when you come back to it, try consciously to put yourself "in the reader's shoes."

There are basically two kinds of testing, depending on the intended use of the document: (1) testing of expository writing and (2) testing of procedural writing. The first pertains to letters, memos, proposals, reports, and other kinds of writing where the writer's goal is to inform the reader about a topic or argue a case. The second refers to how-to documents such as instructions, tutorials, manuals, and other writing, in which the writer's goal is to get the reader to follow a step-by-step procedure.

Figure 7-1 ❑ ILLUSTRATION OF "HIERARCHICAL WRITING"

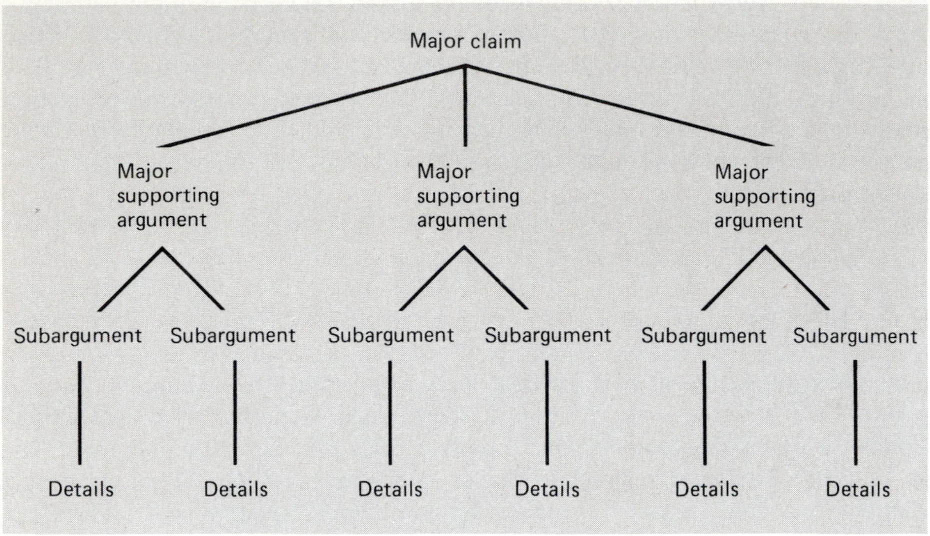

Testing of Expository Writing

Most of the writing you are likely to do as a technical professional is writing that conveys ideas rather than step-by-step instructions. Often these ideas are arranged so as to support a small number of claims, as in many progress reports, proposals, journal articles, sales letters, feasibility reports, etc. When people read this kind of writing, they usually look for such claims. And they expect the rest of the writing—the problem statement, the data, the sub-arguments, etc.—to support those claims. Expository writing has a hierarchical structure: details supporting subarguments, subarguments supporting arguments, arguments supporting major claims (see Figure 7-1). It typically has only a small number of main points, and so a reader who is uninterested in the details can skip through the text searching just for those main points. In contrast, procedural writing does not lend itself to this kind of selective reading. Instead, the reader usually has to follow a full sequence of steps.

The writing in this chapter, up to now, has been expository. We have made four major claims—(1) that testing and revising are an essential part of writing, (2) that some documents need testing and revising more than others, (3) that you should try to test a document from the reader's perspective, and (4) that there are basically two kinds of testing, expository and procedural—and we have supported those claims with arguments. We felt we had to convince you of these points before giving you step-by-step suggestions;

otherwise the suggestions would not have much value. Now we will switch to a more procedural mode of writing in order to give you detailed advice.

To test expository writing, follow the procedure outlined below:

First, give your readers a copy of the document to be tested and let them "mark it up" in private. Encourage them to be critical and do not bias them in any way. Readers often differ on what they look for: one reader may concentrate on ideas; another may focus more on misspellings and other surface details. Give them free rein.

When you get the document back, look at your readers' comments and make sure you understand them. Ask them for clarifications or elaborations, but do not try to defend yourself against any criticism. Don't become defensive and don't take their criticism personally; if you do, you will defeat the purpose of testing. If you have set up the testing conditions properly, it is quite possible that your friends are responding the same way your target audience will. Therefore, your friends' honest criticism is giving you an opportunity to correct mistakes before they reach the target audience.

Finally, ask questions about specific features of the document. We recommend at least the following, in this order:

1 *Are the main points clear?* Since hierarchical writing emphasizes certain claims, these claims must be clear to the reader. You might write down your main points beforehand and compare them against what your friends say are the main points.

2 *Are the main points well supported?* If you are making a case, is it a convincing one? Does your logic hold up? Do you have adequate evidence for your claims? Chapter 4 contains a number of criteria you can apply here.

3 *Are there any "holes" or any mistakes in the details?* The case you're building, or the description you're making, should be as solid as is necessary for the given situation. If there are gaps in your argumentation, your overall case will be weakened. Likewise, if you make errors of detail, your credibility will be damaged.

4 *Is there anything about the tone of the document that might offend readers?* Is it too casual? Too pushy? Too stiff? If a job letter is too informal, for example, it could disqualify you from consideration even if you have good, clearly stated credentials. If a sales letter is too cold and impersonal, it might deter potential buyers (see Chapter 27).

5 *How about the overall "look" of the document?* Does it look professional? Is it nicely formatted, with a reasonable amount of white space? Is the printing clear? Do the headings help the reader make sense of the document as a whole?

6   *Does it read easily?* Does it flow from sentence to sentence? Is the language not too technical? Are any sentences hard to read?

7   *Did you notice any misspellings or grammatical errors?* Although grammar should not be a major concern at this stage, there is no harm in taking note of such mistakes. You'll have to take care of them sooner or later.

Testing of Procedural Writing

Sometimes you may have to write operating instructions, a user guide, a training manual, or some other type of document that contains step-by-step procedures. Procedural writing consists of a list of steps, each of which is to be read and acted on before moving to the next.

   If the document you have written contains instructions for carrying out a step-by-step procedure, you should test it by having someone try to use it for that purpose. This is often called usability testing, and it refers to all kinds of "how-to" writing. Whether it's a tutorial, a training manual, or assembly instructions, you want to make sure that it covers all of the steps and that it is completely clear to the end user. For that reason, *it is especially important that the test subject be someone who resembles the end user, that is, someone who is unfamiliar with the procedure you are describing.* If you make the mistake of using knowledgeable test subjects, your testing will not be worth much because these subjects will "fill in" whatever gaps or flaws there are in your instructions.

   There are basically two forms of procedural testing: (1) visual monitoring and (2) protocol analysis. The former is faster and cheaper, but the latter yields more detailed information.

Visual Monitoring

Visual monitoring is a form of testing in which you simply observe subjects as they try to use your instructions. You have a copy of the document with you, and you note any places where the subjects have trouble. As soon as they are finished, you ask them why they had trouble at those particular places. Visual monitoring is a relatively simple and inexpensive technique that is appropriate for many situations; however, it does not yield as much information as protocol analysis, described below.

Protocol Analysis

Protocol analysis goes one step beyond visual monitoring. You ask subjects to "think aloud" while trying to use your instructions. You tape-record this verbal "stream of consciousness" and then later transcribe and analyze it, looking for trouble spots. This form of testing, if done properly, provides the richest, most fine-grained information of any testing procedure. But it also has certain drawbacks. First, it is time- and labor-consuming, and therefore relatively expensive. Second, it can intrude on subjects' concentration and cause them to perform somewhat differently than they might under normal conditions. Thus, protocol analysis should be used only for documents that are important enough to warrant it, and it should be used only to gain insights, not to "prove" anything.

If you elect to use protocol analysis, here is some advice:

1 Tell your subjects to say whatever's on their mind and not to "edit" their speech. The strength of protocol analysis is that it taps the subject's immediate reactions to written text. If you let your subjects reflect on their thoughts, you lose this immediacy.

2 Give your subjects a warm-up or two to make them comfortable with the procedure.

3 Do not let subjects engage you in "conversation" during the test. Distance yourself physically.

4 Make sure your subjects keep talking. If they lapse into silence, ask them something like "What are you thinking now?"

For more information about the technique of protocol analysis, see the Additional Readings to this chapter, especially Olson, Duffy, and Mack (1984).

7.2 REVISING

Once you've gathered information about your first draft, you'll probably want to use it to do some revising. Testing offers you a great opportunity to revise your draft, and revising a draft is certain to make it better, often much better.

The key principles in revising are:

1 Make major repairs before you make minor ones.
2 Fix up the content before you fix up the form.

It sometimes happens that minor problems disappear once major repairs are made, but the reverse is never true. Thus, you could be wasting your time if you tinker prematurely with details. Even instructions, directions, and other kinds of writing that emphasize step-by-step procedures should be analyzed in their entirety before being analyzed for details. Likewise, the content of your writing is more important than the form. Don't stick with a beautiful sentence just because it's beautiful; if it doesn't say what you want it to say, get rid of it.

The advent of word processors has made revising much easier than it used to be. In the past, you would draw X's and arrows all over your rough draft and try to work from that, imagining what the result would look like.

You would also write relatively few drafts. Now, by doing all the cutting and pasting electronically, you can try one version after another and actually *see* the result. Word processing makes it easier to make major changes, such as moving paragraphs around or inserting new sections. Thus, writers today commonly create many drafts of a document before being satisfied with it. Indeed, it is so easy to "tinker" with a document that you can end up spending a lot of time making small changes and failing to make needed big ones. To avoid this trap, make sure you (1) make major repairs before minor ones and (2) fix up content before you fix up form.

If you are using a word processor, we recommend that you use both electronic text and hard (printed) copy during the revising process. The viewing screen on a personal computer is usually too small for you to see more than a paragraph or two. That's fine for revising words or sentences. You can see enough on the screen to detect problems and try out different solutions without leaving the electronic mode. But if you need to make changes involving larger pieces of text, you should use hard copy from time to time so that you can see "the big picture." Using pages of hard copy will make it easier for you to detect document-level problems and come up with document-level solutions.

Don't forget to use the SAVE command regularly and to keep backup copies!

If you are writing as part of a team, revising becomes a little trickier, particularly if you're not working side by side with your coauthors. As you produce one revision after another, you may find it difficult to keep track of different versions. Thus, "version control" becomes especially important. In order to know which version is which, you should devise some method of numbering or dating your drafts—and then stick to it.

It's a good idea to make different "passes" through a document, rather than trying to revise it all at one time. We suggest that you follow the checklist given on pages 127–128. First look at your *main points* and make sure they're clear. Then check your *supporting points* and *details*. Next consider the *tone* of the document. Then check the *format*. Then focus on *readability* (see Chapters 21–27 for help here). Finally, there's *proofreading* (see Chapter 28). Of course, if your testing has not revealed any problems at some of these levels, you could skip those checklist points.

Testing and revising work together in a "cyclic" sort of way. After you have tested a draft, you should try to make appropriate revisions. And after you have made those revisions, you should test it again. Then, if necessary, you should revise it again. And so on. The amount of testing and revising you decide to do depends, as we mentioned earlier, on how important the document is. We all have constraints on our time and cannot thoroughly test and revise everything we write. But since testing and revising often produce major improvements in a document, you should try to do as much testing and revising as you can, at least for those documents that merit it.

❑ EXERCISE 7-1

Read Situation 7-1 below and try to respond constructively to it. Do you think the reviewer's comments are well taken? If so, try to make the appropriate revisions. If not, do Exercise 7-2 or 7-3.

❑ EXERCISE 7-2

Have two friends read "The Clinic Case" scenario (Situation 2-1, pages 46–51). Then have them each read your draft report and mark it up. When you get their comments back, make sure you understand them. Ask your friends for clarifications or elaborations, but do not try to defend yourself against any criticism. Be constructive. Then, using their criticisms, revise your draft.

❑ EXERCISE 7-3

Form a small group with three or four of your classmates. Give each member of the group a copy of the other members' draft reports from Exercise 6-1. Have everyone in the group independently critique every report, including making written comments in the margins. Then hold a group discussion about each report, with the writer taking notes for revision. Using those notes, rewrite your draft.

REFERENCE

1 Susan Kleimann, "Negotiating to a New Text: Document Cycling in One Government Agency," paper presented at the 40th Annual Convention of the Conference on College Composition and Communication, Seattle, WA, March, 1989.

ADDITIONAL READING

Thomas Duffy and Robert Waller (eds.), *Designing Usable Texts*. (Orlando, FL: Academic Press, 1985.)

Linda Flower, John R. Hayes, and Heidi Swarts, "Revising Functional Documents: The Scenario Principle," in *New Essays in Technical and Scientific Communication: Theory, Research, and Practice*, edited by Paul V. Anderson, John Brockmann, and Carolyn Miller. (Farmingdale, NY: Baywood, 1983), pp. 41–58.

Carol Bergfeld Mills and Kenneth L. Dye, "Usability Testing: User Reviews," *Technical Communication*, Fourth Quarter 1985, pp. 40–45.

G. M. Olson, S. A. Duffy, and R. L. Mack, "Thinking Aloud as a Method for Studying Real-Time Comprehension Processes," in *New Methods in Reading and Comprehension Research*, edited by D. E. Kieras and M. A. Just. (Hillsdale, NJ: Lawrence Erlbaum, 1984).

Situation 7-1 ❑ SEQUEL TO THE CLINIC CASE
(By Mary Sue Garay. Used with permission.)

Tuesday, June 6, 9:15 A.M.

You finished the draft of the Westheimer report in time for your assistant Julie to read it overnight. The two of you have just had a meeting about it. Always a good reader, Julie said that she had read the draft as if she were a Westheimer doctor. From that viewpoint, she saw two problems.

First, the report did not go far enough in *interpreting* the survey findings. Although the Westheimer Clinic had expressly forbidden MBA to make recommendations, Julie thought that the Westheimer doctors would want a report that *made sense* of the survey findings rather than one that just *described* what patients had said in answer to each survey question. Not all the survey questions—and not all their answers—were equally important to helping the doctors improve their clinic.

The Westheimer doctors wouldn't have access to the survey, with its many telling comments. The clinic's promise of anonymity to the patients who had completed the questionnaire made that impossible. Thus, *MBA's unstated role in this project was that of analyst, middle-man, interpreter.* If you could act as an interpreter, MBA could maintain its low profile status and not violate the "no recommendation" stipulation of its contract.

Julie felt that your draft could be revised into such a document. She suggested that you think about *the major implications in the data and focus your report on them.* The report shouldn't be exhaustive—some of the doctors wouldn't have time for that—but it should be an *accurate distillation* of the patients' feelings about their clinic. The report should *summarize* those feelings so that the doctors could see the clinic from their patients' perspective. Simply knowing patient opinion might cause Westheimer to change those operations which needed changing.

Julie also felt that you needed to present your ideas more effectively. The report needed *to be clearer, better focused, and more tightly organized.* If you didn't make the report easy to read, the doctors would have to work too hard to get the message. Julie felt that the Westheimer doctors, like other businesspeople, wanted "the bottom line"—and they didn't want to have to read the report twice to find it.

If MBA failed as an interpreter—because the report either missed crucial points or failed to present those points effectively—the Westheimer Clinic would be likely to continue its decline. And MBA, paid to help the clinic find its problems, would look bad. If, on the other hand, your report could prod Westheimer's doctors into making some changes at the clinic, Westheimer was likely to prosper—and a prosperous Westheimer might become a permanent MBA account.

You push your ultimate goal for the report—the Westheimer as a permanent MBA account—temporarily out of your mind and return to your immediate concern, revising that report. You pull out your first draft of the report, and then state Julie's two concerns as questions to guide your revision:

"What sense can I make of the Westheimer's patients' opinions?

What *main ideas* lurk there and in my draft?"

"How can I make those ideas clear and organize the document so that it is easy to read?"

With those things in mind, you start revising.

Your Task: Write the revised Westheimer report (see Chapters 12 and 13 for advice on structuring). Then format and proofread it carefully (see Chapter 28), so that you can turn it in as a finished report.

PART THREE

VISUAL ELEMENTS

8

Selecting Visual Elements

So far, this book has talked about the production of words, either written or spoken. There are times, however, when words alone are not sufficient to transfer information or points of view, times when words need to be combined with visual aids, formatting (the use of white space and indenting), or other visual elements. For instance, as suggested later in this chapter, appropriate formatting can make a technical report much easier to read—so much easier that the formatting becomes *necessary* given the limitations on the time and attention of an audience. The same can often be said of other visual elements, such as drawings, figures, charts, and graphs, which can quickly summarize an important point or present it in a different way.

Consider the following situation. Figures 8-1 and 8-2 present the same information in different forms: Figure 8-1 in prose text and Figure 8-2 in visual form. The discussion in Figure 8-1 requires 322 words and forces its reader to learn its points linearly—that is, one at a time as they appear in the discussion. In contrast, the chart in Figure 8-2 presents the information quickly and holistically; that is, it presents the whole in a clearly organized pattern that can be seen at a single glance.

Psychologists have shown that you can increase the strength and memorability of a message simply by repeating it or, even better, by repeating it in a different form. Thus, when a visual presentation is added to a verbal one, the combination can produce a *much stronger and more easily remembered*

AUTHORS' NOTE: We would like to acknowledge the efforts and influence of our colleague Lisa Barton on the evolution of this chapter. She was a major contributor to the thinking behind Sections 8.1 and 8.3 and coauthored an early lecture on visual aids from which those sections are derived.

Figure 8-1 ❏ SAFETY RECORDS OF SURFACE AND UNDERGROUND MINING, PRESENTED IN PROSE TEXT FORM
[From Edmund A. Nephew, "The Challenge and Promise of Coal," *Technology Review* 76(2):27–28 (December 1973).]

Safety: Room for Improvement

The safety record of surface mining—in terms of both deaths and nonfatal injuries per ton of coal mined—is spectacularly better than deep mining, which provides at best a hostile, hazardous environment for the miner.

The human costs of moving from surface to underground mining take the form of higher injury and death rates and of greater occupational hazards in general. In 1971, 86% of all coal-mining fatalities occurred in underground mines. In the same year, only half of the total coal production came from deep mines. Over the 7-year period 1965 to 1972, 1412 lives were lost in the underground coal-mining industry in the production of 2335 billion tons of coal. This amounts to an average of 0.606 deaths per million tons of coal—a fatality rate more than five times greater than that of the coal surface mining industry. In recent years, falls-of-roof have accounted for 40% of the deaths in underground mines and coal haulage accidents for about 20%. Annual fatalities from dust and gas explosions fluctuate greatly, but over the years there has been a declining trend.

A similar safety disparity between surface and deep mines holds for nonfatal injuries as well.

During the time period from 1968 to 1971, the safety performance in deep mining of the nation's top 10 coal producers ranged from 0.28 to 1.52 deaths per million man-hours. The differences are even more marked for the category of nonfatal injuries—2.72 to 72.13 injuries per million man-hours. Since the passage of the Coal Mine Health and Safety Act of 1969, most companies have greatly strengthened their safety programs, and this increased emphasis on safety may bring significant improvements in fatality and injury rates. Indeed, the wide range of safety performances cited above makes it clear that—even without a technological breakthrough—much can be done to narrow the safety gap existing between deep and surface mining.

message than either presentation alone. Further, a visual aid can present a *compact summary* of the main points of a verbal text. (There is truth to the expression "A picture is worth a thousand words.") Finally, a visual element can often summarize information in *a more memorable form* than words alone can.

Given these advantages of visual aids, a communicator ought to be able to use them effectively. This involves knowing

1 How to make a visual aid effective

2 When to use the visual aid

3 How to select the best type of visual element in a given situation (e.g., pie chart, bar graph, line graph)

4 How to integrate the visual aid into the text

Figure 8-2 ❑ SAFETY RECORDS OF SURFACE AND UNDERGROUND MINING, PRESENTED AS A VISUAL ELEMENT
[From Edmund A. Nephew, "The Challenge and Promise of Coal," *Technology Review* 76(2):28 (December 1973).]

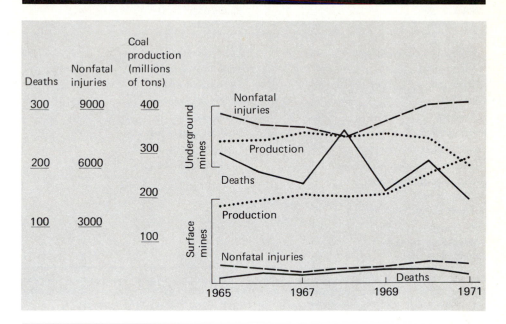

8.1 MAKING A VISUAL AID TRULY VISUAL

Take 2 to 5 seconds to look at Table 8-1 and then cover it up. Do not look at any of the following tables or discussions. Now try to write down the main points made by the table. When you have finished, look at the presentation of the same information in Table 8-2 and see if you can quickly add any more main points to your list. Do this before you continue.

Typically, people who read only Table 8-1 note (1) that job satisfaction declines in each of the two main groups of occupations. These readers will *sometimes* notice (2) that there is a large difference in job satisfaction between the two groups—that is, that most of the first group is relatively satisfied (93 to 75 percent satisfied), whereas most of the second group is much less satisfied (only 52 to 16 percent satisfied). Very few readers of only Table 8-1 will notice (3) that the job satisfaction of skilled printers is higher than that of nonprofessional white-collar workers. These last two observations (points 2 and 3) are very hard to "see" in the format used in Table 8-1.

In contrast, most readers of Table 8-2 easily and quickly note all three

Table 8-1 ❑ PROPORTIONS OF OCCUPATIONAL GROUPS THAT WOULD CHOOSE SIMILAR WORK AGAIN

Professional and White-Collar Occupations	Percent	Skilled Trades and Blue-Collar Occupations	Percent
Urban university professors	93	Skilled printers	52
Mathematicians	91	Paper workers	42
Physicists	89	Skilled auto workers	41
Biologists	89	Skilled steelworkers	41
Chemists	86	Textile workers	31
Firm lawyers	85	Blue-collar workers	24
School superintendents	85	Unskilled steelworkers	21
Lawyers (average)	83	Unskilled auto workers	16
Journalists (Washington correspondents)	82		
Church university professors	77		
Solo lawyers	75		
White-collar workers (nonprofessional)	43		

SOURCE: Based on a study of 3000 workers in 16 industries, conducted by the Roper organization; on Wilensky's study of Detroit workers and professionals; and on a study of Massachusetts school superintendents by Neal Gross, Ward Mason, and W. A. McEachern. From *Psychology Today,* February 1973, p. 39.

observations, as well as a few other, more subtle ones, simply because of the format of the table. Notice that Table 8-2 makes it *visually* quite clear that the job satisfaction ratings of the two groups overlap and that the skilled trade and factory workers as a group are less satisfied than the professionals.

This comparison illustrates an important point: to be most effective a visual aid should present information in a truly *visual* form. Table 8-1 presents information in a chart, but it does not use most of the chart's visual possibilities. It nicely arranges the information in two groups and in descending order within the two groups. However, it does not visually indicate the descending order. To really understand the chart, readers have to do a lot of mental work. For instance, among other things they have to notice

1   That for the left column the first four categories have satisfaction levels of 93%, 91%, 89%, and 89%

2   That these four levels are all quite close to one another

3   That the next seven levels go from 86% to 75%

4   That these are pretty close to one another but lower than the first group

Table 8-2 ❏ ALTERNATIVE ARRANGEMENT FOR TABLE 8-1

Professional and White-Collar Occupations	Percent	
Urban university professors	93	_____
Mathematicians	91	_____
Physicists	89	_____
Biologists	89	_____
Chemists	86	_____
Firm lawyers	85	_____
School superintendents	85	_____
Lawyers	83	_____
Journalists (Washington correspondents)	82	_____
Church university professors	77	_____
Solo lawyers	75	_____
White-collar workers (nonprofessionals)	43	_____

Skilled Trades and Blue-Collar Occupations	Percent	
Skilled printers	52	_____
Paper workers	42	_____
Skilled auto workers	41	_____
Skilled steelworkers	41	_____
Textile workers	31	_____
Blue-collar workers	24	_____
Unskilled steelworkers	21	_____
Unskilled auto workers	16	_____

5   That the lowest category is only 43%

6   That this is pretty low and not close to the next-lowest category

8.2 DECIDING WHEN TO USE A VISUAL AID

Communicators often wonder *when* they should use a visual aid in a communication. Three suggested principles for deciding are to use a visual aid

1   When words alone would be either impossible or quite inefficient for describing a concept or an object

2   When a visual aid is needed to underscore an important point, especially a summary

3   When a visual element is conventionally or easily used to present data

Let us consider each of these uses in turn.

The Visual Aid for Describing or Clarifying

A communicator often has to describe an object or concept which is hard to describe efficiently in words alone. For instance, read the following passage and then try to sketch out the image created in your mind by the passage. Do not look at Figure 8-3 before making your sketch.

Skylab Description

The Skylab cluster is an assembly of modified and specially built space hardware which together provides over 12,750 ft of working volume and weighs in at 199,750 lbs. The Workshop is a Saturn IVB stage modified and outfitted for manned habitation; it contains living quarters, food storage and preparation and waste management facilities and the attitude control thrusters, and it carries the two ill-fated solar arrays (one shows clearly in its extended position in this drawing) whose electric input is routed to the power system in the Airlock Module (AM). The latter, serving as a passageway between the Workshop and the docking facilities, is the focus of many of Skylab's technical systems: atmospheric and thermal controls, power control and distribution, and communications and data handling. The Multiple Docking Adapter (MDA) has two docking ports—the axial port shown in use here and a contingency radial port at the bottom in this diagram; it also contains some space research equipment, including the Earth Resources Experiment Package. The Apollo Telescope Mount (ATM) is a sophisticated solar observatory—the first U.S. manned scientific telescope in space. Here as well are attitude and experiment pointing controls for the entire Skylab cluster and a solar array and associated battery system adequate to power the ATM's equipment. The drawing shows the Apollo Command and Service Module (CSM), used to transport the crew to and from orbit, docked to the MDA.[1]

As suggested earlier, try to sketch out the image created in your mind by the passage before you look at the drawing of the Skylab in Figure 8-3. When you are finished, compare your sketch with the drawing. If you are like most readers, your sketch captures some of the features of the Skylab: it probably shows that the *Airlock Module* comes between the *Workshop* and the *docking facilities* and that the *Apollo Command and Service Module* is attached to the *Multiple Docking Adapter*. However, it may not show that the *Multiple Docking Adapter* is the same as the *docking facilities*, and it probably will not show the appropriate sizes and shapes of the various parts

Figure 8-3 ❏ THE SKYLAB CLUSTER
[From William Schneider and William D. Green, Jr., "Saving Skylab," *Technology Review* 76(3):44 (January 1974).]

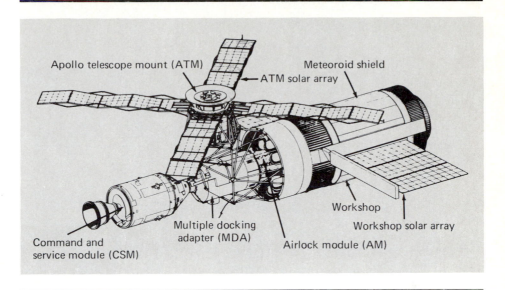

of the Skylab system. Clearly, the formal drawing presents lots of information about sizes, shapes, details, and arrangements which are not presented in the words and which *could not* be presented there, given the limited number of words in the original description and even the extra space occupied by the visual aid. Such a situation obviously calls for a visual aid.

Sometimes you will need to describe a concept or an object which, even if you have an unlimited number of words, seems impossible to capture in words alone. Consider, for example, the passage below describing the pterosaur, a prehistoric flying reptile, and its near relatives.

> Pterosaurs evolved in the early part of the Mesozoic era. Their appearance preceded that of the earliest birds by about 50 million years and followed that of the earliest dinosaurs by about 20 million years. It is hypothesized that the precursors of the pterosaurs (and of the birds and of both orders of the dinosaurs, the ornithischians and the saurischians) were the reptiles called the thecodonts, which for their part evolved from the early, lizard-shaped reptiles known as eosuchians. Among the pterosaurs the rhamphorhynchoids first appeared some 50 million years before the pterodactyloids. The last pterodactyls died out with the dinosaurs 64 million years ago.[2]

If you are having trouble placing all the prehistoric creatures just described, see if their relationships are clearer in Figure 8-4. In this case, the

Figure 8-4 ❑ THE EVOLUTION OF THE PTEROSAURS
[From Wann Langston, Jr., "Pterosaurs," *Scientific American* 244(2):126 (February 1981).

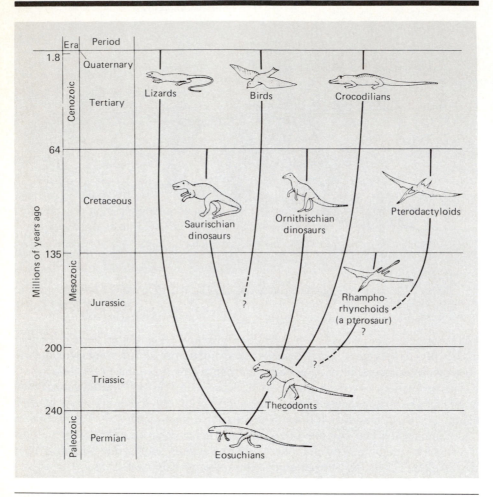

conceptual system described is too complex for most readers to "see" by words alone but is quite easily grasped with the help of the visual aid.

The Visual Aid for Highlighting Important Points

There are many times when you need a visual aid to bring out an important point or to summarize data or a line of argument. For instance, Chapter 5 presents arguments about Introductions and Forewords: that they are based upon psychological principles and that they can be represented by a formula consisting of a problem statement in a certain format, a question or an

Figure 8-5 ❑ OUTLINE OF AN INTRODUCTION OR FOREWORD BY PROBLEM STATEMENT

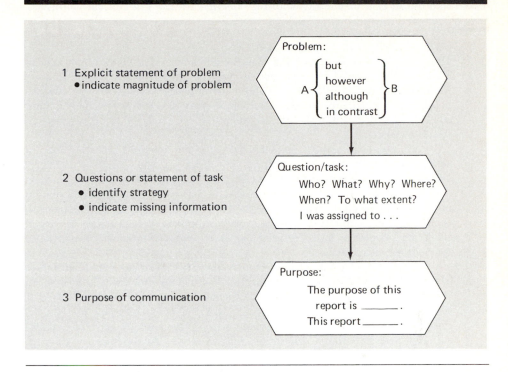

assignment, and a thesis. These arguments included discussions and several examples and ended with a summary of the Introduction. This summary is shown in Figure 8-5, which usually recalls the information in Chapter 5 even for those readers who have not recently reviewed it.

The Visual Aid for Conventional or Easy Presentation of Data

Some types of data are almost always presented in a visual aid. These include cost summaries, budget calculations, frequency spectra, electrical circuits, architectural plans, and the relative positions of geographical areas. As a professional, you ought to be aware of the typical visual forms in your field and use them when appropriate.

Consider the sound spectra in the following extract, a short overview of an interesting research project. In just over 500 words, a problem is introduced, the project's research model and technique are explained, the assumptions are set out, and the findings are reported. Note in this example that using the visual aid is not only more conventional but also more efficient than using words alone: an equally detailed verbal description of the cry spectra would take much more space. Note also *where* the visual aid is placed in this passage: it gives the detailed backup data needed to substantiate a scientific claim. It would be hard to imagine another place in this brief overview where the extra proof and detail of a visual aid would be needed to prove a central claim.

Babies' Cries Give Clues to Diseases

The long, loud, crystal-shattering sounds coming from a newborn infant indicate his good health according to current research by Howard L. Golub and Kenneth N. Stevens of M.I.T.'s Research Laboratory of Electronics. Although scientists in this country and abroad have failed for 20 years to relate cry profiles to infant ailments, their recent tests with a computer cry model showed surprisingly accurate results.

Mr. Golub developed the model by using the accepted acoustic theory of adult speech production and modifying it to fit babies. Since adults and children differ in anatomical proportion, ratios like pharynx length to mouth length, nasal tract length to vocal tract length, and central nervous system disparities had to be adjusted accordingly. Then, using the tape-recorded protests of 55 apparently healthy newborns during a blood test (called the P.K.U. heel stick), Mr. Golub subdivided the cries into 88 variables, such as pitch, intensity, fundamental frequencies, and resonances.

He then used a computer to compare the cries of healthy babies with those of 43 babies having known or suspected health abnormalities. Recordings of these latter cries were supplied by Dr. Michael Corwin, resident in pediatrics at the Upstate Medical Center, Syracuse, N.Y. Of these 43 cases Mr. Golub's computer analysis located 19 of 21 infants with severe jaundice and 9 of 10 babies suffering from respiratory difficulties. Fifteen healthy babies from Dr. Corwin's group were also correctly identified from their cries alone.

Mr. Golub's research team found that a specific ailment is associated with a characteristic cry pattern. Assuming that most infant pathologies alter the acoustically relevant structures, it follows that some aspects of the cry will be correspondingly changed.

For example, if the infant has respiratory distress one would expect a shorter cry due to a change in the dynamics of respiratory muscles responsible for oxygen intake, or a change in vocal tract resonances resulting from a constriction of the pharynx—a possible explanation for the tragic sudden infant death syndrome. In such a case one might also expect a higher frequency sound since the vocal tract acts much like a pipe organ acoustically—a narrow or squeezed pipe giving a higher frequency sound.

Interestingly, Golub and Stevens found precisely these effects, as illustrated by the following two charts [on page 147].

The chart at the left shows the short-time (25 milliseconds) spectrum of the mid-portion of the cry of a normal infant. The chart at the right plots the short-time (25 milliseconds) spectrum of the mid-portion of the cry of an infant that later died of sudden infant death syndrome. Notice the very high first resonance and large amount of noise at the high end of the spectrum, possibly indicating a constriction of the vocal tract near the pharynx.

As a sidelight of his research on jaundice cases, Mr. Golub discovered that some levels of bilirubin—a by-product of the breakdown of hemoglobin— previously considered "safe" are likely harmful. Early treatment for such a condition could prevent brain damage. Other abnormalities likely detectable with the model include bacterial meningitis and deafness.[3]

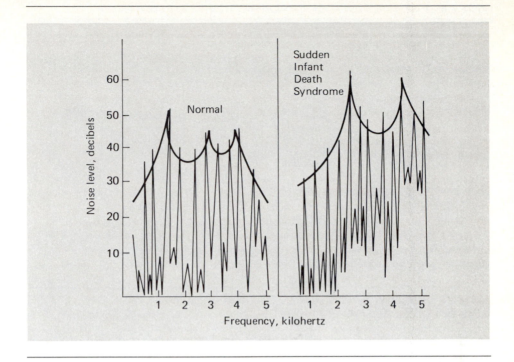

8.3 SELECTING THE BEST TYPE OF VISUAL AID IN A GIVEN SITUATION

When you design a particular visual aid, you are consciously or unconsciously making certain decisions. You are deciding that the particular type of aid you choose (a line graph, bar chart, pie diagram, photograph) is the best *type* to make your point and that the arrangement and highlighting of material on the page is, again, the best to make your point. Unfortunately, there is little information available on which to base such decisions. If you are like most writers, you probably choose one type of visual aid over another simply because it is the first thing you think of using.

The purpose of this section is to sketch out some better or more conscious reasons for choosing visual aids. The section will first identify some conventions of visual perception in Western cultures and then examine several common types of visual aids to see what they do and do not show well.

Conventions of Visual Perception

Given the way Western societies view the world and read, there are a number of general statements we can make about our expectations of visual information. First, we expect written things to proceed from left to right. Note that in

Figure 8-6 ❑ PREFERRED LOCATION OF INDEPENDENT VARIABLE ON A GRAPH

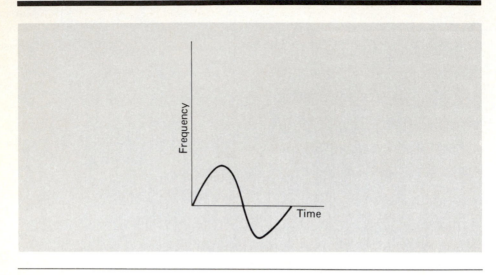

scientific and technical graphs, we place the independent variable on the x axis so that the more important variable moves from left to right. For instance, we plot time on the x axis and frequency on the y axis, as illustrated in Figure 8-6. This pattern is so pervasive that Figure 8-7 looks at best odd and at worst disturbing.

Figure 8-7 ❑ UNCONVENTIONAL LOCATION OF INDEPENDENT VARIABLE ON A GRAPH

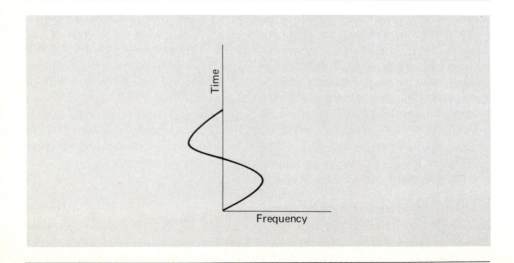

Figure 8-8 ❑ THE SHAPING OF EXPECTATION BY VISUAL ARRANGEMENT: A SAMPLE FOR ANALYSIS

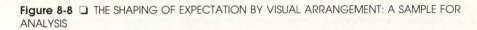

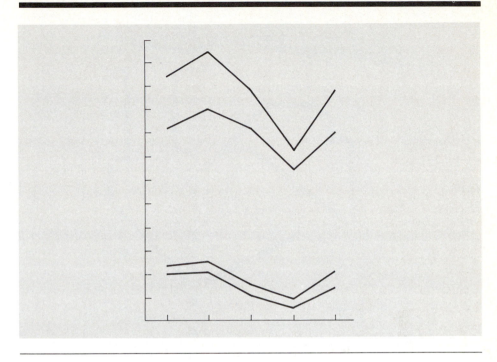

Second, we expect things to proceed from top to bottom, and, third, we expect things in the center to be more important than things on the periphery. Fourth, we expect things in the foreground to be more important than things in the background; fifth, we expect large things to be more important than small things; and sixth, we expect thick things to be more important than thin things. Note that type that is larger, thicker, or bolder than the surrounding type is usually more important: a heading, a title, or an especially important word in a passage. Seventh, we expect areas containing a lot of activity and information to contain the most important information. Eighth, we expect that things having the same size, shape, location, or color are somehow related to one another. (Notice the way the curves in Figure 8-8 are grouped; all labels have been deleted to free you to "see" this point.) Finally, we see things as standing out if they contrast with their surroundings because of line thickness, type face, or color. (You should note that warm or hot colors—red, yellow, and orange—stand out more than cool colors—blue and green.)

Notice how many of the above expectations are verified in Figure 8-9. The dark lines tracing the bus, streetcar, and rail routes clearly stand out from the lighter print and the white background. Conceptually, San Francisco Bay and the Pacific Ocean are merely background for the lines of the transit

Figure 8-9 ❑ THE FIVE LINES OF THE SAN FRANCISCO MUNICIPAL TRANSIT SYSTEM

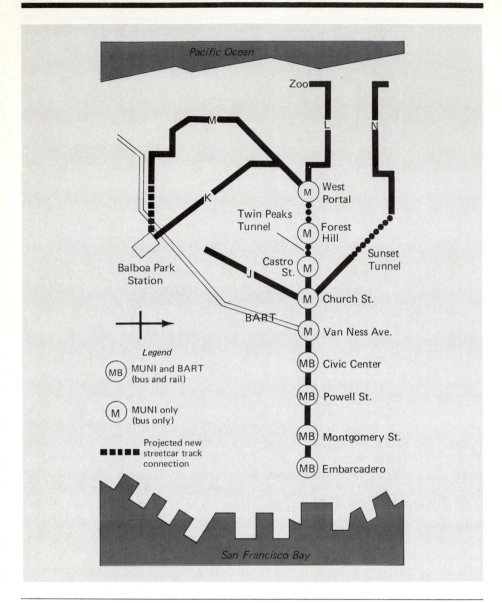

system; both the bay and the ocean are on the edge. In contrast, the most important part of the visual aid, the transit lines, is in the center in dark, thick lines. You might look for such features in other visual aids and try to explain why the designer made the choices you see.

Figure 8-10 ❑ RIVER FLOW BEFORE (1963) AND AFTER (1977) CONSTRUCTION OF THE
ASWAN HIGH DAM ON THE NILE RIVER
[From Julie Wei, "Aswan and After: The Taming and Transformation of the River Nile,"
Research News 31(7):21 (The University of Michigan, July 1980).]

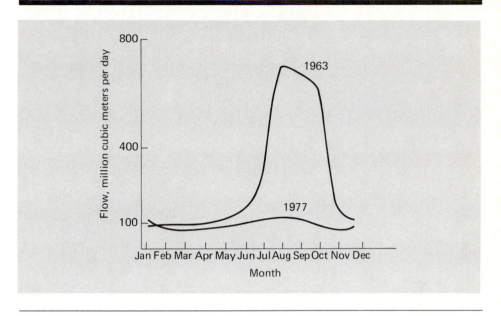

Some Types of Visual Aids and Their Uses

There are six main types of visual aids with which a technical professional
should be familiar: (1) line graphs, (2) bar graphs, (3) pie diagrams, (4) tables,
(5) photographs, and (6) line drawings. Each of these types has particular
strengths and weaknesses, and to use any one appropriately, you must decide
what point you are trying to make and then select the type of visual aid which
makes that kind of point well.

Line Graphs

Line graphs *show well* continuity and direction as opposed to individual or
discrete points, direction as opposed to volume, and the importance of a nodal
point, if there is one. These characteristics are illustrated in Figure 8-10. Line
graphs *do not show well* the importance of one particular point which falls off
a node, the relationship of many lines, or the intersection of three or more
lines. If it is important to be able to trace each line on a graph, you should
probably not put more than three or four on a single graph, especially if they
intersect frequently, or you may produce a graph as hard to follow as the one
in Figure 8-11.

Figure 8-11 ❏ PREFERENCE OF FAMILIES FOR GIRLS VERSUS BOYS IN SIX COUNTRIES

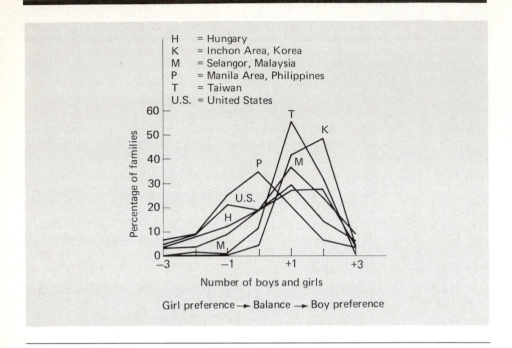

Bar Graphs

Bar graphs *show relatively well* the discreteness or separateness of points as opposed to their continuity, volume as opposed to direction, the relationships among more than three or four items at a time, the contrast between large and small numbers, and the similarities and differences between similar numbers (you can see both that 17 percent and 18 percent are about the same and that 18 percent is a bit larger than 17 percent). These characteristics are evident in the variants of the bar graph presented in Figure 8-12 and in Figure 8-13. Bar graphs *do not show well* the absolute values of the items measured, though you can indicate absolute values with labels, as shown in Figure 8-13.

Pie Diagrams

Pie diagrams *show relatively well* the relationships among three or four items which total 100 percent, the contrast between large and small percentages, and the similarity among relatively similar percentages (they show well that 27 percent and 29 percent are about equal). Pie diagrams *do not show well* the small difference between two similar percentages (you can't usually see the differences between 27 percent and 29 percent). They also do not show well absolute values (unless you label the parts of the pie) or the relationships among more than five or six parts; with too many parts, it is hard to see the

Figure 8-12 ❏ GROWTH OF SURFACE MINING IN U.S. COAL PRODUCTION
[From Edmund A. Nephew, "The Challenge and Promise of Coal," *Technology Review* 76:22 (December 1973).]

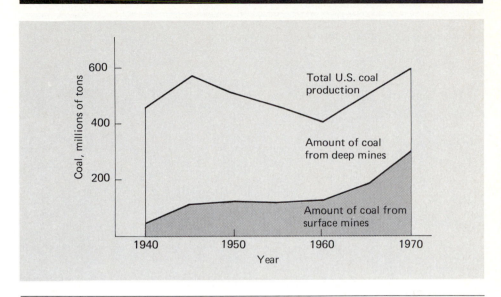

relationships of part to part and part to whole. These strengths and weaknesses are illustrated in Figure 8-14.

Tables

Tables are convenient for presenting lots of data and for giving absolute values when precision is very important. However, since they present items one at a time in columns, they emphasize the discrete rather than the continuous and make it very difficult to show trends or direction in the data.

Tables are not predominantly visual: the reader's mind must translate each number into a relationship with every other number, as described in the job satisfaction example at the beginning of this chapter. Thus, for maximum *visual* impact, tables should probably be a last choice as a visual aid and used only when it is important to provide a great deal of information with precision in a very small space. As an illustration, consider Table 8-3.

Photographs

Photographs *are useful* when you do not have the time, the money, or the expertise to produce a complicated line drawing; when you are trying to produce immediate visual recognition of an item; when you are emphasizing the item's external appearance (as opposed to its internal structure or a cross section); and when you are not concerned with eliminating the abundant detail

Figure 8-13 ❏ COMPARISON OF CESIUM-137 FALLOUT IN THE LAKE HURON AIR-WATER SURFACE, WATER, AND SEDIMENT
[From Julie Wei, "Towards Cleaner Water," *Research News* 29(8/9):22 (The University of Michigan, August–September 1978).]

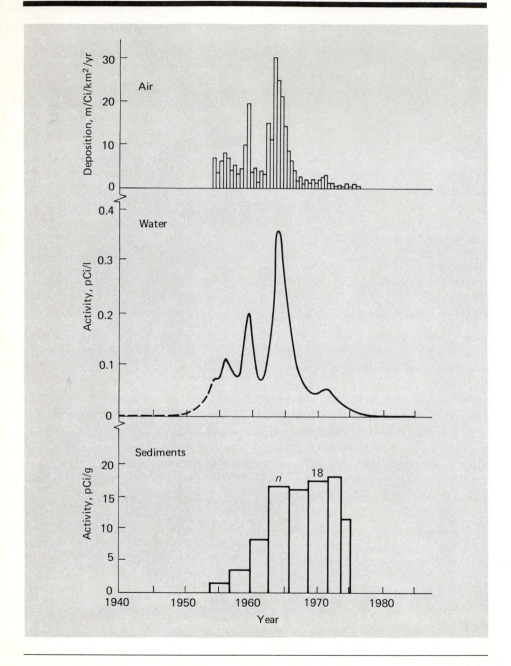

Figure 8-14 ❑ DISTRIBUTION OF FATALITIES IN 181 FATAL CAR-TRUCK CRASHES
[From Daniel J. Minahan and James O'Day, "Fatal Car-into-Truck/Trailer Underride Collisions," *HSRI Research Review* 83:7 (The University of Michigan Highway Safety Research Institute, November–December 1977).]

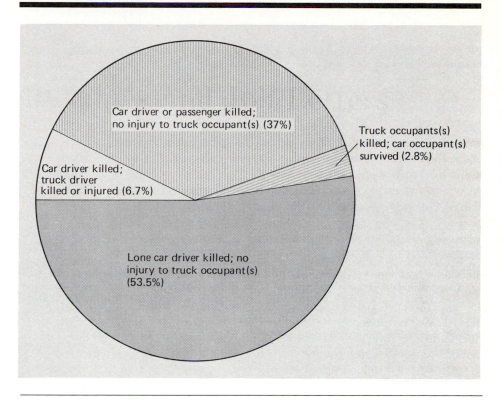

a photograph provides. Although photographs can be airbrushed to eliminate some undesired detail, they still are *not preferred* when you need to focus on some one aspect by eliminating a lot of detail and when you have the time and resources to produce a good line drawing.

Line Drawings

The term *line drawing* includes several types of drawings which focus on external appearance, physical shape, function, or relationship. These include "simplified photos," maps, anatomical drawings, parts charts, and drawings of models (such as atomic or molecular models) or objects from any field of science or engineering. Also included are flowcharts, organizational charts, schematic charts, block diagrams, architectural plans, and blueprints.

Although there are many types of line drawings, all of them share certain functions. They allow you to show things which you can't normally see in a

Table 8-3 ❑ ANNUAL ENERGY SAVINGS FROM SOLAR ENERGY

Year	Solar Space and Domestic Hot Water Systems		Solar Total Energy Systems		Solar Base-Load Electric Power Plants	
	Number of Dwellings (millions)	Annual Energy Savings (10^{15} Btu)	Floor Area of Buildings ($10^4 M^2$)	Annual Energy Savings (10^{15} Btu)	Installed Capacity (10^6 kw)	Annual Energy Savings (10^{15} Btu)
1980	0.3	0.04	—	—	—	—
1985	3	0.4	52	0.24	—	—
1990	8.4	1.2	200	0.92	18.7	1.4
1995	14	1.9	400	1.9	69	5.3
2000	20	2.7	610	2.8	137	10.4
2005	27	3.6	850	3.8	208	15.2
2010	34	4.7	1090	5	284	22
2015	41	5.9	1350	6.2	363	30
2020	49	7	1620	7.6	445	37.6

Assuming the most favorable development, annual energy savings of 52.2 x 10^{15} Btu—over 1160 million tons of oil—may be realized by the use of solar energy by the year 2020. The estimates for dwellings are based on the use of approximately 10^8 Btu annually and a heating plant efficiency of 70%; those for total energy systems on consumption of 6.67 million kwh/yr in a plant of 10,000 m^2 floor area, with an efficiency of 50%; those for power generation on the construction of solar plants with total capacity of 12.5 x 10^6 kwh each year beginning in 1995, the plants having 40% efficiency.
SOURCE: *Technology Review* 76(2):39 (December 1973).

photograph because of size, location, or excessive detail. They also allow you to easily highlight a particular shape, part, or function.

Consider the photograph and drawings of a eukaryotic cell and its compartments shown in Figure 8-15. Notice that the photograph, an electron micrograph, has so much detail that at first it is hard to identify the parts of the cell. Further, because of their size, it is hard even with the drawings for reference to clearly see either the cisternae and ribosomes identified in the lower right corner or the boundary of the nucleus. Obviously, the drawings included with the photograph are much clearer, allow highlighting and focusing impossible in the photograph, and can present important details missing in the photograph.

Computer-Based Experimentation with Visual Aids

If you have a computer and want to experiment with different types of visual aids, you can use a drawing or charting program on your computer to speed up your experimentation. With such a program, you input the data from your table and then choose some form to display it in, and the computer will produce

Figure 8-15 ❑ A EUKARYOTIC CELL AND ITS COMPARTMENTS
[From Albert L. Lehninger, *Biochemistry* (New York: Worth, 1975), pp. 32–33. Used with permission of Glenn L. Decker.]

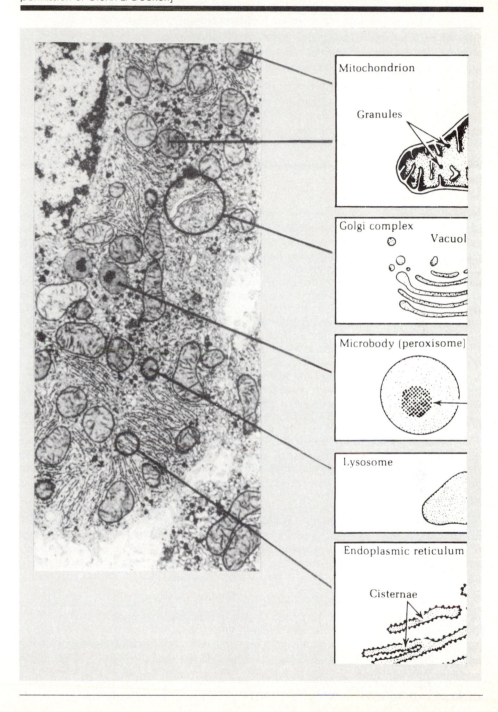

a visual aid in your chosen form from the data. This allows you to quickly and easily produce a variety of presentations of your data: you simply choose a different form for representing the data and the computer will generate it. You can then print out the various forms and compare them or analyze each one while it is one the screen. You can duplicate this process by sketching out visuals by hand, but the computer allows you to produce visuals with quite accurate scales, which is often harder to do in a hurry by hand.

Situation 8-1 ❑ GIVING CREDIT WHERE CREDIT IS DUE (By David H. Balzotti. Used with permission.)

Background

You are a junior engineer working for Ultra-Signal, a company whose product line includes ultrasonic equipment used in the nondestructive testing of various materials. In essence, nondestructive testing is a technique that determines the amount of stress a particular material may be able to withstand, without actually subjecting the material to a stressful element such as wind, friction, or heat. With the aid of transducers, which convert electrical energy into sound, ultrasonic pulses are sent through the material and, by measuring the time it takes the pulses to pass through the material, the thickness and density of the material can be determined. This, combined with other known factors, allows the material's ability to withstand various stresses to be approximated. You are aware that nondestructive testing can be used to test the strength of welds on pipes and bridges and the strength of airplane and ship bulkheads, as well as to check such mundane things as wrenches, hammers, golf clubs, and baseballs coming off a production line.

Your company, Ultra-Signal, has developed several products employing ultrasonic technology for nondestructive testing, but its forte is in the development of transducers. Indeed, Ultra-Signal recognizes that, as science develops new materials, new transducers must be developed that can successfully send a measurable ultrasonic pulse through those materials. Thus, Ultra-Signal has developed a large line of transducers for that purpose.

However, while your company has concentrated on transducer technology, other companies, such as Kyoto-TeKei in Japan, have concentrated on developing equipment for various purposes, including nondestructive testing. These companies have often entered into reciprocal agreements with your company, sharing their new developments and in turn developing more marketable products.

With this in mind, over the past several years Ultra-Signal has sought to broaden its own product base by developing several nondestructive test instruments itself, the most innovative being ULTRA-7734, a desktop unit costing several thousand dollars. But ultrasonic technology is rapidly evolving, and the instruments that use it are getting smaller and smarter. Your company realizes that, even though it is recognized as a leader in the industry producing transducers, it has not quite made it in the instrument business. Thus Ultra-Signal continues to look for new ways to make an impact.

In 1987, Kyoto-TeKei introduced a new miniature hand-held nondestructive test unit, the HUT-1, and offered Ultra-Signal the chance to join in a partnership, marketing the unit in the United States and Europe.

You have heard that the agreement took many months to hammer out and was tenuous at best. Ultra-Signal was forced to charge what your sales manager called "an obscene price" of $4000. Furthermore, no one at your company was quite sure that Kyoto-TeKei wouldn't pull out of the agreement at any moment because the instrument was so untested that it might fail to live up to expectations. Ultra-Signal felt that the risk was worth taking, because the 3-year agreement would give the company time to develop its own miniature instrument as well as to work closely with and monitor any "flaws" in an existing instrument.

Ultra-Signal has an immediate problem to address, however: Kyoto-TeKei's instrument is very powerful and complex to use, and the documentation provided by the Japanese company has been translated, but not into adequate English. It is a 50-page manual consisting of specifications, a few illustrations, some brief explanations, and the BASIC code for a sample program used to format test data uploaded to a personal computer.

The first customers that Ultra-Signal sold to complained loud and long about the condition of the manual. Ultra-Signal's representatives around the country felt that they would not be able to sell the instrument. Even your fellow engineers at Ultra-Signal could not easily determine how to use many of the functions of the machine.

The Problem

You are assigned to write a comprehensive user's manual based on the existing 50-page manual. You sit down with an instrument and grab the engineer who has the most experience with the machine. Together you try to comprehend every aspect of the HUT-1 in order to produce a good manual.

Part of your job is to provide illustrations. However, Ultra-Signal's Drafting Department, the usual source for illustrative matter, is overloaded with work. Furthermore, the company will not hire an outside illustrator. Your boss, Rich, has recently purchased a desktop publishing system, including a scanner with a graphics editing program, to facilitate the Technical Publications Departmental needs. He has always felt that "if we scan in illustrations from drafting, we can edit them as necessary and incorporate them with appropriate callouts."

However, because your plan for the HUT-1 revision requires many illustrations, Rich tells you to "wait until the very end and maybe we can have Drafting do them anyway."

The text nears completion and still nothing has been decided about illustrations. You begin getting pressure from Rich and his boss Bill, the Chief Engineer. You promise to release the manual for a technical review within the next 2 weeks. You go to Rich and ask, "Can you get Drafting to do just a few illustrations so that I can scan and edit them appropriately?"

Rich tries to enlist the aid of the Drafting Department again, but learns that it is still backlogged. However, since he is more than anxious to show off his new desktop publishing system, he suggests, "Why don't you scan the existing drawings from the original manual and edit them as you wish?"

Never having done this, you wonder aloud, "Is this legal?"

Rich assures you that "you can use the illustrations without any problem—besides, they're only line art. But if you're really uncomfortable, check with Bill, because he will know about our relationship with Kyoto-TeKei."

Bill assures you that "you *should* use the illustrations, and there cannot be any problem." "Furthermore," he points out, "K-T didn't even copyright its manual."

So you copy all the illustrations out of the manual and scan, edit, and drop them into the desktop publisher along with the text.

Finally, after 6 long months of intense work, you have made a 200-page manual out of the original 50 pages, with reference material and detailed instructions. You send the manual out for full technical review. The manual is reviewed and returned with the usual corrections, but on the whole the feedback is very positive. Only one thing might slow up production of the manual: Bill would like to see a BASIC program, similar to the one in the original Kyoto-TeKei manual. He instructs you to take it to Ralph, the Software Engineer, so that he can use it as a basis for his own program.

Ralph is flustered by the additional work, saying, "Why don't you just put the damn thing in the manual as it is?" You dissuade him from this, and he agrees to look it over. Within a couple of days, he returns the Kyoto-TeKei program for changes.

You notice that he has not changed the sample program very much, except to refer to the IBM PC where the Kyoto-TeKei had a penchant for the NEC. Also, he has converted metric units into nonmetric units. You decide that the changes may not be a problem, because you can always credit Kyoto-TeKei with providing the original sample after incorporating the program into your manual.

As you are about to proceed, you realize that the illustrations you edited may need to be credited as well, so you set out to determine just if and how you should do so. However, Bill tells you that crediting isn't necessary, since "we have a mutual agreement with Kyoto-TeKei and we can do whatever is necessary to make a marketable product." "Furthermore," he points out again, "Kyoto-TeKei never copyrighted the material anyway."

You check the manual and find that, indeed, Kyoto-TeKei did not copyright it. In any case, you try to determine if there is a way for you to obtain permission from Kyoto-TeKei to use the illustrations and the sample program. But Rich tells you that "it isn't necessary to go any further with this—I need you to get on with your next assignment."

You do as he requests, and the rewritten manual for the HUT-1 goes out the door to the printer. Several days later, you find a memo from Bill in your mailbox asking you to register the copyright of your rewritten manual with the United States Copyright Office. "We've never registered a copyright before," you protest the next time you see Bill.

"I know," he replies, "but the powers-that-be feel we need to protect our investment of time and effort in a new manual. If our deal with Kyoto-TeKei falls through, well ... they ain't gonna get our manual to sell their product."

Your Task:

1 Would you sign the materials to register the copyright? Does it bother you that you are being asked to register the copyright? Would it be different if you weren't being asked to sign the materials?

2 If it turns out that permission is needed from Kyoto-TeKei to use the original material to produce a new manual, whom do you believe to be ultimately responsible for the failure to get permission: you, Rich, Bill, or Ultra-Signal? Who is responsible for not crediting the source of illustrations and the sample program within the manual?

3 If the illustrations were a generic type of line art that anyone with a straight edge and a few French curves could reproduce, and if the sample program were a standard public-domain product, would your feelings change? For more complex illustrations, if you had altered the art dramatically, would you feel that you needed to credit the original? If not, how dramatic is dramatic?

REFERENCES

1 William Schneider and William D. Green, Jr., "Saving Skylab," *Technology Review* *76*(3):44 (January 1974).

2 Wann Langston, Jr., "Pterosaurs," *Scientific American 244*(2):126 (February 1981).

3 *Technology Review 81*(8):79 (August–September 1979).

ADDITIONAL READING

Rudolf Arnheim, *Visual Thinking* (Berkeley: University of California Press, 1969).

Rudolf Arnheim, *Art and Visual Perception: A Psychology of the Creative Eye* (Berkeley: University of California Press, 1974).

Ben F. Barton and Marthalee S. Barton, "Toward A Rhetoric of Visuals for the Computer Era," *The Technical Writing Teacher 12*(2):126–145 (Fall 1985).

Jacques Bertin, *Semiology of Graphics: Diagrams Networks Maps*, translated by William J. Berg (Madison: University of Wisconsin Press, 1983).

William S. Cleveland, *The Elements of Graphing Data* (Monterey, CA: Wadsworth, 1985).

R. N. Haber, "How We Remember What We See," *Scientific American 222*(5):104–112 (May 1970).

Robert Lefferts, *Elements of Graphics: How to Prepare Charts and Graphs for Effective Reports* (New York: Harper & Row, 1981).

Michael Macdonald-Ross, "How Numbers Are Shown," *AV Communication Review 25*:4 (Winter 1977).

Michael Macdonald-Ross, "Graphics in Text," in *Review of Research in Education*, vol. 5, edited by L. Shulman (Itasca, IL: Peacock, 1978).

A. J. MacGregor, *Graphics Simplified: How to Plan and Prepare Effective Charts, Graphs, Illustrations, and Other Visual Aids* (Toronto: University of Toronto Press, 1979).

Cheryl Olkes, "Typography/Graphics," in *Document Design: A Review of the Relevant Research* (Washington, DC: American Institutes for Research, 1980), pp. 103–110, 163–166.

Mary Eleanor Spear, *Practical Charting Techniques* (New York: McGraw-Hill, 1969).

Edward R. Tufte, *The Visual Display of Quantitative Information* (Chesire, CT: Graphics Press, 1983).

Patricia Wright, "Presenting Technical Information: A Survey of Research Findings," *Instructional Science 6*:93–134 (1977).

9

Creating Visual Elements

9.1 DESIGNING THE VISUAL AID

Once you have decided *where* a visual aid is needed and *what type* it should be, you must design it so that it is as relevant, clear, and truthful as possible. This will usually be at least a two-stage process: designing a rough copy and then producing the finished copy. If you work for a company which has an Art or Illustration Department, you may be able to get a technical illustrator to produce the finished copy for you and to counsel you in the design stage. However, even if you have such help, you should be the real designer of the visual aid: you have the best knowledge of the subject and best know the purpose of the aid and the context in which it is being used.

Making a Visual Aid Relevant

Since you place a visual aid in a text to make a point, you should be sure that it makes the point you intend. For instance, suppose that you are discussing expected energy savings from the use of solar energy in the future. You have posed three possible sources of savings—residences, total energy systems such as industrial parks and shopping centers, and solar-based electric power plants—and have broken down the specific savings as illustrated in Table 9-1.

Now that you have your data, you want to construct a visual aid to show the growth in savings and the relative contributions of each source. You construct five possible versions of a visual aid, presented in Figures 9-1 through 9-5, and now have to choose the one most appropriate to your point.

Table 9-1 ❑ EXPECTED ANNUAL SAVINGS FROM SOLAR ENERGY

Annual Savings (10^{15} Btu)

Year	Residences	Total Energy Systems	Solar-Based Electric Power Plants
1985	0.4	0.24	—
1990	1.2	0.92	1.4
1995	1.9	1.9	5.3

On what basis do you choose? What are the differences among the five visual aids?

First let us consider the bar graphs. Among the bar graphs, Figure 9-1 presents the most information in the smallest space and the clearest vision of total growth; however, compared with the other charts, it obscures the comparisons for items in the same year and for the same item in different years. Figure 9-2 obscures the total growth but makes the comparisons already mentioned much clearer, especially for the same item in different years. On the other hand, Figure 9-3 clarifies the comparisons for items in the same year but obscures comparisons between years. The line graphs in Figures 9-4 and 9-5 have the same strengths and weaknesses as their respective bar

Figure 9-1 ❑ ANNUAL ENERGY SAVINGS FROM SOLAR ENERGY: VERSION 1

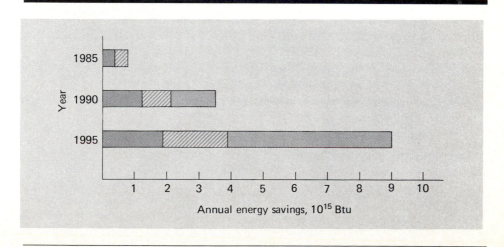

Figure 9-2 ❑ ANNUAL ENERGY SAVINGS FROM SOLAR ENERGY: VERSION 2

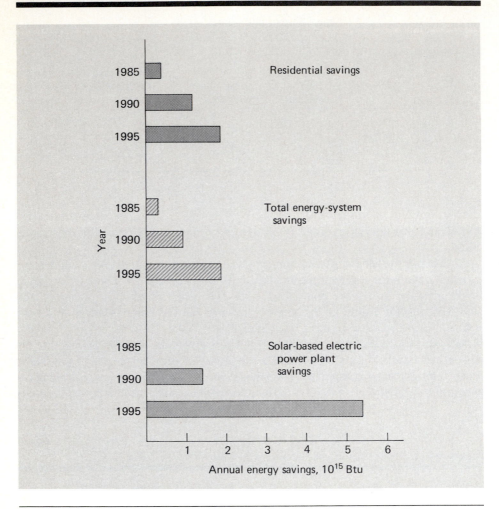

graph counterparts, but they also bring out more strongly the idea of direction and rate of change.

So how do you choose one (or two) from among the group? You pick the one which best matches the focus you wish to take in your report or talk. If you are not much concerned about total growth but want to focus on the contribution of each area for savings, then you might consider Figure 9-2. If you mainly want to focus on the comparison of the contribution of each area in 1995, but want to make a secondary comparison of the contribution of each area in 1990 and a minor comparison of the changed situation between 1990 and 1995, then you might want to consider the presentation in Figure 9-3,

Figure 9-3 ❑ ANNUAL ENERGY SAVINGS FROM SOLAR ENERGY: VERSION 3

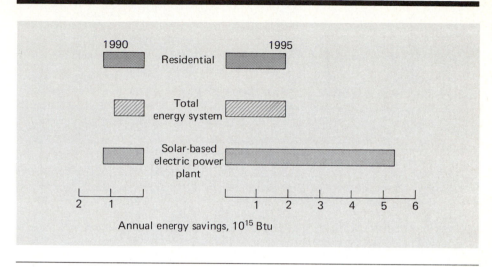

though it is a quite unconventional presentation of the data. If you want to focus on the rate of change of the three areas or the growth of the contribution of each area, you would probably choose Figure 9-4. If you want to focus on the increase in total savings, you would probably choose Figure 9-1 or 9-5.

Figure 9-4 ❑ ANNUAL ENERGY SAVINGS FROM SOLAR ENERGY: VERSION 4

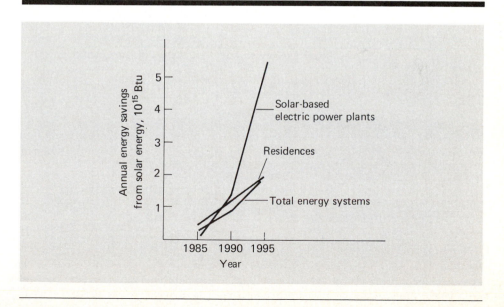

Figure 9-5 ❑ ANNUAL ENERGY SAVINGS FROM SOLAR ENERGY: VERSION 5

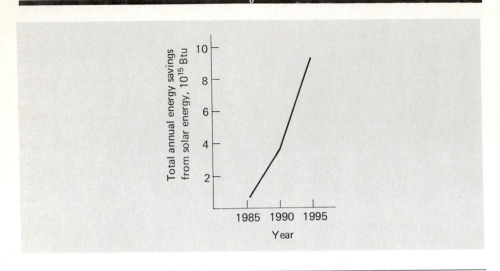

Figure 9-6 ❑ PRODUCTION POTENTIAL FOR AVERAGE SHALE AND MEDINA WELLS: VERSION 1

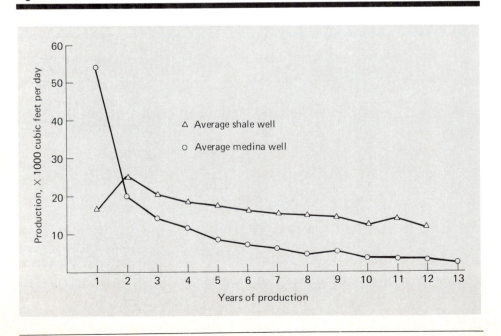

Since relevance and clarity are so important, let us consider one more example illustrating them, the visual aid in Figure 9-6. This visual appeared in a report outlining resources and long-term investment possibilities for a small city. The aid was supposed to show that significant production on a long-term basis could be achieved from shale wells and medina wells (two types of natural gas wells) and that such wells would make good long-term financial investments. Unfortunately, a reader could easily notice the sharp initial drop in the medina well curve and see the flat parts of both curves as low relative to the initial high production of the medina well. Such a comparison *visually* suggests that production is low after 2 years and that neither well will be a good long-term investment, although the medina well might be a promising short-term investment. However, this is the direct opposite of the intended point.

It turns out that these types of wells are financially sound investments if they produce an average of more than 4000 cubic feet per day. If we add this information to the visual as a reference line, it dramatically changes the visual's impact. The modified visual, Figure 9-7, clearly (and visually) reveals the wells' adequate productivity and long-term investment potential.

Figure 9-7 ❏ PRODUCTION POTENTIAL FOR AVERAGE SHALE AND MEDINA WELLS: VERSION 2 [Adapted from Moody and Associates, Inc., *Natural Gas Development Feasibility Study: Report for the School District of Erie, Pennsylvania* (Meadville, PA, October 1980), p. 6. Used with permission.]

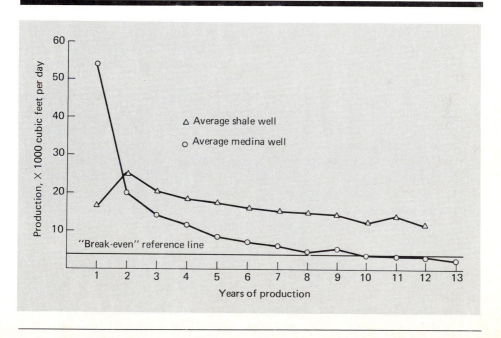

Making a Visual Aid Clear

Making a visual aid clear involves two separate activities: making it conceptually clear and making it technically clear. Making it conceptually clear means having a clearly defined and relevant point and a good form for the point. Conceptual clarity is discussed above. Technical clarity involves having an informative title, appropriate headings and labels, and enough white space so that an audience has the best possible chance of finding the "right" meaning for the visual aid. For example, suppose Figure 9-6 had been titled "Productivity Decline of Medina and Shale Wells." Given the lack of a breakeven reference line, this title would have focused the reader's attention even more strongly on the downward slope of the medina production curve and on the comparatively low long-term production rates. Presumably, this would have made the visual even less effective in communicating its real point.

To really see the benefit of proper labeling and sufficient white space, look at the series of graphs presented in Figure 9-8. Graph *a* is an extremely bad example of a visual aid, since it has *none* of the labeling information usually presented. Graphs *b* and *c* present more information, but still not enough to really get the message across. (Notice that graph *c* lacks enough information even though it provides everything except the title and two critical labels.) Graph *d* provides an adequate title and labels, but the grid in the background is so obtrusive that a reader can hardly see the important lines and labels. Finally, graph *e* provides adequate information and enough white space to let it be seen; from this, a careful and hardworking reader can probably figure out the message. (You should note that graph *d* is typical of most student reports, which are done quickly and checked mainly for accuracy rather than readability.)

Another way to use white space to unclutter a visual aid is illustrated in Figure 9-9. Figure 9-9*a* has no need for three-dimensional bars; in fact, they add distracting detail. Don't you find Figure 9-9*b*, with its two-dimensional bars, easier to read?

You should note that color can be as misused as the graphic flourishes described above. If you use a color other than black or white, you should be able to justify the inclusion of each additional color and of each word printed in a color other than black.

A way to use white space to show structure in a visual aid's data is illustrated in Figure 9-10. In this figure, the consumption-absorption units for various cities and areas are grouped together and extra white space is inserted between the city and area units. This helps to define the structure of the data for the reader. (Note this same use of white space in Figure 9-12 below.)

Making a Visual Aid Truthful

Making a visual aid truthful is important for at least two reasons. First, a visual aid which falsifies information or misleads your audience may hurt your reputation, since it may lead people to see you as dishonest. (However, if you

Figure 9-8 ❏ THE NECESSITY OF LABELS, HEADINGS, AND TITLES IN VISUAL AIDS

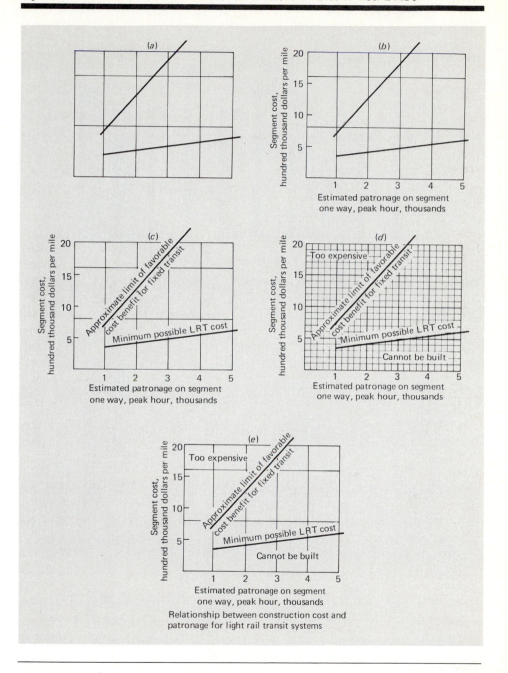

Relationship between construction cost and
patronage for light rail transit systems

Figure 9-9 ❏ ELIMINATION OF UNNECESSARY DETAIL IN VISUAL AIDS

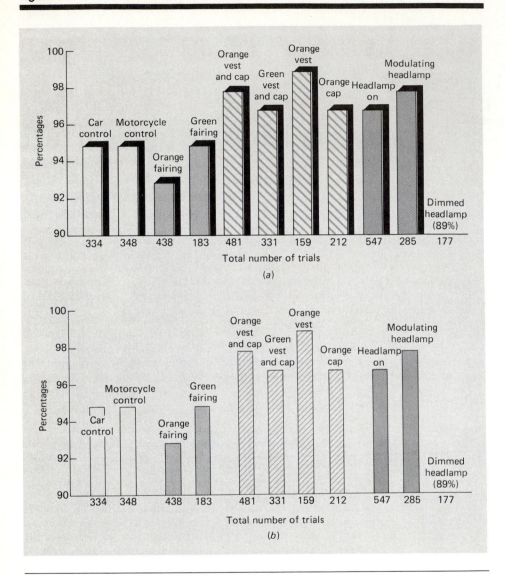

are honest, you probably want to be *seen* as honest.) Second, a false or misleading visual aid may hurt your argument. If members of your audience are upset or dissatisfied with your argument, they may try to discredit it, and one of the easiest ways to do that is to point out a false or misleading spot. How can you believe the arguments of someone who is demonstrably false or misleading?

Figure 9-10 ❑ USE OF WHITE SPACE TO SHOW STRUCTURE IN VISUAL AID DATA
[From David C. White, "The Energy-Environment-Economic Triangle," *Technology Review* 76(2) (December 1973), p. 18.]

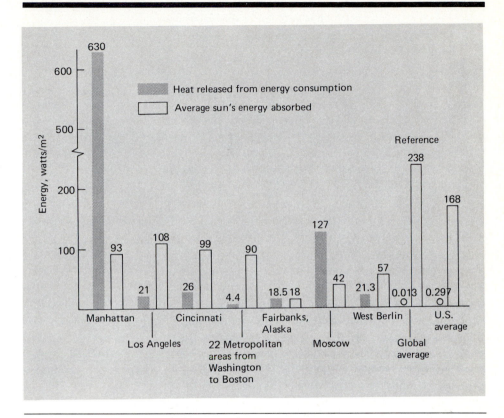

There are several ways to create a false or misleading visual aid. You can inadvertently make a poor choice in the type of visual used for presenting your information and thus obscure your point or make a point quite different from what you intended. This is probably the most frequent cause of misleading visual aids.

In a more serious falsification, you can distort the data. This is sometimes done by obscuring significant differences through inappropriate scale, as illustrated by Figure 9-11. Compare Figure 9-11 with Figure 9-10. Notice that Figure 9-10 preserves the 0 point for reference and indicates the missing part of the scale by a cut in the y axis and in the Manhattan bar; however, it does not obscure the significant differences in energy consumption and absorption among all the other cities and areas, as Figure 9-11 does.

Data are sometimes distorted by exaggerating insignificant differences with a suppressed 0, as illustrated by Figure 9-9. Compare the y axis of

Figure 9-11 ❑ DISTORTION OF SIGNIFICANT DIFFERENCES BY INAPPROPRIATE USE OF SCALE

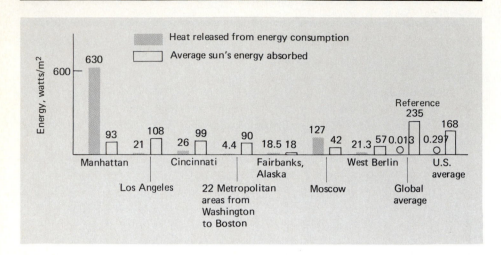

Figure 9-12 with the y axes in Figure 9-9. Notice that Figure 9-9 suppresses the 0 point and greatly exaggerates the differences among the bars by the limited scale range on the y axes. It suggests that the audience responded very differently to the various safety devices. In contrast, Figure 9-12 creates a very different effect: without the suppressed 0, it suggests that the audience responded similarly to the various devices.

9.2 INTEGRATING THE VISUAL AID INTO THE TEXT

Once you have decided to use a visual aid in a particular spot in the text, you must incorporate it so that it seems to belong there. This is easier said than done.

The visual aid needs to be tied to the text and explained, since it appears in the text and must make sense to readers. In addition, if the communicator does not explain the importance of the visual aid—its main point, limitations, assumptions, and implications—readers will have to provide these pieces of information for themselves. As a rule, when readers are put in this position, they will—at least sometimes—see points or implications different from those the communicator wants them to see or perhaps will even completely miss the communicator's point.

The easiest way to integrate a visual aid with the text is to explain its main points and any special implications a reader should note. You might refer back to the passage on cry spectra in Chapter 8, Section 8.2, for a good example of such an explanation.

Figure 9-12 ❏ PRESERVING TRUTHFULNESS BY APPROPRIATE USE OF SCALE
[Adapted from Paul L. Olson, Rich Halstead-Nusslock, and Michael Sivak, "Means of Making Motorcycles More Conspicuous," *HSRI Research Review* 10(2):9 (The University of Michigan Highway Safety Research Institute, September–October 1979).]

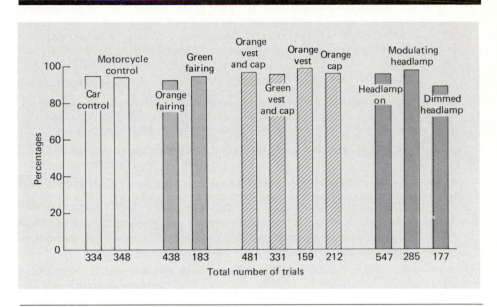

Making a Visual Aid Independent

In addition to being tied to the text, a visual aid also needs to be independent, since it will sometimes be read by itself, without any commentary. This occurs whenever a reader leafs through a book or article just looking at the graphs and tables, or when a listener does not pay close enough attention to a speaker and suddenly finds only the visual aid left from which to find the speaker's point. It also occurs when a speaker needs to make an oral presentation, remembers a particularly fine visual from some report, and has it photocopied for the presentation. In this situation, the visual aid is totally independent of the text, and if it has not been properly titled and labeled, it may be difficult to understand.

A communicator can make a visual aid independent by following the advice given in Section 9.1: make a relevant, important point and use enough headings, labels, and white space. If the visual aid is designed well, the communicator has done his or her job.

❏ EXERCISE 9-1

Field Manufacturing Company produces three major waste products during its manufacturing process: Waste A, Waste B, and Waste C. The company has developed a 10-year program for treating and refining these wastes and then reusing them in the manufacturing process.

A   You are the manager of the treatment program for Waste A. The company has now completed 4 years of the 10-year treatment program, and you need to request increases in your budget for the remaining 6 years of the program. You are scheduled to present your year-end summary and request for future funding next week and are designing your talk and the visual aids you will use in it.

   To aid in designing your visual aid, you have a summary of the cost of treating and refining each waste product which appeared in a recent company report. (This summary appears in Table 9-2.) As part of your argument, you want to compare the funding for your waste treatment program with the funding granted to the other two programs.

   Now you have to design the best visual aid to use in arguing that your program is underfunded in relation to the others. You should first decide what particular point you will use the visual aid to make.

B   You are the manager of the treatment program for Waste B, you are in the same situation as the manager for Waste A, and your job right now is to design the best visual aid to use in arguing that the Waste B program is underfunded in relation to the others. You also have the summary in Table 9-2. You should first decide what particular point you will use the visual aid to make.

C   You are the manager of the treatment program for Waste C, you are in the same situation as the managers for Waste A and Waste B, and your job right now is to design the best visual aid to use in arguing that the Waste C program is underfunded in relation to the others. You also have the summary in Table 9-2. You should first decide what particular point you will use the visual aid to make.

D   You are the head of the Waste Treatment Department at Field Manufacturing. You must present a financial status report on the waste treatment programs to the company's president. The report must describe the economic history of the programs, outline the current economic status of each, and make predictions for the future. Design the best visual aid(s) for the presentation.

❏ Exercise 9-2

Construct visual aids for the information in Tables 9-3 and 9-4. Define a generalization or claim you want the aid to support and make the aid as *visually* supportive as possible.

❏ Exercise 9-3

Read through at least one report you have written and evaluate its use of visual aids. Make sure that you have used a visual aid in each necessary or

Table 9-2 □ COSTS OF REFINING WASTE PRODUCTS A, B, AND C (In Thousands of Dollars)

Waste Product	Expenditures 1987	Increase in Expenditures 1987–1988	Expenditures 1988	Increase in Expenditures 1988–1989	Expenditures 1989	Increase in Expenditures 1989–1990	Expenditures 1990	Projected Total (10-year) Cost of Project
A	153.4	16.6 (10.8%)	170.0	19.0 (11.8%)	189.0	20.5 (10.8%)	209.5	2661.3
B	48.5	0.5 (1.03%)	49.0	2.5 (5.1%)	51.5	3.5 (6.8%)	55.0	1237.0
C	178.0	9.0 (5.05%)	187.0	6.0 (3.2%)	193.0	1.2 (0.62%)	194.2	5169.7
Total (A + B + C)							458.7	9068.0

Table 9-3 ❏ GROSS SALES FOR YOUGAN DAIRY PRODUCTS, 1985–1989

Year	Fresh Milk Products (%)	Dry Milk Products (%)	Diversified Products (%)	Dollars (Thousands)
1989	56.7	15.5	27.8	2074
1988	54.0	17.0	29.0	1750
1987	67.0	20.0	13.0	1229
1986	72.0	23.0	5.0	986
1985	74.0	24.0	2.0	932

appropriate spot, that you have chosen the best type of visual aid for your purpose, that your visual aid is as *visual* as possible, that it does not distort your data, and that it is appropriately integrated into the text. Add or revise (or delete) any visual aids as necessary. When you have finished, trade papers with a friend and evaluate each other's work.

9.3 FORMATTING CONVENTIONS THAT MAKE READING EASIER

To get a good idea of how helpful simple formatting conventions can be, try to quickly read the unformatted text presented in Figure 9-13. This is a version of the Discussion section of a technical report. Give yourself 1 minute to read it and quickly try to summarize the points that the section is making. Then give yourself about 30 seconds to read the formatted version of the report in Figure 9-14 and quickly summarize the main points it is making. Do you agree that formatting makes the version in Figure 9-14 more functional, that is, easier to read and understand?

Table 9-4 ❏ RELATIVE STEEL PRODUCTION IN SOME EUROPEAN ECONOMIC COMMUNITY COUNTRIES

	1983	1984	1985	1986	1987	1988
W. Germany	26.564	32.563	31.597	37.339	36.821	35.316
France	13.442	17.234	17.554	19.781	19.599	19.591
Italy	6.076	9.757	10.157	9.793	12.681	13.639
Belgium	6.367	7.351	7.525	8.725	9.162	8.916
Luxembourg	—	3.456	4.010	4.032	4.559	4.390
Holland	1.000	2.087	2.354	2.659	3.145	3.309

Figure 9-13 ❏ UNFORMATTED VERSION OF THE DISCUSSION SECTION
OF A TECHNICAL REPORT

One of the main problems facing liquid crystal displays (LCDs) is their strong sensitivity to temperature and moisture. These factors can change the behavior of an LCD rather drastically and result in gradual degradation of the device. Two basic categories can be recognized in this case: damages due to a combination of high temperature and high humidity and damages due to high temperature only.

The most damaging environment that an LCD is likely to experience is produced by the combination of high temperature and high humidity. Depending on how the display is constructed, three failure modes can occur under this condition: failure of the polarizers, failure of the adhesive on the polarizer and reflector, and, finally, failure of the liquid crystal mixture. The most common failure occurs with the polarizers. Regular polarizing material will begin to degrade after a short period of time (4 days) when subjected to a temperature of 50C and 95% relative humidity. Once moisture penetrates the plastic backing of the polarizing material, it affects the polarizing properties. The digit area of the display will start to lose contrast, gradually turning from black to brown until it finally disappears completely. To provide protection against the effects of high temperature and humidity, the use of a polarizing material known as "K-sheet," introduced by Polaroid Corporation, is recommended. This material has polarization properties which are relatively unaffected by moisture. Moisture can also attack the adhesive on the polarizer and reflector and cause it to begin peeling away from the glass. If the alignment is lost, dark patches will appear in the viewing area like those as already described. To prevent this kind of failure, the use of a truly hermetic seal is recommended to protect the adhesive. If any of the components in the liquid crystal mixture (such as Schiff-base compounds) are attacked by moisture, the liquid crystal molecules will start to break down, thereby lowering the clearing point of the mixture and the effective upper operating temperature of the display. Due to the ionic by-products of the chemical reactions, the electrical currents of the display will also show a very significant increase. Liquid crystal mixture failure can be prevented in the same way as adhesive failure. Failure can occur when an LCD is exposed to high temperatures only. The twisted nematic display requires the glass plates to be specially treated to impress a uniform alignment on the liquid crystal molecules next to the glass surface. The quality of this alignment can be degraded by high temperature because of decomposition of the interface layer. Failure due to high temperature can be prevented in the same way as adhesive failure.

Problems can arise due to loss of contact between the connector and the indium oxide leads on the LCD. The user will often blame the LCD for an electrical failure such as an open segment, when in fact it is the connector which is responsible. This problem was solved by the introduction of the DIL displays with substrate chips by Liquid Crystal Displays in 1975. They offer a display that can be soldered directly in a circuit board to reduce the possibility of any disconnections.

There are many formatting features that make technical writing look different from most writing we see in newspapers, books, and personal letters. Look, for instance, at Figure 12-1 (p. 238) or Figure 9-16 below. You will notice that each has some very interesting formatting features:

1   To, From, Subject, and Date headings at the beginning

2   Single-spacing

Figure 9-14 ❑ FORMATTED VERSION OF FIGURE 9-13
(Used with permission of Khalil Najafi.)

_____

1. FAILURE DUE TO TEMPERATURE AND
MOISTURE AND PROPOSED PREVENTION
MEASURES

One of the main problems facing liquid crystal displays (LCDs) is their strong sensitivity to temperature and moisture. These factors can change the behavior of an LCD rather drastically and result in gradual degradation of the device. Two basic categories can be recognized in this case: damages due to a combination of high temperature and high humidity and damages due to high temperature only.

1.1   Damages Due to a Combination of Temperature and Moisture
The most damaging environment that an LCD is likely to experience is produced by the combination of high temperature and high humidity. Depending on how the display is constructed, three failure modes can occur under this condition: failure of the polarizers, failure of the adhesive on the polarizer and reflector, and, finally, failure of the liquid crystal mixture.

1.1.1   Failure of the Polarizers
The most common failure occurs with the polarizers. Regular polarizing material will begin to degrade after a short period of time (4 days) when subjected to a temperature of 50C and 95% relative humidity. Once moisture penetrates the plastic backing of the polarizing material, it affects the polarizing properties. The digit area of the display will start to lose contrast, gradually turning from black to brown until it finally disappears completely.

1.1.1.1   Preventing Failure of the Polarizers
To provide protection against the effects of high temperature and humidity, the use of a polarizing material known as "K-sheet," introduced by Polaroid Corporation, is recommended. This material has polarization properties which are relatively unaffected by moisture.

1.1.2   Failure of the Adhesive
Moisture can also attack the adhesive on the polarizer and reflector and cause it to begin peeling away from the glass. If the alignment is lost, dark patches will appear in the viewing area like those as already described.

1.1.2.1   Preventing Failure of the Adhesive
To prevent this kind of failure, the use of a truly hermetic seal is recommended to protect the adhesive.

1.1.3   Failure of the Liquid Crystal Mixture
If any of the components in the liquid crystal mixture (such as Schiff-base compounds) are attacked by moisture, the liquid crystal molecules will start to break down, thereby lowering the clearing point of the mixture and the effective upper operating temperature of the display. Due to the ionic by-products of the chemical reactions, the electrical currents of the display will also show a very significant increase.

1.1.3.1   Preventing Failure of the Liquid Crystal Mixture
Liquid crystal mixture failure can be prevented in the same way as adhesive failure (see 1.1.2.1).

1.2   Failure Due to High Temperature Only
Failure can occur when an LCD is exposed to high temperatures only. The twisted

Figure 9-14 ❏ *(Continued)*

---

nematic display requires the glass plates to be specially treated to impress a uniform alignment on the liquid crystal molecules next to the glass surface. The quality of this alignment can be degraded by high temperature because of decomposition of the interface layer.

1.2.1   Preventing Failure Due to High Temperatures
Failure due to high temperature can be prevented in the same way as adhesive failure (see 1.1.2.1).

2.   FAILURE DUE TO LOSS OF CONTACT
Problems can arise due to loss of contact between the connector and the indium oxide leads on the LCD. The user will often blame the LCD for an electrical failure such as an open segment, when in fact it is the connector which is responsible.

2.1   Preventing Failure Due to Loss of Contact
This problem was solved by the introduction of the DIL displays with substrate chips by liquid crystal displays in 1975. They offer a display that can be soldered directly in a circuit board to reduce the possibility of any disconnections.

---

3   Short paragraphs

4   Lists

5   Section headings (bold or underlined section titles)

6   Numbers to mark the various paragraphs

7   Liberal use of white space

All these features occur frequently in scientific and technical writing because they are functional: single-spacing saves space, and the other devices make a text easier to read, especially for busy and inattentive readers. A heading clearly announces the contents of a section so that busy readers can skip that section if they don't need its details. Short paragraphs and white space make a report easy on the eye, even though it may be single-spaced. Numbering, indentation, and lists provide clues to the organization of the report: they help the reader see what the generalizations or claims are and what sections provide proof for the generalizations. They also allow a reader to skip freely from section to section without reading everything.

Short paragraphs are common in much good technical writing. This may seem surprising to those who were taught that a "good" paragraph should be

Figure 9-15 ❏ SAMPLE MEMO—UNFORMATTED VERSION

SPELLBUCH UNIVERSITY
COLLEGE OF ENGINEERING

October 29, 1989

MEMORANDUM

To: Department and Program Heads
             College of Engineering

From: College Curriculum Committee

Subject: Developing Students' Problem-Solving
             Ability: Request for Information

      The Curriculum Committee is quite concerned about the emphasis given to developing the students' capabilities in problem solving and synthesis. Virtually all the descriptions of new and revised courses submitted to the Committee for approval identify the technical material to be disseminated but do not identify techniques to be used in problem solving in particular. The Curriculum Committee feels that probably the amount of open-ended problem solving in the College should be increased, but the Committee has limited data on the nature and extent of problem solving now incorporated in College courses.

      To more accurately assess the current situation, the Committee is requesting you to provide by November 29, 1989, a list of undergraduate courses (i.e., through 400 level) in

at least 100 to 200 words long to be fully developed. Notice, however, that short paragraphs are not necessarily undeveloped paragraphs. You can spread out a 200-word proof over several short indented paragraphs and create a fully developed section, as illustrated in Figure 9-14. Such a section is much easier to read and probably more meaningful than the long-paragraph version presented in Figure 9-13.

9.4 FORMATTING CONVENTIONS THAT MAKE WRITING CLEARER

In addition to making reading easier, the two visual elements discussed in Section 9.3, white space and formatting, can also make writing clearer. For instance, look at the memo in Figure 9-15 and try to write down the list of items the reader is asked to provide. Then compare your list with that in the version of the same memo presented in Figure 9-16. They might not match.

      The paragraph structure of the first version leads many readers to miss

Figure 9-15 ❑ (Continued)

your program which incorporate a significant amount of design and open-ended problem solving or which discuss problem-solving techniques for either open-ended or closed problems. An example of such a course is attached. "Significant amount" should be interpreted as at least one or two weeks devoted to the subject, including design, analysis, synthesis, and evaluation stages rather than simple assignment of an occasional short problem. Please make a rough estimate of the fraction of effort in these courses which is devoted to open-ended problem solving and design. In addition, please identify which of the listed courses are required in your program and estimate how many are taken by the "average" student.

Please let us have the names of faculty teaching the courses and of faculty who are particularly interested in this topic. The definition of a course as problem solving is obviously not clear-cut, but please make a reasonable interpretation of this classification.

The Committee would also appreciate your giving thought, and possibly some answers, to the following (open-ended!) questions:

What resources could the Curriculum Committee provide to help promote problem-solving courses? Possible examples include organizing seminars, compiling a bibliography of literature related to problem-solving technique and pedagogy, and enlisting the help of the Center for Research on Learning and Teaching. Would involvement be greater if help, release time, and advice in developing these courses were available? Are there special computer needs for these courses?

We would appreciate it if we could have your response to the above questions by November 29, 1989. Thank you for your consideration and help on this questionnaire.

HSF:mb

Attachment

---

one or more of the items. However, in the second version, the formatted list and the extra white space separating the items in the list clarify the number of items and force the reader to see them individually. They also force the writer to decide more carefully how the information is to be presented. For instance, does the writer want all the information specified in the second paragraph of Figure 9-15 presented together, in a "clump," or does the writer really want the information segmented into categories 1, 2, 3, and 4, as suggested in Figure 9-16? If the writer wants segmented information, he or she may need to do a lot of analysis and reclassification of the data received from readers of Figure 9-15; presumably this extra work would be avoided by sending out the version shown in Figure 9-16.

Further, a quick or careless reader is more likely to notice that information is requested in Figure 9-16 because the second paragraph states this in a very obvious way.

In many situations, formatting information into obvious lists or breaking it up into clearly labeled sections helps a reader. When formatting, white

Figure 9-16 ❑ SAMPLE MEMO ILLUSTRATING THE USE OF
FORMATTING, WHITE SPACE, AND HEADINGS
(Used with permission of Scott Fogler.)

SPELLBUCH UNIVERSITY
COLLEGE OF ENGINEERING

October 29, 1989

MEMORANDUM

To: Department and Program Heads
           College of Engineering

From: College Curriculum Committee

Subject: Developing Students' Problem-Solving
           Ability: Request for Information

The Curriculum Committee is quite concerned about the emphasis given to developing the students' capabilities in problem solving and synthesis. Virtually all the descriptions of new and revised courses submitted to the Committee for approval identify the technical material to be disseminated but do not identify techniques to be used in problem solving in particular. The Curriculum Committee feels that probably the amount of open-ended problem solving in the College should be increased, but the Committee has limited data on the nature and extent of problem solving now incorporated in College courses.

To more accurately assess the current situation, the Committee is requesting you to provide by November 29, 1989, the following information:

1. Courses: a list of undergraduate courses in your programs which incorporate a significant amount of design and open-ended problem solving or which discuss problem-solving techniques for either open-ended or closed problems. We realize that the definition of a course as having a "significant amount of problem solving" is not clear-cut, but we have appended an example of such a course and request that "significant amount" be interpreted as at least one or two weeks devoted to problem solving and its techniques (including focus on design, analysis, synthesis, and evaluation) rather than as the simple assignment of an occasional short problem.

2. Effort: a rough estimate of the fraction of effort devoted to open-ended problem solving and design in these courses.

3. Required Courses: the identification of those courses listed in item 1 above which are required in your program.

4. Average Courses: an estimate of how many of the courses listed in item 1 are taken by the "average" student.

5. Resource Persons: the names of faculty teaching these courses who could serve as resource persons within the College.

6. Faculty: the names of faculty particularly interested in problem solving and its techniques.

Figure 9-16 ❑ (*Continued*)

The Committee would also appreciate your giving thought, and possibly some answers, to the following (open-ended!) questions:

1. What resources could the College Curriculum Committee provide to help promote problem-solving courses? (Possible examples include organizing seminars, compiling a bibliography of literature related to problem-solving technique and pedagogy, and enlisting the help of the Center for Research on Learning and Teaching.)

2. How could the College or the departments involve more faculty in expanding problem solving in our courses? Would involvement be greater if help, release time, and/or advice in developing these courses were available?

3. What, if any, are the special computer needs for problem-solving courses in your department?

We would appreciate it if we could have your response to the above questions by November 29, 1989. Thank you for your consideration and help on this questionnaire.

HSF:mb

Attachment

space, and labels are used, the reader and writer are more likely to understand each other and to respond appropriately.

ADDITIONAL READING

Rudolf Arnheim, *Visual Thinking* (Berkeley: University of California Press, 1969).

Rudolf Arnheim, *Art and Visual Perception: A Psychology of the Creative Eye* (Berkeley: University of California Press, 1974).

Ben F. Barton and Marthalee S. Barton, "Simplicity in Visual Representation: A Semiotic Approach," *Iowa State Journal of Business and Technical Communication* 1(1):9–26 (January 1987).

Jacques Bertin, *Semiology of Graphics: Diagrams, Networks, Maps,* translated by William J. Berg (Madison: University of Wisconsin Press, 1983).

William S. Cleveland, *The Elements of Graphing Data* (Monterey, CA: Wadsworth, 1985).

R. N. Haber, "How We Remember What We See," *Scientific American* 222(5):104–112 (May 1970).

Robert Lefferts, *Elements of Graphics: How to Prepare Charts and Graphs for Effective Reports* (New York: Harper & Row, 1981).

Michael Macdonald-Ross, "How Numbers Are Shown," *AV Communication Review* 25:4 (Winter 1977).

Michael Macdonald-Ross, "Graphics in Text," in *Review of Research in Education*, vol. 5, edited by L. Shulman (Itasca, IL: Peacock, 1978).

A. J. MacGregor, *Graphics Simplified: How to Plan and Prepare Effective Charts, Graphs, Illustrations, and Other Visual Aids* (Toronto: University of Toronto Press, 1979).

Cheryl Olkes, "Typography/Graphics," in *Document Design: A Review of the Relevant Research* (Washington, DC: American Institutes for Research, 1980), pp. 103–110, 163–166).

Mary Eleanor Spear, *Practical Charting Techniques* (New York: McGraw-Hill, 1969).

Edward R. Tufte, *The Visual Display of Quantitative Information* (Cheshire, CT: Graphics Press, 1983).

Patricia Wright, "Presenting Technical Information: A Survey of Research Findings," *Instructional Science* 6:93–134 (1977).

PART FOUR

SPECIFIC APPLICATIONS

10

Résumés and Job Letters

Writing a good job letter and résumé is difficult unless you know what you want to do, what you can offer, what the company or program does, and what it offers. Thus, it involves doing some serious thinking and research.

To find out what you want to do, you need to consider your short-term and long-term goals, your previous training and interests, and your financial, geographical, or personal constraints. Do you eventually want to be a technical person or a manager? (Managers are often technical people who, because of the demands of their jobs, have developed managerial skills but have grown away from their technical skills.) What type of training and experience do you need to reach your goal? Is there a conflict between your short-term and long-term goals? If so, what is the best way to resolve the conflict? What constraints are there on your plans and goals? Be sure to consider financial and geographical constraints, as well as such personal constraints as limits on your time and energy. You may not be able to answer all these questions without consulting other people, but if you need some outside advice, go get it. You are making important decisions and should do so with the best knowledge you can get.

To assess your previous training and interests, you might refer to the matrix in Table 10-1.

This table is largely self-explanatory: it stimulates you to consider the accomplishments, skills, knowledge, and recognition you have gained in the course of working, studying, and playing, and finally just living in the time and places you have lived. For instance, if you have lived in France while most of your classmates and colleagues have not, this experience may have given you qualifications which you might not have already noticed: a knowledge of French and greater sensitivity to the customs and culture of the French and other Europeans.

Table 10-1 ❑ A MATRIX TO HELP YOU ASSESS YOURSELF AND YOUR INTERESTS

Activities	Results			
	Products or Accomplishments	Skills	Knowledge	Rewards or Recognition
Working				
Studying				
Playing				
Living				

Be thorough and immodest at this stage; you are generating information for yourself, not for anyone else. And don't forget your special talents and accomplishments; if you speak a language besides English or play the violin, write that down. Many people do not have such skills, and what is unexceptional to you may be very exceptional and interesting to an employer. For example, the ability to speak two or more languages might be a big plus if you wanted to work for a company or agency with international operations. Note that not everything you write down in the matrix will appear in your letter and résumé. You are writing to ensure that you don't leave out something important; you still have to select the information you will finally put in the letter. Be bold and thorough.

To find out what the company or program does, you need to check for information in the library such as company publications, reviews of products, write-ups in financial or business sources such as the *Wall Street Journal* in the United States, and annual reports from the company. If you know someone now at the company, you can try to interview that person for information about the company. If all of these sources fail to produce enough information, you can try to call the company and speak to someone in the personnel office or a manager in one of the technical departments you might want to work in. If you do this, you should explain that you are potentially interested in applying to the company and are trying to get more information before you write a letter of application.

10.1 WHAT MAKES A GOOD APPLICANT?

If you were hiring an employee, you would probably want someone who filled a need in the company and who was (not necessarily in this order) well trained, technically competent, smart, hardworking, reliable, honest, well organized, personable, helpful, resourceful, and a good communicator. Therefore, when you are applying for a position, your cover letter and résumé need to say as many of these things about you as possible.

Most of all, though, the cover letter and résumé must demonstrate good writing skills. The cover letter and résumé seldom establish your technical qualifications for a job, no matter how impressive your training may be. Determining your qualifications is usually left to other parts of the hiring process: interviews, tests, letters of reference, etc. The cover letter and résumé are not so much technical specifications as they are *examples of how well (or how badly!) you write*. This was demonstrated in an experiment conducted by Davida Charney and Jack Rayman of Penn State University.[1] They had 18 campus recruiters evaluate 72 résumés from fictitious mechanical engineering students. The résumés were deliberately varied according to four variables: relevance of previous work experience, elaboration of independent coursework, stylistic quality, and mechanical correctness. Charney and Rayman found that mechanical correctness was the most important of these variables. In fact, résumés of well-qualified applicants that had errors of grammar, spelling, punctuation, parallelism, etc., were rated lower than résumés of less-qualified applicants that were better written. Thus, to write a good job letter or résumé, treat it primarily as a demonstration of your communication skills. Here are some major principles to keep in mind.

1   *Use a conventional format*. Different audiences have different expectations. After you've written a draft of your letter or résumé (perhaps following one of the examples given in this chapter), show it to someone who is familiar with the intended audience or an audience like it and ask for feedback. If you are in school, visit your placement bureau and ask for help. If you are already working, turn to a friend or colleague or to your professional society. Play it safe. Don't try an unconventional "look" unless you yourself know the intended audience well and want to take a chance.

2   *Stress what you can do for the company or program, not what it can do for you*. Look closely at the wording of the job announcement. Try to find out more by phoning the company. In your letter, show that you are aware of the company's needs by using some of the same language the company uses. If you think you meet these needs, say so! Show that you want to be part of the team.

3   *Address the person having the authority to arrange an interview or to admit you into a program*. If you don't know who that person is, try to find out. Don't waste your letter on someone who can't help you.

4   *Stress your accomplishments and responsibilities*, since these will show off many of the desirable qualities listed above. Use dynamic verbs like *devised, organized, researched, supervised, improved, created,* and *developed*. Try to back up your claims with evidence (e.g., quantitative data, personal references).

5   *Make the letter and résumé look beautiful*. This is no place for careless spelling mistakes, typographical errors, or computer-generated formatting

flaws. Your readers will be judging you using whatever evidence they can find, including the physical appearance of these documents. If you want to convey a professional image, you should make your letter and résumé *look* professional. And that means careful proofreading! (See Chapter 28.)

In general, if you need to write a letter of application and a résumé, you should take pains to do them well. If they are impressive enough, you may get an interview with a prospective employer or admission to a desired graduate school or other training program. If they are done poorly, you will probably be screened out of the pool of applicants and into the ranks of those still looking. (Most companies and programs have more applicants than positions and thus are looking for ways to eliminate applicants; a bad letter or résumé makes elimination easy.)

10.2 DESIGNING THE LETTER OF APPLICATION

To write a successful application letter, you need to know the conventions for such writing. In most cases, they are quite straightforward. The letter usually is only one page long and has four main sections:

1   The heading, which includes the writer's address, the reader's address, and the date.

2   A first paragraph, which introduces you (if possible by citing an impressive recommender) and then establishes the company's need and your ability to fill it (or your reason for writing and what you are requesting of the reader).

3   A second paragraph or series of paragraphs which establish your *most relevant* experience and qualifications and which stress your accomplishments, responsibilities, and work quality, not just your activities. This section should include some proof that you did a good job, by citing either a reference ("X will provide a letter of reference") or some objective measure of a job's quality. In contrast to the more complete version of your activities in your résumé, this section should be very selective and focus on only your *most relevant* qualifications.

4   A closing paragraph, which gives any other pertinent data, asks for an interview, and provides your telephone number and the hours you may be reached.

Two sample letters are shown in Figures 10-1 and 10-2. The letter in Figure 10-1 was written by an engineering senior and has been nicely presented at many levels. It is addressed to the person in charge of screening applicants, is written specifically to the company, and was prepared on a high-quality printer. This gets the letter to the right reader(s) and shows that the writer cares enough about the job to prepare the letter carefully. It first cites a reference known and respected by the reader and then defines the job being sought and the writer's general qualifications for the job. The two indented paragraphs in the middle of the letter give the writer's main qualifications, focusing on his accomplishments and on proof that he did a good job. This section is relatively straightforward, since the writer comes from a good school and has some obvious qualifications for the job. Finally, it provides useful information if the employer wants an interview and then it ends politely.

The letter in Figure 10-2 presents a more difficult problem. It is a letter of application to an excellent graduate program, written by a good student but one with apparently marginal qualifications for that program: swimming, work as a camp counselor, and an engineering degree as preparation for one of the most competitive business programs in the country. Since the letter was successful in a difficult situation, it is useful to see why.

First, the student has a good reference in the first sentence. (If no one with an impressive title or reputation has offered to recommend you, then politely ask someone to do so; the worst you can get is a "no.") Second, the student has assessed his qualifications and activities and reinterpreted them from the point of view of the reader. (What are the characteristics an admissions officer would look for in applicants to business school? In future managers? What kinds of things have I done to demonstrate these characteristics?) Obviously, a writer is limited by the facts of a given case, but she or he can organize those facts in more or less effective ways. This student has chosen a more effective way and highlighted it. He has focused on his qualifications by using indented paragraphs which draw the reader's eye. Further, he has used underlining and parallel construction (the repetition of *ability to*) to highlight the qualities he wants the reader to note most carefully: ability to solve problems, ability to lead and motivate others, ability to achieve desired goals, and ability to maintain diverse interests.

When you are ready to write such a letter, try to avoid having a lot of empty space at the bottom of the page. You don't want to suggest that your life has been a blank. Also, try to produce a balanced and visually attractive format. If you have only a little information, spread it out and use wide margins. Finally, make your letter and résumé as concise as possible, models of efficiency.

10.3 DESIGNING THE RÉSUMÉ

The résumé is a summary of *all* your activities and experience. It will repeat some (or all) of the information in the letter of application, and it may be more inclusive than the letter. However, since the letter and résumé may get separated, the résumé should be as strong and impressive as the letter and arranged to highlight your most important qualifications. Sample résumés appear in Figures 10-1 and 10-2.

Figure 10-1 ❏ LETTER OF APPLICATION FOR EMPLOYMENT, AND RÉSUMÉ
(Used with permission.)

6601 Bursley-Lewis
1931 Duffield Street
Fort Collins, CO 80525

May 16, 1989

Mr. Lewis R. Tassellski
Technical Group Coordinator
Pacific Power & Light Company
1634 S.W. Columbia
Portland, OR 97201

Dear Mr. Tassellski:

At the suggestion of Mr. S. Feldman, Instrumentation Supervisor, Fermi II Project Pioneer Corporation, I am writing to apply for a position in the Instrumentation and Control Group at the Astoria Point Power Project. My work experience in both instrumentation and start-up groups should be of value to the project, as should my engineering experience at Colorado State University in automatic control and power plant systems. My credentials include:

Employment as an engineering aide with the Detroit Edison Company at the Enrico Fermi II Nuclear Power Station. While there, I independently researched in-service inspection problems of certain ASME Class II welds, reporting to the production department lead superintendent. I also worked on the initial rough draft of the in-service lubrication manual for the plant. (Ref. R.J. Szcotnicki, Technical Group Supervisor)

Employment as a student engineer with Daniels International Corporation at the Enrico Fermi II Nuclear Power Station. I researched material for input into the component control system, checking data for authenticity and inputting the data to update the system. (Ref. T.R. Bietsch, I. & C. Lead Superintendent)

I expect to receive my Bachelor of Science degree in Mechanical Engineering from Colorado State University in May 1989, and my future goals include a desire to advance in the area of technical management. I am very interested in discussing my credentials with you at your convenience. If this is possible, I can be reached at (303) 555-1857 during the afternoon and evening.

Sincerely,

Steven J. Kaercher

Encl.

Figure 10-1 ❑ *(Continued)*

PERSONAL DATA RÉSUMÉ

Steven J. Kaercher
1931 Duffield Street
Fort Collins, CO 80525
(303) 555-1857

Career Plans and Objectives

I am seeking employment in the consumer power industry, especially in the area of instrumentation and control of power systems. I feel that my experience and education will enable me to contribute to such projects now; later I plan to advance into management.

Qualifications and Experience

Summer 1988 — Engineering aide (student) with Detroit Edison Company at Enrico Fermi II Nuclear Power Station. While working for the Production Department, I (1) researched potential problems in the in-service weld inspection program and (2) worked on the plant lubrication manual, documenting equipment and specified procedures.

Summer 1987 — Student engineer with Daniels International Corporation at Enrico Fermi II Nuclear Power Station. While there, I worked on a component control system for the Instrumentation and Control Group. This involved monitoring all sources of information and doing in-plant inspections to keep the system updated.

College 1985–89 — Employed by Colorado State University, Fort Collins, during all terms in school. I worked at the School of Public Health as a coder one year and at the Bursley Hall Cafeteria for the last three years. This employment paid for 25% of my college costs.

Academic Experience

Education — Bachelor's degree in mechanical engineering, College of Engineering, Colorado State University, expected in May 1989. Emphasis on instrumentation and control of power system equipment.

Societies — American Society of Mechanical Engineers, student member.

References
References and transcript will be supplied upon request.

717 Dartmoor
Ann Arbor, MI 48103
(313) 555-9872

February 15, 1988

Dean Arjay Miller
Graduate School of Business
Canston University
Bolton, CT 06431

Dear Dean Miller:

At the suggestion of Mr. Richard E. Spaid, I am writing to request your support for my application to the Canston Graduate School of Business. My qualifications include a Bachelor of Science in Engineering from the University of Michigan and the following abilities:

Ability to Solve Problems: I have concentrated in mechanics, a strongly problem-oriented field, while maintaining a diverse background in engineering and science. The human aspects of engineering have received the majority of my attention; I have independently researched current advances in the fields of stress-related anxiety, materials for implantation in humans, and nerve conduction velocities in the hand. (Ref. Prof. D. A. Sonstegard)

Ability to Lead and Motivate Others: I had 24-hour direct counseling responsibility for 20 teenage boys at the National Music Camp at Interlochen (summer 1986). I used my motivational and leadership skills to keep the boys interested in their musical and physical activities as well as to deal with homesickness, personality conflicts, group problems, and personal problems. Additionally, I supervised many of the waterfront, recreational, and evening activities. (Ref. L. E. Dittmar)

Ability to Achieve Desired Goals: I have been able to simultaneously attain my goals in swimming and school over the past 12 years through self-motivation, my competitive spirit, and my organizational abilities. I stand in the top fifth of my class scholastically, and I have won three varsity letters as a member of the University of Michigan swim team. (Ref. A. P. Stager)

Ability to Maintain Diverse Interests: I have kept my interests diversified through such activities as being rush chairman of my fraternity and joining Tau Beta Pi, National Engineering Honorary Society. I have been able to share my interests with others as a counselor, scholastic tutor, swim instructor, and music student.

It is essential that the manager master these abilities if he is to contribute to the business world. My perspective of these abilities differs from that of the typical applicant with a B.B.A., and I believe it is a perspective that would be valuable to the MBA program at Canston. I would be happy to provide any further information you might wish. Thank you for your consideration.

Sincerely,

Brian D. Wylie

Figure 10-2 ❑ *(Continued)*

Brian D. Wylie
717 Dartmoor
Ann Arbor, MI 48103
(313) 555-9872

Career Goal

My career goal is to become an effective and conscientious corporate general manager. My strong engineering background, a projected MBA, and my diverse activities will have formed a solid foundation for this goal.

Education and Work Experience

Current
Bachelor's degree in engineering science from the University of Michigan (expected completion: May 1988). Emphasis on biomechanics and industrial human performance. Independent studies on stress-related anxiety, materials used for human implantation, and nerve conduction velocities in the hand.

Summer 1987
Assistant Pool Manager, Travis Pointe Country Club, Saline, Michigan. Responsibilities included maintaining effective relations with the membership, organizing the physical plant, and teaching group and private swim lessons.

Summer 1986
Cabin Counselor, National Music Camp, Interlochen, Michigan. I had 24-hour counseling responsibility for 20 teenage boys with homesickness, personality conflicts, group problems, and personal problems. This job enabled me to use my motivational and leadership skills to keep the boys interested in their musical and physical activities. I also had responsibility for the waterfront, the recreation program, and special events.

Activities and Honors

Activities
Michigan swim team, three varsity letters (1984–1988)
Michigan varsity club water polo (1984–1987)
Psi Upsilon fraternity (1986–1988), rush chairman (1988)

Honors
Tau Beta Pi National Engineering Honorary Society (1987–1988)
Dean's List (three semesters); Regents' scholar (1984)
Academic scholarship (1984–1985, 1986–1988)
Athletic scholarship (1985–1988)

References

Richard E. Spaid, President, Hillman Manufacturing Company
David M. Sonstegard, Ph.D., Associate Professor of Applied Mechanics, U.M.
Larry E. Dittmar, All-State Boys' Director, National Music Camp
Augustus P. Stager, U.M. Varsity Men's Swim Coach

What should the résumé look like? First, it ought to be easy to read: not too long, not too much material crammed on a page, easily visible headings, and, if you prefer, short phrases rather than full sentences. Second, it ought to give your vital statistics: your name, address, and telephone number and, if necessary, a permanent address and telephone number if you intend to move soon. (Résumés used to include such information as height, weight, sex, age, and marital and military status, but this information is often irrelevant and prohibited by law, so it is not included here. Obviously, you can include it if you particularly want to.) Third, the résumé ought to suggest where you are headed professionally (perhaps in a section called "Career Goals") and where you have been (in a section called "Qualifications and Experience," including education and work experience, or two sections titled "Work Experience" and "Education").

The ordering of information is important in the section on qualifications and experience. Generally, you want to put your most relevant and impressive qualifications first. If you have a lot of relevant work experience, you should list that before your educational experience. If you have only a little work experience, you will have to emphasize your education and its special features. What makes you different from any other student with your degree? Have you had any research or design courses which simulate a job situation? Do you have a number of honors and extracurricular activities? You might want to highlight them in a separate section titled "Honors and Activities," since such features show that you are organized enough to handle several activities at one time.

Finally, you need a section titled "References," which either states that references are available on request or lists your references' names (and addresses) if these are particularly impressive or if you need to use up some extra space. (Before you list someone as a reference, ask the person if he or she is willing to serve in that role. It is impolite and potentially disastrous to list people without their approval. They may not like being taken for granted and they may take it out on you by writing a less than flattering letter.)

One note about the placement of information on the page. As with your letter of application, try to avoid a lot of empty space at the bottom of the page. Use wide margins if necessary, spread the information out evenly on the page, and give the full names and addresses of your references to use up more space. If possible, you want to avoid the sense that you haven't done enough to fill up the page. If you can't quite fill up the last page, don't worry but do what you can with formatting without overdoing things. (You don't want 3 inches of text down the middle of a page with margins almost 3 inches wide on each side.)

10.4 OTHER EXAMPLES

Whereas résumés are normally written for a variety of situations, cover letters should be tailored to one particular situation. Figure 10-3 illustrates this point well. When Enrico Gomez began applying for employment during his senior year in college, he prepared a single all-purpose résumé for a variety of prospective employers. However, for each employer he wrote a different cover letter. One of these involved a Japanese company in Tokyo. Enrico's father, who had been to Japan and done some business there, showed him some letters he had received from Japanese executives. Each began with some casual pleasantry before getting down to business. When his father confirmed that this was a standard feature of Japanese business-letter-writing style, Enrico decided to follow that style himself.

As you can see from Figure 10-3, instead of following the American style of going straight to the point, Enrico displays awareness of the Japanese preference for indirectness and modesty. He begins his letter with some inconsequential comments about the weather, and he closes it by thanking the reader for his consideration. In general, he uses a somewhat less assertive style than he might have used for American readers. By respecting Japanese cultural norms, he is showing respect for his Japanese readers and is thus enhancing his chances of getting a sympathetic reading. Enrico also gears his letter to his readers in other ways. He cites someone (Versen) who he knows is known to the addressee. He emphasizes his native-speaker command of English, knowing that such a skill is highly valued in Japanese companies. He mentions other qualifications that would interest LEC.

Figure 10-4 presents a different case. Whereas the other applicants we've looked at all had considerable work experience to cite, Mei Kwan is applying for her first "real" job. She is a sophomore biology major who spotted an announcement in a local newspaper for a summer internship as a lab technician trainee. She sees this as a good stepping-stone toward her long-term career goal as a research scientist. She senses that although she does not have extensive experience in laboratory work, she does have enough to qualify for a summer trainee position. And she is apparently committed to a professional career in biology. In her letter, she emphasizes this commitment. She points out that, despite her youth, she has been "on track" to a biology career for several years. Since her main credential is that of being a good student, Mei Kwan appropriately emphasizes her academic accomplishments, even going back to high school. But she also draws on other aspects of her life to support specific claims. For example, a lab technician must be careful and trustworthy, so she cites her experience running the family grocery store as a way of documenting her sense of responsibility.

❑ EXERCISE 10-1

Evaluate and revise as necessary the letter and résumé combination in Figure 10-5. Note the order of categories and information in both the letter and the résumé. Do the most important qualifications come first? Are they appropriately highlighted? Is the writing clear, correct, and concise?

Figure 10-3 ❑ RÉSUMÉ AND LETTER OF APPLICATION FOR WORK OVERSEAS
(Used with permission.)

Box 606
5115 Margaret Morrison St.
Pittsburgh, PA 15213

8 November 1989

Hiraku Shima
Senior Technical Writer
Leading-Edge Communications, Inc.
Hachiko Bldg., 1-34-6 Takadanobaba
Shinjuku-ku, Tokyo 160, Japan

Dear Mr. Shima:

We are enjoying a beautiful autumn here in Pittsburgh, as I hope you are in Japan.

At the suggestion of Dr. Jean Versen, Assistant Professor of Technical Writing at Carnegie Mellon University, I am writing to inquire about possible job openings in your company for native-English-speaking Americans with training in technical writing. I am a senior in Carnegie Mellon's Technical and Professional Writing Program. As my attached résumé shows, I have taken all the core courses in the program plus computing (Pascal), cognitive psychology, applied statistics, biology, physics, chemistry, and other courses of value to a technical writer. I have worked as a reporter for the campus newspaper. This past summer I worked for IBM as an assistant technical writer; I helped write a user manual for IBM's 3083 series computers and learned much about the entire process of document design and production. I believe that my combined education and experience would make me a valuable asset to LEC.

I will be graduating with a Bachelor of Science degree in Technical Writing in May 1990 and will be available for employment immediately thereafter. If you would like to see my writing portfolio, I will be most happy to oblige.

Thank you for your consideration.

Sincerely yours,

Enrico Gomez

Att.

Figure 10-3 ❑ *(Continued)*

<hr>

ENRICO GOMEZ

School Address
Box 606
5115 Margaret Morrison Street
Pittsburgh, PA 15213
(412) 555-4098

Home Address
221 Flamingo Drive
Vero Beach, FL 33451
(305) 555-3350

OBJECTIVE:	A position as technical writer in an international company that specializes in high-technology document design and production.
EDUCATION:	Carnegie-Mellon University, Pittsburgh, PA. BS degree in technical writing expected May 1990.

Relevant Courses:

Prof. & Tech. Writing I, II, III	Rhetoric in Social Interaction
Computing (Pascal)	Fiction Writing
Cognitive Psychology	Concepts of Engineering
Probability & Applied Statistics	Economics
Physics I & II	Chemistry
Biology	Textual Interpretation
Applied Math I & II	Discursive Practices

EXPERIENCE:

5/89–8/89	Assistant technical writer (summer internship) at IBM's Boca Raton, Florida, facility. Helped write a user manual for the 3083 series computer. (Supervisor: Betty Burnett)
11/87–pres.	Reporter for *The Tartan* (CMU campus newspaper). Have written 23 news articles and 4 reviews. Have also learned how to do editing, typesetting, layout. (Ref.: Robert Couture)
COMPUTING SKILLS:	Pascal, BASIC, UNIX, Wordstar, Macwrite, Visicalc, Minitab, Andrew, Sun Workstation, DEC VAX, IBM PC/AT, MCS 8400 typesetter.
ORGANIZATIONS:	CMU Computer Club, Society for Technical Communicators.
ACTIVITIES:	Skiing, hiking, stamp collecting, church.
OTHER:	Intermediate proficiency in Spanish. Have traveled to four countries in Central America.

References and writing samples available upon request.

Figure 10-4 ❑ RÉSUMÉ AND LETTER OF APPLICATION FOR A FIRST JOB
(Used with permission.)

Mei Kwan
2449 Lurting Avenue
Bronx, NY 10469

March 22, 1989

Dr. Pierre Richard
Director, Central Testing Laboratory
Schneider Biotics, Inc.
Englewood, NJ 07631

Dear Dr. Richard:

I wish to apply for the position of laboratory technician trainee you are advertising in *The Bergen Evening Record*. I am a second-year biology major at Hunter College, City University of New York, and I am confident that I could perform the work with distinction.

Attached is my résumé. It shows that although I am only 19 years old, I have already gained considerable experience in biological testing. As a Dean's List student, I was chosen to be a laboratory assistant for Dr. Ruth Wilson of the Hunter College Biology Department. In this role I help teach freshmen a variety of experimental procedures and evaluate their lab reports. Meanwhile, I have been taking courses in genetics, cell biology, experimental techniques, and other biology courses and getting high marks in them (3.7 average). I also had good training and teaching experience before college, at the Bronx High School of Science, where I was a member of the National Honor Society and tutored underprivileged students in math and science. I have developed computer programming skills as well and can program in both PL/1 and FORTRAN. Having been entrusted with several positions of responsibility in my life—including handling cash register receipts, inventory, and stocking at Kwan's grocery—I can assure you that I would be a careful and reliable employee.

I am eager to talk with you about the trainee position because it is exactly what I am looking for at this stage in my career. I do have the summer free and could begin working at the end of May. Since I am at school and out of reach during the day, I will call you sometime next week to see if an interview can be arranged.

Very truly yours,

Mei Kwan

Att.

Figure 10-4 ❑ *(Continued)*

━━━

MEI KWAN
2449 Lurting Avenue
Bronx, NY 10469
212-555-3200

CAREER OBJECTIVE	Research and development in some area of applied biology, either at a university or in a corporation.

EDUCATION Hunter College, City University of New York
Bachelor of Science expected June 1992
Major: Biology GPA: 3.7 in major (4-point scale)
3.2 overall

Major courses taken
Biology I, II Experimental Techniques in Cell Biology
Genetics Experimental Techniques in Genetics & Molec-
Cell Biology ular Biology

HONORS Dean's List, Fall Semester 1988, Hunter College
National Honor Society (high school)

WORK EXPERIENCE Lab assistant, Hunter College Biology Department, CUNY, September 1988 to present. Supervise first-year biology students, provide tutorial help, maintain lab safety, evaluate reports. (Supervisor: Dr. Ruth Wilson, tel. 212-555-6300)

Tutor, OUTREACH Program, Bronx High School of Science, Bronx, NY, September 1986 to June 1987. Gave voluntary help in math and science to underprivileged students in area high schools.

Store clerk, Kwan's Grocery, Bronx, NY, summers 1984–1987. Served customers, handled cash register, did inventory and stocking. Was sometimes left in charge of entire store.

SPECIAL SKILLS PL/1 and FORTRAN computer languages; bilingual Chinese and English; violin.

OTHER INTERESTS Reading, seeing foreign films, going to concerts, designing clothes.

References available upon request.

309 Packard Road
Embry, OH 44107
September 22, 1988

Dr. Rodger L. Krakau
Biomedical Engineering Research
Johns Hopkins University
Applied Physics Laboratory
8621 Georgia Avenue
Silver Spring, MD 20910

Dear Dr. Krakau:

Pursuant to the advice of Dr. Roberta Shofer, Department of Neurology, Einstein Medical College, New York, I am writing to you concerning potential employment. I will be graduating in May 1989 from the University of Ohio, College of Engineering, with a degree in electrical engineering and with an emphasis in biological sciences and bioengineering. I believe that I could be a valuable addition to your research staff.

For five summers I have been employed as a refrigeration and air-conditioning mechanic and technician. I hold an Unlimited Refrigeration Journeyman's License. While not directly related to my field of study and interest, this employment (and the invitations to return each year) are, I believe, indicative of my integrity and initiative in a commercial situation and my ability to solve technical problems through a logical diagnostic approach. I have letters of recommendation from my two employers, which I would be happy to forward should you so wish.

During the course of my education at the University of Ohio, I have studied in both the College of Engineering (Department of Electrical and Computer Engineering) and the College of Literature, Science, and the Arts (Department of Zoology). I believe that my background in zoology, supplemented with coursework in bioengineering, gives me a solid and adequate basis for work in biomedical engineering.

Outside of the classroom, I have been active on many fronts. I am currently serving as treasurer of the U of O chapter of IEEE. I served on the executive board of the local Democratic party and did research on issues and policy for an unsuccessful bid for a congressional seat (Ohio 2nd district) in 1981. I have also performed in Washington, DC, for the President with an Ohio-sponsored vocal jazz ensemble.

I would like very much to discuss your research projects and my potential employment and am confident that a convenient date and time can be arranged. At present, I am planning to be in the Maryland area for the Thanksgiving and Christmas recesses, or I could arrange to come earlier should it prove necessary. As I am difficult to reach, I will contact your office shortly to arrange a suitable and convenient meeting.

I thank you for your time and consideration, and am looking forward to meeting you.

Sincerely yours,

Lisa C. Caron

Figure 10-5 ❑ (*Continued*)

PERSONAL DATA RÉSUMÉ

Lisa C. Caron
309 Packard Road
Embry, OH 44107
(614) 555-8932

Career Plans and Objectives:

I am seeking employment in biomedical engineering research and development. My education and background in electrical engineering, biology, and biophysics will enable me to contribute significantly to the growth of knowledge in this vast and growing field and thereby eventually help to improve the quality and technology of medical care.

Work Experience:

Summer 1984, 1985, 1986, 1987
Bishop Equipment Company, Fairfax, Virginia

Summer 1988
The Alma Corporation, College Park, Maryland

I worked for these five summers, to finance my education, as an air-conditioning mechanic, beginning as an apprentice and working my way up to master mechanic. I worked in all phases of residential and commercial air conditioning.

Education:

University of Ohio
College of Literature, Science, and the Arts
Department of Zoology 1984–1986

University of Ohio
College of Engineering
Department of Electrical and Computer Engineering 1986–1989
B.S.-E.E. May 1989

Language fluency: French

Societies and Honors:

IEEE 1987–1989; presently serving as treasurer, U of O chapter
Invitation to join Eta Kappa Nu

Professional License:

Unlimited Refrigeration Journeyman's License

References will be supplied upon request.

❏ EXERCISE 10-2

Find an announcement of a job or educational opportunity for which you qualify. (Places to look include the classified ads in a newspaper or trade journal, professional society newsletters, and your school's placement bureau.) ~~Write~~ a résumé and letter of application for the position.

Taylor

Situation 10-1 ❏ THE ETHICS OF HEADHUNTING*
*[Adapted from William B. Werther, Jr., and Keith Davis, *Human Resources and Public Relations,* 3d ed. (New York: McGraw-Hill, 1989), pp. 171–173.]

Darrow Thomas worked as a professional placement specialist for LA&D, Inc., an executive search firm. For the last three months Darrow had not been very successful in finding high-level executives to fill the openings of LA&D's clients. Not only did his poor record affect his commissions, but the office manager at LA&D was not very pleased with Darrow's performance. Since Darrow desperately needed to make a placement, he resolved that he would do everything he could to fill the new opening he received that morning.

The opening was for a Director of Research and Development at a major food processor. Darrow began by unsuccessfully reviewing the in-house telephone directories of General Mills, General Foods, and Quaker Oats. Finally, he stumbled across the directory of a small food processor in the South. In the directory he found a listing for Suzanne Derby, Assistant Director of Product Development. He called her, and the following conversation took place:

Suzanne: Hello. PD Department. Suzanne speaking.

Darrow: Hello. My name is Darrow Thomas, and I am with LA&D. One of my clients has an opening for a Director of Research and Development at a well-known food processor. In discussions with people in the industry, your name was recommended as a likely candidate. I was . . .

Suzanne: Who recommended that you call me?

Darrow: I'm awfully sorry, but we treat references and candidates with the utmost confidentiality. I cannot reveal that name. But rest assured, he thought you were ready for a more challenging job.

Suzanne: Well, okay.

Darrow: Good. How many people do you supervise?

Suzanne: Three professionals, seven technicians, and two clerks.

Darrow: Approximately how large a budget are you responsible for?

Suzanne: Oh, it's about three-quarters of a million dollars a year.

Darrow: What degree do you hold, and how many years have you been Assistant Director?

Suzanne: My undergraduate degree and master's are in nutrition science. After I graduated in 1978, I came to work as an Applications Researcher. In 1983, I was promoted to Chief Applications Researcher. In 1988, I was appointed Assistant Director of Product Development.

Darrow: Good career progress, two degrees, and managerial experience. Your background sounds great! This is a little personal, but would you tell me your salary?

Suzanne: I make $51,000 a year.

Darrow: Oh, that is disappointing. The opening I have to fill is for $70,000. That would be such a substantial jump that my client would probably assume your past experience and responsibility are too limited to be considered.

Suzanne: What do you mean?

Darrow: Well, the ideal candidate would be making about $62,000 a year. That figure would indicate a higher level of responsibility than your low salary. We could get around that problem.

Suzanne: How?

Darrow: On the data sheet I have filled out, I could put down that you are making, oh, say, $65,000. That sure would increase my client's interest. Besides, then my client would know a salary of $70,000 was needed to attract you.

Suzanne: Wow! But when they checked on my salary history, they'd know that $65,000 was an inflated figure.

Darrow: No, they wouldn't. They wouldn't check. And even if they did, companies never reveal the salary information of past employees. Besides, they are anxious to fill the job. I'll tell you what, let me send them the data sheet; I'm sure they'll be interested. Then we can talk about more of this. Okay?

Discussion Questions:

1 Although headhunters do not necessarily engage in the practice of "inflating" an applicant's wage, it does happen occasionally. What would you do in Suzanne's place? Would you allow your name to be used?

2 If Suzanne goes along with Darrow's inflated salary figure and she is hired, what possible problems may she face?

REFERENCE

1 D. Charney and J. Rayman, "The Role of Writing Quality in Effective Student Résumés," *Journal of Business and Technical Communication* (1989).

ADDITIONAL READING

J. Eisenberg, "Guidelines for Writing an Effective Résumé," *Journal of the American Dietetic Association 36*:1401–1403 (1986).

H. S. Field and W. H. Holley, "Résumé Preparation: An Empirical Study of Personnel Managers' Perceptions," *Vocational Guidance Quarterly 24*:229–237 (1976).

A. Helwig, "Corporate Recruiter Preferences for Three Résumé Styles," *Vocational Guidance Quarterly 34*:99–105 (1985).

E. McDowell, "Perceptions of the Ideal Cover Letter and Ideal Résumé,"*Journal of Technical Writing and Communication 17*:179–191 (1987).

V. Oliphant and E. R. Alexander III, "Reactions to Résumés as a Function of Résumé Determinateness, Applicant Characteristics, and Sex of Raters," *Personnel Psychology 35*:829–842 (1982).

J. Penrose, "A Discrepancy Analysis of the Job-Getting Process and a Study of Résumé Techniques," *Journal of Business Communication 21*:5–15 (1983).

11

The Business Letter

There are a number of activities which a business does frequently. These include requesting information or equipment, ordering supplies, praising or thanking someone for a job well done, complaining about a job badly done, and responding to someone else's request, order, praise, or complaint. Although some of these activities may be accomplished orally or by filling out a form, many of them require writing a letter. A letter provides a record of the activity for someone's file, it allows the writer to provide more context or explanation than is usually possible on a form, and it helps the audience remember what is to be done. Although letters may be written more frequently in some jobs than in others, all technical people should be able to write a good letter when it is needed.

Writing a good letter is an art. It basically requires a writer to produce a one-sided conversation with the reader. Of course, all writing does this to some degree in that the writer anticipates the reader's questions and provides answers to those questions where the reader might ask them. However, letters differ from most other forms of writing: they are often more personal, even emphasizing the reader/writer relationship with the generous use of such pronouns as *I*, *we*, and *you*.

Despite this frequent personal sense, letters share several of the organizational features described in Chapters 12 and 13 for the less personal technical report. Like reports, letters need to first orient the reader to the topic at hand, then explain why the writer is writing, and then provide enough information so that the reader can easily understand what he or she is to do. Notice that providing this information requires you to generate ideas, analyze your audience, decide what you need to say, and define your problem as discussed in Chapters 2–5.

The orientation of the reader is especially important when the reader does not expect a letter or might not remember the subject the letter refers to. For instance, consider the letters in Figures 11-1 and 11-2. The first letter has no orientation to the topic being discussed. The writer jumps right into the topic as if she were simply continuing the phone conversation which initiated it. Unfortunately, the conversation occurred almost 3 weeks before the letter was written, and the reader may well have a hard time remembering which shroud is being discussed in the first paragraph and what he needs to know about it. In contrast, the version of the letter in Figure 11-2 gives a good orientation to the topic; it reminds the reader of the conversation and the main questions raised. It also provides a rationale for the organization of the letter's details: they are grouped as answers to the three questions.

11.1 BASIC LETTER FORMATS

Even though letters do not have headings and labels such as the *To, From, Subject,* and *Date* lines in a report (see Chapters 12 and 13), they do have conventional places where readers look for such background information. Figure 11-3 indicates these places, and Figure 11-4 gives an example of a short letter arranged in a conventional format for business letters. Notice that the subject of the letter is mentioned in the first sentence to orient the reader.

The conventional format illustrated in Figure 11-4 is one of three quite common formats for business letters: the unblocked format, the semiblocked format, and the blocked format. In the unblocked format shown in Figure 11-5, the first line of a paragraph is indented a few spaces (as it is in this paragraph) and the following information is indented about two-thirds of the way across the page: the writer's address, the date, the closing (such as *Sincerely yours,*), the writer's signature, and the typed version of the writer's name and job title (if present). In the semi-blocked format shown in Figure 11-6, the first line of a paragraph is lined up with the left margin and there is an extra blank line between paragraphs to signal the start of a new paragraph, but the writer's address, date, closing, and signature information are indented as in the unblocked format. In the blocked format shown in Figure 11-7, the first lines of paragraphs and all of the other address, date, closing, and signature information are lined up with the left margin; there is an extra blank line between paragraphs.

These formats are presented in skeletal form in Figures 11-5 to 11-7 for letters written on plain paper—that is, not on letterhead stationery. It should be noted, however, that when a writer is representing a company or organization, the writer should use the organization's letterhead stationery for any correspondence with people outside the organization. When such letterhead is used, the only change in the formats presented in Figures 11-5 to 11-7 involves the location of the writer's address, city, and state. These are usually given, along with the organization's name and telephone number, in the letterhead printed at the top of the page, as shown in Figure 11-1, for example.

Figure 11-1 ❏ SAMPLE BUSINESS LETTER WITHOUT A GOOD ORIENTATION TO THE TOPIC

ATI
ADVANCED TECHNOLOGIES, INC.
40 Technology Park
Milford, MA 01757
(617) 555-4553   October 4, 1989

Mr. Frank E. Lee
Equipment Department
Harrison Radiator, Inc.
1451 Murray Avenue
Pittsburgh, PA 15217

Dear Mr. Lee:

The shroud you called about on September 15 has the same shape, construction, and functions as the shroud previously purchased by Harrison Radiator Division from ATI. However, that machine was designed to balance a series of blower assemblies having only one fan. With the present balancer, you wish to have the capability to balance a new blower assembly having two fans, one on each end of the motor shaft. This new blower assembly is longer than all previous single-fan blower assemblies. The only dimensional change needed to account for this is to widen the shroud from 10″ to 16½″. Also, a new clamping fixture is needed to hold the double-fan assembly, necessitating that a 3″-diameter hole be added to allow a clamping cylinder to protrude. These changes are illustrated on the enclosed print.

I have found that the change of the focal point of the fan speed sensors, due to the new blower fan's position within the shroud in relation to the sensors, is well within the limits of the adjustments designed into the original shroud, as is shown on the print. This should not present a problem.

The estimated weight of the shroud is 6½ lbs, only ½ lb heavier than the previous one. This results in a cost difference of an additional $5.75 for the new shroud.

I refer you to the enclosed print; it might answer additional questions.

Sincerely,

Barbara C. Benson
Engineering Department

BCB/rs

Enclosure

Figure 11-2 ❑ SAMPLE BUSINESS LETTER WITH A GOOD ORIENTATION TO THE TOPIC

ATI

ADVANCED TECHNOLOGIES, INC.

40 Technology Park
Milford, MA 01757
(617) 555-4553   October 4, 1989

Mr. Frank E. Lee
Equipment Department
Harrison Radiator, Inc.
1451 Murray Avenue
Pittsburgh, PA 15217

Dear Mr. Lee:

In our phone conversation of September 15, you wanted to know:

- What the differences will be between the original shroud for Harrison Radiator Division's D-25-PS balancing machine and a similar one purchased from ATI on September 3 (Ref.: Sales Order #N5682)

- How these differences affect the cost of the shroud

- If the introduction of our double-fan blower assemblies will create any problems with the fan speed sensors

I have listed this information below and enclosed a blueprint of our new shroud for your reference (#25508).

Basically, the two shrouds have the same shape, construction, and functions. However, there are three major differences between the Harrison and ATI shrouds:

1) The ATI shroud has a new double-fan blower assembly.

2) The ATI shroud is 6½″ wider.

3) The ATI shroud has an additional 3″-diameter hole in the top of the shroud.

These changes will result in an additional cost of $5.75. Also, the new blower assemblies will not present any problems with regard to the fan speed sensors.

The original Harrison shroud was designed to work only with a series of single-fan blower assemblies. The introduction of longer double-fan blower assemblies (fans on both ends of the motor shaft) on the ATI shroud requires two things.

The shroud must be lengthened from 10″ to 16½″. Also, a new cradle fixture is required to hold the blower assemblies, resulting in the addition of a 3″-diameter hole in the tip of the shroud, to allow a clamping cylinder to protrude.

Since the two shrouds are the same construction—same material, same method of assembly—the only cost difference will result from the greater weight of the new shroud.

Figure 11-2 ❑ *(Continued)*

Widening the shroud 6½″ adds ½ lb to the total weight. This translates to an additional cost of $5.75.

As can be seen on the enclosed print, the focal point of the new blower fan is well within the range of the electro-optical fan sensors, as provided for by the sensor adjustments designed into the original shroud.

I refer you to the enclosed print. It may answer additional questions.

Sincerely,

Barbara C. Benson
Engineering Department

BCB/rs

Enclosure

If a letter requires more than one page, the additional pages are called continuation pages. They are typed on plain paper, not letterhead. A sample continuation page appears in Figure 11-8. Notice that it contains enough information at the top of the page to identify it if it gets separated from the first page.

In Figures 11-5 to 11-8 the number of blank lines separating the various sections of the letter are indicated in the left margin. One blank line separates most sections, but one to three blank lines separate the writer's address at the top of the page from the reader's name and address, and three or four blank lines separate the closing (*Sincerely yours*) from the typed version of the writer's name. The latter spacing should leave plenty of room for the writer's signature. You may want to adjust the spacing between sections to fit everything on one page or to make a better-looking page.

The letters in Figures 11-1 to 11-8 all have some additional bits of information, the function of which may not be as obvious as that of other parts of the letter. This information appears at the left of the page below the typed version of the writer's name. Such lines should be added to a letter whenever they are appropriate; if any one of these lines is omitted, the remaining lines should be moved up, with one blank line separating them from each other and

Figure 11-3 ❑ LOCATION OF BACKGROUND INFORMATION IN A LETTER, UNBLOCKED FORMAT
(On letterhead stationery, the writer's address will already be printed.)

no name ⟶ Writer's Street Address
Writer's City, State   Zip Code
Date

write out on letter

18 January 1994

Reader's Name
Reader's Job Title
Reader's Organization/Company
Organization's Street Address
Organization's City, State   Zip Code

Subject: *May use*

Dear Name of Reader: *Ms, not misses if unknown*

┌───
│   Subject of letter and reference to past correspondence, if any
│
│
│
└───

┌───
│
│
└───

Sincerely, ~~yours~~,

Writer's Signature ⟵ *Fairly Legable*

Writer's Typed Name

WRITER'S INITIALS: typist's initials
(if writer did not type letter)

Enclosure (if appropriate)
cc: Name of Recipient (of carbon copy of letter, if appropriate)
        or
xc: Name of Recipient (of photocopy of letter)

Figure 11-4 ❑ EXAMPLE OF A SHORT BUSINESS LETTER IN THE UNBLOCKED FORMAT

205 State Street
Livonia, AL 35851
June 20, 1989

Mr. R. B. Sparks
Sales Representative
Computer Connection
384 Grand Drive
Harrington, GA 30018

Subject: BASIC Software for Apple II System

Dear Mr. Sparks:

The Computing Club of Brewster High School will be sponsoring a series of computing activities during the fall of 1989. Since our current computing equipment is inadequate for our projected needs, we are considering buying an Apple II system to expand our equipment line.

Last week I wrote to you requesting information on the cost of the Apple II Plus system and its peripheral equipment as well as information on the Pascal software available with the system.

I would now like to request additional information on

1) The BASIC software currently available with the system, especially for plotting graphs such as that enclosed with this letter

2) Any BASIC software expected to be released in the next year for the Apple system

Thank you for your attention. We are looking forward to hearing from you in the near future.

Sincerely yours,

Ann Jones
President, Computing Club

AJ:vg

Enclosure

cc: Mr. Peter Murray
    Ms. Mary Corvina

Figure 11-5 ❑ UNBLOCKED FORMAT FOR BUSINESS LETTER

205 State Street
Livonia, Alabama 35851
June 20, 1989

(2)*

    Mr. R. B. Sparks
    Sales Representative
    Computer Connection
    384 Grand Drive
    Harrington, Georgia 30018

(1)

    Subject: BASIC Software for Apple II Systems

(1)

    Dear Mr. Sparks:

(1)

_____

_____

_____

_____

_____

(1)

_____

_____

_____

_____

_____

(1)

_____

_____

_____

_____

_____

_____

(1)

                             Sincerely yours,

(3)

                             Ann Jones
                             President, Computing Club

(1)

    AJ:vg

(1)

    cc: Mr. Peter Murray
        Ms. Mary Corvina

* Number of blank lines between sections.

Figure 11-6 ❑ SEMIBLOCKED FORMAT FOR BUSINESS LETTER

205 State Street
Livonia, Alabama 35851
June 20, 1989

(2)*

Mr. R. B. Sparks
Sales Representative
Computer Connection
384 Grand Drive
Harrington, Georgia 30018

(1)

Subject: BASIC Software for Apple II Systems

(1)

Dear Mr. Sparks:

(1)

_____
_____
_____
_____
_____

(1)

_____
_____
_____
_____
_____

(1)

_____
_____
_____
_____
_____
_____

(1)

Sincerely yours,

(3)

Ann Jones
President, Computing Club

(1)

Enclosure

(1)

cc: Mr. Peter Murray
    Ms. Mary Corvina

* Number of blank lines between sections.

Figure 11-7 ❏ BLOCKED FORMAT FOR BUSINESS LETTER

205 State Street
Livonia, Alabama 35851
June 20, 1989

(2)

Mr. R. B. Sparks
Sales Representative
Computer Connection
384 Grand Drive
Harrington, Georgia 30018

(1)

Subject: BASIC Software for Apple II Systems

(1)

Dear Mr. Sparks:

(1)

_____

_____

_____

_____

_____

(1)

_____

_____

_____

_____

_____

(1)

_____

_____

_____

_____

_____

(1)

Sincerely yours,

(3)

Ann Jones
President, Computing Club

(1)

AJ:vg

(1)

Enclosure

(1)

cc: Mr. Peter Murray
    Ms. Mary Corvina

* Number of blank lines between sections.

Figure 11-8 ❏ SAMPLE CONTINUATION PAGE, BLOCKED FORMAT

Mr. R. B. Sparks — *← readers name*   - 2 -   June 20, 1989

(3)*

_____

_____

_____

_____

_____

_____

_____

(1)

Sincerely yours,

(3)

Ann Jones
President, Computing Club

(1)

AJ: vg

(1)

Enclosure

(1)

cc: Mr. Peter Murray
     Ms. Mary Corvina

---

* Number of blank lines between sections.

from the typed version of the writer's name. The following information is included in these lines:

1 *Initials line.* The first line under the writer's name is the initials line (*AJ:vg*), which indicates that someone other than the writer typed the letter. The capital letters are the initials of the writer, the lower-case letters the initials of the typist. If the writer types the letter, the initials line is omitted.

2 *Enclosure line.* The second line under the writer's name (*Enclosure*) indicates that something has been enclosed with the letter and alerts the reader to look for this item. The enclosure may be a whole document, a single page, a picture, a drawing, or anything else. If nothing is enclosed, the *Enclosure* line is omitted.

3 *Copy line.* The third line under the writer's name (*cc: Mr. Peter Murray*) indicates that a carbon copy or photocopy of the letter has been sent to the person (or persons) whose name appears after the *cc:*. Sometimes *xc:* is used to indicate that a photocopy has been sent. If no one has been sent a copy of the letter, the *cc:* line is omitted.

Sometimes the writer doesn't know the name or marital status of the reader and thus has a problem writing a salutation (Dear _____:). This problem has given rise to a distinctly new letter format called the AMS simplified letter style. In this style, the salutation is simply omitted, as is the closing. The subject line replaces the salutation and is typed in all capital letters. In other respects this format resembles the blocked format, with every line beginning from the left margin. Figure 11-9 illustrates this style, which is somewhat less personal than that of other styles.

The AMS style is becoming more and more popular in business circles because it is efficient. But it is quite unconventional and might offend people who are used to the more traditional formats. Therefore, be certain of your audience before you take a chance on this format.

11.2 FORMS OF ADDRESS

Forms of address (salutations) are an important part of letter writing. Except for new formats like the AMS simplified letter style (described above), letters normally begin with some kind of salutation, and the choice of salutation indicates, right away, how the writer sees his or her relationship with the reader. There is a big difference in formality between "To Whom It May Concern" and "Dear Ms. Schmidt," or between "My Dear William" and "Hi, Bill!" Make sure you put some thought into how you address people. If your salutation fails to display the right degree of politeness or the right degree of friendliness toward your reader, your letter may be doomed from the start.

In formal correspondence, it is customary in English-speaking countries and in many other countries to use the addressee's title and last name: "Dear Dr. Maksoud" or "Dear Major Kuroda." If you think the person does not have a title, simply use "Mr." or "Ms." In unusual cases—e.g., when you are writing to a government or religious official—you may want to consult your dictionary: most good dictionaries have a special Forms of Address section in the back. Keep in mind, though, that these conventions are not universal. When writing to someone in a culture different from your own, you would be wise to check with members of that culture to see how they write salutations.

In the American business world, and in certain technical and professional environments, it is becoming increasingly common to address people by their first name, even people whom the writer hardly knows. Though meant to be a sign of goodwill, such casual use of first names can easily be seen as disrespectful. Therefore, unless you are confident that addressing someone by his or her first name will not offend, it is better, we feel, to err on the side of politeness. Keep in mind also that few other business cultures tolerate this practice. If you are corresponding with someone overseas, you would be wise to use a standard formal salutation—unless, of course, your correspondent is really a personal friend.

Avoiding sexism in a salutation can be a little tricky. The widespread use of "Ms." has eliminated the problem of determining whether a female addressee is married. But there is still the problem of how to address someone whose gender you don't know. Suppose you are writing to someone named

Figure 11-9 ❑ AMS SIMPLIFIED FORMAT FOR BUSINESS LETTER

205 State Street
Livonia, AL 35851
June 20, 1989

(3)*

R. B. Sparks
Sales Representative
Computer Connection
384 Grand Drive
Harrington, GA 30018

(3)

Pascal Software for Mac II Systems

(3)

_____
_____
_____
_____
_____

(1)

_____
_____
_____
_____
_____
_____

(1)

_____
_____
_____
_____
_____
_____

(3)

Ann Jones—President, Computing Club

(3)

AJ:vg

(1)

Enclosure

(1)

cc: Mr. Peter Murray
    Ms. Mary Corvina

* Number of blank lines between sections.

J. L. Williams and you don't know whether this person is male or female. How do you write the salutation? If you write "Dear Mr. Williams," you are discriminating against the possibility that Williams is a woman. Likewise if you say "Dear Sir." If you take the opposite tack and write "Dear Ms. Williams," you will not win any points if Williams is a man.

In today's world it is important to avoid any hints of sexism in the way you deal with people. So here are some (albeit imperfect) options to consider:

1   Use the complete name: "Dear J. L. Williams." This is an economical solution, but it's quite cold and impersonal.

2   Use both titles: "Dear Mr. or Ms. Williams." This is a little more personal but also more wordy.

3   Use a memo format: "To: J. L. Williams
                              From: Your Name"
    This changes the entire nature of the communication, embedding it in an organizational environment (see Chapter 13).

Given the different (and perhaps unwanted) effects of each of these options, you may want to deal with the situation—if it's important enough—by taking the trouble to call and find out what the addressee's gender is.

11.3 LETTER OF TRANSMITTAL

When a document of more than two or three pages—a report, article, or proposal—is sent to a reader, it is often accompanied by a letter of transmittal. The purpose of this letter is to identify the document and to orient the reader to it. It may help the reader decide whether to read the document, file it, or send it on to someone else. Thus, the letter of transmittal should be short— no more than one page if possible—and should identify the document being sent, the project about which it is written, and any especially important points being made about the project which the reader does not already know.

The form of the letter of transmittal may vary with the audience. For instance, suppose you have been working with a partner on a proposal for the city of Grand Rapids to develop an industrial waste pretreatment program and a modification to a facilities plan your company submitted to Grand Rapids last spring. You and your partner have written the proposal together, and you have agreed to incorporate some last-minute details into a final draft and have that typed. When the final draft is ready, you might put the following short letter on the report and send it to your co-author:

Ann—
> Here is the final draft of our Grand Rapids proposal. How do you like it?
>   Roger

Obviously, this short letter assumes that the reader knows which proposal you are talking about when you mention "our Grand Rapids proposal" and what points are especially important. Since the reader is a coauthor of the proposal, this is probably a safe assumption.

On the other hand, if you were sending this proposal to your supervisor for review and approval before you sent it out to the city of Grand Rapids, your letter of transmittal might look like this:

> Dear Phil:
>
> You assigned Ann and me to prepare a proposal for the development of an industrial waste pretreatment program for Grand Rapids as an addendum to the facilities plan we submitted to Grand Rapids last spring. Here's a final draft of our proposal for both subjects. As you suggested earlier, this proposal is based on a cost-plus-fixed-fee-with-a-maximum contract. It does not include the city's involvement in the work other than identifying equipment to be purchased by the city.
>
> Think it's ready to go?
>
> Roger

This letter is longer and provides more information than did the version to your partner. It assumes somewhat less knowledge on the part of the reader about the particular project and thus points out certain assumptions behind the proposal which would have been obvious to your partner. On the other hand, it assumes that the supervisor has a relatively high level of technical knowledge in waste pretreatment programs and facilities plans. It assumes, for instance, that the supervisor will see the appropriateness of basing a proposal of this sort on a cost-plus-fixed-fee-with-a-maximum contract and of not including the city's involvement in the work other than identifying the equipment it will purchase.

Finally, if you were sending the final proposal to the city of Grand Rapids, you would need to provide much more context and information than was provided in the two previous letters. Those letters were written to people who had knowledge about the project, who understood the implications of important technical statements, and who were aware of the experience and credentials of your company in developing industrial waste programs for waste pretreatment and facilities modifications. Much of this knowledge needs to be spelled out in more detail for someone outside the company, someone less familiar with the company's credentials and the details of the development plans. Thus, a letter of transmittal accompanying the proposal might look like that presented in Figure 11-10. This letter, written to the city engineer but potentially read by anyone else to whom the document will be routed, identifies the proposal it accompanies and then spells out some of the information the city engineer and city planners might need to know about the proposers. It emphasizes the proposers' willingness to cooperate with the city and stresses their qualifications in the area of the proposed work. This is probably the most important point to stress, since any community about to spend a large amount of money wants to believe that the money will be well spent, that the community's problem will be solved for the specified amount. Most probably,

Figure 11-10 ❑ SAMPLE LETTER OF TRANSMITTAL TO AN AUDIENCE
OUTSIDE THE WRITER'S ORGANIZATION
(Used with permission.)

McNamee, Porter and Seeley
3131 South State Street - Ann Arbor, Michigan 48104 - (313) 665–6000
Consulting Engineers

June 15, 1989

Mr. John L. Hornbach, P.E.
City Engineer
300 Monroe Avenue, N.W.
Grand Rapids, MI 49503

Re: Request for Proposals

Dear Mr. Hornbach:

We are pleased to submit four copies of our proposal for the development of an Industrial Waste Pretreatment and Nondomestic Users Program and the preparation of a 201 Facilities Plan Addendum.

Our proposal is based on a cost-plus-fixed-fee-with-a-maximum type of contract and is our best estimate at this time to complete the work outlined in your request. We are prepared to make an oral presentation of its contents when requested, and, should we be selected as your engineer, we are prepared to discuss our proposal in greater detail.

Our proposal does not include the city's involvement in the work other than identifying equipment that is to be purchased by the city. We believe that there will be considerable labor saving on the city's part resulting from our understanding of your needs; our team approach; our knowledge of your wastewater collection system, your wastewater treatment facilities, and your NPDES requirements; our ongoing working relationship with the Michigan DNR; and the fact that we did prepare and had Michigan DNR approval of a Step 1 Grant Amendment for this work.

We are aware of the slowdown in the U.S. EPA grant program and the possibility of some of the work not qualifying for federal and state assistance. We are prepared to discuss a scaledown of this work if so directed by the city.

We value our long association with the city of Grand Rapids and look forward to its continuance. Please contact us if you have any questions on our proposal.

Sincerely,
McNAMEE, PORTER AND SEELEY

By _____
Philip C. Youngs

the problem will be solved only if the company funded to do the work is knowledgeable, effective, and experienced.

Whatever the audience, the function of the transmittal letter is to orient the audience to the document. This can be done well only if the writer of the letter is sensitive to the needs, interests, and knowledge of the reader.

11.4 LETTER OF COMPLAINT

Sometimes in our dealings with individuals or businesses we find that we have not received what we were promised or what we expected. We may have received a part different from what we ordered, one that is defective, or one that does not perform as advertised. We may have been badly treated by someone, or we may be exasperated by a long-term dangerous situation. Whenever these sorts of things happen, whenever we feel cheated or victimized, especially in important ways, we want to complain. We want to get things "off our chests" and to correct unpleasant situations.

This desire to complain and correct gives rise to the letter of complaint. The goal of such a letter is usually to improve the situation about which one is complaining. To do this, the writer of a letter of complaint should politely but firmly

1   Identify the nature and seriousness of the problem

2   If possible, request or suggest a solution to the problem

This procedure is illustrated in Figure 11-11, a sample letter of complaint.

Notice that this letter of complaint asks the reader to provide time, effort, and perhaps money to address the complaint. Since this is typical of many letters of complaint, you can imagine that such letters should be polite but firm. They probably shouldn't threaten the reader, at least not on the first exchange of letters, since they shouldn't alienate the person who may be able to solve the problem. For the same reason, if the letter of complaint suggests a solution to some problem, the solution should be reasonable. From a purely practical standpoint, reasonable solutions have a better chance of being accepted than unreasonable ones. Further, they show the writer to be fair and helpful even in a difficult situation, and thus they may encourage people to do business with the writer in the future.

11.5 RESPONSE TO A LETTER OF COMPLAINT

Responding to a letter of complaint poses its own set of problems. The writer of such a response needs to look knowledgeable, helpful, concerned, and appropriately apologetic. As with other letters, the writer needs to

Figure 11-11 ❑ SAMPLE LETTER OF COMPLAINT

HARRISON RADIATOR, INC.
1451 Murray Avenue
Pittsburgh, PA 15217
(412) 555-4356
September 15, 1989

Ms. Barbara C. Benson
Engineering Department
Advanced Technologies, Inc.
40 Technology Park
Milford, MA 01757

Dear Ms. Benson:

Two weeks ago we purchased a shroud #22508 from Advanced Technologies, Inc., for our Harrison Radiator Division's D-25—PS balancing machine (reference sales order #N5682). Your sales representative told us that your shroud would fit our Harrison balancing machine. Unfortunately, when we tried to install the shroud, it did not fit.

I am writing now to request instructions for adapting your shroud to our Harrison balancer. If you have such instructions, we would appreciate receiving them as soon as possible. We need to use our balancer within two weeks, if that is possible, and we cannot operate it until we have a suitable shroud.

If we cannot adapt your shroud to our Harrison balancer, we would like to return the shroud for full credit and find a suitable substitute from another supplier.

Thank you for your prompt attention to this matter.

                                        Sincerely,

                                        Frank Lee

FL:aa

___

1 Identify the purpose of the letter, reminding the reader of the source of the complaint as well as of any suggestions the reader originally made for dealing with the complaint

2 Deal with the complaint, outlining whatever the writer can do to help the reader

3 Assure the reader of the writer's goodwill and attention to the problem

These characteristics are illustrated in Figure 11-12. Notice that the writer of this letter stresses the speed with which she is responding and then carefully provides the help the reader needs: she outlines the characteristics of the two shrouds and the few simple modifications which need to be made to the problematic shroud. Finally, the letter ends on a positive, helpful note. It does not grovel, but it stresses the company's commitment to satisfying its customers.

11.6 LETTER OF REQUEST

Writers sometimes need to request information or equipment. Since such requests usually cost the reader time or resources in responding, the writer should very clearly

1 Orient the reader to the topic of the letter

2 Indicate why the writer should be willing to respond

3 Indicate exactly what the writer is requesting the reader to provide

Orienting the reader to the topic was discussed earlier in this chapter and in Chapter 5 and so will not be treated again here. Indicating why the writer should be willing to respond may be done along with the orientation or in addition to it. An example of a letter which provides a motive for responding along with an orientation appears in Figure 11-13; one which provides a motive for responding in addition to the orientation appears in Figure 11-11, a letter of complaint as well as a letter of request.

Routine requests for information should be handled in routine fashion. You can make life easier for your reader if you go straight to the point and do not insert unnecessary personal information. If you do not know the addressee, use a standard salutation like "Dear Sir/~~Madam~~" or "~~To Whom It May Concern.~~" Give only enough background information as is necessary to explain the reason for your request. Then make the request. Politely thank the addressee for his or her help. Close with "Sincerely" or "Yours sincerely" and your signature.

A study of 168 routine letters of inquiry by Brenda Sims at the University of North Texas found that letters from nonnative speakers tended to contain unnecessary personal information.[1] Furthermore, many had a tone of exaggerated courtesy, with adverbs and adjectives like *greatly, eager, esteemed,* and *most grateful* and sentences like "I humbly request you to take into consideration my ardent desire to continue my studies." And they were significantly (and unnecessarily) longer than the letters written by native speakers. Sims attributes these practices to cultural differences. She points out that these writers thought they had to persuade the reader to send them information, not realizing that the reader was more or less obligated to send

Figure 11-12 ❑ SAMPLE RESPONSE TO A LETTER OF COMPLAINT

ATI

ADVANCED TECHNOLOGIES, INC.

40 Technology Park,
Milford, MA 01757
(617) 555-4553   September 19, 1989

Mr. Frank E. Lee
Equipment Department
Harrison Radiator, Inc.
1451 Murray Avenue
Pittsburgh, PA 15217

Dear Mr. Lee:

On September 15, you wrote to me about a problem you were having fitting our shroud #22508 on your Harrison balancing machine. You asked for instructions, as quickly as possible, which would allow you to adapt our shroud to your balancer.

I have just received your letter and am sending by return mail the instructions you requested. I am also enclosing, for your reference, a blueprint of our shroud #22508.

Basically, our shroud and the original Harrison shroud have the same construction, shape, and functions. However, there are three major differences between our shroud and the Harrison model:

1) Our shroud has a new double-fan blower assembly.

2) Our shroud is 6½ inches wider.

3) Our shroud has an additional 3″-diameter hole at the top of the shroud.

These differences are probably causing your problem with fit.

Fortunately, it will be very simple to accommodate our shroud to your Harrison balancer. The original Harrison shroud was designed to work only with a series of single-fan blower assemblies. Our shroud introduces longer double-fan blower assemblies (fans on both ends of the motor shaft), and this requires only two small modifications.

1) Our shroud must be lengthened from 10" to 16½".

2) Our shroud will require a new cradle fixture to hold the blower assembly. You can insert a cradle fixture by drilling an additional 3″-diameter hole in the top of the shroud, inserting the fixture, and allowing the clamping cylinder to protrude.

If you make these simple modifications, I am sure that your new shroud will fit perfectly.

Figure 11-12 ❏ *(Continued)*

If there is anything else we can do to help you in this matter, please feel free to call us. We firmly believe that customer satisfaction is our most important goal.

Sincerely yours,

Barbara C. Benson
Engineering Department

BCB/rs

Enclosure

---

it to them. If you are a nonnative speaker requesting routine information from a Western university, agency, or other organization, keep these cultural differences in mind; make your request as simply and directly as possible.

11.7 RESPONSE TO A LETTER OF REQUEST

The response to a letter of request, like any of the other letters already described, should

1 Orient the reader
2 Identify the purpose of the letter, reminding the reader of his or her request and any conditions upon it
3 Provide the information requested

If the reader requested a physical object that will be sent in a separate package or at a later date, the writer should include any available information on the package's status and projected date of arrival. If the reader requested information which is now available, the letter writer should try to include all the relevant information the reader requested without overwhelming the reader with irrelevant or unnecessary detail. A sample response to a letter of request appears in Figure 11-14. Notice that in form this letter resembles a very short technical report as presented in Chapters 12 and 13, minus the To, From, Subject, Date format and headings to indicate the Foreword and Summary.

❏ EXERCISE 11-1

Write a letter of request for information that can be provided by one of your classmates, friends, or colleagues. Then ask that classmate, friend, or colleague to write a letter responding to your letter of request.

Figure 11-13 ❑ SAMPLE LETTER OF REQUEST

ANDERSON COLLEGE
400 Haggerty Road
San Francisco, CA 94105
(415) 555-6400

May 28, 1989

Mr. R. B. Sparks
Sales Representative
Computer Connection
384 Grand Drive
Harrington, GA 30018

Dear Mr. Sparks:

The Computing Science Club of St. Mary's School will be sponsoring a new course in the Pascal programming language for the fall term of 1989. Because of the overcrowded conditions of our present computer system, we are going to purchase several microcomputers rather than add Pascal software to our present system.

We are considering purchasing ten of your Apple II Plus systems for our Pascal course. However, before we can make a final decision, we must review the following:

1) Information on the cost of the Apple II Plus system and its peripheral equipment

2) Information on the Pascal software available with the system

3) Specifications for the computer system and its microprocessor

Could you please send us whatever information you have which addresses these questions?

Thank you for your attention. I am looking forward to hearing from you in the near future.

Very truly yours,

C. Beth Norther
Instructor, Data Processing

CBN/vr

Figure 11-14 ❏ SAMPLE RESPONSE TO A LETTER OF REQUEST

Pine Dunes Country Club
Orchard Dunes, NJ 07081

August 16, 1989

Mr. Thomas F. Wilson
Club Manager
Pine Dunes Country Club
Orchard Dunes, NJ 07081

Dear Mr. Wilson:

After receiving numerous complaints from some of the club members about the unsafe conditions of the wooden boat docks on the club's west shoreline, you asked me to investigate the condition of the docks. In particular, you asked me to find out what parts were broken or missing from the dock (if any) and what it would cost to make the necessary repairs. I am now writing to report my findings and actions.

I inspected the boat docks on August 12. In general, they seemed to be in good condition, but they did have a few minor trouble spots. Six crossarm brackets were missing bolts, and three stanchion braces had pulled away from the dock and required some minor welding.

After receiving authorization from Virginia Lette, Beach Director, I bought the required bolts for $4.75, replaced the missing bolts in the crossarm brackets, and welded the three stanchion braces. The docks are now in good working order.

Details on Repairs

On August 13, the docks which were causing the problem were removed from the water. The missing bolts from the crossarm brackets were replaced with new bolts (stock #83443.7AG) that were bought from Commerce Dock Supply for $4.75. They were tightened to the recommended torque of 40 ft-lb as stated in the installation instructions.

The three stanchion braces that needed welding were brought over to the Maintenance Department, where they were rewelded by Roger Sands. No problems were encountered during the welding, and the stanchion braces seemed to be completely repaired. There was no cost for the welding.

On August 14, the repaired dock sections were reinstalled and the dock system is now in good working order.

Sincerely,

Steven M. Berggruen
Assistant Beach Director

❑ EXERCISE 11-2

Write a letter of request or complaint to someone you do not know. Try to pick a situation about which you really need information or about which you really want to complain. Send your letter and see if it is effective.

❑ EXERCISE 11-3

Write a letter of transmittal for a report you have produced, found in the library, or encountered on the job. If you are a student, you may choose a report you have produced in one of your technical classes or in your technical writing class. Define the audience and situation for your letter, and write at least one version that might accompany your report to total strangers.

11.8 CROSS-CULTURAL DIFFERENCES

When writing to international audiences, be aware that cultural norms differ from one country to another and try to adapt your letter to the norms that the reader expects. For example, as we noted in Section 11.6, American readers expect routine letters of inquiry to be short and to the point. If you are writing a letter of inquiry to an American audience, that's how you should write it. If you are unsure how to do it, the models given in this chapter may be of some help.

In other countries, it is often more appropriate to start off with some pleasantries before getting down to business and to conclude the letter with more pleasantries. Such "small talk" is seen as promoting the business relationship. One of these countries is Japan. The Japanese preference for politeness and indirectness is well illustrated in Figure 11-15, as compared with Figure 11-16.[2] These two letters have similar content but clearly differ in style and tone. Tests with American and Japanese readers showed that American readers consistently preferred Letter B, while Japanese readers consistently preferred Letter A. Letter A has a warmer, more personal tone. Although some of its language may be formulaic, it contains emotive words like *lovely, prosperity, happiness, delighted,* and *warmest*; it makes a gentle transition to the main topic (with *by the way*); and it talks about things other than the business at hand. Letter B, by contrast, dispenses with these preliminaries and goes straight to the point. It uses no language more emotive than the conventional phrases *we hope* and *we would be happy.* Americans would characterize it as having a more "businesslike" tone.

If you are an American writing a business letter to a Japanese audience, or a Japanese writing to an American audience, you should keep these different preferences in mind. More generally, whenever you are writing to readers in another country, try to familiarize yourself with the customary letter-writing style of that country. Even if you don't know the language of that country

Figure 11-15 ❑ LETTER OF TRANSMITTAL DESIGNED FOR A JAPANESE AUDIENCE

Letter A

Communications Design, Inc.
2401 Highland Avenue
Pittsburgh, Pennsylvania 15206, USA

April 30, 1989

Mr. Akami Tadashi
Senior Manager
Takase Corporation
Tokyo, Japan

Dear Sir,

Greetings from Communications Design, Inc. It's a lovely time of year in this part of the United States, and we are enjoying the warmer weather. We wish you and your company continued prosperity and happiness in the year ahead.

By the way, we offer a unique service to Japanese companies. We simplify documents such as instruction manuals so that they can be easily understood by English-speaking people. Recent research has found that there is a strong relationship between well-written manuals and marketing success. We hope you can take a moment to look at the materials we've enclosed.

Please know that we are delighted to write to you and accept our warmest greetings.

Sincerely,

William M. Powell

Encl.

(and thus have to write in English, say, to a Chinese audience), you can make your letter more effective by observing the appropriate protocol. The best way to learn about cultural norms is to consult someone from that country; lacking that, you may be able to find a book that treats the subject in greater depth than is possible here.

Figure 11-16 ❑ LETTER OF TRANSMITTAL DESIGNED FOR AN AMERICAN AUDIENCE

Letter B

Communications Design, Inc.
2401 Highland Avenue
Pittsburgh, Pennsylvania 15206, USA

April 30, 1989

Mr. Akami Tadashi
Senior Manager
Takase Corporation
Tokyo, Japan

Dear Sir,

We hope you can take a moment to look at the materials we've enclosed describing the work we do at Communications Design, Inc.

At Communications Design, we offer a unique service to Japanese companies. We simplify documents such as instruction manuals so that they can be easily understood by English-speaking people. Recent research has found that there is a strong relationship between well-written manuals and marketing success. We hope you can take a moment to look at the materials we've enclosed.

We would be happy to answer any questions you might have about our services and look forward to meeting you.

Sincerely,

William M. Powell

Encl.

REFERENCES

1 Brenda Sims, "Discourse Community and Business Communication: Problems in International Business Letters." Paper delivered at the 40th Annual Convention of the Conference on College Composition and Communication, Seattle, March 1989.

2 Figures 11-15 and 11-16 were designed by Chris Freeble, who also conducted the tests on them. Used by permission.

ADDITIONAL READING

Roger Cherry, "Politeness in Written Persuasion," *Journal of Pragmatics 12*:63–81 (1988).

Susan Jenkins and John Hinds, "Business Letter Writing: English, French, and Japanese," *TESOL Quarterly 21*(2):327–350 (June 1987).

12

Basic Features of Reports

As noted in Chapter 1, experienced scientists and engineers rate writing and speaking skills as being very important or even critically important to job performance. Written communications cited for their particular importance and frequency include project proposals, progress reports, technical descriptions, memos, and short reports.

Memos and technical reports, including short informal reports and progress reports, are read by a variety of audiences: by technical experts in the field as well as by managers and other nonexperts—the busy, inattentive, and perhaps uninterested reader described in Chapter 3. Thus, the memo or report should be accessible to all these readers and responsive to their needs and reading habits. In part, responsiveness may be increased through two important structural features: (1) the Foreword and Summary (or their analogues—the Introduction and Summary, or the Executive Summary, or the Executive Overview) at the beginning of the memo or report for managerial and nonspecialist readers, followed by a Discussion or Details section for specialist readers, and (2) the placement of generalizations or claims before their support in the part of the memo or report written for specialist readers, including the placement of supporting but nonessential information in Appendixes. Although the Foreword and Summary may actually be written after other parts of the report, they appear first in the finished report and thus will be discussed first here.

This overall structure is an argumentative structure and can be contrasted with the more narrative or chronological structure found in the laboratory report, the research report, or the journal article. The more narrative structure presents information in the chronological order in which it occurred and is conventional in such types of writing as laboratory reports or journal articles. While the narrative structure is easier to write since it closely follows the order of the process of investigation, it is usually less functional for managerial readers: since such readers are typically more interested in the results, conclusions, recommendations, and implications, which all occur at the end of the narrative document, these readers have to hunt around in a narrative document to find the information of most interest to them. In contrast, the argumentative structure presented here makes it very easy for managers to see the information they most need, since it all occurs "up front" in the report format.

12.1 THE FOREWORD AND SUMMARY: ORGANIZING MAIN POINTS FOR NONSPECIALIST READERS

Technical reports may be aimed at specialists, at nonspecialists, or at both at the same time. They are usually read by a wide variety of audiences— managers, salespeople, technicians, financial analysts, specialists in the field, and sometimes even the public. Thus, the beginning of such a document must orient audiences of varying backgrounds, experiences, values, and assumptions. This orientation is done by beginning the report with a Foreword and Summary. The Foreword and Summary are placed at the beginning of a memo or technical report to be accessible to the busy managerial or nonspecialist reader.

The Foreword orients the reader and shows that the report addresses a problem or issue important to the reader. The Summary provides a compact statement of results, conclusions, and recommendations to help managerial readers make decisions: it tells what the writer discovered, the implications of the discovery, and the recommendations for action based on the writer's special knowledge—all in terms that the nonspecialist manager will find useful. (Other terms for the Foreword and Summary include the Introduction and Summary, Executive Summary, Executive Overview, and Synopsis.) By placing the Foreword and Summary first in a memo or technical report, the writer allows the managerial reader to stop after the first page or so and still have all the necessary information. To make sure that all readers understand the problem set up at the beginning, the writer must write to the least knowledgeable reader. This is the nonspecialist, especially the manager unfamiliar with and uninterested in the details of a particular project.

The Foreword or Introduction

The Foreword orients the reader by identifying and defining the problem being addressed in the memo or report, as generally discussed in Chapter 5. As with other types of problem-defining introductions, the purposes of the Foreword are

1 To catch the audience's attention: to place the report in a context so that the reader can see how it fits in with other communications and with overall goals, to tell *why* the project was done and *why* it is important, to define what problem was addressed

2 To quickly orient the audience to the subject of the report: to define *what* was done, to indicate the missing information the report provides (these are often covered with an explanation of the writer's technical assignment)

3 To define the purpose of the report, to predict what kinds of information the report will present: for example, to tell the reader whether this is a proposal to solve the problem identified, an evaluation of someone else's solution, a cost study of a solution proposed earlier, or some other kind of information. (This information will restate information in the "Subject" line.)

For illustrations of Forewords and the way they fulfill these three purposes, you should review the many Forewords presented and discussed in Chapter 6. Here are two of them:

Foreword 1
During my visit to your nursery on September 15, 1989, you and I observed the high mortality of the 2-year-old red pine seedlings in bed 19. As you stated, this would lead to a production shortage in 1990. You requested my help in finding the cause of this problem so that it could be corrected in the near future. The purposes of this report are (1) to explain that parasitic nematodes are the probable cause of the high seedling mortality and (2) to recommend a solution to the problem.

Foreword 2
Slowdowns and necessary repairs have occurred in the incineration area of the Sewage Treatment Plant, costing $2 million last year alone. These problems may have occurred because of inaccurate data on the amount of sludge cake being produced and delivered to individual furnaces. Thus, Dennis Moore, the Chief Production Engineer, requested that I investigate our data on sludge cake production and suggest improvements in monitoring it if those are necessary. The purpose of this report is to document the unreliability of our present monitoring technique and to recommend a better one.

Although Chapter 5 has amply discussed the first two purposes of the Foreword, we would like to expand here on the third: defining the purpose and focus of the report. Consider the sample Foreword which appears in Figure 12-1 along with the rest of the report which it introduces. As expected, the Foreword sets up a problem, the Summary identifies the solution to the problem, and the Details or Discussion section provides the detailed support for the proposed solution.

However, notice that the problem set up in the Foreword drives the structure of the rest of the report. The problem is cast as the need for a system which meets two criteria—performance effectiveness and cost superiority. These criteria become two major structures around which the whole report is organized. To stress their importance, they are highlighted by appearing in the indented list. These criteria drive the organization of the Summary: the Summary is written to show that the aerated lagoon meets the criteria and meets them better than the other options on the basis of performance effectiveness and cost superiority. The indented list in the Summary highlights the meeting of the two criteria (in the same order as they appear in the Foreword). Finally, the two criteria drive the structure of the rest of the report, which provides details on performance superiority and cost superiority of aerated lagoons. An effective Foreword, then, really provides the driving force and overall organizational structure of the memo or report.

The Summary

The Summary follows the Foreword at the beginning of a memo or technical report. It operates with the Foreword to directly address the busy managerial or nonexpert reader. The purposes of the Summary are

1 To quickly present the main results of the project

2 To quickly present the important recommendations and implications of the project

The Summary provides a compact statement of results, conclusions, and recommendations to help managerial readers make decisions: it tells what the writer discovered, the implications of the discovery, and the recommendations for action based on the writer's special knowledge—all in terms that the nontechnical manager will find useful.

The purposes of the Summary and their relation to the information in the Foreword are summarized in Table 12-1.

Here are some of the general questions that a managerial reader will have about a project or study.

1 What is the project's *importance* to the company? Its *scope* of application?

2 What will it *cost*?

3 Are any *problems* projected because of it?

4 Are there any *implications* to the company from it? Any *more work* to be done? Any *resources* required or involved (people, facilities, equipment)? Any *priorities* to be added or changed?

5 Are there any important *dates* or *deadlines* associated with it?

6 Are there any important *recommendations* for future action which the writer could make on the basis of his or her special expertise?

Figure 12-1 ❑ SHORT INFORMAL REPORT: STRUCTURAL ORGANIZATION ARISING FROM THE FOREWORD, MOVEMENT FROM GENERALIZATIONS AND CLAIMS TO DATA AND SUPPORT

To: Ms. J. Jones, Director
Orange Grove Products
1135 Halifax Building
Orlando, Florida 32105

From: Leo Ming Chen, Environmental Engineer
Camp, Dresser and McKee Inc.
One Center Plaza
Boston, Massachusetts 02108

Date: October 29, 1989

Subject: Citrus Processing Waste: Treatment System Proposal

Foreword

In your letter of July 7, 1989, you asked me to suggest a treatment process for the wastewater from your new citrus processing plant. You stated that any treatment process selected should

1.   exhibit performance effectiveness under average and adverse flow conditions

2.   exhibit cost superiority in terms of initial cost and yearly spending

Consequently, I have compared several treatment alternatives using the data you have supplied and your criteria as a basis for comparison. The purpose of this report is to recommend a process for economically and efficiently treating citrus processing wastes.

Summary

An aerated lagoon is recommended as the most efficient and economical method for treating citrus processing wastes. Several treatment processes were considered in the selection. These include the activated sludge process, the anaerobic lagoon, and the aerated lagoon. The advantages of the aerated lagoon over the other treatment processes are as follows:

1.   The aerated lagoon is the only alternative which could meet the federal pollutions standards under adverse flow conditions. It exhibits significantly better performance under all conditions through more consistent BOD reduction and higher organic loading potential.

2.   The aerated lagoon affords significantly lower initial and yearly costs due to its ease of construction, operation, and maintenance. Per lagoon, the estimated initial cost is only $114,000 and the annual operating cost $22,800, approximately half as expensive as the more economical of the other two options.

1 Performance Superiority of Aerated Lagoons

Aerated lagoons consistently produce a better-quality effluent than do activated sludge processes or anaerobic lagoons. Aerated lagoons exhibit better BOD reduction and higher

Figure 12-1 ❑ *(Continued)*

organic loading potential under both average and adverse flow conditions than do either of the other treatment schemes.

1.1 Superior BOD Reduction by Aerated Lagoons

The standard for BOD, as published in the *Federal Register* of July 1, 1979, states that all discharges into receiving streams shall contain no more than 30 mg/liter of BOD. Table 1 shows aerated lagoons with 95% BOD reduction potential to be capable of producing an effluent in compliance with federal standards under both average and adverse flow conditions. Activated sludge processes and anaerobic lagoons, on the other hand, can only effectively treat wastewaters of average BOD values.

Table 1 BOD REDUCTION COMPARISON OF TREATMENT PROCESSES

Secondary Treatment System	BOD Reduction (%)	Effluent BOD of Average Strength Flow (mg/liter)	Effluent BOD of Maximum Strength Flow (mg/liter)
Aerated lagoon	95	9.5 (OK)	28.2 (OK)
Anaerobic lagoon	88	22.8 (OK)	79.2 (unacceptable)
Activated sludge	85	28.5 (OK)	99.0 (unacceptable)

1.2 Superior Organic Loading Potential of Aerated Lagoons

Seasonal shockloads typical of citrus processing plants are easily handled by aerated lagoons but tend to pose problems for activated sludge processes and anaerobic lagoons. Production within the plant will be a one-shift-a-day operation and may shut down completely on weekends and holidays. The volume of wastewater therefore will fluctuate through the harvesting season, which begins in October and ends in June. A brief performance comparison of how these three systems react to such periods of high loading will show the distinct advantage which aerated lagoons have over the other processes.

A. **Aerated Lagoons**: Aerated lagoons, because they are not continuous-flow processes, can store wastewater until it can be treated. These lagoons can therefore be designed to handle a shockload of unlimited quantity.

B. **Anaerobic Lagoons**: Anaerobic lagoons also exhibit the same storage capabilities as aerated lagoons. However, because of the nature of anaerobic organisms, obnoxious odors are produced at high organic loadings. Hydrogen sulfide and mercaptans are produced at loading values exceeding 240–320 kg BOD/1000 cubic meters, a value very typical of the size of plant in question.

C. **Activated Sludge:** Activated sludge processes can only tolerate organic loading values not exceeding 260–340 kg BOD/1000 cubic meters. Higher loading rates were followed by bulking primarily caused by Sphaerotilus organisms.

Figure 12-1 ❑ *(Continued)*

---

2 Cost Superiority of Aerated Lagoons

Table 2 shows that aerated lagoons cost less than activated sludge systems and anaerobic lagoons in terms of initial capital as well as operational and maintenance (O&M) costs. The initial capital and O&M costs are usually reflective of the amount of equipment required for each system. Aerated lagoons require only turbine-type aerators and therefore cost very much less to build and maintain than activated sludge systems, which need several settling basins and numerous pumps. Anaerobic lagoons, although not needing any equipment, have a higher operating cost than aerated lagoons because of the chemicals required in stabilizing the anaerobic bacteria. The estimated cost of these various systems is based primarily on average flow and BOD loadings. More detailed cost tables appear in the Appendix.

Table 2 COST COMPARISON OF TREATMENT PROCESSES

Secondary Treatment System	Cost (Thousands of 1989 Dollars)	
	Capital	Annual
Aerated lagoon	114.0	22.8
Anaerobic lagoon	212.0	44.8
Activated sludge	725.0	39.0

3 Conclusions and Recommendations

My objective in this report was to recommend a process for economically and efficiently treating the citrus wastes from the new Orange Grove Products plant. The use of an aerated lagoon is recommended if the ultimate process is to produce an effluent which consistently complies with federal standards. Such a system will also save initial capital as well as annual operational and maintenance costs.

A more detailed list of these questions has been prepared by Richard Dodge[1] and is presented in Table 12-2. Writers should try to answer these questions in the Summary because they are particularly qualified to do so. They often know more about the answers than anyone else. Further, they are hired to analyze and evaluate data, not just to present it. If they make no effort to analyze their project in the larger context of the organization, they are not really doing their job. Finally, by answering such questions for managers, writers can look exceptionally competent. Relatively few technical people properly address these issues, and those who do will seem unusually useful to the organization.

Table 12-1 ❏ OUTLINE OF THE FOREWORD AND SUMMARY
IN THE MEMO AND TECHNICAL REPORT

Foreword		Summary	
Purpose	Type of information	Purpose	Type of information
Define context of report to catch audience's attention; tell *why* project was done and *why* it is important	[A but] B problem statement	Present main results for managers	Results
		Present information needed by managers to decide and act	Recommendations and implications
Indicate missing information; define subject of report; define *what* was done	Assignment or question; criteria		
Define *purpose* and focus of report	Purpose and overview of contents of report		

What does a Summary that addresses these questions look like? For a sample, you might look at the Summary in Figure 12-1. It comes right after the Foreword, which has defined a problem: Orange Grove Products needs a method for treating wastewater from its new citrus processing plant, a method which meets criteria of effectiveness and cost. The Summary provides a solution: an aerated lagoon will meet both criteria better than other potential solutions. The Summary also gives general performance and cost data, the kinds of additional information a manager needs to make a decision. (A manager would not approve a solution which was ineffective or too expensive.)

Framing Summaries for Particular Audiences

Once you have defined the audiences for a report, you need to decide how to convey your information so that the audiences can understand and use it. This can be tricky when different audiences need different kinds of information or the same information presented in different ways.

For instance, consider the following situation. You are a geologist working for a township council composed of two businesspeople, one farmer, one schoolteacher, and one lawyer. The township has so attracted new industry and residents for the past 8 years that water shortages have developed and future growth will be impossible without an expanded water supply. You have been hired to locate, if possible, new wells to increase the township's water supply. After some effort, you have succeeded and must inform the council of your success. Which version of your results would be most appropriate for the township council?

Table 12-2 ❏ WHAT MANAGERS WANT TO KNOW

Problems

What is it?
Why undertaken?
Magnitude and importance?
What is being done? By whom?
Approaches used?
Thorough and complete?
Suggested solution? Best?
Consider others?
What now?
Who does it?
Time factors?

Materials and Processes

Properties, characteristics, capabilities?
Limitations?
Use requirements and environment?
Areas and scope of application?
Cost factors?
Availability and sources?
What else will do it?
Problems in using?
Significance of application to company?

Field Troubles and Special Design Problems

Specific equipment involved?
What trouble developed? Any trouble history?
How much involved?
Responsibility? Others? Company?
What is needed?
Special requirements and environment?
Implications to company?
Most practical solution? Recommended action?
Suggested product design changes?

Tests and Experiments

What tested or investigated?
Why? How?
What did it show?
Better ways?
Conclusions? Recommendations?
Who does it? Time factors?

New Projects and Products

Potential?
Risks?
Scope of application?
Commercial implications?
Competition?
Importance to company?
More work to be done? Any problems?
Required labor, facilities, and equipment?
Relative importance to other projects or products?
Life of project or product line?
Effect on company's technical position?
Priorities required?
Proposed schedule?
Target date?

From Richard W. Dodge, "What to Report," *Westinghouse Engineer* 22 (4–5):108–111 (1962).

Summary 1
Six test wells were drilled, four of which penetrated suitable sand and gravel.

1 The relationship of geologic units encountered in Test Wells 1 through 6 is interpreted and presented in Plate 1. Two aquifers were identified; an extremely thick aquifer was encountered in Test Wells 2, 5, and 6 ranging in thickness from 78 to 125 feet, and a deep aquifer was encountered in Test Well 1 extending from 150 to 185 feet. There is theorized to be limited hydraulic connection through fine-grained units between the two aquifers. Some degree of connection between the deep aquifer and upper aquifer was established by the rapid rise in water levels in Test Well 1 observed during a heavy rain in September 1989.

2 A conservative estimate of the yield from the upper aquifer is 2.0 mgd. Aquifer data indicate this could be produced by pumping any combination of two wells among Test Wells 2, 5, and 6 (assuming manganese in Test Well 2 proves acceptable). Ultimate production from a French Creek well field could prove to be in the 3.0 to 5.0 mgd range.

Summary 2
We conducted studies and six test drillings to locate an area in which a reliable, high-quality municipal water well field could be developed. We have located such a site.

Results: Four of the six wells drilled were productive and penetrated sand and gravel suitable for development of municipal wells. Each of the four productive wells showed very good water-producing capability, yielding 300 to 500 gallons per minute.

Conclusions: Production wells completed at the test well sites should more than supply the township's current and future needs. Analysis of test data indicates that such production wells could produce over 2 million gallons per day, twice the township's projected requirements. Further, if the water-bearing sand and gravel are as extensive as we project them to be, yields up to 5 million gallons per day could conceivably be developed.

Unless the reader is a geologist, Summary 1 will probably not make any sense. It contains unfamiliar terms such as *aquifer* and *mgd*. It also contains statements which sound as though they ought to convey information but don't. For instance, Section 2 states, "A conservative estimate of the yield from the upper aquifer is 2.0 mgd." Even if readers guess that *mgd* means *million gallons per day*, they still don't know if 2.0 mgd is a sufficient water supply.

In contrast, Summary 2 gives the information in a form that nonspecialists can understand. It explicitly states the generalizations the nonspecialist reader needs to know:

1 The geologist has found water.

2 The geologist has found enough water to supply the township's current and future needs.

Figure 12-2 ❏ SAMPLE REPORT FOR ANALYSIS

Date: June 6, 1989

To: Meg A. Lith, Manager
              Research Division
              Scoria Mining Company
              Menhir, MI 48003

From: Michael A. Newton, Research Assistant
              Phoenix Memorial Laboratory
              47-28 Technology Mall
              Atlanta, GA 30081

Subject: Sediment Sample Analysis: Report of Investigation

Foreword: On 14 May you requested an analysis of two sediment samples: TG-1 and TG-2. You were especially concerned about the amount of cobalt in each. In response to your request, we have performed a neutron activation analysis on each of the samples to determine the major constituent elements in each. This letter concerns the results of the analysis.

Details: The final analysis is based upon the collected results of two separate procedures: one for short-lived isotopes and one for long-lived isotopes. Although sample preparation and gamma ray spectra analysis for both of the procedures are identical, the process of irradiation is different. The process of analysis is detailed below.

1. Preparation: Sample preparation was performed by obtaining two 1-gram samples from each of the sediments. These samples were then sealed into individual quartz tubes for ease in handling and irradiation.

2. Irradiation: One sample tube from TG-1 and one from TG-2 underwent short half-life activation. The other sample tube from TG-1 and TG-2 underwent long half-life activation.

    2.1 Activation of Short-Lived Isotopes: Sample activation was performed by exposing the sample tube to a neutron flux for a period of 10 seconds. The sample was allowed to sit for a period of 30 seconds to ensure that it had decayed to a safe handling level. The sample was then available for analysis.

    2.2 Activation of Long-Lived Isotopes: Sample activation was performed by exposing the sample tube to a neutron flux for a period of 10 hours. The sample was allowed to sit for a period of 5 minutes to ensure that it had decayed to a safe handling level. The sample was then available for analysis.

3. Analysis: Sample analysis was performed by individually measuring each tube for gamma ray emission. The measurement was performed with a Ge(Li) detector in conjunction with a multichannel analyzer that was controlled by a PDP-8 computer. The resultant gamma ray differential pulse height spectrum was analyzed to determine the following:

    3.1 The elements present in the sample by referring to the photopeaks present on the spectra.

Figure 12-2 ❑ *(Continued)*

3.2   The concentration of each element present in the sample by referring to the magnitude of the gamma ray emission rate relative to a known sample's emission rate. (The concentration of each element in the known sample had been previously determined.)

Sample composition was then determined by combining the results of the short- and long-lived isotope analysis. Those results were reduced to include only the six major elements of each sediment sample.

Results: The final results of the analysis are shown below for the six major elements in each of the samples. These measurements represent the relative concentration of the elements to within 0.01 part per million.

	TG-1	TG-2
Na-23	31,600 ppm	30,200 ppm
Mn-56	763 ppm	260 ppm
Fe-59	18,800 ppm	46,100 ppm
Co-60	4.40 ppm	15.40 ppm
Sc-46	3.63 ppm	12.30 ppm
Pa-233	25.60 ppm	6.98 ppm

Note: ppm denotes concentration in parts per million.

If any questions arise as to the analysis of the samples, please feel free to contact me.

❑ EXERCISE 12-1

A   Write a Summary for the report shown in Figure 12-2. When you have finished, be sure to check your Summary for the important results, conclusions, implications, and recommendations.

B   Compare your written Summary with those written by a small group of your classmates. If you do not all have the same information in your Summaries, try to justify your choices to one another and then to produce a "Consensus Summary" from the group.

❑ EXERCISE 12-2

If you have recently written any other reports, go back now and check their Summaries for appropriateness and completeness. Since Summaries are so important, you should practice writing and rewriting them whenever you have the opportunity.

Figure 12-3 ❑ STRUCTURE OF THE SHORT INFORMAL REPORT

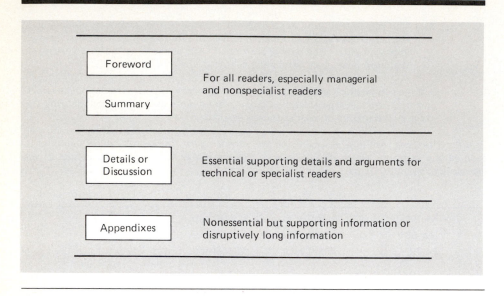

12.2 STRUCTURING PROOFS AND TECHNICAL DISCUSSIONS: ORGANIZING DETAILS FOR SPECIALIST OR INTERESTED READERS

As just discussed, a report has a short Foreword and Summary which set up a problem and give a proposed solution and cost in general terms for a manager or nonspecialist reader. They are followed by a Details or Discussion section which provides details to support the proposed solution. These are details which the manager could read or skip depending upon a need to know, but which a specialist would find essential in evaluating the argument. These essential details are followed by Appendixes which provide other supporting details: either (1) nonessential details, details which support the arguments in the Details section but which are not so central to the argument that they need to be there, or (2) such lengthy sections of details that they would disrupt the line of argument in the Details section if they were presented there. This structure is summarized in Figure 12-3.

Thus, the overall report structure moves from generalizations or claims (the Foreword and Summary) to support (the Details).

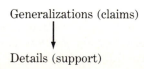

Generalizations (claims)

Details (support)

Figure 12-4 ❏ ORGANIZATION OF INFORMATION FOR EXPERT READERS

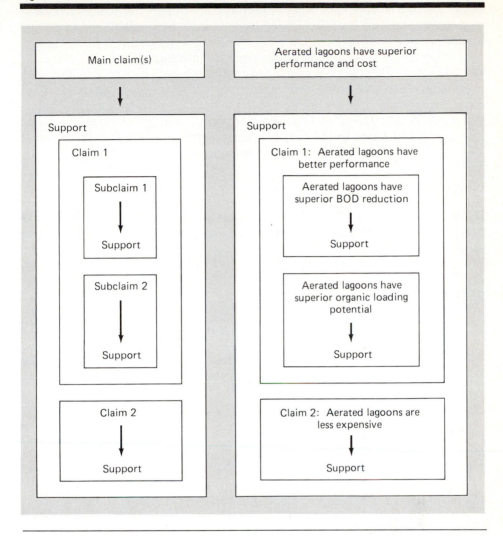

The section of the report giving details is written for experts, readers with the technical background and interest to follow the details and fine points of the technical argument. This section is usually called the Discussion in a long report and the Details or Discussion in a short report. Such a section is illustrated in the Details section of Figure 12-1.

As does the structure of the overall report, the structure of the Discussion and its individual paragraphs moves from generalizations and claims to support and details. For example, examine the first main section after the Summary in Figure 12-1: "Performance Superiority of Aerated Lagoons." This section begins with a heading and then a first paragraph that states the claim being made: Aerated lagoons have superior performance to activated sludge processes or anaerobic lagoons because aerated lagoons show better BOD reduction and higher organic loading potential. Notice that the main claim (*aerated lagoons have superior performance*) is visible from the heading alone.

The main claim is then supported by the subsections which follow, with each subsection proceeding from claims (or subclaims) to support. In each case, the subclaim appears in the subsection heading—"Superior BOD Reduction by Aerated Lagoon" and "Superior Organic Loading Potential of Aerated Lagoons"—and again in the first sentence of the section. This structure is summarized in Figure 12-4 on the previous page.

❏ EXERCISE 12-3

Rewrite the report in Figure 12-2 to strengthen the movement in the Details Sections from generalizations or claims to data and support. Don't forget to consider the headings.

REFERENCES

1   Richard W. Dodge has prepared an extended list of questions managers want answered for various types of projects. The list appeared in "What to Report," *Westinghouse Engineer* 22(4–5):108–111 (1962) and was based on information provided by James W. Souther, "What Management Wants in the Technical Report," *Journal of Engineering Education* 52(8):498–503 (1962).

ADDITIONAL READING

J. C. Mathes and Dwight Stevenson, *Designing Technical Reports* (Indianapolis: Bobbs-Merrill, 1976).

Henry Mintzberg, "The Manager's Job: Folklore and Fact," *Harvard Business Review*, July–August 1975, pp. 49–61; *The Nature of Managerial Work* (New York: Harper & Row, 1973).

13

Memos, Short Informal Reports, and Progress Reports

As noted in Chapter 1, experienced scientists and engineers rate shorter types of technical and professional communication as being among the most important for success on the job: memos or short informal reports, including progress reports. The memo or short informal report is a one- to four-page document, usually single-spaced and usually written to someone within the writer's own company or organization. It may note the existence of a problem, propose some course of action, describe a procedure, or report the results of a test or investigation. The memo may take the form of a letter or a short informal report. The organization and format of letters are discussed in Chapter 11, and those of the short informal report are discussed in this chapter. The term *informal* here does not mean sloppy or casual or carelessly done; an informal report should be as carefully prepared as possible—thoughtfully written, neatly typed or word-processed, and thoroughly proofread for errors. It may be short or long; short informal reports (or memos) will be discussed in this chapter, and long informal reports are discussed in Chapter 15. The term *informal* refers to the format in which the report is presented, and that will be described below. As with other informal reports, progress reports may be long or short, depending on the size of the project being discussed. Short progress reports will be discussed in this chapter as a type of short informal report. Long progress reports are a type of long report and are treated in Chapter 15.

13.1 THE STRUCTURE OF MEMOS AND SHORT REPORTS

Memos, short informal reports, and progress reports all share the basic features of reports discussed in Chapter 12. They are all read by a variety of

249

Figure 13-1 ❑ FOREWORD, SUMMARY, AND DETAILS SECTIONS OF A SHORT INFORMAL
REPORT; MOVEMENT FROM GENERALIZATIONS AND CLAIMS TO DATA AND SUPPORT

Watson Telephone Laboratories, Inc.
Lincoln Hill, NJ 07123

To: Dr. J. C. Smith
         Manager, Low-Energy Research Group

From: Catherine Doe
         Mechanical Systems Engineer
         Positron Research Group

Subject: Positron Beam Experiment: Reducing Delays

Date: February 5, 1989

Foreword: In your letter of January 23, 1989, to Dr. D. Newman, Director of the Positron
Research Group, you expressed concern about the numerous delays in the positron beam
experiment. Dr. Newman attributes most delays to the 10 hours required to pump the
vacuum system down from atmospheric pressure to an acceptable operating level and has
asked me to investigate means of reducing these delays. This report recommends a solution
for reducing the delays.

Summary: The purchase of a gate valve for the Vac-Ion pump was found to be the
optimal solution to the delay problem in the positron beam experiment. Addition of the
gate valve will reduce evacuation time to 3 or 4 hours, cost no more than $600, and present
no installation problems. The use of a larger fore pump was seen as the only alternative
approach, but this would require extensive modification of the vacuum system, cost
approximately $1100, and give no improvement in the evacuation time.

Details: The excessive time required to obtain a suitable vacuum, 10^{-7} torr or less, is
primarily due to contamination of the ion pump. When the vacuum system is opened to
the atmosphere, the entire inner surface of the system and the ion pump itself become
contaminated. The contaminated surface then outgasses considerably upon evacuation,
keeping the pressure high for more than 10 hours. The two possible solutions to this
problem are to avoid the contamination altogether or to decontaminate more quickly.

The proposed solution would avoid contamination by the installation of a gate valve on
the Vac-Ion vacuum pump. This would allow the ion pump to remain in operation even
while the rest of the system is open to the atmosphere, thus preventing contamination of
the pump. This would be a major improvement, since only the relatively small surface
area of the rest of the vacuum system would become contaminated, and evacuation time
would be reduced to 3 or 4 hours. A suitable gate valve costs $500 to $600.

The rejected alternative, the purchase of a larger fore pump, would speed the decontami-
nation of the vacuum system, but cost too much. Replacing our present 120 ft³/min fore

Figure 13-1 ❑ (*Continued*)

pump by an 800 ft³/min pump would reduce the evacuation time to about 4 hours. However, an 800 ft³/min fore pump costs approximately $1100, and its installation would require substantial modification of the vacuum system.

Thus, I recommend the purchase and installation of a gate valve as the best method for reducing delays in the positron beam experiment.

If you have any questions, I would be happy to answer them.

---

audiences: by technical experts in the field as well as by managers and other nonexperts—the busy, inattentive, and perhaps uninterested readers described in Chapter 3. Thus, the memo or report should be accessible to all these readers and responsive to their needs and reading habits.

As part of their accessibility, all these reports share the basic structural features described in Chapter 12 and summarized in Figure 12-3. They all have a Foreword (the statement of the problem) and Summary (the main results and other important information a manager needs to know) written for managerial and nonexpert readers. (A short company report to a small, knowledgable audience sometimes may combine the Foreword and Summary into one section which may be either unlabeled or labeled with a local label such as Overview.) All of the reports move from general to particular throughout the report—that is, they order generalizations or claims before data and support. Thus, short reports are organized so that a busy managerial reader (a reader who wants to read no more words than are absolutely necessary) can stop after the Foreword and the Summary. The Discussion or Details section gives the extra information needed by technically involved readers: support for the claims in the Summary or extra details needed to implement or fully understand the solution proposed.

To see how a short report illustrates these features, examine the report in Figure 13-1. As did the reports in Chapter 12, the main text of this report begins with a Foreword, which poses a problem: there have been frequent delays in an experiment because of the excessive time required to pump down a vacuum system. Since these delays are undesirable, the writer has been asked to reduce them. This Foreword is followed by a Summary which reports the writer's solution: buy a gate valve for the vacuum pump. The Summary also tells why this is a good solution—it will reduce evacuation time to acceptable levels, cost no more than $600, and present no installation problems. Notice that these advantages address the manager's concerns. Before a manager could approve the purchase of the gate valve, she or he would need to know the cost (only $600), the benefits (reduction of delay), and the implications (no installation problems). In addition, the phrase "no installation problems" implies other useful information: that there should be no more problems with the unit and no unusual work or resources required.

13.2 THE FUNCTIONS OF MEMOS AND SHORT REPORTS

The view of reports and memos in Chapter 12 stresses their persuasive nature—that they are basically arguments of fact or policy (described in Chapter 4) which the writer is presenting to the reader. For instance, a progress report claims that the writer (or the writer's group) has made certain progress and then supports that claim with evidence and documentation of the progress. A final report presents the results of an investigation (the writer's claim about some state of affairs) along with the data to prove that the investigation was done carefully and appropriately—that is, with the proof that the writer's vision of "reality" is true or defensible. In all the reports/memos in Chapters 12 through 14, the writers are making arguments of fact or policy and providing support.

However, the writers do more than provide support for current readers. They are also providing crucial documentation for future readers, for those who may need to continue, extend, test, or defend the writers' work. This function is crucial in all reports but is especially pressing in the progress report. Progress reports are written before the entire project is completed. The documentation of what has been done is important, since the only other sources of information about current work on the project are usually notes and workbooks in a form only the writer can understand. If something happens to the writer—an accident, an untimely death, a job change—then those who must continue the project need to have the earlier work available in a usable form. Thus, the progress report becomes a convenient vehicle for spelling out conclusions and for assembling data at regular intervals. If the writer has finished some library research and discovered something important for the project, she or he should spell out the discovery and provide a relevant Bibliography. If the writer has designed a part or developed a process flowchart or derived an important mathematical relationship, that work should be documented by including the relevant material either in the body of the report or in an Appendix. That work shouldn't have to be done again if something should happen to the writer.

Short reports can serve a wide variety of functions, as illustrated by the short reports in Figures 13-2 through 13-6. The two reports in Figures 13-2 and 13-3 serve complementary functions and provide two views of the origin of a report addressing an organizational problem. The memo report in Figure 13-2 illustrates a response to a request for problem solving. In this situation, someone other than the writer (here a supervisor) has identified a problem and requested the writer to deal with the problem: someone other than the writer has initiated the situation to which the writer is responding. In the response, the writer outlines the problem generating the request for action in the first paragraph, recommends a solution to the problem in the second paragraph (the unlabeled Summary paragraph), and provides some arguments to support its recommendation in the rest of the report. (The attachments mentioned in the report are not included in Figure 13-2.)

In contrast, in the report/memo in Figure 13-3, it is the writer who identifies a set of problems which no one else has noticed before (a drain of the time of usability specialists) and who thus initiates the writing of the report. The report first outlines the problems and proposes a solution: hire a usability researcher. The report then goes on to provide the information that a manager would need to approve and process the request without having to wait for another memo on the duties and qualifications of the new usability researcher: the report provides a job description for the new position and a list of the qualifications needed by the new researcher. Thus, this report is a complete little package totally initiated by the writer: it identifies the problem, identifies the solution, and identifies the information needed to implement the solution. It does everything but actually hire the applicant!

The report in Figure 13-4 is a short test report which gives the results of a series of tests, along with the methods and criteria used in carrying out the tests. Such test reports are important because they provide empirical data for managers to use in making often difficult decisions with technical and financial implications. They also provide an important source of material for lawyers in product liability suits who are trying to justify (or attack) the company's decisions or for managers who are reviewing earlier decisions. Unfortunately, test reports are often written so frequently and quickly that they become compressed into the "bare bones" of the results, without adequate context setting or specification of methodology or interpretation of the test results needed by later audiences.

Figure 13-5 presents a self-assessment written at the end of the work year. It provides at least some of the information which the writer's supervisor will use in evaluating the writer's performance for promotions and raises. According to the writer,

My strategy in writing this was to inform my supervisor of the kinds of work that I had been doing during the past year. I had the feeling that my supervisor was not really aware of what I had been doing and, because her appraisal of me is important to my future, I wanted to make sure she was not missing something.

By the way, the rating scores that my company uses mean the following: The scale is 1–5. A 5 rating is practically superhuman. A 4 is above the average (consistent 4 ratings put someone in line for a promotion). A 3 means you've done a good job, but not really out of the ordinary. A 2 means there is a deficiency, and a 1 means you'd better watch out because you could lose your job real soon.

Figure 13-6 presents a trip report whose purpose is to document that the writer went on the trip and is justified in having certain expenses paid. Trip reports also allow writers to tell their organizations what they saw, concluded, or recommended as a result of the trip.

❑ EXERCISE 13-1

Compare the reports in Figures 13-2 and 13-3 for similarities and differences in types of information and relative amounts of space devoted to various sections.

MICROSOFT MEMORANDUM

To: Tandy Trower, Head, Usability Testing Department

From: Mary Dieli, Usability Test Designer

Re: The Gratuity Software Policy

Date: 16 November 1988

Dist: Paul Moore, Usability Test Designer

Encl: Attachment 1, Current Gratuity Software List
              Attachment 2, Promotional Software
              Attachment 3, Gratuity Software Breakdown

There have been complaints within Microsoft about the current usability test gratuity software policy. As you requested, the Usability group has reviewed this policy and reevaluated it. The purpose of this memo is to provide (1) a review of the purpose and spirit of the gratuity software policy, (2) a review of the current policy, the problems in it that you mentioned, and a proposal that deals with those problems, and (3) an alternative proposal.

Our recommendation is that we continue the current gratuity software policy with the following modifications:

1) Use a modified list. For a list of products we will offer, see Attachment 3: Gratuity software breakdown.* (The list also shows products we will not offer.) We have listed the product prices for your information only; we would remove them before giving the subjects this list. Columns A–E refer to categories I explain below under Proposal 2.

2) Give out only promotional software. We will ask subjects to list their top three software packages; that way, if a promotional copy of their first choice is not available, we can substitute their second choice, and so on.

The purpose and spirit of the gratuity software policy

Our usability test pool ideally consists of a list of people who, among them, represent the varied population of our target market. To ensure unbiased and useful testing, we don't want to be limited in our subject selection to only certain sections of the market. If we discontinued the gratuity software policy, we don't think the subject pool would completely dry up. However, we do think certain groups of people in our target market would not participate (e.g., busy professionals).

We offer gratuity software as a "thank you" rather than as a payment for services rendered. We see that it has a positive public relations effect and imagine that it, in the long run, increases sales, since use of one Microsoft product often encourages use of another. (It is also important to mention that the internal cost of these products is low.)

---

*This list is intended for use during our standard usability tests, where subjects come to Microsoft singly to participate independently in a usability test of limited duration.

Figure 13-2 ❏ *(Continued)*

Proposal 1: Maintain current gratuity software policy, with modifications

Currently, when subjects participate in a usability test, we offer them a choice of some Microsoft software or books. Although initially there were no restrictions on what was on the list, this summer we did modify the list on the basis of two loosely-defined criteria: (1) limit the list to applications, and (2) don't offer anything "too expensive." (See Attachment 1: Current gratuity software list.)

Potential problems and our responses to them. You and I discussed three potential problems with that policy. Below I list those problems and our responses to them, which we propose as modifications to the current policy.

1) One problem was that there is some concern in other parts of Microsoft that Usability is giving away "free" software. To help mitigate this concern, we suggest giving subjects promotional software only. We have researched this option and found that it is feasible. (See Attachment 2: Promotional software.) Furthermore, we think it is consistent with the spirit of the gratuity software policy: promotional software, although a complete package, is intended for promotional purposes only, not for sale.

2) Another problem we discussed is one of seeming inequity: because a subject's choice of product is in no way limited, one subject may choose a product that retails for, say, $99, while another subject in the same test may choose one that retails for $495. We've discussed this at length and don't see it as a problem because subjects can choose the software they want including software with a high retail value.

3) The third problem we discussed is that if we were paying people cash for their participation in these tests, we would not pay them $495. This is true, although often our target market contains high-level professionals whose time is valuable in the monetary sense. However, since the spirit of the gratuity software policy is not payment for services rendered, and since it is not presented to subjects in that way, thinking about it as such is inappropriate.

Proposal 2: An alternative gratuity software plan

In our discussions, the following plan emerged as our second choice: categorize software according to retail price and then categorize usability tests similarly according to time they take and level of subjects' effort involved. (In this plan, we would also offer only promotional software.) Columns A–D on Attachment 3 represent our first attempt at categorization. Proposal 2 is to offer subjects a choice of products from Columns A and B only for a shorter/easier test and from Columns A–D for a longer/more difficult test.

Categorization has its own problems, some of which are listed below:

1) It is not easy to categorize usability tests. The extremes (e.g., very long or very short) are easy; it's the middle areas that are difficult.

2) What is now a positive public relations activity may turn into a negative if subjects are told they can't have a certain product that they want or that they know their colleagues got when they participated.

3) The subject pool may become biased and/or limited if subjects will only sign up for certain tests on the basis of their product choices.

So, again, we recommend Proposal 1.

Figure 13-3 ❑ SAMPLE SHORT MEMO/REPORT: IDENTIFICATION OF A PROBLEM AND PROPOSAL SOLUTION
(Used with permission of Microsoft Corporation.)

MICROSOFT MEMORANDUM

To: Tandy Tower, Head, Usability Testing

From: Mary Dieli, Usability Test Designer

Re: Usability Group Position Needed

Date: 14 November 1988

Dist: Usability Testing Department Personnel

During the last few months, I've noticed some problems in the usability test process that we have not yet addressed. One problem is that, at the beginning of each test, the usability specialist needs to spend time researching whether or not the usability test questions at hand have been answered before, either at Microsoft or elsewhere. Even when the research shows that the question has been answered and we learn something useful, that activity is not the best use of the specialist's time.

Another problem is that we get a steady stream of requests for information about usability, like the following:

"Do you have any studies showing that users tend to use online materials more as they become more accustomed to them?"

"Just recently, I had a writer express some concern about one of our book designs. My question is one of usability. Do you have or know of data that might address the user's reaction to page number placement?"

"I'm also interested in readability research (the Fog Index, and so on). If you could simply suggest a few directions to pursue, that would be very helpful."

Right now, we share responsibility for responding to such requests, but it takes time away from the scheduled usability tests.

A third problem is more of an omission than a problem: we are not pulling results from a series of tests into periodic reports about what we have learned about any particular product across time.

Recommended Solution

Therefore, I'd like to propose a new position for the usability group: *Usability Researcher.* The rest of this memo outlines the purpose of the new position, its primary activities, and the qualifications we would seek in applicants for the position.

Figure 13-3 ❏ *(Continued)*

The purpose of this position would be:

1) To save time by making sure the answers to the usability questions we are asked to test haven't already been answered in the literature or by other usability groups.

2) To allow usability specialists to stay focused on the tests on which they're working by dealing with Microsoft's requests for research.

3) To periodically compile results of tests into technical reports for dissemination within Microsoft. This would also help educate the Microsoft community about usability, about the types of issues we deal with, and about our process.

Primary activities

1) Do pretest research of usability issues. Use the research results (a) in place of some usability test (if the usability question has already been answered) or (b) as the basis for some usability tests (if the question hasn't been answered).

2) Develop a database for entering usability test data and tracking results.

3) Design a technical report series. Group existing reports, summarize results, compile technical reports, and distribute them.

4) Follow up on and disseminate "research request" results.

Qualifications

- Extensive research experience, particularly focused in library science or a related field
- Excellent writing skills
- Knowledge of usability testing methods and practices
- Knowledge of Human-Computer Interaction (HCI) issues
- Experience in a corporate environment
- Familiarity with personal computers (e.g., Macintoshes, IBMs)
- Ability to solve problems quickly and independently, with a minimum of instruction
- Good interpersonal skills

In the interim, while we are waiting to get approval for this position, I would like to see if I can get some help from some University of Washington graduate students. Obviously, the person in this position will be more useful as he or she is here longer and learns the history of Microsoft usability. Help from UW would just alleviate the load temporarily.

Figure 13-4 ❏ SAMPLE SHORT MEMO/REPORT: TEST REPORT

INTERNAL TEST REPORT FORM

Date: July 18, 1988

To: A. M. Phillips

From: S. Mason

Subject: Snow Intrusion Testing: Air Induction System

PURPOSE

This memo documents development testing of air induction snow intrusion testing done on 1950 and 1275 engine cars. The cars studied included the 1989 Devil Phase "C"; the 1989 Star Phase A; and the 1989 Raider 1950 Phase A. It lists our conclusions, followed by the general test procedure, criteria for passing the test, and test results.

CONCLUSIONS

1. As currently released, the 1989 Devil Phase C fails snow intrusion testing. It will pass minimum snow intrusion requirements with the addition of the left radiator side baffle from the 1988 Devil. Snow intrusion would be virtually eliminated by adding a closeout panel above the headlamps.

2. The 1989 Star 1275 Phase A fails the snow intrusion test. It passes the test with the addition of a simplified thermac system or with revision of the air intake location to above the tie bar.

3. The 1989 Raider 1950 Phase A passed with no changes. It ingested virtually no snow into the induction system.

GENERAL TEST METHOD

These tests were run at our facility in Toledo, Ohio, in February of 1988. The test uses a "lead" vehicle (usually a pickup truck) which drives through the loose snow on the test track, making a cloud of snow which the test car drives in. The test car follows the lead car as closely as possible at a speed of 30 MPH. This simulates following a truck in a snowstorm, but is more severe due to the very close following distance. The test lasts for 1 hour or until the test car loses power because of snow packed in the air induction system. The induction system is checked every 15 minutes for snow buildup, but is *not* cleaned out.

TEST CRITERIA

The goal for our designs is to have no snow get into the induction system during this test. However, the design is considered acceptable for production if the test car can follow the lead truck for 1 hour at 30 MPH.

Figure 13-4 ❑ (Continued)

1989 DEVIL PHASE C PASSES WITH CHANGES

The 1989 Devil ingested virtually no snow when it was modified by adding a close-out panel above the headlamp area and by installing a left radiator side shield. (Radiator side shields are already used on the 1988 Devil.) Without the headlamp close-out panel, cars do pass the test, but the air cleaners are half full of snow. Without the radiator side shields, they fail the test at 45 minutes (3/4 through the test). These two changes will be implemented for production by Pete Ring at our request. The radiator side shield will make SOP [standard operating procedure], and the headlamp close-out panel will be an already approved change.

1989 STAR 1275 PHASE A PASSES WITH CHANGES

The 1989 Star 1275 Phase A fails the test in unmodified condition, but passes with either of two changes. The first alternative is a simplified version of thermac. At the end of the test, this version had about 1 cup of water and slush in the air cleaner. The second alternative is a different air intake location. With the inboard air intake moved from "through the tie bar" to "over the tie bar," the vehicle had no snow intrusion into the air cleaner. Both of these alternatives will require further study. Thermac is expensive and adds complexity. The air intake revision will affect water and dust intrusion. Both will affect engine air temperature and possible air meter tables. Some change will be required, because without it the car failed the test 30 minutes into the testing period.

1989 RAIDER 1950 PHASE A PASSES WITHOUT CHANGES

The 1989 Raider 1950 Phases A passes without changes. It had only trace snow intrusion into the air cleaner at the end of the test.

Figure 13-5 ❏ SAMPLE SHORT MEMO/REPORT: END-OF-YEAR SELF-EVALUATION

TextProduction INTERNAL MEMORANDUM

To: Carol Dunn, Dearborn Manager

From: Craig Hurst, Technical Writer

Date: December 8, 1988

Subject: 1988 Performance Review/Self-Appraisal

Goals

Nearly every day that I have worked at TextProduction since joining the company has been devoted to the Product A project. Thus, my goals have been nearly identical with those of that project. These goals have been

1. To learn enough about Product A that I could rewrite and reformat all Jones Agricultural Products documentation according to TextProduction standards. In particular, I am responsible for Manual 1 and Manual 2.

2. To rewrite these manuals, where necessary, for user-friendliness, accuracy, and overall quality.

3. To edit these manuals for brevity and reduce page count.

The project is due to be completed near the end of the first quarter of calendar year 1989. At this point, all three of the above goals are well on the way to being reached.

In addition, I have had the following personal goals:

4. To work on the problems of reducing page count and of integrating online and print documentation.

5. To contribute to a positive working environment.

Though I have had virtually no time to devote to the fourth goal, I have participated in discussions with other members of the TextProduction documentation staff about this problem. I have worked on a proposal with others for a seed project related to this goal, though I know that it is unlikely that I will have the time to participate in such a project in the foreseeable future.

A General Rating

I believe that I have contributed significantly to the improvement of Product A documentation and will, given the opportunity, contribute even more to company and departmental goals. I give myself a general rating of 4.

Figure 13-5 ❏ *(Continued)*

Objective 1—Learning about the product

My objective was to understand the material contained in the Product A manuals and to obtain some proficiency in using the product. I rate myself a 4 on this objective.

I have gained significant knowledge of Product A. This knowledge translates to an ability to write database queries to test the truth of a particular statement in Manual 1. In addition, I can contribute to a discussion of the language and understand its place in Product A.

I have not learned as much about the C programming functions and macros contained in Manual 2 as I would like. Mostly, this is due to significant time constraints during the project when attending C programming classes became impossible.

I did contribute significantly to the product education of others by organizing a class taught by an expert trainer. The class went very smoothly. In preparation for this class, I procured equipment (computer chips); found, scheduled, and collected audiovisual equipment; arranged for technical support to install software; and attended to other details.

Objective 2—Editing and completing a first draft of Manual 1 and Manual 2

I rate my performance a 4 on this objective.

I completed a first draft/rewrite of Manual 1, meeting the deadline for all chapters and very significantly increasing readability and coherence of this manual. With Manual 2, I reformatted the manual and made mostly minor editorial changes owing to severe time constraints.

As part of the editing process, I fought to obtain the electronic files for these manuals as soon as possible and suggested strongly the abandonment of the piecemeal approach of applying minor comments to the manuals for Jones Agricultural Products and Carson Medical Supplies.

I set a high standard for the types of changes that needed to be made, and implemented them in Manual 1. I succeeded in improving its coherence, cohesion, usability, and organization. I believe that the type of changes I made had a positive effect on our editor, who also abandoned the piecemeal approach and began to make serious improvements in clarity and organization. Some sections have been cut out and wordiness has been reduced.

I worked closely with the Graphics Department in coming up with approximately 20 diagrams to be used in Manual 1.

Comments on this manual from Joe Smith were that "it was the best introduction to databases that he has seen." On the level of style, Sally Jones, our proofreader, said that Manual 1 was stylistically better written than many other manuals that she had proofread at TextProduction.

Figure 13-5 ❏ *(Continued)*

Objective 3—Reformatting according to A-T conventions

I rate myself a 4 here.

Although the bulk of the work of setting the notational conventions was done by the editor, there were occasions when I made suggestions to solve a particular problem. For instance, I suggested a solution to the problem of how to represent in-text statement variables which was implemented.

In addition to my participation in the many discussions of notational conventions and style regarding Manual 1, I also had to format Manual 2. In conjunction with the editor, I determined that this manual would require different notational conventions from the other three Product A manuals. Little time was allotted for editing and formatting this manual, but I did manage to do both on schedule.

I coded both manuals for their first draft in a very short period of time.

Objective 4—Completing a technical review

I rate myself a 3 here.

A technical review of Manual 1 has been completed. I entered over 600 comments into the technical review database on this manual and conducted a meeting on the basis of these comments. I keystroked the entire manual, in addition to the keystroking done by the Technical Support and Test groups. I believe that this manual has been thoroughly reviewed, though there are still issues that I would like to address, given the time.

Manual 2 is in the process of being reviewed and a technical review meeting is scheduled before the end of 1988. Technical reviewers are compiling programs from electronic files I have supplied. Significant programmer's knowledge of C is required to adequately review this manual, and this knowledge is lacking among some of the reviewers. On the whole, I would like a more thorough review of this manual, but it has received less attention than any other manual owing to time constraints.

Objective 5—Contributing to a positive working environment

I rate myself a 4 here.

Generally, I feel that I have established excellent and pleasant working relationships with nearly everyone on this project, including members of the Test, Support, Graphics, and Editorial groups. These positive relationships create a beneficial environment for the exchange of information so necessary to producing a good product.

Figure 13-6 ❏ SAMPLE SHORT MEMO/REPORT: TRIP REPORT
(Used with permission of Intel Corporation.)

INTERNAL CORRESPONDENCE
ISO TECHNICAL PUBLICATIONS

TO:	**ISO Tech Pubs Manager**	**DATE:**	**May 20, 1987**
FROM:	**Marcia Petty, x2316, EY2-06**		
RE:	**Trip report from STC conference in**		
	Denver, 5/10 - 5/13	**CC:**	**List**

The Value of a Professional Conference

This is a brief trip report to discuss the general value of sending an Intel employee to a professional conference and the specific value of sending a writer, editor, or illustrator to the annual conference of the Society for Technical Communication. I just returned from this year's conference in Denver; as I summarize my conference experience, I see value returned in the following ways:

- Renewed sense of the value of our work
- Shared ideas
- Renewed enthusiasm
- Reminder of individual responsibility

Renewed sense of the value of our work

Sometimes individuals feel isolated and frustrated as the pressures of engineering and marketing schedules force us to accept lesser documentation quality. But a gathering of peers reminds me of the value of our communication interface between Intel's products and the customers. In light of the continual fight for professional recognition that writers have within Intel, it's good to remember that we really do make a contribution!

Shared ideas

Getting together lets us share ideas and solutions. Many of the problems we face, others have already faced and solved. From the sharing, I bring back ideas to help our department. There's always a balance: we've done things other people haven't, others have done things we haven't. It's particularly valuable to pick the collective brains of the huge IBM contingent. IBM funds a lot of communication R & D; the rest of us, lacking such funding, can at least listen and ask and learn from the best.

Figure 13-6 ❏ SAMPLE SHORT MEMO/REPORT: TRIP REPORT
(Used with permission of Intel Corporation.) (*Continued*)

(Conference value, continued)

Renewed enthusiasm	My professional enthusiasm gets renewed from attendance at an STC conference. Sometimes, slogging along day-to-day, I lose sight of fresh horizons, the new and different. For a fresh perspective, you need a bit of professional distance from your work occasionally. A vacation is great, but it doesn't necessarily bring new ideas about work; we go on vacation to forget work, not improve it.
Reminder of individual responsibility	This should go without saying, but the conference reminds me of the following: each writer bears individual responsibility for continuing to train and improve in writing and graphics. If you want to improve, you must expend effort and show commitment toward polishing your own communication skills. Read the experts and ask and observe. Our department is so small, with each of us heavily loaded with our own projects, that no one can serve as a 100% editor for others' work.
How did I choose what to attend?	There were over 150 sessions to choose from during three days. To help those who might attend future conferences, I'll list the questions I asked myself when choosing:

- Is the topic relevant to my department's needs?

- Is the session designated for experienced people?

  I have learned from past conferences that sessions labeled "novice" and mislabeled "mixed" are indeed valuable for novices, and sessions labeled "experienced" are more relevant for experienced communicators. They're not kidding when they assign the labels.

- Is the session paper available in the *Proceedings*?

  If my selection process left several alternatives for a time slot, I tried to attend a session for which no paper was available. This didn't always work out; for example, most workshops didn't have published papers, but were so popular that they were hard to get into.

Figure 13-6 ❑ SAMPLE SHORT MEMO/REPORT: TRIP REPORT (*Continued*)

Insights and Examples	The rest of this report presents insights and examples from sessions I attended.

Usability–is the item physically easy to read?	I think the terms "usability" and "readability" often get used interchangeably. For the sake of this report, I'll define "usability" to mean that the item is physically easy to use and is useful for the reader's circumstance. That is, a quick reference guide is pocket-sized for a repairman; a keyboard template is provided for machine-independent S/W applications; or literate graphics are provided when words would be too slow for the task (the exit instruction card on a airplane, for example).
Page layout, hanging heads	Usability is improved through the use of header columns with "hanging heads" that make it easy to skim the material quickly (this report uses hanging heads). In the competition, approximately 70% of award-winning user, training, and reference manuals used hanging heads.
Convenient spiral binding	For user, training, and installation guides, plastic spiral binding made quick-reference books easy to spread flat. One speaker stated that we should use **white** plastic so the reader's eye movement isn't disturbed by a strip of color down the gutter.
Cover over spiral binding provides spine for title	With a plastic spiral binding, include a cardstock cover that covers the binding from the outside. This way the reader not only has a manual that lies flat, but also has a manual spine with the title on it for quick reference.
Related items on facing pages	To repeat a universal truth: as much as possible, have text and tables or text and illustrations on facing pages. One manual in a 3-ring binder even had dotted line arrows leading across the gutter from text on the left page to a screen illustration on the right page.
Narrow gutter	In the competition, downsized manuals tended to have a very narrow gutter margin, 3/4" or less. This gives a wider text column and thus allows more information per page, meaning less page turning for the reader. Let's face it, downsized manuals already have special formatting constraints.

Marcia Petty, 5/20/87, STC Trip Report   Page 3

Figure 13-6 ❑ SAMPLE SHORT MEMO/REPORT: TRIP REPORT (*Continued*)

(Usability, continued)

Commands organized alphabetically	Some communicators recommend that in operating system reference manuals, we list commands alphabetically from beginning to end, not in separate sections by type of command. This way, the person who doesn't know the type can look up the command by name.
Some redundancy is O.K., she repeated	Include some redundancy. Much as we try for brevity and conciseness, sometimes it's better to include a piece of information twice rather than make the reader search for it.

Readability–is the item easy to comprehend?	For this report, I'll define "readability" as a measure of how easy it is for the reader to comprehend the material, both text and graphics.
Job aids	The easiest document for a user to understand may not be a manual. One speaker gave an excellent argument for "job aids" that are not full-blown manuals. Examples are keyboard templates, wall posters of important warnings, and quick reference, downsized guides. The point is that good readability means giving the magical right amount of information...not too much, not too little. A job aid may have just the right information for the task in hand; a manual may result in overload. (Note that for two years we've been writing Technical Updates and 6-8 page installation guides to supplement manuals; I guess you could call these job aids.)

One competition manual had a tear-out, cardstock reminder list for "Managing Your System" right after the front cover. The list included page number references. |
| Enough background information | Make sure you provide or assume the proper background for understanding. This was referred to by one speaker as the given-new contract: If I provide or correctly assume the given information, then the reader can more easily comprehend the new information. |
| Audience analysis | Determining the correct level of readability depends on correct analysis of the audience. This will continue to be a shortcoming if our audience is never clearly defined. I think we wind up with a schizophrenic view of our integrated system products. How should we respond if our questions get the following answers? |

Figure 13-6 ❑ SAMPLE SHORT MEMO/REPORT: TRIP REPORT (*Continued*)

(Readability, continued)

> **Question:** Is this document for the novice or for
> the experienced person?
> **Answer:** Yes.
>
> **Question:** Do we sell to VARs or end users?
> **Answer:** Yes.

One answer is to continue to provide multi-manual sets with different reader levels addressed in each manual. And to provide job aids.

Testing readability

The importance of testing manuals for readability is a common theme. IBM tests manuals all the time, with carefully selected subjects, one-way mirrors, and videocameras. The rest of us try to benefit from their general results. For specific results, we have to rely on our reviewers before a product is released and on customer comments after a product is released. Heck of a way to run a railroad! But our schedules don't allow the extra time needed for detailed readability tests.

Physical organization can increase readability

Although this may sound more relevant to a manual's usability level, physical structure can increase or decrease readibility. This refers to general and specific levels:

- At the level of overall manual organization, for example, make sure the material appears in the correct sequence. Make sure the learning chunks aren't too big.

- At the level of specific page format, if we help readers gain a sense of organization, we help them find what they need more quickly. I quote, "...typographic format and white space create a sense of organization for a reader...." For example, headers are signposts to organization, and that's why we use different typefaces and weights to indicate hierarchy. This argument also supports hanging heads, which help you skim the organizational elements more quickly. Other organizational devices include rules, tables, boxes, and bulleted lists.

Marcia Petty, 5/20/87, STC Trip Report   Page 5

Figure 13-6 ❑ SAMPLE SHORT MEMO/REPORT: TRIP REPORT (*Continued*)

(Readability, continued)

Readable forms and contracts

Two authors made an argument for applying readability standards to such mundane items as contracts and forms. An excellent example was a contract that went to a customer in "term-paper-format": 8½x11 page, text the full width of the page (excluding narrow margins), headers barely distinguished from text. The customer thought it was a draft and marked it up! Needless to say, the company that wanted the contract signed redid the format so that it became much more readable.

We have a "good" bad example right in house: we need to reorganize the disclaimer page in H/W manuals. The page gives too much information at once and is barely readable. Maybe we don't care if users actually read it, it's just legal backup; if so, let's put all that stuff at the back of the manual! For example, one competition manual had the FCC statement at the back, but listed in the contents...odd, but workable.

Miscellaneous

A session on "visual literacy" reminds us that having an advanced text and graphics tool does not automatically create an experienced communicator. Training is still needed to use the features of any tool, and I don't mean just specific machine-dependent training. I refer you back to my paragraph headed "Individual Responsibility." The individual needs to learn the fundamentals of writing and graphics as well as the use of advanced electronic tools.

Summary

I hope Intel will continue to fund participation in professional conferences. The company, the department, and the individual all benefit. It may take time afterwards to put new ideas into place, but in the meantime, those ideas have a daily impact in our communication process.

13.3 THE FORMATTING OF MEMOS AND SHORT REPORTS

In addition to their structure, the reports in Figures 13-1 through 13-6 demonstrate two other notable features about memos and short informal reports: the format of the heading information and of the text of the report.

Formatting the Heading of the Memo and Informal Report

The memo and the short informal technical report begin with a heading segment, which typically gives the following information. (*Distribution, Enclosures,* and *References* are optional.)

To: Name of reader, job title
           Reader's department
           Reader's organization
           Address of reader's company or organization (sometimes included)

From: Name of writer, job title
           Writer's department
           Writer's company or organization
           Address of writer's company or organization (sometimes included)

Subject: Title of report (~~Subject may be replaced by Re or RE, for Regarding~~)

Date: Date

Dist: Distribution list of other people receiving the report (omitted if there is no
           distribution list)

Encl: Enclosures; other documents which are included with the report (omitted if
           there are no enclosures)

Ref: References; list of particularly important background documents (omitted if
           there are no such documents)

Often, memos and informal reports are written on company letterhead, special company stationary that has the company's name—and usually the address and phone number—already printed on it. If the memo or report is written on letterhead and if any of the above heading information appears in the letterhead, then it is not repeated in the *To, From,* or *Distribution* items. Thus, a typical heading on letterhead paper would include the following information for the *To* and *From* parts of the heading. The letterhead information is in bold.

Name of Company or Organization
Name of Department (if included)
Address and Phone Number of Department or Company

```
To:      Name of reader, job title
         Reader's department (if not in letterhead)

From:    Name of writer, job title
         Writer's department (if not in letterhead)
```

Note that the heading segment is easy to read because of its format: liberal use of white space and aligned columns. It is also functional, because the first information it gives is the information readers need first.

It is obvious why writers need to put the subject and date on the report and why they need to give the name of the reader and writer. They give the subject to define the purpose of the report and to predict what kinds of information the report will present, and they give the date to indicate when the report was written. It may be less clear why they need to give the job title, department, and organization of the reader and writer. These are provided because, at one level, documents are written not to people but to and from the occupants of particular jobs. Thus, it needs to be clear what job or position a person holds. Note that none of the information is presented more than once at the beginning of the report. If the names of the organization and/or the department are included in the letterhead, then these do not reappear under the *To* and *From* headings. If they do not appear in the letterhead, then they must appear under the headings.

Figures 13-7 and 13-8 present two different heading segments with appropriate organization of information. Note that Figure 13-7 is written on organizational letterhead—no department or division is specified in the letter-head—and so *Optoelectronics Department* must appear under the names of the reader and writer. In contrast, Figure 13-8 is written on departmental letterhead, and so *Optoelectronics Department* does not need to be repeated. If the reader had been in a different department—say, in the Finance Department—then the appropriate heading segment would be that of Figure 13-9. Notice that the writer's department appears in the letterhead but not under his or her name, whereas the reader's department appears under his or her name and not in the letterhead. The rest of the information remains the same.

Formatting the Text of the Short Informal Report and Memo

Each main section of the report/memo—the Foreword, the Summary, and the Discussion (or Details)—follows the previous section, with only a double or triple space between sections. A clearly visible heading indicates the beginning of each section and announces the contents and, if possible, the point of the

Figure 13-7 ❏ HEADING SEGMENT FOR SHORT INFORMAL REPORT/MEMO
ON ORGANIZATIONAL LETTERHEAD
(No Department or Division Specified)

AEI American Electronics Incorporated

To: Mr. John Nicol
            Chief Optoelectronics Engineer
            Optoelectronics Department

From: Khalil Najafi
            Assistant Optoelectronics Engineer
            Optoelectronics Department

Date: October 30, 1989

Subject: Liquid Crystal Displays: Analysis of Failure and Recommendations for
            Solution

Dist: Mr. Edward Jones, President

---

section. If there are Appendixes, they begin on a new page at the end of the
report/memo. Formatted short informal reports and memos appear in Chapter
12 as well as in this chapter, and the overall structure of an informal report/
memo appears in Figure 13-10.

Section headings should be as informative as possible. Notice that the
headings in Figures 12-1, 13-2, 13-4, and 13-5 are especially informative. For
instance, the headings in Figure 12-1 not only announce the type of information
to be found—performance information and cost information—but they also tell
the reader what point is being made about these sections: aerated lagoons
have superior performance and aerated lagoons have superior cost characteristics (are cheaper). Such informative headings make it very easy for the
reader to read quickly. Indeed, they allow a reader to know the main point
of a section without having to read it. As with other reports, short reports
and memos should use white space and formatting to indicate structure.

❏ EXERCISE 13-2

Examine the section headings in the figures in Chapter 13. If the headings
are very general (such as Details), see if you can suggest more informative
headings.

Figure 13-8 ❏ HEADING SEGMENT FOR SHORT INFORMAL REPORT/MEMO
ON DEPARTMENTAL LETTERHEAD
(When Reader and Writer Are in the Same Department)

AEI American Electronics Incorporated
Optoelectronics Department

To: Mr. John Nicol
           Chief Optoelectronics Engineer

From: Khalil Najafi
           Assistant Optoelectronics Engineer

Date: October 30, 1989

Re: Liquid Crystal Displays: Analysis of Failure and Proposed Solutions

Dist: Mr. Edward Jones, President

Figure 13-9 ❏ HEADING SEGMENT FOR SHORT INFORMAL REPORT
ON DEPARTMENTAL LETTERHEAD
(When Reader and Writer Are in Different Departments)

AEI American Electronics Incorporated
Optoelectronics Department

To: Ms. Karen Smithfield
           Chief Financial Analyst
           Finance Department

From: Khalil Najafi
           Assistant Optoelectronics Engineer

Date: October 30, 1989

Re: Liquid Crystal Displays: Analysis of Failure and Proposed Solution

Dist: Mr. Edward Jones, President

Figure 13-10 ❏ FORMAT OF AN INFORMAL REPORT/MEMO WRITTEN ON LETTERHEAD PAPER

Name of Company or Organization
Name of Department (if necessary)
Address and Phone Number of Organization or Department

To:
From:
Subject:
Date:
Dist:

Foreward

Summary

Discussion

Appendixes

Attachments

Situation 13-1 ❏ INFORMATION AND COMMUNICATIONS SYSTEMS PLANNING AT PURE-PAC: RESOLVING WHO DOES WHAT
(by Barbara Couture and Jone Rymer Goldstein*. Used with permission of Little, Brown.)

"That's very, very good news," you think, smiling to yourself as you put down this morning's edition of the *Wall Street Journal*. You just read an article announcing that Pure-Pac will be paying high dividends this quarter because of increased profits. That news promises to make your job as a communications systems planner[1] for the Forever-Seal Division Expansion project a lot easier. Pure-Pac managers will be more eager to buy communications equipment to accommodate future growth when they are confident that the company will show a profit.

At Pure-Pac, your most important task is to convince Harvey Hamburgh (Forever-Seal Division Manager) and James Schreiber (President of Pure-Pac) that the communications systems you plan for the new Forever-Seal Division offices will grow with the company. (See Organization Chart.) Planning for growth means designing a system which integrates all communications equipment that may be needed for the foreseeable future by the various Forever-Seal Division offices. The list includes telephone service; video display service for security and employee training; a network of data communications services, such as computer-to-computer, terminal-to-computer, and word-processing systems; and speciality communications services, such as radio paging, intercom systems, and teleconferencing capabilities—not to mention all building management systems, such as fire and security alarms and temperature control systems.

The job of assessing all these needs seems overwhelming at times, but you are working with a team of dedicated and competent people who are managing the entire Forever-Seal Expansion project. You have nothing but respect for your boss, Dennis Jaynes, Director of the Division Expansion Office. Jaynes coordinates the work of an internal team of architects, engineers, planners, and interior designers who are working with outside engineering and architectural firms on the expansion.

The planning and designing of the Forever-Seal expansion have gone quite well, particularly because Jaynes has the full support of Division Manager Hamburgh. Also, Jaynes gets along famously with the Forever-Seal Director of Operations, Jason Firebaugh, to whom your office reports. Jaynes's work would be a lot easier, though, if Adam Updike, the division's Director of Finance, would trust the leadership of those under Hamburgh. Updike sits at Hamburgh's right hand, and, since he watches the purse strings, he feels that he ought to control every part of the Forever-Seal operation in some way. When he does not succeed, he becomes standoffish. Jaynes believes that Updike would like to totally revamp the Division Expansion Office. In fact, your boss has said that Updike would reorganize every office in the whole division given the chance. That is one of Updike's problems as an administrator; he always has his fingers in too many pies. The irony is that you were hired because of this very behavior.

Adam Updike is responsible for planning the information systems for the expansion project. He recommended that the Division Expansion Office hire a separate communications systems planner; moreover, he supported *your* candidacy wholeheartedly because you

---

*From Barbara Couture and Jone Rymer Goldstein, *Cases for Technical and Professional Writing* (Boston, MA: Little, Brown, 1985), pp. 75–82.
[1] A communications systems planner directs the selection of equipment that is used to transmit all kinds of information throughout an organization, including telephones, computers, paging systems, security systems, and other devices. In order to recommend adequate communications systems, the communications planner works closely with an information systems planner who determines what kind of information must be transmitted and what systems will manage the sorting, processing, and delivery of information. For example, in planning and implementing a company payroll system, an information systems planner determines what data are needed by the payroll officers (for example, withholding tax, wage rates, information from the corporate financial system) and specifies how this data will come together in the payroll. A communications systems planner analyzes what kind of equipment is needed to store the data, make changes, transfer data, and produce the ultimate output.

convinced him of your commitment to designing a communications system that will accommodate any expansion of the information systems and procedures. Since you have been with the company, however, Updike has given you the cold shoulder. Jaynes has suggested that this is because you took hold of the job so thoroughly, certainly with more gusto than Updike had expected. During the two months you have been at Forever-Seal, you have already established a Communications Systems (CS) Planning Team[2] and have held regular meetings to assess state-of-the-art communications equipment and to assess how new products are being used by other companies. By all rights, Updike should be pleased by that. But for some reason he is not, and he has made no move to work with you. That has seemed strange to you, given his tendency to want to run everything. But Jaynes explained why that might be.

"You see," Jaynes said, taking you aside after a cool reception by Updike at a meeting of central management and the Division Expansion Office, "since Adam is planning the information systems for the new Forever-Seal plant, I think he's afraid you may make him look bad. He hasn't been half as organized as you've been in getting his plans together. That CS Planning Team you put together really got the jump on him."

You asked Jaynes what to do about the situation. He suggested that you "do nothing" at the moment and wait for Adam to approach you instead. You accepted his advice but were a little concerned about the lack of communication (can you believe it!) between the communications and information systems people. The problem could have severe consequences if not resolved soon. Inevitably, your state-of-the-art research has kept you busy enough. But you do tend to mull over what to do about Adam Updike whenever you have nothing specific scheduled—right now, for instance.

You are making a few notes on how to approach Updike when the phone rings. For a moment, you think you are psychic, for it is none other than Updike on the line. He is unusually friendly, so you wonder what is going on. He starts telling you about some meeting of the top officers of Pure-Pac. Apparently last weekend the President, all division managers, and the Vice President and Chief Financial Officer for the corporation were away at a management retreat. Hamburgh could not attend, and he sent Updike in his place. Following a presentation, President Schreiber got especially interested in communications systems planning for the Forever-Seal Division expansion and wanted to know how short-range and long-range planning for both information systems and communications systems would be carried out. Sure enough, he cornered Updike at the meeting to ask him about it. At this point in his tale about last weekend's powwow, Updike hesitates a bit.

"So what did you tell him?" you ask, anxiously.

"Well, that's why I called you," Adam says, getting to the point. "I told him that I planned to hire a consultant to assist me with information systems planning for the new division offices, and that you were handling communications systems planning."

"And what did you tell him about the communications plans?" you ask.

"Well, I thought it would be best not to speak for you," Updike continues. There is a brief silence on his end of the line. "Perhaps you ought to drop him a line and explain how your office is handling things. Just to let him know you have everything under control."

Cutting short his conversation with you, Adam does not give you time to ask questions. You are totally puzzled and not sure what to do about this situation, so you decide to drop in on Jaynes and discuss it with him.

"Looks like Adam got nailed," Jaynes tells you. "And so now he's trying to get back at us. We've got to do something about this right away. You better write a memo to Schreiber telling him about our plans before he asks us himself."

"Well, what should I tell him?" you ask. "We haven't done much planning because we haven't gotten together with Updike."

"I know that. First, we've got to let Schreiber know that we will design a totally flexible

[2]Members of the Communications Systems Planning Team include user representatives from each of the 12 departments in the Forever-Seal Division. The user representatives discuss their future needs for communications equipment in the light of what you have found will be available to purchase.

facility to accommodate any future needs for communications equipment. Then we'll tell him you've got research on needed communications equipment in progress—you know—mention the CS Planning Team. Whatever else we say, we can't let him think we have our heads in the sand over here," Jaynes adds with emphasis.

"What do you mean by that?"

"I mean that we must let him think we are working well with other corporate offices," Jaynes explains. "We should mention that you are cooperating with the corporation-wide Telecommunications Planning Committee . . ."

"Of course," you break in, "I meet with that group once a month. Every Pure-Pac division is represented there.[3] Communications people in all the divisions raise questions about how to use and purchase computers, televideo display systems, the whole bit. My contact with that group has helped me do equipment research faster and more efficiently."

"That's right," Jaynes says. "And it could be important to bring that up now. We know Updike has not gone out of his way to communicate with us, but he may feel that we're the ones who don't want to cooperate. What I fear is that he might give Schreiber that impression. Your work with the Telecommunications Committee would show Schreiber we're team workers." Jaynes steps toward you. "But in that memo to Schreiber, you better mention Updike's 'good works.' Tell Schreiber that Updike's office is in fact planning information systems now, with the help of an outside consultant as he says. Say that we are gathering as much data as we can about communications systems to handle everything from word processing to printing payroll checks, so we'll be ready to suggest alternatives when Updike tells us what he needs."

"So, you want me to emphasize cooperation, but not to say anything too explicit? I see," you say. "But what can I use as a premise for writing this memo in the first place?"

"Be straightforward," Jaynes answers. "Tell Schreiber we're aware of the retreat and heard he's curious about the communications systems planning for the new plant."

Still somewhat uncertain about the situation, you nevertheless say, "Okay," and start out toward your own office.

"Wait a minute." Jaynes stops you. "Before you start that memo, I'm going to give a call to Harvey Hamburgh." Jaynes takes a deep breath. "It just occurred to me, this communication should come from him. If we go straight to Schreiber, it might look like Hamburgh isn't in control of our operation."

So, you wait while Jaynes phones Hamburgh. After Jaynes reaches him, he soon learns that Updike never told Hamburgh of his conversation with Schreiber. However, Hamburgh gives your office the go-ahead to write the memo to Schreiber for him. When Jaynes hangs up, he asks you to draft the memo to Schreiber immediately, reminding you that it must represent the division as united in its efforts. "We must keep any hint of our feelings about Updike's lack of cooperation out of this," Jaynes emphasizes. "Actually, it probably wasn't so bad that this whole event happened. Now, we'll have to deal directly with the problem of coordinating Updike's efforts with ours. A good way to make it all get off on the right foot is to assure Schreiber we're one big happy family."

Feeling satisfied that everything will be resolved satisfactorily, Jaynes shows you out of his office. Now you must take everything into account, including all social and ethical implications, in an effort to solve the problem.

---

[3]The Telecommunications Planning Committee was formed by Vice President Steven Lindstrom eight months ago, before you joined the company. You were appointed to the committee when you were hired. Lindstrom established the group to encourage research in telecommunications at Pure-Pac so that computers and other equipment could be purchased and used efficiently and effectively throughout the company. At committee meetings, company representatives from each division pose problems on the use of telecommunications equipment in their divisions. Vendors are sometimes called in to explain products.

Your Tasks:

1   Draft the memo which Dennis Jaynes asks you to write to James Schreiber, President of Pure-Pac Packing Company. The memo will be from Harvey Hamburgh, Manager, Forever-Seal Division. Steven Lindstrom, Adam Updike, and Dennis Jaynes will be on the distribution list. (Write your name on the upper-right corner of this paper to identify it for your instructor.)

2   In mulling over what Jaynes has asked you to write to Schreiber, you become uncomfortable with what you have been asked to do. You think you have a better strategy for dealing with the problem. Write a memo to Jaynes which briefly explains your alternative plan and requests him to discuss it with you. Append a draft of any document required by your new strategy (for example, a memo to Schreiber which reflects your approach to the problem.)

3   Assume that you have completed Assignment 1. Harvey Hamburgh has approved your memo to Schreiber, but he is not happy about the mishap which prompted it. He wants Jaynes to provide written documentation of the events that led to your memo to Schreiber. Write a memo from Jaynes to Hamburgh explaining background events.

4   Write a memo inviting Updike to attend the next meeting of your Communication Systems Planning Team. (You indicate the date.) You will want Updike to brief the team on his plans for gathering data on needed information systems. The memo will come from you, and Jaynes will be on the distribution list.

ORGANIZATION CHART FOR PURE-PAC PACKING

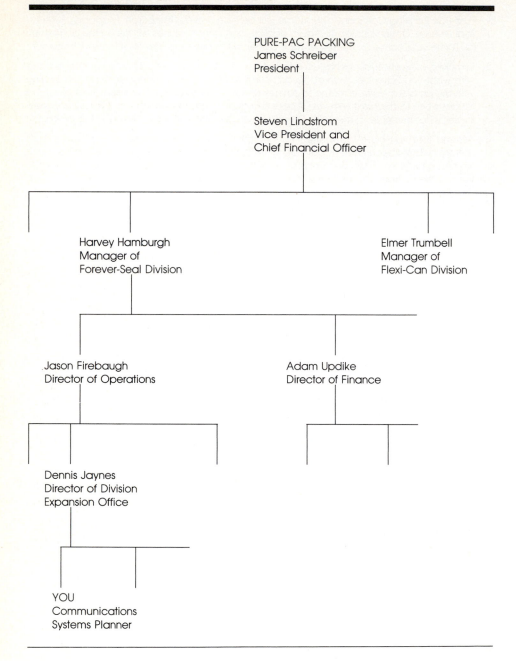

PURE-PAC PACKING
James Schreiber
President

Steven Lindstrom
Vice President and
Chief Financial Officer

Harvey Hamburgh
Manager of
Forever-Seal Division

Elmer Trumbell
Manager of
Flexi-Can Division

Jason Firebaugh
Director of Operations

Adam Updike
Director of Finance

Dennis Jaynes
Director of Division
Expansion Office

YOU
Communications
Systems Planner

14

Feasibility Reports

In most report-writing situations, someone has noticed a problem or undesirable situation and then proposed or asked someone else to propose a solution. Sometimes, however, it may not be clear that there is a solution or that any given solution would be worth what it would cost in time, money, and effort. At such times, it is useful to conduct a feasibility study. A *feasibility study* analyzes the problem and its implications, evaluates alternative solutions, and (usually) recommends a particular course of action. The results of such a study are written up in a *feasibility report*.

The term *feasibility* has two somewhat different meanings. In its more limited sense it means technically capable of being done, executed, or effected. We shall call this *technical feasibility*. If we were considering building a tunnel through a mountain, this more limited sense of feasibility would refer to such questions as the following:

1 Can we build the tunnel? Do we have the technology to keep the soil from caving in?

2 Do we have enough people available for the job?

3 Do they have the required expertise?

4 Can we get the required building supplies and building permits?

In addition to this sense of technically capable of being done, *feasibility* has the wider, more inclusive sense of suitable or reasonable. To return to the tunnel example, this wider sense of feasibility encompasses such additional questions as the following:

1 Assuming that we can build this tunnel (that it is technically feasible), should we? Will it produce enough benefits to be worth the cost?

2 Is the cost reasonable?

3 Can we afford it, even if it is reasonable?

4 Will building or having the tunnel produce any undesirable side effects?

5 Will it produce any desirable side effects, such as increased employment in the area?

In this larger sense, the notion of feasibility covers the five general criteria discussed in Chapter 4 that are used to evaluate any proposed solution to a problem:

1 *Effectiveness:* Is the solution effective? Will it solve the problem posed? Why? How do we know?

2 *Technical Feasibility:* Can the solution be implemented? Does it require technology or resources that are unavailable? How do we know?

3 *Desirability*: Would we want to implement the proposed solution? Does it have any undesirable effects? Does it have desirable effects? Why? What are they?

4 *Affordability*: What will the solution cost to implement? To maintain? Is this cost reasonable? Is it affordable? Will it reduce costs in the future? Why?

5 *Preferability*: Is the solution better than or preferred over any other possible solution? Why?

This larger notion of feasibility also covers any special criteria necessary or appropriate to the given situation—for instance, important issues of safety, environmental pollution, conformity to particular laws or traditions, precedent setting, or justice. As a technical expert, you must be able to deal with all the relevant issues and criteria in a feasibility study.

When all these criteria enter into the notion of feasibility, they provide the basis for a full-fledged argument of fact or policy. For instance, when you judge a tunnel as being effective, technically feasible, desirable (including safe), affordable, and preferable, and then argue on the basis of these judgments that the tunnel is feasible, you have constructed an argument of fact supported by several subarguments of fact as discussed in Chapter 4. If you go one step further and argue that the tunnel should be built, you have constructed an argument of policy supported by several subarguments of fact, one of which is the argument of feasibility.

Let us examine the arguments being developed in a sample feasibility report to further illustrate this point. The report concerns flight tests on airplanes—in particular, flight flutter tests. A flight flutter test studies the vibratory motion and resulting instability of aircraft wings during flight. When a plane is in flight, air passing over and around the wings creates some wing motion. If this motion is too great, the wings will vibrate catastrophically, or "flutter." If this flutter becomes too severe, the wings can actually break off. The Foreword to the report sets up the problem being addressed in a feasibility study.

Foreword

Flight flutter tests are inherently dangerous. To maintain a margin of safety, we must have fast and accurate in-flight analysis of the vibratory motion in the wings created by the test flight conditions. However, the computer program currently used for in-flight analysis is outdated, and its inadequacies present safety problems. Thus, a new analysis technique must be implemented to improve safety and to maintain our reputation in this area of flight testing.

For this purpose $10,000 was allotted, and a survey of the field was made to determine possible choices for the new analysis technique. I was assigned to review the choices for feasibility, select the best one and, if it seemed feasible, identify initial development and implementation concerns. This report presents my solution to our problem and provides initial information on the development and implementation of that solution.

While doing the analysis and evaluation necessary for this report, the writer realized that any proposed solution to the problem—that is, any computer program which would produce an adequate in-flight analysis—would have to meet the following criteria:

1 *Effectiveness*: The solution must be capable of providing accurate damping estimates of up to four superimposed vibration modes from the accelerometer data available.

2 *Technical feasibility*: The solution must be capable of working with the present computer (HP-5451C Fourier Analyzer), with minimal modifications. The solution should be adaptable by our staff, given their expertise and available time.

3 *Desirability*: The solution should have some flexibility in its software structure to allow modifications to the existing package should a particular test call for them.

4 *Affordability*: The solution purchase, development, and implementation costs should be less than the allotted $10,000.

5 *Preferability*: A solution that provides conservative damping estimates and requires the minimum analysis time is preferred.

The writer then evaluated three possible alternatives—the Random Decrement (Randomdec) Technique, the Time Domain Package, and the Flight Test Data

Reduction System—against these criteria, finding that two of the alternatives were ultimately infeasible:

1 The Time Domain Package was rejected for failing to meet the desirability criterion. The package is proprietary, and the producer refuses to disclose the details of the software structure. Therefore, the flexibility in modifying the package is absent.

2 The Flight Test Data Reduction System was rejected for failing to meet the effectiveness criterion under certain conditions. Accurate damping estimates were difficult to obtain when two or more modes fell within a frequency range of $\Delta 5$ Hz. Since this condition is quite possible during testing, the package was rejected.

The remaining alternative, the Randomdec Technique, was feasible: it met all the criteria and was ultimately recommended as the solution to the problem of inadequate in-flight testing.

The writer now was in a position to present his results in a feasibility report. The feasibility report essentially follows the basic structure described in Chapter 12 for technical reports and illustrated in Figure 14-1. It has a Foreword and Summary on page 1 for managerial and nonspecialist audiences, followed by a Details section aimed at more specialist audiences. The Details or Discussion section provides the argument, backup data, conclusions, and references to support the claims stated in the Summary. This general pattern is reflected in the writer's outline for the feasibility report on the flight flutter tests:

Foreword
Summary
Details
    I   The Problem and Its Background
   II   Argument for the Randomdec Technique
       A   Criteria for Judgment
       B   The Randomdec Technique as the Recommended Solution
       C   Rejection of Alternatives
  III   Development and Implementation of the Randomdec Technique
       A   Brief Explanation of the Technique
       B   Cost and Time Required
  IV   Conclusions
   V   References

For this feasibility report, the Details section first outlines the problem, and then presents arguments for the Randomdec Technique. Then it addresses the issue of development and implementation (a part of the writer's assignment). Finally it presents the writer's conclusions. Given this structure to work from, the writer produced the report shown in Figure 14-2 (the

Figure 14-1 ❑ STRUCTURE OF THE FEASIBILITY REPORT

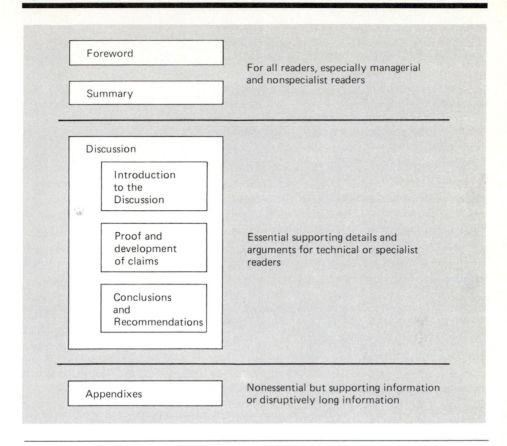

Attachment to the report is not included). A second feasibility report, an informal feasibility report, is presented for analysis in Figure 14-3.

The feasibility report is much like other kinds of technical reports: it is based on a problem; it presents an analysis of one or more possible solutions; and, if possible, it recommends the best solution given the circumstances involved. Such a recommendation is based on an argument of fact (this is or seems to be feasible) and it may constitute an argument of policy (you should adopt this solution). If a feasibility report does not build all the way to recommendations and an argument of policy, it should at least make, as clearly as possible, the various subarguments of fact for the people who will ultimately have to decide whether to fund or approve the project being studied. If these decision makers are better qualified than you are to make recommendations, then leaving the recommending to them may be appropriate. However, in many cases they will know less about the issue than you will and will look to you to help them decide what to do. In such cases, you should not hesitate to make clear recommendations. Remember: writers of feasibility reports are

Figure 14-2 ❑ SAMPLE FEASIBILITY REPORT

Date:	December 5, 1987
To:	Regina J. Petty Branch Chief Dynamics Branch
From:	Louis J. Bogart Aerospace Engineer Dynamics Branch
Subject:	DEVELOPMENT AND IMPLEMENTATION OF A FLUTTER ANALYSIS PACKAGE: A FEASIBILITY STUDY
Attachment:	Functional Diagram of the Random Decrement Analysis Procedure
cc:	R. J. Jones, Chief, Safety Office T. R. Wilson, Chief, Flight Operations

FOREWORD

Flight flutter tests are inherently dangerous. To maintain even a margin of safety, we must have fast and accurate in-flight analysis of the wings' vibratory motion created by the flight test conditions. However, the computer program currently used for in-flight analysis is outdated, and its inadequacies present safety problems. An incorrect damping estimate obtained during in-flight analysis could result in the loss of an aircraft due to a wing fluttering off. Thus, a new analysis technique must be implemented to improve safety and to maintain our reputation in this area of flight testing.

For this purpose $10,000 was allotted, and a survey of the field was made to determine possible choices for the new analysis technique. I was assigned to review the choices, select the best one, submit it to a feasibility test, and, if it seemed feasible, identify initial development and implementation concerns of the selected technique. This report presents my solution to our problem and provides initial information on development and implementation of that solution.

SUMMARY

In an attempt to find a replacement for our inadequate flutter analysis package, a survey of the field turned up three alternatives: the Random Decrement (Randomdec) Technique, the Time Domain Package, and the Flight Test Data Reduction System. Each alternative was studied for effectiveness, technical feasibility, desirability, affordability, and preferability. As a result of this study, I recommend the Randomdec Technique as our solution: it meets all of the criteria, whereas the others fail to do so. An initial cost estimate of developing and implementing this technique is $8,000 ($2,000 under budget). Implementing this technique will increase the safety of our flight flutter tests and provide a desired margin of safety; it will also allow us to maintain a prominent position in this area of flight testing.

Figure 14-2 ❑ (*Continued*)

DETAILS

I The Problem and Its Background

The problem arises out of the fact that our software package for flutter analysis is outdated. The Power Spectral Density (PSD) Package presently in use was developed over 10 years ago, and the level of accuracy it provides is no longer acceptable. The accuracy of the damping estimates obtained by the PSD method is erratic, and several superior analysis techniques have been developed in recent years.[1]

The importance of having damping estimates that are as accurate as possible can readily be seen from the potential danger involved in flight flutter testing. An incorrect damping estimate obtained during in-flight analysis could result in the loss of an aircraft due to a wing fluttering off.

An updating of our software is needed to reduce the risk involved and to maintain a prominent position in this area of flight testing. Funds ($10,000) have been allotted to improve our capabilities in this area, and a survey of the analysis techniques presently being used in the field has been made. The survey resulted in three analysis techniques being chosen as the most promising alternatives to our present situation. They are

1 The Randomdec Technique—broadly used by a number of aircraft manufacturers

2 The Time Domain Package—specific package developed by the Aerotest Corporation and used by the Air Force at Edwards Test Center

3 The Flight Test Data Reduction System—specific package developed and used by the Northrop Corporation

Since our present PSD Package has already been modified to its limits with unsatisfactory results,[1] the solution to our problem lies in choosing one of the above alternatives and implementing it into our flight testing system.

II Argument for the Randomdec Technique

A Criteria for Judgment

1 Effectiveness: The solution must be capable of providing accurate damping estimates of up to four superimposed vibration modes from the accelerometer data available.

2 Feasibility: The solution must be capable of working with the present computer (HP-5451C Fourier Analyzer), with minimal modifications.

3 Desirability: The solution should have some flexibility in its software structure to allow modifications to the existing package should a particular test call for them.

4 Affordability: The solution purchase, development, and implementation costs should be less than the allotted $10,000.

Figure 14-2 ❑ SAMPLE FEASIBILITY REPORT (*Continued*)

5 Preferability: A solution that provides conservative damping estimates (Attachment 1) and requires the minimum analysis time is preferred.

B   The Randomdec Technique as the Recommended Solution

I recommend the Randomdec Technique as the solution to our problem because it meets or exceeds all of the given criteria.

1 Effectiveness: This technique provides accurate damping estimates of up to four superimposed modes.

2 Feasibility: This technique can utilize a software structure which is well suited for use on the available computer. We have the personnel and expertise in house to develop and implement the necessary software.

3 Desirability: This technique is well documented and available for use by the general public. A proprietary software package need not be purchased. The necessary software can be developed in house and tailored to our specific needs with the flexibility desired.

4 Affordability: This technique is estimated to cost under the allotted figure. Our survey and in-house discussions gave an initial estimate for development and implementation costs of $8,000 (200 hours of labor @ $40/hour).

5 Preferability: This technique is mathematically derived to provide conservative damping estimates, thus allowing some safety margin (Attachment 1). Initial estimates of analysis time were also found to be acceptable.

C   Rejection of Alternatives

1   The Time Domain Package was rejected for failing to meet the desirability criterion. The package is proprietary, and the producer refuses to disclose the details of the software structure. Therefore, the flexibility in modifying the package is absent.

2   The Flight Test Data Reduction System was rejected for failing to meet the effectiveness criterion under certain conditions. Accurate damping estimates were difficult to obtain when two or more modes fell within a frequency range of $\Delta 5$ Hz. Since this condition is quite possible during testing, the package was rejected.

III   Development and Implementation of the Randomdec Technique

A   Brief Explanation of the Technique

The Randomdec Technique was developed by Cole[2] and has been used successfully by a number of aircraft companies[3,4]. The proposed functional diagram of the

Figure 14-2 ❏ (*Continued*)

analysis procedure can be seen in Attachment 1. To start the procedure, the Randomdec signature is extracted from the initial vibration data (time history) using the methods detailed.[2] The signature is then truncated, since only the beginning is needed for further analysis. A curve fitting of the signature is then performed to obtain the frequency and damping characteristics of the individual modes. However, first a power spectral density of the signature is taken to supply the curve-fitting program with initial frequency estimates of the modes. After the curve fitting, the individual modes are separated and output with their respective frequency and damping estimates.

B   Cost and Time Required

The familiarity of the engineers in the Dynamics Branch with both the computational procedures involved in the Randomdec Technique and the capabilities of our computer system should facilitate a quick and efficient development and implementation of the new package. Having worked extensively in programming the computer in the past, I would estimate that a maximum of 200 hours would be necessary to complete the development, implementation, and checkout phases. This represents $8,000 (@ $40/hour), which is below our allowable budget of $10,000.

IV   Conclusions

I recommend the development and implementation of the Randomdec Technique as the solution to our problem of an inaccurate and outdated flutter analysis package. Use of the Randomdec Technique will reduce the risk involved in the potentially dangerous area of flight flutter testing and help us to maintain a prominent position in this flight test field.

V   References

1   DFRC Memorandum 81-0611-2: Inadequacies of the Power Spectral Density Flutter Analysis Package. June 11, 1981.

2   Henry A. Cole Jr.: On-Line Failure Detection and Damping Measurement of Aerospace Structures by Random Decrement Signatures. NASA CR-2205, 1973.8

3   W. J. Brignac, H. B. Ness, and L. M. Smith: The Random Decrement Technique Applied to the YF-16 Flight Flutter Tests. AIAA Paper No. 75-776, May 1975.

4   R. V. Doggett Jr. and C .E. Hammond: Determination of Subcritical Damping by Moving-Block/Randomdec Applications. In "Flutter Testing Techniques," NASA SP-415, 1975, pp. 622–46.

Figure 14-3 ❏ SAMPLE INFORMAL FEASIBILITY REPORT
(Used with permission of Intel Corporation.)

Intel

**INTERNAL
CORRESPONDENCE**

**Boards and Systems
Technical Publications**

DATE: **Oct. 12, 1988**
TO: **Management Review Committee (MRC)**
FROM: **Marcia Petty, x6-2316, HF3-72, ccmail**
RE: **End of Pilot for Electronic Publishing**

CC: **List**

ONE LAST LOOK   We are almost at the end of the electronic publishing pilot program, and you will soon be making a decision about the vendor's implementation of EP. Throughout the lengthy selection and evaluation process, I have seen technology factors consistently outweigh the human factors that must also be considered when we try new equipment or a new process. Our desire for innovation is real, because "innovation" implies that things get better, not worse.

But at this point, the product-under-test looks as if it will impose less efficient and less user-friendly processes and tools than many of us are using now. Less efficient means less productive; less productive means a higher cost to Intel.

The Writer's Viewpoint   As an individual writer, I'd like to offer a few observations about the status of the pilot. Here's the basis for my professional judgment: I am a senior technical writer at Intel. I have been a technical writer and editor for 16 years, with 10 of those years spent in computer companies. I've used a variety of publishing tools and processes to create final output of manuals, brochures, data sheets, fliers, and advertising. The tools of my trade have included the following (from oldest to newest):

1

Figure 14-3 ❏ (*Continued*)

10-12-88, EP Pilot (M. Petty)

- Electric typewriters
- Typesetting and illustrator-drawn art
- UNIX-type troff commands to format files sent to a dot matrix printer
- Word processing software on PCs
- Dedicated, stand-alone word processors
- Integrated on-line text and graphics sent over a network to a laser printer: electronic publishing (my current process)

I've been a member of the Tech Pubs Electronic Publishing Pilot (TPEP2) committee since February of this year; I've monitored and helped evaluate a number of the pilot acceptance criteria (which were already established before major writer involvement).

Remember that the paragraphs that follow are my observations as a professional writer. I am not providing pages of financial, technical, or statistical information here; we do not have enough data from the pilot to provide valid financial, technical, or statistical information.

The Changing EP Concept

What the pilot has attempted to do in the past few months is not what the pilot group was set up to do. To quote a member of the original MRC for the project, the original concept was to use local print capability to save money on printing costs mostly via reducing scrap. The EP system was to be a back-end system that would have no impact on the tools the writers use or on manual content.

Change without Adequate Involvement

Unfortunately, I have watched the concept change, gradually making inroads into both manual content and the tools with which we do our jobs. I have seen the emphasis change from testing a back-end system for demand printing, to attempting to create a front-end system not designed to do what we need. Admittedly concepts evolve, but in this case the changes have occurred without enough input from those affected: the people with the knowledge about what is needed for publishing.

As the process has focused on technology and has ignored human factors and the creative process, we have seen a growing pressure to make manuals fit the limitations of the selected EP pilot equipment. This seems contrary to what we professional communicators try to do...make our manuals fit the needs of our customers, then find equipment that lets us create those manuals to the best of our ability.

2

Figure 14-3 ❏ SAMPLE INFORMAL FEASIBILITY REPORT (*Continued*)

10-12-88, EP Pilot (M. Petty)

The Desire for *New* Technology, not Old

I'd like to have newer, better tools to work with, not a regression to older technology and processes. For example, the trend in the industry is to provide distributed processing to allow each user group to handle its work as appropriate to local needs. Local needs are those of a particular product and its customer audience.

Centralized Processing Is Older

The current EP implementation with the vendor's equipment calls for a return to centralized processing. If we centralize the publishing process, we may lose sight of local customer needs; we may lose the flexibility to meet those needs, which are not identical from one Intel product line to another.

Centralized Processing Is Costly

A return to centralized processing is very likely to cost, not save, money for Intel. Based on information from the pilot, with the equipment being tested, the writers will see a 10 to 20% increase in time required to do tasks that will be needed because of centralized processing.

For example, it's an old process loop that we see between the writer and any centralized production resources (electronic typesetter, illustrator, formatter). Multiple passes always have to be made to produce accurate final output. How does the pilot's equipment vendor fit in this picture? Well, one acceptance criterion calls for 90% accuracy after a file is converted to the vendor's file standard. Can we live with file conversion errors and send manuals that are 90% accurate to an air traffic control system installation or a nuclear power plant?

Obviously 90% accuracy isn't good enough. Thus, we must revive the proofreading and correction loop from the days of typesetting. Either the writers or else proofreaders must recheck each entire manual after conversion, character by character. This translates into time cost for the writers (and fewer manuals written) or increased headcount for Intel (newly hired proofreaders). Either way, it costs money.

THE CREATIVE PROCESS AND THE CUSTOMER

I already use electronic publishing to create manuals here at Intel, and I really like EP. In the right implementation, with a user-friendly interface, EP lets me do my job more easily and creatively. If the mechanics of my job are easier, I have more freedom *to make my reader's job easier.* If I make my reader's job easier, I help the customer and help keep the customer.

3

Figure 14-3 ❑ (*Continued*)

10-12-88, EP Pilot (M. Petty)

But if the mechanics of my job are harder than before, if the machine interface is user-hostile (as the vendor's writer workstation has been demonstrated to be), I will have a harder time just putting together the basics of a manual, much less thinking about how to convey information clearly and directly. The tools I have now are better for the writing/publishing task than what is being offered to writers through the vendor's implementation of EP.

SUMMARY

I hope you will reject the implementation that we tested during the pilot, but that you will let us move forward to find a true EP solution.

I've already stated that I think EP is a good concept, but I don't think the pilot implementation will serve us well. I think that because the concept changed in mid-stream, there was a failure to involve writers early enough to shape the acceptance criteria to meet real-world needs. Many of the criteria were set to do minimum work, not the complicated tasks we do here in multiple departments.

We have learned much from the pilot, both what we want and what is unacceptable. This knowledge can be valuable to help us shortcut our time and costs as we evaluate other systems.

Also, it is stated that by virtue of the pilot, we have learned of production cost savings that can be made by implementing process changes separate from any additional purchases. These savings alone can justify the money Intel has spent for the pilot.

4

hired for their interpretive ability as well as for their fact-gathering ability. In many instances, the writer of the feasibility report is the person best qualified to make the needed recommendations.

❏ EXERCISE 14-1

One of your duties as Assistant Manager of Raney's Manufacturing is to check the plant's suggestion box. Recently a number of complaints have appeared about the unsightly, unsanitary mess in the washroom. Some employees have been careless about using and disposing of paper towels. Each dispenser has a built-in trash basket and even a sticker to remind employees to place used towels in the basket. However, the problem persists.

As a solution, you are considering replacing paper towels with hot-air hand dryers. At your request, Haworth Inc., makers of the Jetaire Hand Dryer, sent a salesperson to evaluate your needs. The salesperson made a cost analysis of your situation and also presented an estimate of the cost for changing to the Jetaire Hand Dryer. Raney's has 240 employees, and each employee visits the washroom an average of 4 times a day. The average number of paper towels used each visit is 2.5 towels, at a cost of $0.003 per towel. Assuming 22 working days a month, a 3-year cost figure for towels alone was derived. Haworth's salesperson also pointed out that there are "intangible" costs involved with paper towels. These include time spent in filling out purchase orders and the cost of mailing the orders; cost per square foot of storage of the paper towels; and plumbing expenses for toilets and sinks clogged with paper towels. These intangible costs usually amount to 50 percent of the 3-year total cost of paper towels for any business.

According to the salesperson, each of the four washrooms in your plant should have three Jetaire units. The units costs $150 apiece and together use about $4.50 in electricity per month. There is also an installation fee of $10 per Jetaire unit.

Because the estimated acquisition cost of the Jetaire Hand Dryers is over $1000, it is necessary to receive authorization from Steve Batter, the Plant Superintendent. He is a very busy person and often does not have the time to sort through a lengthy cost analysis report. He would like to know immediately what the costs and savings are to the company.[1]

A Write a feasibility report which brings the towel problem to Mr. Batter's attention and explores the feasibility of replacing the towels with hot-air hand dryers. Be sure to consider all relevant information. If your class structure allows this, try brainstorming and organizing the structure of the report with a small group of classmates.

B Once you have created a draft of your feasibility report, submit it to a small group of other students or colleagues for review and evaluation. If you brainstormed and organized with one group, get a different group to review and critique the report. Ask your classmates or colleagues to review it as if they were Mr. Batter and any other possible readers of the report, and to evaluate it for the clarity and appropriateness of the Foreword, Summary, structure, and formatting choices; the completeness with which it addresses the feasibility criteria outlined earlier in this chapter; and the adequacy of the editing and proofreading.

REFERENCES

1 The situation was adapted slightly from Herta A. Murphy and Herbert W. Hildebrandt, *Effective Business Communications*, 5th ed. (New York: McGraw-Hill, 1988), pp. 488–489.

15

Long Reports

15.1 THE LONG INFORMAL REPORT

The long informal report is quite similar to the short informal report. Both are written for readers within the writer's organization; both use the *To, From, Subject, Date* heading segment; and both are formatted so that each section follows the previous section with only a heading and a double or triple space between sections. However, the long informal report is different from the short report in two important ways: (1) the long report is longer, and (2) because of its length, its Discussion section is organized differently to make a greater amount of technical material accessible.

If you examine the table of contents of the long informal report at the end of this chapter, the similarities and differences between the long report and the short report are obvious. Let us focus on the differences. As illustrated in Figure 15-1, the Discussion section has three main subsections: the Introduction to the Discussion, the main Proof or Development section, and the Conclusions and Recommendations section. As a whole, the Discussion restates the material in the Foreword and Summary, but for a technically competent and interested reader. The Discussion gives the technical details, technical reasoning, and data that the technical reader needs—in contrast to the general orientation to the subject that the managerial reader needs. How does the Discussion do this?

The Introduction to the Discussion

The Introduction to the Discussion is the first part of the Discussion; it restates the Foreword in terms meaningful to the technical reader. The Foreword

294

Figure 15-1 ❏ STRUCTURE OF THE LONG INFORMAL REPORT

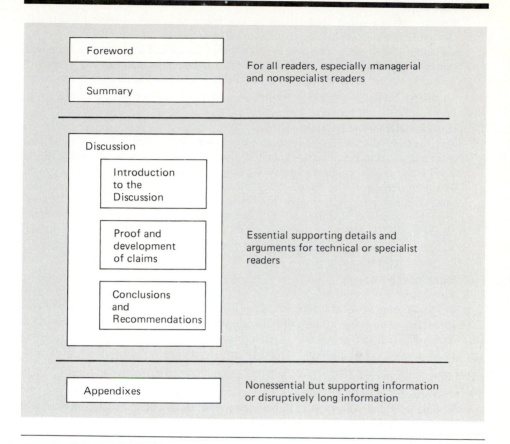

focuses on the implications of the problem for the managerial reader by presenting the problem in terms of safety, cost, efficiency, quality control, labor relations, or compliance with legal regulations. In contrast, the Introduction to the Discussion focuses on the implications of the problem for the technical reader: it defines the technical nature of the problem or outlines technical assumptions, values, subproblems, background, or resources relevant to the problem.

For example, consider the following two examples. The first was written by an electrical engineer at a company making heaters, fans, and portable lights, and its Introduction to the Discussion focuses on the technical nature and details of the problem. The second was written by a harried inventory controller, and its Introduction to the Discussion focuses on the technical background.

Example 1
Foreword
Because of increasing competition in the home appliance market, Electronic Design Division is evaluating the operating convenience and marketability of its less successful products. As part of this effort, I was asked to evaluate a proposed control circuit for our combined heater, light, and fan; to document any problems with this circuit; and, if necessary, to design a new control circuit. This report presents my evaluation and new design.

Introduction to the Discussion
Ms. Durkis of Marketing has requested a simpler control switch for our combined heater, light, and fan. I have evaluated a preliminary design for a circuit which could turn on all three units by flipping a single switch and then remove power from one or more of the units by depressing another switch. The design for the control circuit was built around three relays and claimed to operate by picking these three relays in different combinations to produce the desired switching capabilities. The following sections detail the preliminary design of the circuit and its problems, a revised design, and a cost analysis of the revised design.

Example 2
Foreword
During Photon's recent companywide expansion program, the inadequacy of our inventory numbering system became apparent. Thus, I was asked to design a more efficient and systematic numbering system based on our departmental needs and expected growth. This report provides an overview of the system and its response to our needs.

Introduction to the Discussion
The present inventory (parts) numbering system has been in use for almost 5 years. At the time of its conception, Photon did not have the computer capabilities it has now, nor did it envision the growth it has recently experienced. Because of these factors, the current parts numbering system is quite basic and limited. As the quantity of parts increased over the years, the numbering system could not efficiently keep track of them, nor could it assign new part numbers on a systematic basis. (In fact, new part numbers were assigned on a purely random basis.)

Last month, Photon underwent one of its most comprehensive expansion programs to date and suffered critically from the present numbering system. Often customers had to wait for us to reorder unexpectedly depleted supplies, and our Inventory Department logged hundreds of hours of overtime. Thus, management decided to do away with the present system and develop a new one. This report documents the various departmental needs identified, the computer-assisted inventory control system I have designed, and features of the system directly responding to the needs.

The different emphases of the Foreword and the Introduction to the Discussion are reflected by these examples. For instance, in the second example, the Foreword states the problem generally, devotes only one sentence

Table 15-1 □ DISTRIBUTION OF CONTENTS IN THE FOREWORD AND THE
INTRODUCTION TO THE DISCUSSION

	Foreword	Introduction to the Discussion
Managerial problem	35–70%	0–20%
Technical problem or assignment	5–35%	60–80%
Purpose/Forecast of contents	10–30%	0–25%

to it, and presents it in terms relevant to managerial concerns: the inventory system is inadequate. In contrast, the Introduction to the Discussion states the problem in more detail, devotes the first one and one-half paragraphs to it, and focuses on what particular technical reasons made the inventory system inadequate: for instance, "the numbering system could not efficiently keep track of them, nor could it assign new part numbers on a systematic basis. (In fact, new part numbers were assigned on a purely random basis.)"

These different emphases of the Foreword and the Introduction to the Discussion are reflected in Table 15-1, which presents the relative percentages of each devoted to the managerial problem, the technical problem or assignment, and the forecast of contents. Notice that the Foreword devotes proportionally more of its space to the managerial problem, whereas the Introduction to the Discussion devotes more of its space to the technical problem.

□ EXERCISE 15-1

A A student complained that she was having trouble with the following Foreword, that it was more like an Introduction to the Discussion than a Foreword (and she was right). Turn it into a Foreword by eliminating details unnecessary for a managerial reader. Feel free to rewrite sentences or to move information around.

Foreword
Flutter is a phenomenon in which structural instabilities arise from aerodynamic loads. In aircraft, these instabilities are often characterized by divergent oscillations of the wings, resulting in structural failure and possible loss of the aircraft. Flight flutter testing involves the tracking of damping estimates of the excited vibration modes at different flight conditions.

The violent nature of flutter makes safety an important concern during flight flutter testing. As with any flight test program, cost is also a major concern. Our objective at Smith is to provide safe testing at the lowest cost. Costs can be lowered by decreasing flight time. This requires utilizing the fastest analysis technique that will meet the accuracy requirements that safety demands.

Presently, we have available two software packages for data analysis during flight flutter testing. They are

1 the Power Spectral Density (PSD) Package

2 the Random Decrement (Randomdec) Package

The main objective of each package is to determine damping estimates of the structural modes. I was assigned to do a comparison study of the two packages looking at two questions:

1 How accurate are the damping estimates?

2 How much time is required for the analysis?

These questions are concerned with safety and cost, respectively. The purpose of this report is to present the findings of my study and to make some recommendations concerning future testing.

B When you have finished with your revision, compare it with the Foreword in Figure 14-2 and to the revisions done by several of your classmates. Discuss the differences among the versions.

Conclusions and Recommendations

The Conclusions and Recommendations section appears at the end of the Discussion section and gives information a specialist reader might want in addition to that needed by the manager. It might include the most important technical results which led to the conclusions and recommendations, and it might be technically too complex or difficult for the managerial reader. When possible, the Conclusions and Recommendations section should move from generalizations or claims to data and support. The Conclusions and Recommendations section contrasts with the Summary in that it provides more technical detail and assumes a higher level of relevant technical knowledge.

For instance, consider the Summary and the Conclusions and Recommendations presented below from a report written by a geologist to the Milton Township Council. (The Foreword is given first to orient you.) Unless you are a geologist, you probably won't understand all the information presented in the Conclusions and Recommendations, though you should easily find your needed information in the Summary.

Foreword
Within Milton Township, the greatest industrial and residential development has occurred in areas with limited groundwater resources. This development has already created water shortages and a need to deepen existing wells; more seriously, it has greatly handicapped or prevented the development of new industry. Thus, the township council authorized Moody and Associates to locate, if possible, several high-yield water wells to provide a sufficient municipal water supply. The purpose of this report is to describe the successful location of sand and gravel materials with excellent groundwater-yielding capability.

Summary

We conducted studies and six test drillings to locate an area in which a reliable, high-quality, municipal water well field could be developed. We have located such a site.

Results: Four of the six wells drilled were productive and penetrated sand and gravel suitable for development of municipal wells. Each of the four productive wells showed very good water-producing capability, yielding 300 to 500 gallons per minute. Further, the wells produced water of generally excellent quality, though with a potentially high but treatable concentration of manganese. One test well showed manganese concentration now high enough to require treatment, and there is concern that manganese levels in all the wells might increase to such a level. However, this would not prevent development of the well field, and additional study will be required to determine if manganese removal is necessary at all.

Conclusions: Production wells completed at the test well sites should more than supply the township's current and future needs. Analysis of test data indicates that such production wells could produce over 2 million gallons per day, twice the township's projected requirements. Further, if the water-bearing sand and gravel are as extensive as we project them to be, yields up to 5 million gallons per day could conceivably be developed.

Recommendations: We recommend that the property tested be acquired for development as a municipal well field. The property is unique in its supply capability within Milton Township, and a well field on this property should meet any present and foreseeable water requirements within the township.

Conclusions and Recommendations

Conclusions: Six test wells were drilled, four of which penetrated sand and gravel suitable for developing adequate municipal wells.

1 The relationship of geologic units encountered in Test Wells 1 through 6 is interpreted and presented in Plate 1 [not included here]. Two aquifers were identified; an extremely thick aquifer was encountered in Test Wells 2, 5, and 6 ranging in thickness from 78 to 125 feet, and a deep aquifer was encountered in Test Well 1 extending from 150 to 185 feet. There is theorized to be limited hydraulic connection through fine-grained units between the two aquifers. Some degree of connection between the deep aquifer and upper aquifer was established by the rapid rise in water levels in Test Well 1 observed during a heavy rain in September 1989.

2 A conservative estimate of the yield from the upper aquifer is 2.0 mgd. Aquifer data indicate this could be produced by pumping any combination of two wells among Test Wells 2, 5, and 6 (assuming manganese in Test Well 2 proves acceptable). Ultimate production from a French Creek well field could prove to be in the 3.0 to 5.0 mgd range.

3 It is difficult to estimate the maximum production available from the lower aquifer encountered in Test Well 1 because this unit was not identified in the other five test wells. Based on the data from the pump test of Test Well 1, the yield of the lower aquifer is estimated to be 0.5 to 1.0 mgd.

4   The water quality from all wells is good, with the exception of the anomalously elevated manganese found in Test Well 2. Iron and manganese were present in all other wells but in acceptable concentrations. If there is a source of manganese (none has been identified to date) in the immediate vicinity of Test Well 2, there is concern that due to the extremely high permeability of the aquifers, the high manganese groundwater could migrate toward the other wells with long-term pumping.

Typically, the water is moderately hard to hard, with hardness ranging from approximately 7 grains in Test Well 1 to 10 grains in Test Well 5 to 14 grains in Test Well 2.

Recommendations: Based on results from drilling and sample analysis, the following tests and land purchases are recommended.

1   A short pump test (8 hours) should be conducted on Test Well 2 to recheck the iron and manganese concentrations in this well.

2   In order to stress the aquifer as heavily as possible and simulate municipal well field operation, a pump test should be conducted on Test Well 6 with an attempted pumping rate of 700 to 1,000 gpm for 48 hours.

3   In order to secure the existing test wells, provide a buffer around these wells, and permit some aquifer test drilling, it is recommended that the township purchase property roughly rectangular in form extending 800 feet north and 800 feet south of Test Well 1 and west to the abandoned railroad tracks. This would be approximately 40 acres. If obtaining this acreage is not feasible, then it is recommended that at a minimum the township obtain property extending 800 feet north and 800 feet south of Test Well 1 and 600 feet to the west of Test Well 1. This would be approximately 22 acres. It is also recommended that a long-term option be obtained on the Barco property south of Test Well 2 and southwest of Test Well 3 for potential long-range future well field development.

The Conclusions and Recommendations section of this well report was written for geologists: it assumes a certain level of technical knowledge which a nongeologist would not have and implies conclusions which are not directly stated. For instance, item 1 of the Conclusions assumes that the reader knows what an aquifer is (a water-bearing layer of permeable rock, sand, or gravel). It assumes that the reader will see the facts in item 1 and know that they mean "we found water that we can use." This conclusion is implied, but not directly stated. Item 2 assumes that the reader knows what mgd means (million gallons per day), that a community the size of Milton Township would use only about 1 mgd, and thus that 2 mgd or 3–5 mgd would be more than enough water for the township. This item implies that enough water was found, but it does not directly state it. Item 2 of the Recommendations assumes that the reader known what "gpm" means (gallons per minute).

This kind of assumed and implied knowledge is common in texts written for experts. However, when nongeologists read the Conclusions and Recommendations section of this well report, they encounter many problems of vocabulary, assumptions, implications, and missing connections, and have serious problems understanding the text. Thus, you should not write this way in the sections of the report intended for nonspecialist readers, such as the Foreword and Summary.

We should note that although experts can understand the type of text in which generalizations are only implied, they appreciate the direct presentation of generalizations, such as those in the Summary above:

- We have located such a site (in which a reliable, high-quality, municipal water well field could be developed).

- Four of the six wells drilled were productive and penetrated sand and gravel suitable for development of municipal wells.

- Each of the four productive wells showed very good water-producing capability.

- The wells produced water of generally excellent quality.

Specialists appreciate such directly stated generalizations because they make the text easier for even the specialist to read. The specialist does not have to do the mental work to translate a fact into a conclusion—for instance, to translate the fact that "the yield from the upper aquifer is 2.0 mgd" (from item 2) to the conclusion that we have found enough water for the township. Thus, a direct general-to-particular structure, such as that presented in Chapter 12 with generalizations specified "up front," is recommended even for specialist readers.

❏ EXERCISE 15-2

A   The following Foreword and Summary were written by an engineer at Jur-Kay Carriers, a moving company specializing in long-distance hauling. Unfortunately, the engineer provided far too much detail for the busy managerial readers of this report. Your job is to write a new Foreword and Summary which give a proper (and short) overview of the problem and proposed solution. Try to make each section no longer than 100 to 150 words. Then write an Introduction to the Discussion and a Conclusions and Recommendations section. A Table of Contents for the report follows to outline the report's structure.

B   When you have finished with your revision, compare it with the revisions done by several of your classmates. Try to agree on a consensus version of the assignment in A.

Threaded Fasteners: Failure Analysis
and Proposed Improvement
Table of Contents
1   Foreword
2   Summary

Foreword

On April 3, 1989, a Jur-Kay Carriers semitractor-trailer overturned on I-96 approximately 30 miles east of St. Louis, Missouri, and the driver was seriously injured. There were no other vehicles involved in the accident. The exact cause of the accident is unknown. However, it is believed that one of three bolts in the steering mechanism failed, causing the driver to lose control. The truck, trailer, and contents of the trailer were a total loss.

On July 27, 1989, a Jur-Kay Carriers semitractor-trailer lost control on U.S. 101, 52 miles south of St. Louis, involving three other vehicles in a crash resulting in two deaths. The driver of our vehicle incurred minor injuries. The accident investigation revealed that a bolt in the power steering linkage of our tractor failed, thus causing the driver to lose control. Maintenance Record #1825, dated July 27, 1989, revealed that 5 hours before the accident a fleet mechanic replaced the original SAE-grade-8 linkage bolt with an ungraded bolt. It can be assumed that if the bolt had been replaced with an SAE-grade-8 bolt, the accident would not have occurred.

Jur-Kay Carriers is now in litigation with a suit brought against them by the owners of the three other vehicles involved in the crash.

As a direct result of these two accidents, and because for the last six months our maintenance records have shown an extremely large number of threaded fastener replacements, Paul Milani, Senior Engineer, Jur-Kay Carriers, suspected that a problem existed with threaded fastener usage in our Maintenance Department. On July 29, 1989, Mr. Milani requested that I evaluate this usage of threaded fasteners. The investigation has been completed.

The purpose of this report is to show that our mechanics have a dangerous lack of knowledge about fasteners. The purpose of this report is also to present the needed material for the recommended instruction class, emphasizing:

1   Bolt strength and preload specifications

2   Torque requirements

3   Proper selection of a nut

4   Why fasteners fail

5   What different failures indicate about the inadequacies of the specific nut-and-bolt assembly in question

Summary

A preliminary examination of maintenance records dated since January 1, 1989, was conducted by Mr. Milani. The examination has shown that a large number of threaded fastener replacements and repairs are being made in the Maintenance Department of Jur-Kay Carriers. In addition to this, there have been two serious vehicle accidents in the last four months caused by Jur-Kay Carriers. They were both the result of threaded fastener failures. Because of these facts, it is believed that a definite problem exists and an investigation into threaded fastener usage in our Maintenance Department has been made.

The problem had to lie in one of four areas (or a combination thereof): (1) our fasteners were of inferior quality, (2) there was such a lack of standardization in the fasteners we used that many types of gradings were present, resulting in utter confusion on the part of the mechanics whose job it was to select the replacement part, (3) there was simply a high degree of carelessness in the mechanics' performance of their jobs, or (4) the mechanics were simply not knowledgeable enough to make the proper grade choice when replacing a bolt.

Research revealed that the problem lay in none of the first three areas:

1   Our fasteners are not of inferior quality.

2   There is no lack of standardization in the grading of the threaded fasteners we use.

3   There is no degree of carelessness in our mechanics' performance of their jobs. In fact, the opposite was found to be the case.

The problem was found in the last area:

4   Our mechanics simply do not know the field of fasteners. Indeed, subsequent observations of activities in the garage and conversations with the mechanics verified this conclusion.

Our mechanics lack necessary knowledge in the threaded fastener field. Thus, they are unable to make the correct choices in replacing a nut-and-bolt assembly that has failed. Low-tensile-strength bolts are being substituted for high-tensile-strength bolts, and bolts are not being used for their designed purposes. This has resulted in an unnecessarily large number of failures and replacements, not to mention the two latest vehicle accidents.

I recommend that an instruction class be initiated to inform and educate our mechanics in basic areas of threaded fastener technology. I estimate that five sessions, each an hour in duration, are sufficient to present the necessary material. Such a class would eliminate the problems involving threaded fasteners present in our Maintenance Department. That is, the serious problem of substituting low-tensile-strength bolts would be avoided. Our mechanics would begin using bolts for their designed purposes. We would also solve the recurring bracketing and "bolt loosening" problems caused by reusing bolts that have stretched way beyond their optimum strength.

I volunteer my services as instructor if the need is to be met. Such a class will emphasize:

1   Bolt strength and preload specifications

2   Torque requirements

3   Proper selection of a nut

4   Why fasteners fail

5   What different failures indicate about the inadequacies of the specific nut-and-bolt assembly in question

I also recommend that after our mechanics are taught the basic threaded fastener technology, a number of bolt-grade-marking charts be posted throughout the garage.

15.2 THE FORMAL REPORT

Formal reports are usually relatively long reports written to readers outside the writer's organization. Since such reports represent the organization and spread its reputation, they are sent out with a "dressier" look—often a fancy cover and binding with the organization's seal prominently featured. Instead of the *To, From, Subject,* and *Date* heading of the informal report, a formal report begins with a Title Page, an Abstract (treated in Chapter 18), a Table of Contents, and sometimes a List of Illustrations, each beginning on a new page. Formal reports also begin the Foreword, Summary, and Discussion—or their equivalents—on new pages. In all these senses, the formal report closely resembles a book.

The Title Page, Abstract, and Table of Contents will probably be the last things written, but they will be discussed first here since they come first in the finished version of the report. The Title Page of the formal report identifies the subject and major actors in the project being reported and helps potential readers decide if the report is something they have to read. The Title Page includes all the information you have already seen in the heading of the informal report plus some funding information:

1   Name of receiver and/or receiving organization

2   Name of writer and writer's organization

3   Subject of report

Figure 15-2 ❑ TITLE PAGE FROM A FORMAL REPORT

A DESCRIPTOR VARIABLE APPROACH TO MODELING AND OPTIMIZATION OF LARGE-SCALE SYSTEMS — TITLE

FINAL REPORT
March 1985–February 1989 — TYPE OF REPORT / PROJECT DATES

DONALD N. STENGEL
DAVID G. LUENBERGER
ROBERT E. LARSON
TERRY S. CLINE — AUTHORS

SYSTEMS CONTROL, INC.
1801 Page Mill Road
Palo Alto, CA 94004 — AUTHORS' ORGANIZATION

Date Prepared—February 1989 — REPORT DATE

Prepared for the
DEPARTMENT OF ENERGY
OFFICE OF THE ASSISTANT ADMINISTRATOR FOR ENERGY
TECHNOLOGY
Division of Electric Energy Systems — FUNDING AGENCY

Work Performed under Contract NCS EX-75-0-0-2090
ET-73-0-0-2353 — FUNDING CONTRACT NUMBERS

---

4   Date of report

5   Funding agency (often this is the same as the receiver or receiving organization)

6   Funding contract number(s)

7   Beginning and ending dates of the project

8   File number of the report or document authorizing the project

The specific information on a given Title Page and its format may vary with the requirements of different organizations. You must find out what the specific requirements are for your situation and provide the required information in the required format. A sample Title Page appears in Figure 15-2.

Figure 15-3 ❏ SAMPLE TITLE PAGE AND TABLE OF CONTENTS FOR A FORMAL REPORT

SOUTHEASTERN PENNSYLVANIA TRANSPORTATION AUTHORITY

SEPACT II DEMONSTRATION PROJECT

Ridership Surveys of Pennsylvania Railroad Commuter Lines:

Levittown, Chestnut Hill, Manayunk, and Paoli Lines

and all

Reading Railroad Commuter Trains Operating in Southeastern Pennsylvania

SEPACT II Demonstration Project
Commuter Rail System Study Program
Project No. PA-MTD-4
Contract NO. H-598

Marketing Report No. 3
Philadelphia Metropolitan Area
Report 106 (D)

November 30, 1988

Volume 1

Volume 1 contains the narrative and the findings,
Volumes 2 and 3 have the complete statistical presentation.

Figure 15-3 ❑ (Continued)

Figure 15-3 ❑ SAMPLE TITLE PAGE AND TABLE OF CONTENTS FOR A
FORMAL REPORT (*Continued*)

Note that the first four items on a Title Page correspond to the *To, From, Subject,* and *Date* segment of the informal report. A major difference is in the arrangement of this information on the page. The Title Page spreads the information over the page in a visually pleasing arrangement with lots of white space between items, while the opening of an informal report does not.

In other ways, the formal report duplicates the long informal report. It is long and has a Discussion section which includes an Introduction to the Discussion and a Conclusions and Recommendations section. These sections restate the Foreword and the Summary, as described above, and the main part of the Discussion provides support for the generalizations in the Summary and the Conclusions and Recommendations.

To see how these sections are presented in a finished report, consider the sample Table of Contents for a formal report, along with its title page, which appears in Figure 15-3. The Table of Contents in Figure 15-3 has seven major parts. Part I (Introduction) corresponds to the Foreword or Introduction of the long report. Part II is the Summary of the findings and recommendations. Parts III through VII are the Discussion section of the report: Parts III through VI each have an introduction to the topic covered in that part, and Part VII contains the specific conclusions and recommendations for Parts III through VI. The information in Part VII is presented in more general form for managerial audiences in Part II (Summary). The information presented in Parts III through VI provide details and arguments needed to support the findings and recommendations, including maps and tables of ridership data. As indicated on the title page, the report described by this Table of Contents is only Volume 1 of a detailed three-volume report. Volumes 2 and 3 present the statistical presentation and analysis to support the data and conclusions in Volume 1. In this sense, Volumes 2 and 3 function as Appendixes for Volume 1: they contain supporting information so it will be available if needed, but not interrupt the flow of the argument in Volume 1. Also note that the title page differs from that in Figure 15-2 in that it does not list individual authors or the funding agency which will receive the report. In fact, this report was written for the Southeastern Pennsylvania Transportation Authority, which was both the funding agency and the corporate author. The report is a large marketing report written by the organization for the organization, and it does include all of the other standard information: title, corporate writer (the writer's organization), date, and funding contract numbers. Another Table of Contents for a formal report appears in Appendix B.

16

The Proposal

A proposal requests support—usually money—for work that a proposer wants to do. It could be written to a funding agency and propose to conduct a scientific research project; it could be written to the managers of a company and propose to solve a problem within the company; it could be written to a customer and propose to provide services or products which the customer needs. In any event, the person or group receiving the proposal has its own interests and goals, which may or may not coincide with those of the proposer. Thus, the proposal must convince the person who receives it—the potential funder—that the proposed activity will be a good "investment," that is, that the activity is worthy of support and will advance the funder's goals, produce high-quality results, and do all this better than other activities competing for the same funds. To make things even more difficult, the proposal must make these arguments to busy readers trying to divide too few resources among too many applicants.

To be successful in this environment, you must quickly and clearly answer the questions a potential funder will bring to any request for support:

What does the proposer want to do?

How much will it cost?

Is the problem important and relevant to the funder's interests?

Will the proposed activity solve or reduce the problem?

Can the proposed activity be done? Will it duplicate other work?

Is the method or approach appropriate, clearly defined, and well thought out?

310

Can the results be adequately evaluated?

Is the proposer qualified to do the activity or work? Better qualified than others?

Will the results of the activity be available to others?

Are the proposed schedule of activity and budget reasonable?

Answers to these questions will depend on the particular guidelines, interests, and standards of particular funders. To be sure that *your* proposal meets the expectations of a given funder, you should examine whatever literature you can find on the funder's goals or interests. For example, if you are thinking of submitting an unsolicited proposal to a private company, you might first want to consult published reviews of the company and the company's annual report to see if your proposed project fits into its agenda of activities and then talk to someone at the company about the company's needs.

In the case of solicited proposals, the best initial source of information is usually the Request for Proposals (RFP) given out by the funder. To avoid being inundated by inappropriate proposals and to provide guidelines for reviewers, funding agencies usually put considerable time and thought into the wording of RFPs. Therefore, you should read the RFP very carefully before deciding whether to send a proposal to that particular agency. If your proposed project and the RFP do not match up, you would do well to look for another agency.

Even if there seems to be a good match between a company's needs or an RFP and your project, you never know unless you first check with a relevant company manager or the project officer at the funding agency. It's better to make sure you're on the right track early on than risk losing months of time on a proposal that has no chance of winning. Therefore, at some point early in the proposal process, it's a good idea to call the relevant company manager or project officer and make sure that your project truly fits the company's or agency's interests. In many situations, the manager or project officer plays a major role in the decision-making process—screening submissions, selecting reviewers, presenting cases to the final selection committee, etc. Talking with the manager or project officer allows you to make sure that he or she understands the nature of your project. (Written preproposals serve the same purpose, but not all companies or agencies accept them. If you are seeking support from some institution that does accept preproposals, you would be foolish not to take advantage of such a good opportunity for early feedback.)

The penalties for failing to have the right information are severe: failure in the proposal process and much wasted time. This is illustrated in the following comment aimed at university proposal writers, but appropriate for a more general audience.

> For many of the agencies issuing RFP's [Request for Proposals], the colleges responding with proposals are automatically rejected unless proposal instructions

are completely complied with. Others are more subtle. They simply imply or interpret inability from the proposal's inability to meet the RFP criteria. Either course is negative.[1]

16.1 THE ORGANIZATION OF A FORMAL PROPOSAL

Most proposals follow the general outline presented in Figure 16-1, an outline similar to that given in Chapter 4 for the argument of fact and policy. Indeed, you could argue that a proposal is an argument of policy (the funding agency should fund this proposal) based on several embedded arguments of fact (the problem is important, the activity proposed will alleviate the problem, the proposers are qualified to do the work proposed, etc.).

The general outline should be adapted and modified according to the needs of the readers and the demands of the topic proposed. For instance, long or complicated proposals might well contain all the sections shown in Figure 16-1 in fully developed form, with their own headings corresponding to the sections and subsections of the outline. In contrast, shorter or simpler proposals might contain only the sections noted in Figure 16-2; they might treat a given subsection very briefly or combine several subsections. A very short informal proposal is shown in Figure 16-4. While its heading information (To, From, Subject, Date, etc.) is formatted differently than that for a formal proposal, it well illustrates the compression of sections that can occur in shorter formal proposals.

Notice that since a short proposal is short and simple, it may have no Table of Contents or separate sections for Background or List of References, as indicated in Figure 16-2. Any background information and references may be folded in with the discussion of the problem (in Problem Addressed) or with the other sections. Also, in Figure 16-2, the Plan for Accomplishing Objectives, the Plan for Evaluating Results, and the Schedule for Project Completion are listed as major topics, not subtopics under Description of Proposed Activity as in Figure 16-1, Parts of a Long Proposal.

Title Page

The Title Page provides the basic *To, From, Subject, Date* information found in headings and title pages for other types of technical communications; it also includes financial information relevant to proposals alone. Specific formats for title pages vary from one proposal to another, but most include the following:

1 The title of the proposal (as short and informative as possible)

2 A reference number for the proposal

3 The name of the potential funder (the recipient of the proposal)

Figure 16-1 ❑ PARTS OF A LONG PROPOSAL

Title page

Abstract

Table of contents

Introduction
 Problem addressed
 Purpose or objectives of proposed activity
 Significance of proposed activity

Background

Description of proposed activity
 Plan for accomplishing objectives
 Plan for evaluating results
 Schedule for project completion

Institutional resources and commitments

List of references (for about six or more references)

Personnel
 Explanation of proposed staffing
 Relevant experience of major personnel

Budget
 Budget in tabular form
 Justification of budget items

Appendixes
 Letters of endorsement
 Promises of participation, subcontractor's proposals
 Biographical data sheets (vita sheets)
 Reprints of relevant articles, reports, background documents

4 The name and address of the proposer(s): the project director or project
 manager and other especially important participants (research proposals
 will include the name of the principal investigator and any coprincipal
 investigators)

5 The proposed starting date and duration of the project

6 The total funds requested

Figure 16-2 ❑ PARTS OF A SHORT PROPOSAL

Title page (ideally, with abstract)

Introduction
    Problem addressed
    Purpose or objectives of proposed work
    Significance of proposed work

Plan for accomplishing objectives

Plan for evaluating results

Schedule for project completion

Institutional resources and commitments

Personnel
    Explanation of proposed staffing
    Relevant experience of major personnel

Budget
    Budget in tabular form
    Justification of budget items, where necessary

Appendixes or attachments, if needed

---

7 The proposal's date of submission

8 The signatures of the project director and responsible administrator(s) in the proposer's institution or company

    A sample Title Page is shown in Figure 16-3. It lacks items 7 and 8 because these were included on the institution's cover form, which accompanied the proposal.

    The Title Page provides the basic title information needed by a reader deciding whether or not to read the proposal, and the title and reference number needed for filing it. It also provides an overview of the funding and timing of the project.

Abstract

Like other Abstracts, the Abstract of a proposal is short, often 200 words or less. In a short proposal addressed to someone within the writer's institution, the Abstract may be located on the Title Page; in a long proposal, the Abstract will usually occupy a page by itself following the Title Page.

Figure 16-3 ❏ SAMPLE TITLE PAGE FOR A FORMAL PROPOSAL

PROPOSAL FOR EXTENSION OF
NASA GRANT NSG 1306

MODELS AND TECHNIQUES FOR EVALUATING THE

EFFECTIVENESS OF AIRCRAFT COMPUTING SYSTEMS

IN REPLY REFER TO:
DRDA 81-2096-P1

Submitted to the

NATIONAL AERONAUTICS AND SPACE ADMINISTRATION
LANGLEY RESEARCH CENTER
HAMPTON, VIRGINIA 23365

Submitted by the

SYSTEMS ENGINEERING LABORATORY
DEPARTMENT OF ELECTRICAL AND COMPUTER ENGINEERING
THE UNIVERSITY OF MICHIGAN
ANN ARBOR, MICHIGAN 48109

PRINCIPAL INVESTIGATOR: John F. Meyer

PROPOSED STARTING DATE: 1 July 1990

PROPOSED DURATION: 1 year

AMOUNT REQUESTED: $39,932

The Abstract is a critical part of the proposal because it provides a short overview and summary of the entire proposal; it is the only text in the proposal seen by some readers. The Abstract should briefly define the problem and its importance, the objectives of the project, the method of evaluation, and the potential impact of the project; it normally does not define the cost. A longer treatment of Abstracts appears in Chapter 18.

Table of Contents

The Table of Contents lists the sections and subsections of the proposal and their page numbers.

Introduction

The Introduction of the proposal, like the Foreword of the technical report, orients a nonspecialist to the subject and purpose of the document. It explains why the project is important and should provide the following pieces of information in terms appropriate for a managerial audience.

1 The problem being addressed (perhaps defining what it isn't as well as what it is)

2 The purpose or objectives of the proposed project

3 The significance of the proposed project

If the project is simple, the Introduction may also include some relevant details which would belong in the Background section of a more complicated proposal.

If you are responding to a Request for Proposals (RFP), your approach to the definition of the problem and the purpose or objectives of your proposal should reflect the thrust of the RFP.

Background

As a separate section, the Background allows you to fill in important technical details inappropriate for the nonspecialist readers of the Abstract and Introduction. The Background provides a place to discuss the history of the problem, to survey previous work on your topic (a survey leading up, of course, to some problem or gap in the previous work), and to place this project in the particular context of previous work you may have done on the problem. If the proposal extends earlier work you have done, be sure to show why your previous work needs to be continued and how the proposed work differs from it. Do not spend a great deal of time justifying your earlier efforts and budgets; concentrate on the new work proposed.

If previous work has been done on your project, you need to demonstrate that you are aware of it and understand its importance and limitations. (Funders don't want to pay you to reinvent the wheel.) You usually do this in a subsection called Literature Survey or Previous Work, in which you demonstrate your competence by carefully selecting and evaluating the works you cite. If you can't select the most crucial items for your project and briefly show why they're crucial, you're probably not expert enough to know what you need to do and how best to do it. Also, you won't be able to show how your work fits into the larger scheme of things, how it builds on previous work and goes beyond it, how it is original and contributes to knowledge in the field.

Description of Proposed Activity

The Description of Proposed Activity is the most important section of the whole proposal. It describes what you want to do and how you intend to do it:

1   The plan for reaching the stated objectives

2   The plan for evaluating the results

3   The schedule for completing the work

This section will be evaluated carefully by the proposal's reviewers, who will be knowledgeable in the field. Their job is to eliminate all proposals whose objectives or plans are inappropriate or unclear or not well thought out. Thus, your job as a writer is to convince them that you are doing what needs to be done and are doing it in the most careful and thorough way.

When preparing the Description section, you should assume that you are writing for a critical, hostile audience, for you are. You should provide all the details a knowledgeable critic would need to assess your argument:

1   The assumptions on which your work is based

2   The approach or hypothesis you are following

3   The specific problem(s) or question(s) you are trying to address

4   The particular work and evaluation methods you are using

5   The appropriateness of your methods for the problem proposed

In addition to providing these items, you may need to justify them, especially if there might be any question or controversy about them. It is particularly important to demonstrate the appropriateness of your method for solving the problem posed. If this isn't clear, probably nothing else will matter.

You also need to convince the critical reader that your proposed schedule is appropriate and realistic. You don't want to propose to do too much, given your time and resources, or it will seem that you have a poor assessment of the project and don't really know what you are doing. One way to demonstrate that your schedule is realistic is to spell it out very specifically so that the critical reader can easily see its merits and the care and thoroughness you put into determining it. If you demonstrate that you've really thought of everything, carefully, you're halfway to success.

Institutional Resources and Commitments

If you are proposing a project that requires special equipment, one important factor in your ability to do the proposed work is having access to that equipment. Having this equipment already available at your institution is a big plus for your proposal, since a funder could pay you much less to do the work than it would have to pay someone who had to buy the equipment. Thus, it is to your advantage to list relevant institutional resources. Further, funding agencies often feel that proposers work harder (and institutions monitor them more carefully) if the proposer's institution has resources invested in the project.

List of References

If your references are so extensive that they may interrupt the text if you insert them as you go along, you may want to set them up in a separate section. You may also want to do this if the previous work is especially important and you want your reviewers to see that you have cited all the "right" items. As a rule of thumb, if you have more than six references, you might consider a List of References, placed before the sections on Personnel and Budget. The references may be listed consecutively as they appear in the text, with the author's name in normal order (first name or initials first), or they may be listed alphabetically with the author's last name first.

Personnel

The purpose of the Personnel section is to explain who will be doing what and to demonstrate that the people listed for a proposed activity are competent to do it. This is normally accomplished in two subsections, one outlining the responsibilities of the individual participants and the structure for coordinating their activities and one providing short biographical sheets (usually no more than two pages) for the main participants. The biographical sheets should focus on only the *relevant* qualifications of the participants.

Budget

Like the Personnel section, the Budget section has two purposes: to explain what things will cost and to justify and explain individual expenditures, especially when these are not obvious. The Budget is usually summarized in a table, such as Table 16-1. (A simple proposal may have a much simpler budget.) The typical headings in a Budget are Personnel, Equipment, Supplies, Travel, Computer Time (if relevant), and Indirect Costs. Items typically included under these headings are listed in Table 16-2. Note that not all proposals will need to include all items, but the items listed in Table 16-2 should be treated in a proposal's budget when appropriate.

Table 16-1 ❏ SAMPLE 12-MONTH BUDGET

	Contributed by Sponsor	Contributed by Our Company	Total
Personnel			
Project director, half time	$ 25,000	$ 0	$ 25,000
Project associate, quarter time	0	9,000	9,000
Research assistant, full time	15,000	0	15,000
Clerk-typist, full time	20,000	0	20,000
Subtotal	$ 60,000	$ 9,000	$ 69,000
Staff benefits	9,000	1,200	10,350
(15% of salaries and wages)			
Subtotal	$ 69,000	$10,200	$ 79,350
Consultants			
Warren Duval, $400/day, 2 days	$ 800	$ 0	$ 800
Equipment			
Methometer	$ 2,000	$ 0	$ 2,000
Materials and Supplies			
Miscellaneous office supplies	$ 200	$ 0	$ 200
Glassware	200	0	200
Chemicals	200	0	200
Subtotal	$ 600	$ 0	$ 600
Travel			
Project director consultation with sponsor, Chicago to Washington, DC, and return; 1 person, 2 days			
Airfare	$ 294	$ 0	$ 294
Per diem @ $65/day	130	0	130
Local transportation	40	0	40
Subtotal	$ 464	$ 0	$ 464
Total Direct Costs	$ 72,864	$10,200	$ 83,214
Indirect Costs (69% of salaries and wages, including staff benefits)	$ 48,162	$ 7,038	$ 55,306
Grand Total	$121,026	$17,238	$138,520

SOURCE: Adapted from *Proposal Writer's Guide* (Ann Arbor: University of Michigan, Division of Research Development and Administration, September 1975), p. 9.

Table 16-2 ❑ CHECKLIST FOR PROPOSAL BUDGET ITEMS

Salaries and Wages

1   Academic personnel
2   Research assistants
3   Stipends (training grants only)
4   Consultants
5   Interviews
6   Computer programmers
7   Tabulators
8   Secretaries
9   Clerk-typists
10 Editorial assistants
11 Technicians
12 Subjects
13 Hourly personnel
14 Staff benefits
15 Salary increases in proposals that extend into a new year
16 Vacation accrual and/or use

Equipment

1   Fixed equipment
2   Movable equipment
3   Office equipment
4   Equipment rental
5   Equipment installation

Materials and Supplies

1   Office Supplies
2   Communications
3   Test materials
4   Questionnaire forms
5   Duplicating materials
6   Animals
7   Animal food
8   Laboratory supplies
8   Laboratory supplies
9   Glassware
10 Chemicals
11 Electronic supplies
12 Report materials and supplies

Travel

1   Administrative
2   Field work
3   Professional meetings
4   Travel for consultation
5   Consultant's travel
6   Subsistence
7   Automobile rental
8   Aircraft rental
9   Ship rental

Services

1   Computer use
2   Duplication services (reports, etc.)
3   Publication costs
4   Photographic services
5   Service contracts
6   Special services (surveys, etc.)

Other

1   Space rental
2   Alterations and renovations
3   Purchase of periodicals and books
4   Patient reimbursement
5   Tuition and fees (training grants)
6   Hospitalization
7   Page charges
8   Subcontracts

Indirect Costs

SOURCE: *Proposal Writer's Guide* (Ann Arbor: University of Michigan, Division of Research Development and Administration, September 1975), p. 10.

Appendixes

Appendixes are reserved for necessary supporting documents which, because of their length or type, would disrupt the "flow" of the proposal. The most common Appendix items are biography sheets, letters of endorsement for the proposal, and promises of participation from important participants. Other materials may be pertinent to a given proposal, but the proposal writer should consider carefully any item included in the Appendix and eliminate anything not really needed to support the importance of the topic, the credentials of the proposers, or the ability of the proposers to carry out their work.

16.2 THE ORGANIZATION OF A SHORT INFORMAL PROPOSAL

The outlines for proposals presented in Figures 16-1 and 16-2 are relatively full outlines; they allow writers to argue convincingly to someone who doesn't know them that the stated problem is important, that their proposed activity will alleviate the problem, and that they are qualified to do the proposed work. Sometimes, however, proposers need to write very short, almost routine proposals meant for people in their own unit. Such proposals might concern some small thing to be done, some small problem to be addressed, some small piece of equipment to be purchased. In such a situation, a proposal using a compressed form of the outline shown in Figure 16-2 might be very appropriate.

Such a proposal appears in Figure 16-4. This proposal omits the Personnel and Budget sections. These are, however, implied in Sections 5 and 6: one person, the writer, will work for 4 weeks, and perhaps someone from the Hardware Department will work for a short time to provide the necessary hardware. However, the proposal does contain the other sections in relatively obvious terms. It has an Introduction, which states

1   The problem (we need a catchy demonstration program)

2   The objectives of the proposed work (to provide a demonstration program by simulating the control panel of a nuclear reactor)

3   The significance of the work (to prove the ability of the computer programming language, Industrial Pascal, to communicate with various input/output devices and to test its capabilities to monitor and control several simultaneous tasks of the sort described in the proposal)

The proposal has a plan for accomplishing the objectives, and this is outlined in Sections 2–5. It also has an implied plan for evaluating the results; an observer would simply see if the demonstration program simulates the nuclear control panel by producing the actions described in Section 4. Finally, the proposal has a schedule (Section 6), and it lists resources already available in the unit (Section 2) as well as needed resources (Section 5).

Figure 16-4 ❏ SAMPLE OF A SHORT INFORMAL PROPOSAL

**Chen Computer Systems
Interoffice Memorandum**

To: John Van Roekel
             Software Manager

From: Larry N. Engelhardt
             Senior Programmer

Date: 9 August 1989

Subject: Proposed Demonstration Project for Industrial Pascal: Specifications and
             Description

1   Introduction

The Programming Division has just finished developing an Industrial Pascal* language
system for our new 3935 RacPac minicomputer system. We now need a computer program
that will allow our salespeople to demonstrate the capabilities of the RacPac system
running Industrial Pascal in a "real" environment.

I propose to develop a demonstration program that will control a simple but realistic
representation of a control panel of a nuclear reactor. This demonstration program will
be written entirely in Pascal and will run on a RacPac system in an Industrial Pascal
environment.

The RacPac will be hooked into a 3910A-3920A simulator, and together the RacPac and
simulator will represent the control panel of a nuclear reactor. The program will read and
write digital and analog inputs and outputs to and from the controls and indicators on
the simulator front panel and respond in an interesting and amusing manner. This
demonstration will both prove the ability of Industrial Pascal to communicate with various
input/output devices and test its capabilities to monitor and control several simultaneous
tasks in our environment.

2   Hardware Capability

The hardware configuration on which the demonstration program will run consists of a
simulator built by Bill Lepior and a 3935A RacPac system. The simulator is a cabinet
which features two potentiometers and a meter connected to a 3910A analog I/O controller,
and eight lamps and eight switches connected to a 3920A digital I/O controller. Both
controllers are plugged into the chassis of the RacPac. Furthermore, the RacPac has an
integral plastic keyboard and 5–inch CRT screen controlled by an 1812 interface for

---

*Authors' Note: Pascal is a computer language, and Industrial Pascal (IP) is a version of Pascal used on small
computers in industrial settings to control machinery or to monitor various processes. Software is a computer
program or programs. I/O is input/output, that is, the text or information put into a computer (input) or that given
out by it (output). A CRT screen is a cathode ray tube screen, a screen similar to a television, which prints out
the letters and numbers the operator types and which prints anything the operator commands the computer to
calculate and print. Hardware is equipment—computers (the "hard" parts of computers) and other equipment—
and a PROM (Programmable, Read-Only Memory) is a small part in a computer.

Figure 16-4 ❑ (*Continued*)

operator I/O. This mix of digital and analog I/O devices offers a good sample of control devices.

3 The Demonstration

One assembler language program has already been written to demonstrate the simulator. Of course, a Pascal program could be written to duplicate the functions of the assembler program exactly. However, to be interesting for the customers, the Pascal demo program could be made more realistic. In addition, a dynamic and robust demonstration program is needed to make use of all the I/O devices of the simulator and also to exercise the multitasking features in IP in a real-time control situation. This document proposes that the IP demo program be designed to be a nuclear reactor control panel.

4 Operator Controls

The IP demo program would turn the simulator, keyboard, and screen into a simplistic representation of a control panel of a nuclear reactor. For example, the eight panel switches could move damper rods into and out of the make-believe reactor. The corresponding panel lights would go on when a rod is removed. As damping rods are removed, the core temperature would rise over time. The core temperature can be displayed by the large analog meter. As more rods are withdrawn, the temperature would rise at a faster rate. The "operator" can reduce the rate of temperature rise by increasing the cooling water feed rate with one potentiometer. The second potentiometer could set an alarm point on a temperature which will cause the reactor to "scram" automatically. The goal of the "operator" is to keep the reactor running in a range between letting the core cool off and scramming, or going critical.

The screen of the RacPac can display values for all of the activity of the simulation. The screen can also display helpful but ominous warning messages such as "Warning: Evacuate the state of Pennsylvania!" While the screen duplicates the analog and digital inputs and outputs, the keyboard will accept some simple commands to control execution of the program.

5 Hardware Required

The development of the demonstration program will require a 3800B for editing and compiling. Intermediate debugging could be done by plugging the simulator into a 3800B if it is equipped with an 1850 controller board in addition to its standard 1824. Final checkout will require either down-line loading of the program from a 3800B to the simulator's RacPac or burning of the program into PROMs for installation into the RacPac. The Hardware Department has both a 3800B and an 1850 controller board, and it could burn the program into PROMs and install the PROMs.

6 Schedule

Description	Weeks Required	Total Time
Research drivers	1	1
Design program	1	2
Code	1	3
Debug	1	4

In what it includes, then, this proposal provides the context and information necessary for those reviewing it while still being an appropriately short proposal for a small project.

16.3 EDITING THE PROPOSAL

Once you have written a proposal, it is wise to analyze it for weak spots, areas needing more proof or detail. As a guide, you might refer to the analysis below of the problems detected in 605 proposals rejected by the National Institutes of Health.[2] While this is an analysis of research proposals, it seems to reflect problems with other kinds of proposals as well. More than one item may have been cited in rejecting a proposal.

A PROBLEM (58%)	%
1 The problem is not of sufficient importance or is unlikely to produce any new or useful information.	33.1
2 The proposed research is based on a hypothesis that rests on insufficient evidence, is doubtful, or is unsound.	8.9
3 The problem is more complex than the investigator appears to realize.	8.1
4 The problem has only local significance, or is one of production or control, or otherwise fails to fall sufficiently clearly within the general field of health-related research.	4.8
5 The problem is scientifically premature and warrants, at most, only a pilot study.	3.1
6 The research as proposed is overly involved, with too many elements under simultaneous investigation.	3.0
7 The description of the nature of the research and of its significance leaves the proposal nebulous and diffuse and without a clear research aim.	2.6

B APPROACH (73%)	%
1 The proposed tests, methods, or scientific procedures are unsuited to the stated objective.	34.7
2 The description of the approach is too nebulous, diffuse, and lacking in clarity to permit adequate evaluation.	28.8
3 The overall design of the study has not been carefully thought out.	14.7
4 The statistical aspects of the approach have not been given sufficient consideration.	8.1
5 The approach lacks scientific imagination.	7.4
6 Controls are either inadequately conceived or inadequately described.	6.8
7 The material the investigator proposes to use is unsuited to the objective of the study or is difficult to obtain.	3.8
8 The number of observations is unsuitable.	2.5
9 The equipment contemplated is outmoded or otherwise unsuitable.	1.0

C	INVESTIGATOR (55%)	%
1	The investigator does not have adequate experience or training for this research.	32.6
2	The investigator appears to be unfamiliar with recent pertinent literature or methods.	13.7
3	The investigator's previously published work in this field does not inspire confidence.	12.6
4	The investigator proposes to rely too heavily on insufficiently experienced associates.	5.0
5	The investigator is spreading himself too thin; he will be more productive if he concentrates on fewer projects.	3.8
6	The investigator needs more liaison with colleagues in this field or in collateral fields.	1.7

D	OTHER (16%)	%
1	The requirements for equipment or personnel are unrealistic.	10.1
2	It appears that other responsibilities would prevent devotion of sufficient time and attention to this research.	3.0
3	The institutional setting is unfavorable.	2.3
4	Research grants to the investigator, now in force, are adequate in scope and amount to cover the proposed research.	1.5

Other problems lead to the following suggestions to proposal writers offered by a foundation executive, who reviews proposals and deals with authors.[3]

1   Be realistic; don't promise global changes from your efforts.

2   When you call a funder, be organized; don't use the funder to motivate you to organize your thoughts.

3   Keep the funder informed of your work; don't let the next time the funder sees you be when you are asking for second-year funding.

4   Requests for continued funding are just as important as the original request; don't merely submit a letter asking for more money.

16.4 GETTING THE PROPOSAL APPROVED FOR SUBMISSION

Different organizations have different procedures for handling proposals. Often proposals to someone within the writer's organization can be submitted without much red tape. You may need to get your supervisor's approval before sending out a proposal asking that you work for another department or unit, but a verbal OK may be all the approval you need. (Note, however, that you should at least inform your supervisor if you propose to work on a project outside your unit.)

In contrast to the simple procedures for approving internal proposals, the procedures for approving proposals to funding agencies outside the proposer's organization may be quite involved. Organizations often insist on formally reviewing and approving any proposal to external funders. This allows the organization to monitor commitments made to outside organizations; it also allows the organization to eliminate undesirable competition among members of its own staff who may be competing with one another for funding. The problem with this organizational review and approval—from the proposer's point of view—is that it takes extra time, and the proposer must plan that time into the writing process.

In addition, the proposer may need to get special forms filled out and signed by administrative officers in the proposer's organization. For instance, the sample approval form in Figure 16-5 for proposals seeking external funds had to be filled out and signed by four administrative officers in addition to the proposer. Just finding these people can take some time, and theoretically each person should have some time to review the contents of the proposal he or she is approving.

If the proposer's organization requires such approval, the proposer should be aware of this fact long before the proposal's deadline and should plan to allow sufficient time for each reviewer.

❑ EXERCISE 16-1

Compare the structure of a long and short proposal to the proposal outlines given in Figures 16-1 and 16-2. See if you can identify the various sections outlined in the figures. Compare the treatment of these sections with the comparable treatment of sections in the short informal proposal in Figure 16-4.

❑ EXERCISE 16-2

Write a short formal or informal proposal for some project you want to do. Be sure to include all relevant information, and be prepared to justify the information you have included or excluded.

REFERENCES

1   Tim Whalen, "Grant Proposals: A Rhetorical Approach," *The ABCA Bulletin*, March 1982, p. 36.

2   *Proposal Writer's Guide*, Division of Research Development and Administration, University of Michigan, Ann Arbor, 1975. From E. M. Allen, *Science*, November 25, 1960, pp. 1532–1534.

3   Bill Somerville, "Where Proposals Fail: A Foundation Executive's Basic List of What to Do and Not Do When Requesting Funding," *The Grantsmanship Center News*, January–February 1982, pp. 24–25.

Figure 16-5 ❑ SAMPLE APPROVAL FORM FOR A PROPOSAL SEEKING EXTERNAL FUNDS

Proposal no. DRDA-81-2096-P1
Date 5/21/89

The University of Michigan

APPROVAL OF APPLICATION FOR GRANT OR CONTRACT
(See back of form for instructions.)

X Research   New
___ Instruction   X Continuation of acct. no.: 014524
___ Facilities   ___ Competing renewal
___ Fellowship/traineeship   ___ Noncompeting renewal
___ Equipment/materials   X Supplement
___ Service

To be submitted to: National Aeronautics and Space Administration
(Proposed Sponsor)

Project Director   Prof. John F. Meyer
Department or unit   Elec. & Comp. Eng.
(Limit to One)
Organization code   2160

1. Major fields of study to which project is related:
   Code nos.   11.80.50
   (Please Limit to 3 Codes)

2. PROJECT TITLE (80 spaces max.) Models and Techniques for Evaluating Effectiveness of Aircraft Computing Systems

3. Proposed period (must agree with period for which budget in Item 9 is given):
   From 7/1/89 through 6/30/90   Total months: 12   Total cost: $39,932

4. Does the proposed activity involve:
   a. Use of human subjects Yes ___ No X   If yes, date of committee approval ___
   b. Use of vertebrate animals Yes ___ No X
   c. Lease or purchase of computer equipment Yes ___ No X
   d. Recombinant DNA Yes ___ No X   If yes, date of committee approval ___
   e. Classified research Yes ___ No X   If yes, date of committee approval ___

5. Participating faculty and students:
   a. The following faculty members other than the project director will participate.

Name	Rank and Unit	Nature of Participation

Figure 16-5 ❑ SAMPLE APPROVAL FORM FOR A PROPOSAL SEEKING EXTERNAL FUNDS
(*Continued*)

    b. Number of students to participate: Postdoctoral: _____ Graduate: __2__
       Undergraduate: _____

6. University facilities required:
    a. Will adequate space be available for the period proposed?
       Yes __X__ No_____
       If yes, Bldg.: E. Eng. Room: 2523,2506,1   Approved: _____
       If no, space required: _____ sq. ft. Source: _____ Approved: _____
    b. Space renovation $_____   UM account no.: _____ Approved: _____
    c. Will acquisition of major equipment items require installation and building modi-
       fication at cost to the University?
       Yes _____ No __X__.

7. Is additional office equipment required for this project? Yes _____ No __X__
   Is it provided for in the proposal? Yes _____ No_____
   If not, attach list and indicate University unit to provide: _____

   _____
      (approved)

8. Is work to be performed off Univeristy property? Yes _____ No __X__
   If yes, indicate location and duration: _____

9. Summary of proposed budget:

	UM sources:	Sponsor	Total	Acct. no. for UM sources	Approved
a. Salaries and Wages (S&W)	_____	21,405	21,405	_____	_____
b. Staff Benefits (SB)	_____	3,853	3,853	_____	_____
c. Fellowships/Stipends	_____	_____	_____	_____	_____
d. Tuition	_____	_____	_____	_____	_____
e. Consultants	_____	_____	_____	_____	_____
f. Travel	_____	3,100	3,100	_____	_____
g. Supplies/Materials	_____	_____	_____	_____	_____
h. Equipment*	_____	_____	_____	_____	_____
i. Computer Costs	_____	1,000	1,000	_____	_____
j. Subcontracts	_____	_____	_____	_____	_____
k. Other (itemize under notes)**	_____	600	600	_____	_____
l. Total Direct Costs (TDC)	_____	29,958	29,958	_____	_____
m. Indirect Costs***	_____	16,974	16,974	_____	_____

   17.8% Computer = $178
   58% MTDC = $16,796

Less carryover	_____	(−7,000)	(−7,000)	
n. TOTAL		$39,932	$39,932	

10. NOTES:
     *Itemize items of equipment costing more than $10,000.
     **Publication costs/supplies.
    ***Explanation if other than negotiated rates.

Figure 16-5 ❑ *(Continued)*

(Requested by: Project Director)
J.F. Meyer

(Fiscal/Legal Review)

(Approved by: Dept. or Unit Head)
G.I. Haddad

(Approved for DRDA by: Project Representative)
Mary L. Egger

(Approved for School/College by: Dean)
M.J. Sinnott

(Approved for the University by: Vice President)
C.G. Overberger

ADDITIONAL READING

John H. Behling, *Guidelines for Preparing the Research Proposal*, rev. ed. (1984).

Greg Myers, "The Social Construction of Two Biologists' Proposals," *Written Communication 2*: 219–245 (1985).

William S. Pfeiffer, *Proposal Writing: The Art of Friendly Persuasion* (Columbus, OH: Merrill, 1989).

17

Instructions, Procedures, and Computer Documentation

Writers sometimes need to produce descriptions of standard procedures which must be followed, or instructions for teaching someone how to do something, or other types of documentation. For simplicity, all of these will be called *instructions* in this chapter. Instructions can be written for many types of activities, some very complex and others relatively simple, some confined to the workplace and others found in everyday life. Such activities might include using or repairing a piece of highly technical equipment, running a new videotape recorder, writing a report, enrolling a child in a new school, preparing a child for camp, running a conference, moving an office with its equipment and staff, or cooking a new dish. In each case, if the instructions are not adequately developed, the user of the instructions will not be able to complete the desired task. The adequacy of instructions is especially critical for people who do not know much about the task to be done—for example, new users of a computer system who must rely on the adequacy of the instructions to learn how to set up the computer and how to run it.

There are two basic approaches to writing instructions—organizing the information according to topics, or organizing the information according to the tasks which the user is likely to do. Information organized according to topics tends not to match the reader's needs very well; information organized according to tasks is usually easier for the reader to use.

Thus, in this chapter, the writing of instructions will be presented as a four-stage process of analyzing the situation and tasks; planning the text; writing and editing the text; and evaluating, revising, and testing the text[1]—all to be completed before the instructions are given to the final users. This process may be applied to relatively modest projects and to the individual documents in large, multivolume projects.

17.1 ANALYZING THE SITUATION AND TASKS

The analysis stage occurs before any planning or writing has been done and focuses (1) on the situation *for* which the instructions are to be written—the audience, purpose, and context for the instructions; (2) on the situation *in* which the instructions are to be written—the resources, constraints, and deadlines inherent in the writer's situation; and (3) on the content of the instructions—the tasks which need to be learned. Only when these things are known can the writer adequately plan and develop the instructions.

Analyzing the Situation: Audience, Purpose, and Context

Before you start writing instructions or other documentation, you must carefully define your audience. To do this, you can make use of current research about the audiences for instructions and then analyze the special characteristics of your particular audience. Let us first consider some insights from the research.

Sticht, Fox, Hauke, and Welty-Sapf studied the reading tasks done in Navy technical training,[2] and Mikulecky analyzed reading tasks done by students and civilian employees.[3] Both identified two major classes of tasks: reading-to-do tasks in which information is found and read to perform some action (but not necessarily remembered) and reading-to-learn tasks in which information is read to be understood, learned, and remembered. The studies found that 78% of the work tasks for civilian or military employees were reading-to-do tasks and 15% were reading-to-learn tasks; these percentages are almost exactly reversed for school situations. Unfortunately, as Kern and Sticht have found, writers of training and instructional material are often college graduates using a textbook reading-to-learn model for their instructions,[4] and they produce instructions which do not work well in the workplace[5] and are often too general for use on the job.[6]

These problems are compounded by reading behavior such as that observed by Wright in a study of instructions for a variety of common consumer products. She found that less than 60% of the users said they would read all of the instructions, and that in most instances at least 30% of the users would not read any instructions.[7] Of course, the percentage who read at least some instructions increased as the products became more complex. However, even for complex systems, many people read instructions only sketchily. This is illustrated by Sullivan and Flower's study of users of a new information retrieval system in a library: "They read only sections of the document, they skip the orienting material in introductions, and they read primarily to answer specific questions rather than to learn or do"[8] and they skip relevant information in the process.

Given these reading needs and behaviors, audiences for instructions have predictable demands, as shown in Showell and Brennan's study of users of Army instructions: "The most frequent type of new publication requested was one oriented around job duties, in effect, a how-to-do-it handbook instead of a textbook on everything you always wanted to know about the topic."[9]

Thus, your definition of audience should consider the points just made about audiences for instructions and then focus on the specific audience for your particular document. If possible, the audience definition should be quite specific, listing for each type of person in the audience such issues as level and type of education, expertise on the subject, type of job, and particular needs and interests in your instructions. It may sometimes be hard to do this, since

> writers frequently have no direct contact with the users, the job, or the job setting. The writer's assignment may not specifically identify the user and the purpose or function the manual is intended to serve (Kern & Sticht, 1980; Kern, Sticht, Welty, & Hauke, 1977). The intended user may turn out to be several different groups of users performing different types of job activities and sharing only a general type of relationship to the same piece of equipment or content area. In addition, the manual is frequently required to serve the multiple functions of a general reference text, a training text, and a procedural guide for performance of specific tasks.[10]

Further, if you are writing computer documentation, you want to be careful to distinguish among the various types of knowledge in your target audiences as identified by Rosenbaum and Walters[11]: computer novice, computer expert, subject novice, and subject expert. (See Figure 17-1.)

With a subject matter such as technical writing, obviously any single person could know nothing about computers or technical writing (be a combination of computer novice and subject novice), know a lot about computers but nothing about technical writing (be a computer expert and a subject novice), know nothing about computers but a lot about technical writing (be a computer novice and a subject expert), or know a lot about both computers and technical writing (be both a computer expert and a subject expert). If any or all of these types of people are present in your audience, you must design instructions which accommodate their needs as well as possible.

For instance, suppose you need to write instructions for the secretaries and technical report writers at Company B to teach them how to start using the text editing system on a new computer just bought for their business office. You note the following characteristics about this particular audience.

1   Your instructions will be used by two groups of people: by various engineers and managers who would probably use the system only occasionally to write the content of reports and by secretaries doing typing, editing, and formatting of reports every day.

2   In this particular office, the secretaries and managers do not know much about any computer, but the engineers do know something about computers. The secretaries have had training in formatting documents and are quite particular about the way the finished document looks. Some of the managers

Figure 17-1 ❏ TARGET USER AUDIENCES FOR COMPUTER DOCUMENTATION

SOURCE: Stephanie Rosenbaum and R. Dennis Walters, Audience Diversity: A Major Challenge in Computer Documentation," *IEEE Transactions on Professional Communications* 29(4) (December 1986). © 1986 IEEE.

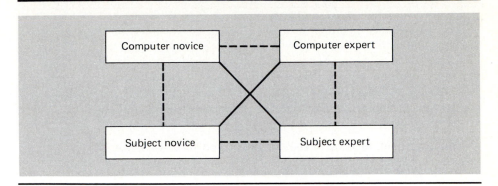

and engineers are concerned about formats but let the secretaries produce the final pages.

3 Initially, all the secretaries, managers, and engineers want to learn a simple set of procedures for getting their work done. However, on the basis of past experience, it seems that later on the secretaries would probably need more expertise on the hard details of the system, since they would be doing a lot of the final formatting and "finishing up" and would use the computer system all the time. The engineers and managers would probably use this particular system only occasionally and would probably worry only about the content of the text, not its look.

Once you have defined all of your audiences as specifically as possible, you need to analyze your situation, analyze the tasks you want your readers to do, and plan the overall structure of the document.

Analyzing the Situation: Resources, Constraints, and Deadlines

It is important for you to understand the situation *in* which the instructions are to be written—the resources, constraints, and deadlines inherent in your situation, your organization's business goals and production practices, and any specifications which your instructions need to meet. If these are not understood, you may later plan and try to write a document whose content and presentation cannot possibly be accomplished with the available resources of personnel, time, and money, or which will not meet required specifications. In defining these resources, constraints, deadlines, and other factors, you should probably be conservative and plan for a worst-case scenario.

Analyzing the Tasks

Instructions tell someone how to do a task or a set of tasks. Thus, before you can write the instructions, you must first determine exactly what tasks need to be done by performing a task analysis. A *task analysis* is a procedure in which you carefully examine the overall task to be learned to see what subtasks are involved, and then break down the subtasks until you have identified each step that must be done. For instance, if you needed to write instructions for getting started on using a text editing system on a computer, you might first describe the overall task as "learning the basic operations for entering and editing text." You might then list the major subtasks as follows:

1   Start (or open or create) the document you want to write.

2   Write your new text in the document.

3   Edit the text you have written where necessary (add and delete and change text).

4   Save the document.

5   Exit from or close the document.

Although this list is a good start on analyzing the overall task, it is a relatively high-level breakdown, that is, each item on the list is a kind of summary of several smaller steps. Thus, to complete the task analysis, you must analyze each subtask to see what its components are and continue doing so until you have reached the lowest level which is needed for your audience. For instance, the first subtask above might include the following steps:

1   Start (or open or create) the document you want to write,

    a   Turn on the computer.

    b   Load in or call up the text editing program.

    c   Start the document you want to write.

Each of these steps may include substeps. For instance, if you are going to use a text editing program which is stored on a diskette, you will need to break down step b as follows:

    b   Load in or call up the text editing program.

        (1)   Put the diskette in the machine.

        (2)   Locate the text editing program on your diskette.

        (3)   Give the command to start up the text editing program.

We will not continue this breakdown any further, but merely note that we have not yet arrived at the step-by-step instructions which would allow a new user to actually do the task of writing a document on the computer. Here is what has been generated so far:

1   Start (or open or create) the document you want to write.

    a   Turn on the computer.

    b   Load in or call up the text editing program.

        (1)   Put the diskette in the machine.

        (2)   Locate the text editing program on your diskette.

        (3)   Give the command to start up the text editing program.

    c   Start up the document you want to write.

2   Write your new text in the document.

3   Edit the text you have written where necessary (add and delete and change text).

4   Save the document.

5   Exit from or close the document.

One common problem which occurs in doing task analyses is that, as certain steps are broken down into their components, it becomes clear that there is more than one option to be considered. For instance, locating the text editing program will be a slightly different task if the computer user has the program on a hard disk (a large disk inside the computer which is not normally taken out) rather than on a diskette (a small disk outside the computer which must be inserted into a disk drive). These various options and their differences must be accurately reflected in the instructions or the instructions will not be adequate for the users.

In doing a task analysis, then, the writer must be sure to consider all of the tasks and subtasks that a user of the instructions will need to do and to define these tasks down to the most specific steps needed by the audience. The steps should be presented in the chronological order in which they must be done, and no steps should be left out.

CAUTION: If you have a large and very complex task, you should probably define the major tasks that need to be done as outlined above, then plan the overall structure of the document as outlined in the next section, then continue with the task analysis to determine the lowest levels and steps needed to complete the task, and finally recheck the plan to see if it is still adequate. If you have a relatively simple task to teach, you can usually complete the task analysis and then move into the planning stage.

❏ EXERCISE 17-1

A   Identify two tasks you would like to teach someone else to do. Try to choose at least one task that is not much more complicated than running a videotape recorder or a compact disc player. For each task, define the audience(s) you would be teaching to do the task, the purpose(s) for which they would need to know the task, and the context in which they would be operating.

B   Choose one of the tasks you defined in A and make a task analysis of it. If the task is quite complex, choose one of the subtasks to analyze completely.

17.2 PLANNING THE DOCUMENT

Once you have done situation and task analyses, you need to plan the structure and presentation of the document(s) which will be created to teach or describe the desired tasks. The first stage of planning is to define the strategy or approach you will be using. This may involve creating one or more documents, as illustrated in the case study below. In this case, by Rosenbaum and Walters, a professional documentation company was asked to design instructions for an accounting program on a computer.

> A manufacturer of a popular home computer system (Company A) approached us with a plan to market an accounting software package along with their microcomputer hardware. The accounting software consisted of a general ledger system, an accounts receivable system, and an inventory control system. The target audience was small businesses, such as "mom-and-pop" retail stores. As we discussed the documentation possibilities with Company A, we identified three major issues.
>
> The first issue related to the software configuration. Although the accounting package consisted of three connected systems, a customer could purchase one without buying the others. The general ledger system accepted data from the other two, but the accounts receivable and inventory control systems could function independently. Therefore, Company A needed not one, but three stand-alone documents.
>
> The second issue concerned the client company's product support. Company A had acquired the software from a third-party developer, and expected to sell it through retail stores and other low-support organizations. Consequently, the documentation for each system had to contain complete procedural and reference information.
>
> The third issue was the user audience itself. Company A's marketing analysis showed the prospective buyer of the system to be both a computer novice and a subject-area novice, with little knowledge of accounting principles or practices. In particular, Company A expected that most of the "user" questions would be questions about accounting rather than the software.

Because the software required accurate charts of accounts, Company A expected that many small businesses would hire a certified public accountant (CPA) to set up the accounts on the system. Some customers would subsequently choose to maintain their own books, but others might still need the accountant. Thus, the CPA—a subject expert but probably a computer novice—would often be a hidden audience of the proposed manuals.

The solution we finally adopted for Company A involved three major elements. Company A helped us address the needs of the accounting-naive audience by commissioning a "Big Eight" accounting firm to prepare a draft of an accounting primer for small businesses, the *Business Manager's Companion Guide*. This document summarized basic accounting principles for businesspeople to whom they are unfamiliar. Our organization took the draft text produced by the accounting firm and rewrote it for clarity and comprehension, inserting examples coordinated with the examples in the user's manuals.

The second major decision was to supply a sample company's books with the software product on diskette, and coordinate the sample company with the tutorial information in the user's manual. Thus, Company A's customers could practice using the accounting software with the sample company's books before they took the serious step of entering and working with their own accounting data.

The third major element of Company A's documentation structure was to divide the tutorial information into two groups of procedures according to frequency of use. All the preliminary procedures—hardware and software installation, start-up, and setting up the customer company's books—were separated from regular operating procedures. Some of the other organizational and style decisions we made to address this user audience were the following:

- Target all user procedures at the small business owner, and place them in the first half of each software manual. Write the instructions in a tutorial style, but assume that the user understands the accounting principles underlying the software.

- Divide the user procedures into three main areas: installation and startup of the hardware and software, day-to-day operating procedures (using the sample company as a training device), and setup of the real chart of accounts.

- Place the reference section in the same book, after the user instructions. Organize reference material alphabetically by function, and summarize the procedures (for the user who has read the tutorial chapters and become familiar with the system).

- Add a glossary and a dozen sample charts of accounts for selected businesses.

- Coordinate the graphic design and production of the entire documentation package, using numerous illustrations and samples of screens and reports.

As the project progressed, we incorporated other elements, such as a binder that would stand upright next to the computer and would hold all the system diskettes.

Figure 17-2 ❏ SUMMARY FOR AUDIENCE ANALYSIS AND OVERALL DOCUMENT DESIGN FOR ROSENBAUM AND WALTERS CASE

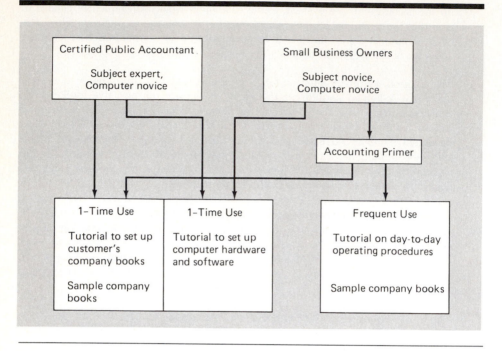

The end result of this project was a series of manuals which exactly suited Company A's needs and, more importantly, the needs of its customers.[12]

A visual summary of this strategy and the results of the audience analysis appears in Figure 17-2. As illustrated by this case, some careful thinking about your overall strategy early in the project will surely lead to a better design in the end and a more efficient process of generating the instructions. Imagine the time that might have been wasted if the writers in the case above had pressed ahead with a single set of instructions instead of the solution they devised, and only later decided that they would have to rewrite their material into three documents.

Planning the Content

Once you have planned the overall strategy for your instructions, the second stage of planning is to organize the various tasks and subtasks into some logical form or outline (such as that presented in the task analysis section for the editing system). This outline defines your approach to the tasks and will strongly guide the instructions you will finally write. The highest level of the outline can serve as a guide to the overall structure of the document. If you have completed the task analysis down to the level of individual steps, these steps may correspond to entries in your outline. However, the outline and steps are only guides at this point, and you must turn them into a well-thought-out design which has a logic and consistency at all levels of the document: at the major structural levels (the "book" level), at the major subsection levels (the "chapter" level), and at the section level (the "paragraph" level). You must also decide what kinds of examples and graphics will be used and how they will relate to the written instructions.

Planning the Examples

Since examples are very concrete, users of instructions frequently rely on examples to learn any procedure which incorporates an example. (Some people apparently rely on the text rather than on the examples.) However, examples can be overused, poorly constructed, and poorly integrated into a document, so they must be designed with care. Further, they must be used appropriately, since research has shown that in some situations examples are not particularly useful and may even be counterproductive.

Unfortunately, there has been relatively little specific and systematic research on the use of examples in technical documentation, but a few studies do examine the use of examples or elaborations (other added material) in a general sense. Some of the major conclusions of these studies are presented below. However, we note strongly that the issues involved here are quite complex and that much more research is needed to provide a solid base for the guidelines needed by writers.

As noted earlier in this book, readers have different purposes for reading: some readers need to learn all the information in a document, some need to learn only part of that information, some need to find one specific item out of a document, and some want to get a sense of the general approach being used in the document, but do not need to learn any specific fact or procedure in it. In reading instructions, some people read to learn or remind themselves of a *procedure*, while others read to learn *facts* presented in the document. It appears that procedure learning is different from fact learning in the way that each makes use of examples.

Effectiveness of Examples for Procedure Learning

Examples and other elaborations seem to help readers learn procedures. They improve comprehension and memory for readers learning to select and execute procedures correctly.[13] However, once subjects know what tasks need to be performed, they benefit little from seeing examples and other elaborations.[14]

One strong implication of these findings is that if documents are set up with very strong task focuses, at least some of the examples and elaborations may not be needed. It appears that with a strong task focus, readers will supply at least some of their own examples and elaborations and perhaps gain little or no additional benefit from those supplied by the writer.

Variable Effectiveness of Examples for Fact Learning

Although examples seem to be useful for learning procedures, research suggests that they are less helpful for learning facts. Indeed, several studies indicate that examples generated by the writer may actually reduce learning and memory in situations that stress fact learning. For instance, Reder and Anderson found that students reading only summaries apparently learned and remembered the material better than did students studying textbook chapters with examples and other elaborations. The advantage for summaries held up over a range of times (20 minutes to 1 year), a variety of study conditions, and various types of tests.[15]

In contrast to the studies above, several studies of author-provided examples and other types of elaboration show some advantages for the learner of facts under specialized conditions:

1 When readers are already knowledgeable[16]

2 When elaborated passages were mixed into the same texts with unelaborated passages[17]

3 When elaborations consisted of added material which increased the importance of the material being tested[18]

In summary, it seems that in situations requiring fact learning, author-generated examples and elaborations may actually reduce learning and memory, especially when these elaborations are placed in texts for nonexperts or novices.

However, reader-generated examples seem to improve comprehension and memory in similar situations. In several experiments, subjects with prior relevant knowledge were more likely to introduce information that was relevant but not presented in the reading.[19] This observation suggests that readers with more relevant prior knowledge are elaborating on material they are reading, that is, they are creating their own examples and other types of elaborations as and after they read. Those readers who have more relevant prior knowledge and who elaborate on it also show significantly better comprehension and memory for the material they have read.[20]

The implication here is that to help fact learning, designers of documentation need to find ways to engage the reader in the generation of examples and other types of elaborations. Unfortunately, it is not yet clear how this is best done.

Planning the Graphics

Once you have defined your content, you need to decide what graphics you will use to illustrate the text and what form they will take. Some issues to be considered have been outlined by Twyman.

a *Purpose* whether, for example, to impart information or to persuade

b *Information content* the essence of the information or other message to be conveyed c *Configuration* the different ways of organizing the graphic elements of language spatially

d *Mode* whether verbal, pictorial, schematic, or a mixture of two or more of them

e *Means of production* from hand-produced to computer-controlled

f *Resources* in terms of available skills, facilities, funds, and time.

g *Users* taking into account such factors as age, abilities, training, interests, and previous experience.

h *Circumstances of use* whether, for example, the user is working in a well-appointed library or under stressful circumstances in a moving vehicle.[21]

Guidelines for choosing and designing graphics appear in Chapters 8 and 9. In addition, if you are producing computer documentation, you need to remember that documentation users often scan through the text—looking for pictures of the screen representing their present state—and choose what to read according to the apparent appropriateness of the pictures.[22]

Planning the Design

At this stage, you should also decide on a format for your instructions. Many of the issues to be planned have been outlined by Smillie.

1 Layout and size

2 Typeface and size

3 Borders

4 Page numbering scheme

5 Indexing

6 Method of tracking changes

7 Placement of warnings, cautions, and notes

8 Paper stock

9 Binding method

For illustrations, requirements have to be established for

1 Quality

2 Level of detail

3 Angle of view

4 Locator illustrations

5 Item enlargement

6 Exploded views

7 Call-outs[23]

Other issues to be planned include the size and position of headings and margins at various levels, and the formats for various types of lists, examples, and graphics. If you can define such things before you actually begin writing, you will save a lot of editing later on.

Finally, in planning the overall design for the document(s), you need to consider that "usability and comprehensibility of completed manuals suffer from the competitive claims of developers' organizational practices and business goals regardless of the goodness of their model of the user."[24] Thus, be sure that you understand how your organizational practices and business goals may affect your design. In particular, you need to make sure that you are meeting any specifications for manuals which your organization has adopted.[25]

17.3 WRITING AND EDITING A DRAFT

Once you have defined the audience and purpose, completed a task analysis, and planned the document, you are ready to write the first draft of your instructions. Note that this is only a first draft, one that will be edited and tested and then revised and retested as many times as needed to allow users to successfully complete the tasks. The first draft should follow the outline set up in your document plan and should address the following guidelines.

Provide Information about Goals

You should give the users of the instructions some information about why they are doing the steps in your instructions, what their goals are in following a particular sequence of steps. Research suggests that if you do this, users will be better able to follow your instructions and to apply them to variants of the specific situations you discuss in examples. Since you generated the steps in a top-down task analysis, the goal for any series of steps will usually be quite visible in the next level up in the outline. For instance, consider the following steps from the task analysis presented above:

(1) Put the diskette in the machine.

(2) Locate the text editing program on your diskette.

(3) Give the command to start up the text editing program.

The user was to do these steps to complete the subtask

  b Load in or call up the text editing program.

This goal was listed immediately above the steps in the outline.

Provide Adequate and Visible Warnings

Writers of instructions and other documentation may be and probably will be legally liable for any damages that occur when a user is following instructions. Thus, if there are warnings that might affect a user's safety, or cautions that might affect the safety of equipment or software or data, the writer **must** place those warnings and cautions where they will be seen and followed. This means that **warnings and cautions must appear at the point of use both in the documentation and on dangerous equipment.** If they are too small or hidden by other material—if they are not visually emphatic—then the writer has not adequately warned the user and is legally liable. If they come too late in a text, after the dangerous situation is created, or if they do not adequately convey the nature and extent of the danger, then the writer has failed to adequately warn the user and is legally liable.

> In writing warnings, the key thing to remember is to address all possible users, to command their attention, and to use every means possible to assure that they do in fact appreciate the risk. As several court decisions indicate, even that may not be enough. (In Thomas vs. Kaiser Agricultural Chemicals, a farmer read the warnings, read the instructions, had used similar equipment in the past, but was still awarded damages as a result of injury. The court held that he did not appreciate the danger.)[26]

What should warnings and warning labels say and how should they be presented? As of March 1989, *no national standards* for warnings within a document or warning labels have been defined by the American National Standard Institute, the International Standards Organization, or any national professional society. This means that there is no standard which defines, for warnings or warning tags,

1 The types of content which must be present,

2 The minimum sizes of print, or

3 The relative size of warning words or icons to informational words or icons.

However, the American National Standards Institute has had a Standard for Accident Prevention Signs (withdrawn in March 1989), which included a classification of dangers, and a Standard for Safety Color Codes for Marking Physical Hazards. Although the classification of dangers was not defined for warnings within texts, the basic classifications plus a biohazard classification appear in most standards dealing with accident prevention signs, accident prevention tags (tags attached to a piece of equipment), and color codes for marking physical hazards. Thus, they may be useful interim guides to the determination of the class of a hazard and color coding of its warnings until a national standard is established by a standards institution or by the courts.

Types of Hazards Needing Warnings

The Standard for Accident Prevention Signs defined six classes of physical dangers to be marked by signs.

4.1 Class I (Danger) These signs indicate immediate and grave danger or peril, a hazard capable of producing irreversible damage or injury, and prohibitions against harmful activity.

4.2 Class II (Caution) These signs are used to call attention to a potential danger or hazard capable of resulting in severe but not irreversible injury or damage. In some instances, the hazards may be those associated with Class I (Danger) signs but are of significantly less magnitude.

4.3 Class III (General Safety) These signs include notices of general practices and rules relating to health, first aid, medical equipment, sanitation, housekeeping, and general safety other than the two above classes. NOTE: The general broad utility, application, or intent of Class III (General) signs sets them apart from the other 2 classes of signs that relate to specific hazards, generally having graver circumstances associated with them.

4.4 Class IV (Fire and Emergency) These signs shall be used only to label or point the way to fire extinguishing equipment, fire escapes and exits, gas shutoff valves, sprinkler drains, and emergency procedures.

4.5 Class V (Radiation) These signs shall be used only to warn against the actual or potential presence of harmful radioactivity, and to identify objects, devices, or materials which emit such harmful radiation.

4.6 Special Signs These signs have been developed for special uses and have been separated from the other 5 major classes of accident prevention signs.[27]

Other classification systems mention a seventh physical danger:

Biohazards

In addition to physical dangers to the user, some systems present

Dangers to aspects of the user's property or environment

These dangers should also be cautioned against. For instance, users of a computer program may unintentionally modify or destroy their databases or other information if the program is not used correctly. This possibility must be pointed out by the writer of the instructions for using the program.

Color Codes for Hazards

The American National Standards Institute has defined standards for the use of colors on warning signs and tags, where a particular color is reserved for a specified use. These colors are defined in terms of the Munsell Notation System, a color identification and specification system,[28] and are specified for "the identification of physical hazards, the location of safety equipment, protective equipment, stationary machinery, portable powered hand tools, signs, and markers,"[29] as follows:

6.1 Safety Red Safety Red shall be the color for the identification of (1) Danger; (2) Stop. The following are examples of application of the color Safety Red.

(1) Emergency Stop Bars on hazardous machines.

(2) Stop buttons or electrical switches used for emergency stopping of machinery.

6.2 Safety Orange Safety Orange shall be the color for designating dangerous parts of machines or energized equipment. The following are examples of applications of the color Safety Orange:

(1) dangerous parts of machines or energized equipment which may cut, crush, shock, or otherwise injure and to emphasize such hazards when enclosure doors are open or when gear, belt, or other guards around moving equipment are open or removed, exposing unguarded hazards.

(2) inside of movable guards, inside of transmission guards for gears, pulleys, chains, etc. Exposed parts (edges only) of pulleys, gears, rollers, cutting devices, power jaws, etc.

6.3 Safety Yellow Safety Yellow shall be the color for designating caution. Solid yellow, yellow and black stripes, or yellow and black checkers shall be used for maximum contrast with the particular background. The following are examples of the applications of the color Safety Yellow.

(1) For marking physical hazards, such as striking against, stumbling, falling, tripping, or being caught between.

(2) Storage cabinets for flammable materials and containers of flammable or combustible liquids. . .

6.3.1 The radiation hazard symbol colors shall be Safety Blue on Safety Yellow.

6.4 Safety Green Safety Green shall be the color for designating safety and the location of first aid and safety equipment.

6.5 Safety Blue Safety Blue shall be the color for designating information such as informational signs and bulletin boards (not of a safety nature). Safety Blue also has specific applications in the railroad area to designate warnings against the starting, use of, or movement of equipment that is under repair or being worked upon.[30]

Some Suggested Guidelines for Writing Warnings

The standards for classification of hazards and for color coding give some starting points for writers of warnings. Other starting points appear in the *Trial Lawyer's Quarterly*, in which a lawyer suggests nine guidelines for writing warning labels for toxic substances.

- Provide the chemical name of the product, including well-known abbreviations.

- Provide a signal word such as **danger**, **warning**, or **caution**.

- Identify the ingredients making up the product, including solvent.

- Explain the hazards.

- Explain the precautionary measures in a clear and specific way.

- Provide instructions in the case of an accident.

- Explain the consequences of exposure to different degrees and different concentrations.

- Use bold and graphics.

- Devote approximately 40% of the surface area of the container to warnings.[31]

Some of these guidelines apply specifically to toxic substances; the rest apply to general warning situations and suggest areas of content which should appear in a warning. Although this list is not a standard, it can serve as a helpful interim guide until national standards organizations provide more specific guidelines.

❏ EXERCISE 17-2

To see how the first guideline for preparing a draft—provide information about goals—interacts with an example, consider the following instructions for deleting text with a particular text editing system. Identify the location of information about goals and the example information.

Deleting Text

1 Use the arrow keys to move the cursor so that it is under the first character of the text you want to delete, say the *l* of *lollipop*.

2 Press the F4 key.

3 Notice that the following "menu" appears at the bottom of the computer screen:

      Move   Copy   Delete   Recall   Save

Also notice that Move is highlighted.

4 Press the right-arrow key twice to highlight

    Delete

5 Press the ENTER key to select Delete.

6 Notice that at the bottom of the screen the computer displays the question:

    Delete what?

7 Move the cursor to the end of the text you want to delete. To do this, type the last character of the text you want to delete, here the *p*

of the last p in *pop*. The cursor will automatically advance to the first occurrence of this character, highlighting all the text it passes through. If this character occurs more than once in the text you want to delete, then type the character again and again until the cursor has finally reached the last occurrence of the character, at the end of the text you want to delete.

8 If you highlighted the wrong text and want to start over, press the ESC key (ESC for ESCAPE). After you press the ESC key, you will have to move the cursor back to the beginning of the text you wish to delete and then press the F4 key to restart the deletion procedure.

9 If you highlighted the correct text, then press the ENTER key to signal that you are finished highlighting.

10 Notice that the text that was highlighted disappears.[32]

Present Conditions before Actions

Users of instructions or documentation often follow the instructions unit by unit. For example, suppose step 9 of Exercise 17–2 were rewritten to read

Press the ENTER key, if you have highlighted the correct text.

In this case, many users will first press the ENTER key and then check to see if they have highlighted the correct text. (Of course, by then the text will be gone—as indicated by step 10—so it will be too late to see if it was correctly selected.) To avoid this problem, and to make it easier for readers to decide what they are supposed to do, you should put the conditions under which an action can be done *before* the action itself. The conditions for action will usually involve some evaluation of a situation and will often begin with *If* or *When*. For instance, in the sample on deleting text in Exercise 17-2, there are two specific conditions: *If you highlighted the wrong text and want to start over* from step 8 and *If you highlighted the correct text* from step 9.

Tell Users What They Should Notice in Response to an Action

New users of instructions or documentation like to know that they are doing the right thing and may get very nervous if they don't feel confident about their actions. Thus, writers should put orienting information into the instructions to reassure the user. The sample instructions on deleting text in Exercise 17-2 present several instances of this kind of information in steps 3, 6, and 10. In each case, the orienting information begins with *Notice*, as in step 3:

3 Notice that the following "menu" appears at the bottom of the computer screen:

     Move   Copy   Delete   Recall   Save

Also notice that Move is highlighted

Be Especially Careful about the Wording

Once you have analyzed the situation and tasks and planned your document, you need to actually produce the sentences. Chapters 21–28 outline a range of guidelines for writing sentences, but there are a few additional ones relevant to instructions and other documentation. These include

- Be accurate and complete.

- Put one step per item in a list of instructions.

- Order steps in chronological sequence.

- Use verbs of command for steps in instructions.

- Keep the language simple enough for poor readers and children.

- Consider the possible language problems of nonnative speakers of English (see Chapters 21 and 28).

- Separate steps in the instructions from explanatory material or other comments (or use formatting to indicate the differences).

The effect of these guidelines is illustrated by considering this version of steps 1–4 in the passage in Exercise 17-2.

Version A
Deleting text is accomplished by pressing the F4 key after highlighting the text to be deleted and then pressing the right-arrow key twice to highlight "Delete" after "Move" is highlighted.

The original passage is reproduced below as Version B.

Version B

Deleting Text

1 Use the arrow keys to move the cursor so that it is under the first character of the text you want to delete, say the *l* of *lollipop*.

2 Press the F4 key.

3 Notice that the following "menu" appears at the bottom of the computer screen:

       Move   Copy   Delete   Recall   Save

Also notice that Move is highlighted.

4 Press the right-arrow key twice to highlight

     Delete

Version A violates several of the guidelines presented earlier. First, it is not completely accurate, because it omits most of the important information in step 3 from Version B, which the user needs to evaluate his or her success in accomplishing steps 1 and 2. Second, Version A does not put one step per item in a list of instructions, making it hard for a user to see all the actions which need to be done. (The first sentence includes two actions: *Press the F4 key* and *Press the right-arrow key twice to highlight "Delete."*) Third, Version A does not put the steps in chronological sequence, with conditions before actions. In fact, it tells the user to do two actions, and in each case a prior action is discussed after the action presented:

- Deleting text is accomplished by pressing the F4 key *after highlighting the text to be deleted*

- Pressing the right-arrow key twice to highlight "Delete" *after "Move" is highlighted*

Fourth, Version A does not use verbs of command for steps—*use, press, notice, move*—which makes it hard for users to identify what they are to do. Finally, Version A uses unnecessarily complex language, such as the word *accomplished.* In contrast, Version B follows all of the guidelines to produce a more usable text.

Edit the Document

Once you have written a first draft of your instructions, you should edit the draft as described in Chapters 21–28 of this text. Edit first for logic of structure, then for accuracy of statements, then for completeness of steps, then for readability, then for punctuation and spelling. Edit, edit, edit.

17.4 EVALUATING, REVISING, AND TESTING

The last stage of producing a set of instructions or a description of a procedure involves evaluating the document through technical and editorial reviews (and revising it if something clearly needs revision), testing it, and finally re-revising and retesting it as needed.

Getting Technical and Editorial Reviews

Once you have finished a draft of the document, you will want to get it evaluated or reviewed by one or more technical experts in the area and by one or more other editors. If you are a technical expert or an editorial expert, you still need to get someone else to review your document because often a writer cannot see problems which other readers can see. Your reviewers should check for technical accuracy and completeness, consistency with company formats and policies, consistency with other documents in the same series if you are writing one document in a series, and adequacy of the writing in terms of structure, readability, and grammar.

Testing the Document with Sample Users

Once you are satisfied that you have created a set of instructions which is as well written as you can make it, you must test it on real users to see if it is usable, that is, if it allows a user to easily complete a specified task. Efficient methodologies for usability testing are still being developed,[33] and indeed the best role for writers in usability testing is still under debate.[34] However, as is suggested in Chapter 7, there are some generally accepted practices which should make your usability test more effective. First, you must test your document on real users. If your instructions will be used by computer novices, you must test it on computer novices. If you test it with computer experts, they may be able to succeed because of their computer knowledge, not because of your instructions. Second, if you can, you should test your document on at least several real users and carefully note where they have problems—where they make mistakes or don't seem to understand easily. Third, you need to keep track of the problems your users have. If you can videotape the sessions in which your test users read and try to follow the instructions, you will have a record you can check later to see what really seemed to be going wrong. (Something or several somethings will surely go wrong!) If you cannot videotape the tests, you will have to devise a method of keeping track of the problems.

We should stress that something will surely be wrong with even the most carefully prepared instructions. One famous example may help convince you of this and of the benefits of testing. When personal computers first came out and diskettes were unfamiliar items, one company with excellent technical writers had prepared instructions for getting started on the computer. At one point, the instructions said *Take the diskette out of its protective cover.* The first test subject came to that instruction and was videotaped taking the diskette out of its paper jacket and then ripping apart the square diskette cover until he proudly held the round interior "disk" from inside the square cover. The next instruction said *Hold the diskette by the top right corner and insert it into the disk drive.* At this point the test subject was videotaped looking at the round "disk" he held in his hand, looking at the ripped-up square cover with its mangled corner, and then trying to hide the whole diskette with its ripped-up cover under the desk. The writers changed that part of the instructions after only one test.

The first few tests will probably catch the most obvious errors, such as the one just described. Careful attention will be required to catch the more subtle errors and problems, but you have a much better chance of finding them if you test with the same kinds of people who will be using the finished instructions.

Revising the Document as Needed

After each test or cycle of tests, you will have to revise the instructions to remove the problems you have identified. As in the first editing stage, the best advice is to edit, edit, edit.

Retesting as Needed

When you have finished revising, you need to retest the document to be sure that you have not introduced additional problems or allowed other problems which were there before to become visible. When you have a series of tests in which your users are successful, you will have probably done as well as you will be able to do. Of course, in the real world, you may run out of time before this occurs, and then you will have to decide if you will send out untested and probably ineffective documentation or try to get some more time. Whatever happens, testing and retesting will probably take much more time than you had ever imagined, but will provide a fascinating view of what a reader does with your documents.

17.5 DOCUMENTING LARGE PROJECTS

A large documentation project is any project requiring more than one or two people and producing more than one volume of documentation. In some instances, a documentation team may have hundreds of contributors and produce many volumes of documentation of many different types. For instance, a large software development project might produce the following documents:

- Management Plan
- System Definition or Product Specification/Description
- Functional Specification
- High-Level System Specification
- Detailed System Specification
- Hardware Specification
- Database Specification
- Test Plan
- Test Report
- User's Guide
- Installation and Maintenance Guide
- Operations Guide

Thus, at the beginning of a large documentation project, the manager of the project should determine

- What types of documentation will be produced
- Who will coordinate the documentation efforts

- Who will actually produce the instructions or documentation (conduct the analyses and write the instructions)

- What the deadlines will be

- Who will be designated as technical and editorial reviewers

- Who will maintain the documentation, that is, who will correct any mistakes or carry out any updating in the documentation

In each case, the production of the documentation should follow the procedures outlined earlier in this chapter: analyzing audiences, purposes, tasks, and resources and constraints; planning content and presentation; writing and editing; and testing, revising, and retesting. In each case, the documents should generally be structured as follows:

- Set up all documentation with a top-down, general-to-particular design such as that described for reports in Chapter 12.

  - Begin each major section and subsection with a description of the function of the software or part or procedure being described.

  - Organize each document and section and subsection so that
    function → design
    what → how

- Provide an orientation at the beginning of each document which explains the function and purpose of the document.

- Break each document down into functional modules.

- For each document, provide an explanation of how the sections and subsections function and describe the information flow through each section and subsection.

REFERENCES

1 More detailed, useful commentaries on many of the issues involved in designing instructional material are found in the following: Thomas M. Duffy and Robert Waller, *Designing Usable Texts* (Orlando, FL: Academic Press, 1985); Jonathan Price, *How to Write a Computer Manual: A Handbook of Software Documentation* (Menlo Park, CA: Benjamins/Cummings, 1984); and R. John Brockmann, *Writing Better Computer User Documentation: From Paper to Online* (New York: Wiley, 1986).

2 T. G. Sticht, L. C. Fox, R. N. Hauke, and D. Welty-Sapf, *The Role of Reading in the Navy*, NPRDC-TR-77-40 (San Diego, CA: Navy Personnel Research and Development Center, September 1977), NTIS No. AD A044 228.

3 L. Mikulecky, *Job Literacy: The Relationship between School Preparation and Workplace Actuality*, Final Report (Bloomington, IN: Indiana University, 1981).

4   Richard P. Kern and T. Sticht, *Guidebook for the Development of Army Training Literature: Rationale and Policy Implications for Interviews with Army Writers*, Interim Report (Alexandria, VA: Human Resources Research Organization, March 1974).

5   T. Sticht, "Understanding Readers and Their Uses of Texts," in *Designing Usable Texts*, edited by Thomas M. Duffy and Robert Waller (Orlando, FL: Academic Press, 1985).

6   M. Showell and M. Brennan, *A Survey of User Attitudes toward Army Training Literature*, Final Report (Alexandria VA: Human Resources Research Organization, 1974).

7   Patricia Wright, "The Instructions Clearly State . . . Can't People Read?" *Applied Ergonomics*, September 1981, pp. 131–141.

8   Patricia Sullivan and Linda Flower, "How Do Users Read Computer Manuals? Some Protocol Contributions to Writer's Knowledge," in *Convergences*, edited by Bruce T. Peterson (Champaign, IL: National Council of Teachers of English, 1986).

9   T. Sticht, "Understanding Readers and Their Uses of Texts," in *Designing Usable Texts*, edited by Thomas M. Duffy and Robert Waller (Orlando, FL: Academic Press, 1985), p. 320.

10 Richard P. Kern, "Modeling Users and Their Use of Technical Manuals," in *Designing Usable Texts*, edited by Thomas M. Duffy and Robert Waller (Orlando, FL: Academic Press, 1985), p. 341. The passage cites R. P. Kern and T. G. Sticht, "The Study of Writing in Functional Contexts: The Army Writer and the Design of a Guidebook for Developing Army Training Literature (HumRRO Special Report), in *Usefulness of Readability Formulas for Achieving Army Readability Objectives: Research and State-of-the-Art Applied to the Army's Problems*, TR 437, edited by R. P. Kern (Alexandria, VA: US Army Research Institute for the Behavioral and Social Sciences, January 1980), NTIS No. AD A086 408/2, p. 341.

11 Stephanie Rosenbaum and R. Dennis Walters, "Audience Diversity: A Major Challenge in Computer Documentation," *IEEE Transactions on Professional Communications*, *29*(4) (December 1986).

12 Rosenbaum and Walters, pp. 50–51.

13 B. Ross, "Remindings and Their Effects in Learning a Cognitive Skill," *Cognitive Psychology 16*:371–416 (1984). J. Pepper, "Following Students' Suggestions for Rewriting a Computer Programming Textbook," *American Educational Research Journal 18*(3):259–269 (1981). L. M. Reder, D. H. Charney, and K. I. Morgan, "The Role of Elaborations in Learning a Skill from an Instructional Text," *Memory and Cognition*, 1985. C. M. Allwood, T. Wikstrom, and L. M. Reder, "Effects of Presentation Format on Reading Retention: Superiority of Summaries in Free Recall," *Poetics 11*:145–153 (1982).

14 Reder, Charney, and Morgan (1985).

15 Lynn M. Reder and John R. Anderson, "A Comparison of Texts and Their Summaries: Memorial Consequences," *Journal of Verbal Language and Verbal Behavior 19*:121–134 (1980).

16 H. Mandl, W. Schnotz, and S. Tergan, "On the Function of Examples in Instructional Texts," *Kognitive Prozesse und Unterricht. Jahrbuch für Empirische Erziehungswissenschaft*, edited by L. Kotter and H. Mandl (Dusseldorf, W. Germany: Schwann, 1983).

17 E. Z. Rothkopf and M. J. Billington, "Passage Length and Recall with Test Size Held Constant: Effects of Modality, Pacing, and Learning Set," *Journal of Verbal Learning and Verbal Behavior 22*:667–681 (1983).

18 Stein and Bransford, op cit.

19 H. L. Chiesi, G. J. Spilich, and J. F. Voss, "Acquisition of Domain-Related Information in Relation to High and Low Domain Knowledge," *Journal of Verbal Learning and Verbal Behavior 18*:257–273 (1979). H. R. Arkes and M. R. Freedman, "A Demonstration of the Costs and Benefits of Expertise in Recognition Memory," *Memory and Cognition 12*:84–89 (1984). R. A. Sulin and D.J. Dooling, "Intrusion of a Thematic Idea in Retention of Prose," *Journal of Experimental Psychology 103*:255–262 (1974). A. L. Brown, S. S. Smiley, J. D. Day, M. A. R. Townsend and S. C. Lawton, "Intrusion of a Thematic Idea in Children's Comprehension and Retention of Stories," *Child Development 48*:1454–1466 (1977). G. H. Bower, J. B. Black, and T. J. Turner, "Scripts in Memory for Text," *Cognitive Psychology 11*:177–220 (1979).

20 F. C. Bartlett , "Remembering: A Study in Experimental and Social Psychology," (Cambridge, England: Cambridge University Press, 1932). D. L. Schallert, "Improving Memory for Prose: The Relationship between Depth of Processing and Context," *Journal of Verbal Learning and Verbal Behavior 15*: 621–632 (1976). R. C. Anderson and J.W. Pichert, *Recall of Previously Unrecallable Information Following a Shift in Perspective*, Technical Report 41 (Champaign-Urbana, IL: University of Illinois, Center for the Study of Reading, 1977), ERIC Document Reproduction Service No. ED 142 974. D. J. Dooling and R.E. Cristiaansen, "Episodic and Semantic Aspects of Memory for Prose," *Journal of Experimental Psychology: Human Learning and Memory 3*:428–436 (1977). C. E. Weinstein, "Elaboration Skills as a Learning Strategy," *Learning Strategies*, edited by H. F. O'Neil, Jr. (New York: Academic Press, 1978). H. R. Arkes and M. R. Freedman, "A Demonstration of the Costs and Benefits of Expertise in Recognition Memory," *Memory and Cognition 12*:84–89 (1984).

21 Michael Twyman, "Using Pictorial Language: A Discussion of the Dimensions of the Problem," in *Designing Usable Texts*, edited by Thomas M. Duffy and Robert Waller (Orlando, FL: Academic Press, 1985), pp. 248–249.

22 Patricia Sullivan and Linda Flower, "How Do Users Read Computer Manuals? Some Protocol Contributions to Writer's Knowledge," in *Convergences*, edited by Bruce T. Peterson (Champaign, IL: National Council of Teachers of English, 1986).

23 Robert J. Smillie, "Design Strategies for Job Performance Aids," in *Designing Usable Texts*, edited by Thomas M. Duffy and Robert Waller (Orlando, FL: Academic Press, 1985), p. 220.

24 Richard P. Kern, "Modeling Users and Their Use of Technical Manuals," in *Designing Usable Texts*, edited by Thomas M. Duffy and Robert Waller (Orlando, FL: Academic Press, 1985), p. 341. The passage cites R. P. Kern and T. G. Sticht, "The Study of Writing in Functional Contexts: The Army Writer and the Design of a Guidebook for Developing Army Training Literature" (HumRRO Special Rep.), in *Usefulness of Readability Formulas for Achieving Army Readability Objectives: Research and State-of-the-Art Applied to the Army's Problems*, TR 437, edited by R. P. Kern (Alexandria, VA: U.S. Army Research Institute for the Behavioral and Social Sciences, January 1980), NTIS No. AD A086 408/2, p. 342.

25 Thomas M. Duffy, Theodore Post, and Gregory Smith, "Technical Manual Production: An Examination of Five Systems," *Written Communication*, *1*(4):378 (October 1987).

26 Dwight W. Stevenson, "The Legal Context for Technical Writing," *Technical Writing Skills and Computer-Based Writing Tools* (Course #8815), July 18–22, 1988 (Ann Arbor, MI: The University of Michigan College of Engineering, 1988), p. S-9.

27 Reprinted by permission from American National Standards Specifications for Accident Prevention Signs, ANSI Z35.1–1972, withdrawn in March 1989. See also ASAE S441–83 Safety Signs, 2 pp.; SAE J 115–87 Safety Signs, Recommended Practice, 2 pp.; American National Standards Specifications for Accident Prevention Tags, Z35.2–1968; USA Standard Specifications for Accident Prevention Tags, USAS Z35.2–1968.

28 American National Standards Method of Specifying Color by the Munsell System, ANSI/ASTM D1525–68 [Z 13815].

29 Reprinted by permission from American National Standard Safety Color Codes for Marking Physical Hazards, ANSI Z53.1–1979, Revision of ANSI Z53.1–1971. See also American National Standards Safety Color Code for Marking Physical Hazards, Errata, 16 pp., ANSI Z53.1–1971; ISO Recommendation R408–1964 Safety Colors (with reference to the assignment of colors to various hazard levels).

30 ANSI 253.1–1979.

31 Abraham, Bidanset, Newman, and Warner, "Labeling Liability in the Toxic Tort Case," *Trial Lawyer's Quarterly*, *15*(4):25–27 (Fall 1983). Quoted in Dwight W. Stevenson, "The Legal Context for Technical Writing," *Technical Writing Skills and Computer-Based Writing Tools* (Course #8815), July 18–22, 1988 (Ann Arbor, MI: The University of Michigan College of Engineering, 1988), p. S-9. See also American National Standards Hazardous Industrial Chemicals—Precautionary Labeling, 59 pp.; ANSI Z129.1–88.

32 Passage adapted from Richard Catrambone, "Specific versus General Procedures in Instructions," Doctoral Dissertation, University of Michigan, 1988.

33 Patricia Sullivan, "On Users and Usability: On Learning about Users," *Proceedings of the SigDoc'88 Conference*, Ann Arbor, MI, October 1988.

34 Mark Haselkorn, "The Future of 'Writing' for the Computer Industry," in *Text, Context, and Hypertext: Writing with and for the Computer*, edited by E. Barrett (Cambridge, MA: MIT Press, 1988), pp. 3–13.

ADDITIONAL READINGS

Susan G. Hadden, *Read the Label: Reducing Risk by Providing Information* (Westview: AAAS, 1986).

Elizabeth Harris, "A Theoretical Perspective on 'How To' Discourse," in *New Essays in Technical and Scientific Communication*, edited by Paul V. Anderson, R. John Brockmann, and Carolyn R. Miller (Farmingdale, NY: Baywood, 1983), pp. 139–155.

Industry Standards Subject Index (Englewood, CO: Information Handling Services, 1989).

18

Theses and Journal Articles

Two of the most important forms of advanced academic writing are the thesis and the journal article. Both are like many other forms of technical communication in that they should address a problem or an issue important to their audience and present sound arguments clearly and coherently. In this respect they should follow the principles of communication presented throughout this book.

Theses and journal articles differ from other types of technical documents, however, in that they are typically written for a much narrower audience: specialists in a field, who share assumptions, knowledge, and backgrounds and who have the need and interest to read carefully. This means that theses and journal articles can eliminate some of the features provided primarily for nonexperts, such as Foreword and Summary sections, and can employ other features designed for experts, such as technical jargon and scholarly references.

18.1 THESES

The master's or doctoral thesis is a major milestone in a student's career. Many students blanch at the prospect of writing one, and develop "writer's block." Probably the best way to avoid such mental paralysis is to realize that, basically, these forms of writing are merely academic exercises, part of the "rites of passage" in academic life.

A thesis is written for a committee of experts who know the field. Its purpose is to convince them that *you* know the field as well. You must show them that your topic represents "an original contribution to knowledge" (no matter how minor). This means, first, that you must *situate* the topic: you

356

must demonstrate familiarity with all of the current knowledge related to that topic and you must identify an interesting issue within that body of knowledge. The review of literature must be extensive, because it has to begin at a very general level and narrow down to a very specific one. (Journal article writers are spared most of this "display knowledge.")

Having identified an issue, you then must frame it in a way that is amenable to examination using methods acceptable to your research community. If the methods are at all innovative or controversial, you are expected to justify their use. The research design, of course, should be cleared by your committee long before you reach the thesis-writing stage.

You then proceed to describe your research. This should be done in very detailed fashion so that your committee can make sure you did everything properly and left nothing out. Do not make the mistake of copying the style of your favorite journal article. Journal article writing assumes that the reader can "fill in" many unstated steps; therefore, it generally eliminates far more material than does thesis writing. Thesis writing is supposed to be pedantic.

After describing the results of your research, you are expected to discuss them in a long concluding section. Many students try to skimp on this part, but it is a mistake to do so. The Discussion is, in a sense, the "mirror image" of the Introduction. Having narrowed down the scope of your work in the Introduction, you are now expected to broaden it out again in the Discussion, that is, to "resituate" it. This is where you can emphasize the significance of your work. You do so partly by relating your findings to the problem statement given in the Introduction, and partly by describing the implications of your work for future research. Finally, you should acknowledge any shortcomings in your research. Nobody expects your research to be absolutely perfect, and if you fail to acknowledge any weaknesses, you will be jeopardizing your credibility. Indeed, this is one case where professing weakness is actually a strength.

A good source of information and advice about dissertation writing is R. Sternberg's *How to Complete and Survive a Doctoral Dissertation* (see Additional Reading at the end of the chapter).

18.2 JOURNAL ARTICLES

The first question to ask yourself in preparing to write a journal article, of course, is "What journal shall I send it to?" You might be tempted to automatically send it to the most prestigious journal you think might take it. But this could be a mistake. The more prestigious journals may not represent your best readers. If you want your work to be maximally appreciated and used, you should probably send it to the journal your professional colleagues pay closest attention to, not necessarily the most glamorous one in the field. A survey by Richard M. Davis of 85 journal editors from professional scientific and technical societies found that the most common reason for rejecting an article is that the subject is not suitable for the journal.[1]

Table 18-1 ❑ MOST COMMON REASONS FOR REJECTING ARTICLE MANUSCRIPTS
(Cited by 85 Editors of Scientific and Technical Journals)

Reason	Number of Respondents
Subject	
Not suitable for journal	63
Not timely	4
Coverage	
Questionable significance	55
Questionable validity	39
Too shallow	39
Too exhaustive	8
Length	
Too long	26
Too short	4
Presentation	
Bad organization	35
Ineffective expression	33
Ineffective or unusable illustrations	11
Failure to follow style guide	4

Davis's editors cited many other reasons for rejecting an article as well: coverage of the topic is of questionable significance, of questionable validity, or too shallow; the paper is badly organized, badly written, or too long. The results are summarized in Table 18-1.

Davis's editors were also asked several open-ended questions, including "What is the most common mistake made by contributors?" and "What general advice would you give to contributors?" Their responses are enlightening. (Number of respondents is given in parentheses.)

Contributors' most common mistake

Organization and Presentation (50): Rambling—do not show problem or significance of results; no summary; failure to make a case; failure to cite previous work; too long—overly detailed information; poor graphics; no mention of uncertainties; technical errors.

Manuscript (21): Failure to follow instructions for authors.

General (15): Unaware of the scope of the journal—look at a few issues and see what we publish; too PR oriented—tooting their own horns; insignificant papers—not up to professional standards.

Expression (8): Lack of clarity, conciseness (try to write clearly, not profoundly); failure to write for the audience—use of highly specialized terms.

General advice to contributors

Manuscript (27): Follow the guidelines in the journal and style manual, submit a clean manuscript.

Expression (20): Write clearly, distinctly, concisely; be specific, avoid esoteric jargon; revise several times before submitting.

Organization and Presentation (18): Think about the audience—its interests; show significance to the reader and the field; emphasize what is relevant; cite appropriate related work—omit unnecessary reference to your own; spend time on organization—state the problem, significance, results; don't try to cram too much detail into an article.

General (10): Don't rush into print—a few good papers are better than many bad ones.

Revision (7): Get colleagues to read and comment, put it in a drawer for 30–90 days—then revise.

Review Process and Suggested Changes (5): Don't pester the editor; follow reviewer comments.

Journal articles vary somewhat in style from one journal to another, even within the same field. If you are writing an article for a particular journal, you should read that journal's instructions on organization and format and then look at sample articles in several issues to become familiar with the journal's style. (Here style includes format for footnoting, ways of setting up headings and visual aids, length of paragraphs, and other special features.)

Although they may differ in style, most articles share a remarkably uniform purpose and structure. The purpose of an article in any field is to advance an argument of fact or policy: (1) an argument of fact that the results reported are valid, that previously reported results are supported (or not), that a given theory is supported (or not), that other observations are necessary to resolve some debate in the field; or (2) an argument of policy that previous results should be questioned or reinterpreted, that a given theory should be abandoned, recast, or extended.

To be genuine contributions to knowledge, these arguments should be *original*, that is, they should be new and different. Ideally, the results being reported on should be somewhat *unexpected*. Our own case-study research indicates that specialist readers do not read journal articles from start to finish but look immediately for the main results. As a writer, you should try to accommodate such readers by "foregrounding" your main results. You can do this in several ways:

- Use a *title* that states or at least implies the major findings.

- State the major findings in an *informative abstract*.

- State the major findings in a *purpose statement* at the end of the introduction.

- Cite a *key visual aid* early in your article so that the journal editor will be induced to position it there.

- Use *informative section headings* instead of traditional ones like Results and Conclusions. (More and more journals are allowing this.)

At the same time, if you are doing experimental research, you should adhere to the traditional ordering of sections for experimental research reports:

1   *Introduction*, which defines the problem and describes its importance

2   *Materials and Method*, which describes how the research arrived at the results

3   *Results*, which describes what was discovered

4   *Discussion*, which analyzes the importance of the results and their implication(s)

Let's examine each of these sections in more detail.

The Introduction

The Introduction is an especially tricky part of the article, since it must present a great deal of information and orientation in a short space. Robert Day, an experienced journal editor and author of a helpful and entertaining book on writing articles, has provided four rules for a good introduction:

> (i) It should present first, with all possible clarity, the nature and scope of the problem investigated. (ii) To orient the reader, the pertinent literature should be reviewed. (iii) The method of the investigation should be stated. If deemed necessary, the reasons for the choice of a particular method should be stated. (iv) The principal results of the investigation should be stated. Do not keep the reader in suspense; let him follow the development of the evidence. An O. Henry surprise ending might make good literature, but it hardly fits the mold that we like to call the scientific method.[2]

Articles aimed at specialists may begin with an Introduction based on either a long-form or a short-form problem statement. (Since specialists share assumptions, methods, and knowledge of their field, a writer can often assume this shared experience and not state it.) In contrast to other types of introductions, article introductions aimed at specialists include technical details and a short review of previous work on the topic. These establish a context against which the author contrasts an observation or an inadequacy in theory or method, as is illustrated in the following example.

The Thermal Conductivity and Specific Heat of Epoxy-Resin
from 0.1 to 8.0 K

A term	The thermal properties of glassy materials at low temperatures are still not completely understood. The thermal conductivity has a plateau which is usually in the range 5 to 10 K, and below this temperature it has a temperature dependence which varies approximately as T^2. The specific
Amplification of or proof for A term	heat below 4 K is much larger than that which would be expected from the Debye theory, and it often has an additional term which is proportional to T Some progress has been made towards understanding the thermal be-

Table 18-2 ❑ A POSSIBLE STRUCTURE FOR ARTICLE INTRODUCTIONS

The Four Moves			Number of Occurrences in 48 Article Introductions
Move One	Establishing the field		43
	A Showing centrality		25
	i By interest	6	
	ii By importance	6	
	iii By topic prominence	7	
	iv By standard procedure	6	
	B Stating current knowledge		11
	C Ascribing key characteristics		7
Move Two	Summarizing previous research		48
Move Three	Preparing for present research		40
	A Indicating a gap	20	
	B Raising questions	14	
	C Extending a finding	6	
Move Four	Introducing present research		46
	A Giving the purpose	23	
	B Describing present research	23	

SOURCE: John Swales, *Aspects of Article Introductions,* Aston ESP Research Report No. 1, University of Aston, Birmingham, England, 1981, p. 22a

(Unstated *however* B term	haviour by assuming that there is a cutoff in the photon spectrum at high frequencies (Zaitlin and Anderson 1975 a,b) and that there is an additional system of low-lying two-level states (Anderson 1975, Phillips 1972). Nevertheless,
Missing information	more experimental data are required, and in particular it would seem desirable to make experiments on glassy samples whose properties can be varied slightly from one to the other. The present investigation reports attempts to do this
Purpose	by using various samples of the same epoxy-resin which have been subjected to different curing cycles. Measure-
Assignment	ments of the specific heat (or the diffusity) and the thermal conductivity have been taken in the temperature range 0.1 to 8.0 K for a set of specimens which covered up to nine different curing cycles.[3]

As suggested by this example, article introductions and especially their problem statements can be quite complex and varied. They have been studied in some detail by John Swales, who has proposed a common structure for them presented in Table 18–2. The results are based on 16 articles each from physics, biology/medicine, and the social sciences.

In Swales's proposed structure, the problem statement often extends into or through Move Three, the statement of missing information (the question or task) appears in Moves Three and Four, and—if it appears at all—the purpose is found in Move Four. (Note that a statement of purpose appears in only half of Swales's examples.) The sample article introduction analyzed above is repeated below, this time with Swales's divisions.[4]

The Thermal Conductivity and Specific Heat of Epoxy-Resin from 0.1 to 8.0 K

1 Establishing the field

The thermal properties of glassy materials at low temperatures are still not completely understood. The thermal conductivity has a plateau which is usually in the range 5 to 10 K, and below this temperature it has a temperature dependence which varies approximately as T^2. The specific heat below 4 K is much larger than that which would be expected from the Debye theory, and it often has an additional term which is proportional to T. Some progress has been made towards understanding the thermal behaviour

2 Summarizing previous research

by assuming that there is a cutoff in the photon spectrum at high frequencies (Zaitlin and Anderson 1975 a,b) and that there is an additional system of low-lying two-level states (Anderson 1975, Phillips 1972). Nevertheless, more

3 Preparing the present research

experimental data are required, and in particular it would seem desirable to make experiments on glassy samples whose properties can be varied slightly from one to the other. The present investigation reports attempts to do this

4 Introducing present research

by using various samples of the same epoxy-resin which have been subjected to different curing cycles. Measurements of the specific heat (or the diffusity) and the thermal conductivity have been taken in the temperature range 0.1 to 8.0 K for a set of specimens which covered up to nine different curing cycles.[5]

The Materials and Method Section

The Materials and Method section is a critical part of your report's argument, since it establishes the validity of your results and allows them to be taken seriously. It demonstrates that you have done everything "the right way": that you have been scrupulously careful and thorough, that you have used an accepted method, that you have made no technical mistakes.

This section also provides the mechanism by which the scientific community can repeat and verify your work. It must contain sufficient detail to allow any relatively experienced researcher in your field to reproduce your results exactly. This means that you must

1   Identify exactly what materials you used to conduct your research—
    reactants, enzymes, catalysts, organisms, experimental subjects (human
    or animal), etc. You should also identify your materials specifically enough
    that another researcher could use exactly what you specify to reproduce
    your results.

2   Identify any special conditions under which you conducted your research—
    special temperatures, irradiation with ultraviolet light, testing with un-
    usually high current or voltage loads.

3   Identify any special criteria you used to select materials, subjects, test
    apparatus, or test method. (If you chose one gluing material or catalyst
    over another, why?)

4   Identify the specific method you used to conduct the research. If you
    followed a standard procedure, you may simply reference it. If you followed
    an unorthodox or new procedure, you need to describe it fully.

5   Justify, where necessary, any of your choices of criteria, materials, method,
    or conditions.

The Results Section

The Results section of an article presents (1) the major generalization(s) you
are making about your data and (2), in compact form, the data supporting the
generalization(s). The generalizations must be clearly and obviously stated:
"This glue has successfully bonded stainless steel artificial joints to human
bone" or "UV-irradiated DNA contains 5,6-dihydroxydihydrothymine as well
as pyrimidine dimers in ratios dependent upon wavelength."[6] The supporting
data must be presented fully enough that the reader can evaluate the strength
of your generalizations but succinctly enough that the data do not overwhelm
the generalizations.

For instance, consider the relationship between generalizations and data
in the following example from an article entitled "The Effect of Graded Doses
of Cadmium on Lead, Zinc, and Copper Content of Target and Indicator
Organs in Rats." The subheadings in the Results section are Zn, Cu, Pb, and
Correlations. Only the text of the section on zinc (Zn) is presented here.

Zn

The Zn contents of the target and the indicator organs are presented in Tables
I and II. The tables also state significant differences ($p \leq 0.001$) between the
control group and the experimental groups for each tissue. As can be observed,
several of the soft tissues had increased Zn levels when the Cd dose had been
above 5 ppm, whereas for Cd doses above 12.5 ppm significantly decreased Zn
contents in epi's and dia's were registered (see Table I). In addition, Cd supply
was associated with a significant decrease in Zn content in incisors and molars
of the majority of the experimental groups (see Table II).

In none of the pooled left molar samples was the Zn content of enamel different from that of dentin. This was valid also for incisor samples through all groups.[7]

A second example of the relationship between generalizations and supporting data in the Results section comes from an article on tensile and fatigue tests on human cortical bone specimens. The example presents only a part of the Results section, the results of the tensile tests; it does not include the two tables of supporting data or the figure referenced in the text.

Tensile Tests

The mean values of the physical characteristics of the test specimens are shown in Table 2. The mechanical parameters of ultimate stress, yield stress, yield strain and modulus of elasticity showed no statistically significant correlation with porosity, ash fraction, wet density, or dry density.

Statistically significant ($P < 0.05$) positive linear regressions were found between ultimate stress and elastic modulus and between yield stress and elastic modulus (Table 3). A weak negative linear regression (not statistically significant) was found between yield strain and elastic modulus. The influence of bone elastic modulus of the ultimate stress, yield stress, and yield strain are shown in Fig. 1. The mean values ($\pm$s.d.) were ultimate stress 140 ($\pm$12) MPa, yield stress 129 ($\pm$11) MPa, yield strain 0.0068 ($\pm$0.0004), and elastic modulus 17.5 ($\pm$1.9) GPa.[8]

The Discussion

The Discussion section explains the implications of your results. It fits the results into the context of the field by relating your results to other work, both theoretical and experimental. Along with the Introduction, it explains why your work is important, how it contributes to the advancement of the field. It is critical that the Discussion be done carefully and thoroughly. As Robert Day has noted,

> *Many* papers are rejected by journal editors because of a faulty Discussion, even though the data of the paper might be both valid and interesting. Even more likely, the true meaning of the data may be completely obscured by the interpretation presented in the Discussion, again resulting in rejection.[9]

If you want to show how your work contributes to the advancement of your field, you might consider what circumstances create advancement. One philosopher, John Platt, believes that some fields advance much more rapidly than others because of a rigorous intellectual methodology:

> These rapidly moving fields are fields where a particular method of doing scientific research is systematically used and taught, an accumulative method of inductive inference that is so effective that I think it could be given the name of "strong inference.". . . The steps are familiar to every college student and are practiced,

off and on, by every scientist. The difference comes in their systematic application. *Strong inference consists of applying the following steps to every problem in science, formally and explicitly and regularly:*

1 *Devising alternative hypotheses;*

2 *Devising a crucial experiment* (or several of them), *with alternative possible outcomes,* each of which will, as nearly as possible, *exclude one or more of the hypotheses;*

3 *Carrying out the experiment so as to get a clean result;* and

4 *Recycling* the procedure, making subhypotheses or sequential hypotheses to refute the possibilities that remain; and so on.[10] [Italics original]

The keys to this system, according to Platt, are logical rigor, the alternative hypotheses, the crucial experiments, and the disproving of hypotheses. These allow scientists in a field to identify those problems crucial to the advancement of the field, then to construct a series of logical options or lines of inquiry for each problem, and then to eliminate quickly unproductive hypotheses and lines of inquiry. This system reduces wasted time because it focuses the theoretical and experimental efforts of a field on the hypotheses and lines of inquiry likely to produce successful results.

Now, given this perspective, what do you say in an adequate Discussion? How do you establish the importance of your work for the advancement of your field? If you can, you argue that your work is a crucial experiment, disproving some hypothesis or supporting another. Notice that this view encourages the reporting and discussion of negative results. "Hypothesis A predicted that I would find B, but I did not. This result questions the validity of Hypothesis A or indicates its inadequacy." Such a statement might save researchers and graduate students from wasting years of needless effort. It may also provide the best kind of statement scientists may make about nature:

> As the philosopher Karl Popper says today, there is no such thing as proof in science—because some later alternative explanation may be as good or better—so that science advances only by disproofs. There is no point in making hypotheses that are not falsifiable, because such hypotheses do not say anything; "it must be possible for an empirical scientific system to be refuted by experience."[11]

In arguing that your work is important, ideally because it is a crucial experiment, you typically consider some or all of the following questions.

1 Were your results expected? If not, why not?

2 What generalizations or claims are you making about your results? How do you interpret these generalizations?

3 Do your results contradict or support other experimental results?

4 Do they suggest other observations or experiments which could be done to confirm, refute, or extend your results?

5 Do your results support or contradict existing theory?

6 Do your results suggest that modifications or extensions need to be made to existing theory? What are they?

7 Could your results lead to any practical applications? What are they?

If you have recently written a dissertation and are unsure about whether to try to publish it or parts of it, a useful source of information and advice is Beth Luey's *Handbook for Academic Authors* (see Additional Reading).

❏ EXERCISE 18-1

A If you expect to write technical articles, analyze the organization and style of articles in one journal to which you might submit an article. Compare the journal's features with those discussed above.

B For a topic you are interested in, write an Introduction to an article on a research project you have conducted or would like to conduct.

C Select a laboratory report from one of your technical courses and write a Materials and Method section for it. Then write a Results section. If you can, write an Introduction and then a Discussion.

18.3 ABSTRACTS

The abstract is a very important document for both specialist and nonspecialist readers. It serves primarily as a *screening device* to give readers an idea of what the article as a whole is about, so that they can decide whether or not they want to read it. This is an especially important function in cases where the article is physically separated from the abstract. For example, it is increasingly common practice in many research fields to publish abstracts separately in both hard copy and electronic databases. When researchers scan such databases, they need to determine, from the abstract, whether or not they should read the entire article. Since looking up an article can be quite time-consuming, a well-written informative abstract can save valuable time and energy.

A second important function of abstracts is their use as *stand-alone texts*. After reading an abstract, a researcher may decide that the subject matter is not important enough to merit reading the article as a whole but *is* important enough to keep in mind for some possible future use. In such a case, the researcher may store the abstract in his or her files and may even use some of the information it contains, without ever seeing the article as a whole.

Third, in cases where the reader does go on to read the article as a whole, a good abstract provides a helpful *preview*. This "frames" the article and prepares the reader for the main points to come.

Finally, abstracts facilitate *indexing*. Indexers at all levels, from small in-house libraries to major institutions like the National Library of Medicine, often depend on author-generated abstracts to help identify key words and phrases for cross-indexing. By writing a good abstract for your article, you will improve your chances of having your article properly indexed, and thus will improve your chances of having it read by the right people.

The bibliographical information needed in an abstract varies depending on the report's purpose, audience, and expected circulation. For instance, a report which will circulate only inside the organization that produced it (an in-house document) might need only:

1   The name of the author

2   The title of the project

3   The date it was written

4   The project under which the document was produced

5   A special organizational number (if appropriate)

On the other hand, a report available to people outside the organization that produced it might require all that information plus:

6   The name of the organization that produced it

7   The name of the organization that commissioned it (if any)

8   The contract number under which the document was produced (if any)

9   A security classification

10   Key words and special numbers for cataloging in an information retrieval system

Items 7 through 10, for instance, are necessary in reports on government-sponsored research. As a report writer, you will have to find out what types of information are required in your abstract. You can do this by (1) looking at similar reports that have already been published to see what kinds of information they include in their abstracts, (2) consulting the "Instructions to Authors" published by the organization for which you are writing, or (3) consulting with your librarian.

The substance of an abstract should be the heart of the article it describes. In experimental research articles, for example, abstracts typically describe the methodology used, the main results, and major conclusions. Occasionally there is also an opening background statement or statement of purpose, although this is likely to be useful only to nonspecialist readers. Usually, the abstract for an experimental research article is *informative* in the sense that it emphasizes major findings and conclusions. Here is an example:

A. J. Boulton, J. H. Bowker, M. Gadia, J. Lemerman, K. Caswell, J. S. Skyler, and J. M. Sosenko (1986). Use of Plaster Casts in the Management of Diabetic Neuropathic Foot Ulcers. *Diabetes Care* 9(2):149–152.

Neuropathic foot ulceration is a major medical and economic problem among diabetic patients, and the traditional treatment involves bed rest with complete freedom from weight bearing. We have investigated the use of walking plaster casts in the management of seven diabetic patients with long-standing, chronic foot ulcers. Although all ulcers healed in a median time of 6 weeks, this therapy was not without side effects. We conclude that casting is a useful therapy for neuropathic ulcers, although several clinic visits, including cast removal and foot inspection, are necessary to avoid potential side effects caused by the casting of insensitive feet.

KEY WORDS: Surgical Casts, Foot Diseases, Diabetic Neuropathies, Skin Ulcer Therapy.

Notice how the abstract goes through four distinct "moves." The first sentence provides some background information, the second gives a brief idea of the specific topic and methodology, the third reports the major findings, and the fourth draws some conclusions. These different functions are clearly marked by the use of different verb tenses. The first and last sentences use the so-called present tense to make broad statements that are not restricted to a particular time frame. The second sentence uses the present perfect tense to connect the opening background statement to the particular study being reported on. The third sentence uses the past tense to describe specific results that occurred at a specific time.

With review-type articles, in which there is often no single main point or finding, abstracts tend to describe only the general contents of the article rather than specific findings. And they are normally written in the present tense. The following is an example:

M. Lord, D. P. Reynolds, and J. R. Hughes (1986). Foot Pressure Measurement: A Review of Clinical Findings. *J. Biomedical Engineering* 8(4):283–294.

In this review, a description of what is known about foot pressure distribution while standing and while walking is followed by sections on clinical findings. Two major clinical areas are treated extensively, namely the diabetic foot and the rheumatoid arthritic foot. Other applications, including the assessment of surgical procedures for orthopedic corrections, are included. A large variety of different techniques for foot pressure measurement have been used; interpretation of the results has to be made on the basis of a firm understanding of the technique employed. Often, quantitative results from different pieces of apparatus are difficult to compare, indicating a need for accurate calibration and a standardized presentation. An up-to-date summary of pressure measurement systems reported over the past five years is included.

KEY WORDS: Rheumatoid Arthritis, Diabetes Mellitus, Foot Diseases, Foot Physiology, Pressure, Biomechanics.

Usually, abstracts are written "after the fact": after the research has been carried out and the paper written up. But there may be cases where you have to write an abstract *before* the paper is written, indeed sometimes even before the work itself is done. For example, if you want to give a presentation at a professional conference and have to submit an abstract 8 to 12 months before it, you may find yourself in the position of wanting to submit a "promissory" abstract. If it is common practice to submit promissory abstracts to such conferences, then there is no harm in doing so. However, keep in mind that your ethical standards and professional credibility are at stake. Do not succumb to the temptation to overstate the material that you will ultimately present. Your abstract is expected to be an accurate reflection of the actual paper you will give. If it isn't, you are in effect guilty of "false advertising."

Experienced researchers are able to (1) anticipate quite accurately the eventual outcome of their research, (2) make interesting claims without using a lot of hard data, and (3) hedge their claims for those unusual cases where the results don't turn out as expected. If you want to write a promissory abstract but are not experienced at it, you would be wise to seek specific advice from a more experienced colleague.

❏ EXERCISE 18-2

Find several journal articles in your field. For each one, do the following: (1) cover the abstract; (2) read the article; (3) write an informative abstract for it; (4) uncover the original abstract and compare it with yours.

REFERENCES

1   Richard M. Davis, "Publication in Professional Journals: A Survey of Editors," *IEEE Transactions on Professional Communication PC-28* (2):34–42 (June 1985).
2   Robert A. Day, *How to Write and Publish a Scientific Paper* (Philadelphia: ISI Press, 1979), p. 24.
3   S. Kelham and H. H. Rosenburg, "The Thermal Conductivity and Specific Heat of Epoxy Resin from 0.1 to 80K," *J. Phys. C.: Solid State Physics 14* (1981).
4   John Swales, *Aspects of Article Introductions*, Aston ESP Research Report No. 1 (Birmingham, England: University of Aston, 1981), p. 16.
5   Kelham and Rosenburg.
6   Philip C. Hanawalt, Priscilla K. Cooper, Ann K. Ganesan, and Charles Allen Smith, "DNA Repair in Bacteria and Mammalian Cells," *Annual Review of Biochemistry 48:*787 (1979).
7   G. B. R. Wesenberg, G. Fosse, and P. Rasmussen, "The Effect of Graded Doses of Cadmium on Lead, Zinc, and Copper Content of Target and Indicator Organs in Rats," *International Journal of Environmental Studies 17:*193–194.

8   Dennis R. Carter and William E. Caler, "Uniaxial Fatigue of Human Cortical Bone: The Influence of Tissue Physical Characteristics," *Journal of Biomechanics* *14*:463–464 (1981).

9   Day, p. 33.

10 John R. Platt, *The Step to Man* (New York: Wiley, 1967), p. 20.

11 Platt, p. 27.

ADDITIONAL READING

American National Standards Institute, American National Standard for the Preparation of Scientific Papers for Written or Oral Presentation, ANSI Z39.16–1972, New York, 1972.

C. Bazerman, *Shaping Written Knowledge: The Genre and Activity of the Experimental Article in Science* (Madison: University of Wisconsin Press, 1988).

V. Booth, *Writing a Scientific Paper*, 4th ed. (London: Biochemical Society, 1978).

CBE Style Manual Committee, *Council of Biology Editors Style Manual: A Guide for Authors, Editors, and Publishers in the Biological Sciences*, 4th ed. (Washington, DC: Council of Biology Editors, 1978).

L. Debakey, *Guidelines for Editors, Reviewers, and Authors* (St. Louis: C. V. Mosby, 1976).

G. N. Gilbert and M. Mulkay, *Opening Pandora's Box: A Sociological Analysis of Scientists' Discourse* (Cambridge, England: Cambridge University Press, 1984).

B. Latour, *Science in Action* (Cambridge, MA: Harvard University Press, 1987).

B. Luey, *Handbook for Academic Authors* (Cambridge, England: Cambridge University Press, 1987).

J. H. Mitchell, *Writing for Professional and Technical Journals* (New York: Wiley, 1968).

G. Myers, "The Social Construction of Two Biology Articles," *Social Studies of Science* *15*:595–630 (1985).

G. Myers, "The Pragmatics of Politeness in Scientific Articles," *Applied Linguistics* *10*:1–35 (March 1989).

R. Sternberg, *How to Complete and Survive a Doctoral Dissertation* (New York: St. Martin's, 1985).

19

Oral Presentations

In many fields of science and technology, the ability to communicate technical information orally is just as important as the ability to write well. Indeed, study after study has found that people in management positions rely more on oral communication than they do on written communication. Phone calls, committee meetings, conversations—these are the primary vehicles by which managers convey and receive information. And they require good speaking and listening skills. A number of studies suggest that oral communication skills are of crucial importance, in fact, in determining who gets promoted to upper management levels in the first place.

But managers are not alone in their need to have good speaking and listening skills: engineers, scientists, and other technical professionals also need these skills. This can be seen, for example, in the results of a survey reported in *Engineering Education*.[1] In this survey, 367 engineers in chemical, civil, electrical, geological, mechanical, metallurgical, and mining engineering were asked to rate 30 specific communication tasks for job-related importance. Some of these engineers were working as managers, but most were not. Of the 24 tasks that were rated important by the respondents, 14 were writing tasks and 10 were oral tasks. Of these, 5 involved the giving of formal oral presentations (project proposal presentations; oral presentations using graphs, charts, and/or other aids; project progress report presentations; project feasibility study presentations; and formal speeches to technically sophisticated audiences). The remainder were of a more informal nature, involving one-to-one talks or small-group discussions.

The main purpose of this chapter is to give you advice on how to give formal and informal oral presentations. It also has some suggestions for getting the most out of meetings and teleconferences.

19.1 GIVING A FORMAL ORAL PRESENTATION

There are several steps to giving a successful formal oral presentation: you must prepare properly, practice adequately, and deliver the presentation with energy and enthusiasm. Your delivery will be much more successful if you pay careful attention to your preparation and practice.

Preparation

The basic principle to keep in mind in preparing any kind of oral presentation is that all listeners have a limited attention span and cannot be expected to follow everything you say. Their attention will probably wander from time to time, even if your presentation is only 10 minutes long. So, if you want to make sure that your listeners will come away from your talk with your main points clear in their minds, you must organize your presentation in such a way that these main points stand out. Here is how to do it:

1 *Analyze your audience, and limit your topic accordingly.* What do your listeners already know about the topic? What do they need or want to know about it? How much new information about it can they absorb? If you tell them what they already know, they'll be bored, but if you go to the opposite extreme and give them too much information too fast, they might be overwhelmed and might simply "short circuit" on you.

2 *Determine your primary purpose.* Is there some main point you want to get across? Is there something you want your listeners to believe, or be able to do? Whatever it is, you should have it clear in your mind so that you can build your presentation around it.

3 *Select effective supporting information.* What kind of information will best support your main point? What kind of information will appeal to your listeners? They will probably be able to remember no more than three or four main supporting points and no more than two or three supporting details for each of these. So choose wisely.

4 *Choose an appropriate pattern of organization.* Often, your supporting information can be presented according to a single dominant pattern of organization. For example, if you're trying to tell your listeners how to do something, you may want to organize your information into a list of instructions, chronologically ordered. On the other hand, if you're describing an experiment, you'll probably want to use the standard descriptive pattern: introduction, materials, procedure, results, discussion. There are many possible patterns of development to choose from, some of which are described in Chapters 4 and 22. Whichever one you choose should be appropriate to the subject matter, to your primary purpose, and to your audience—and it should be followed consistently.

5 *Prepare an outline.* Keep it brief: main points, main supporting points only. Arrange these points according to the pattern of development you chose in step 4. Do not write out a complete text unless the technical content or circumstances warrant it.

6 *Select appropriate visual aids.* Visual aids are virtually indispensable to technical presentations. First of all, they serve as "cue cards," allowing you to remember all your important points and stay on track without reading from a manuscript or from notes. Second, they have tremendous power as attention-getters. If you want to emphasize a point, by all means try to do it visually as well as orally. Studies have shown that people remember the visual parts of speeches far better than they do the verbal parts. Finally, visual aids can help clarify your message. (Would you rather describe a DNA molecule in words or show somebody a drawing of one?)

Pick your visual aids carefully. Generally, it's best to use the type of visual aid your audience expects. For example, if you're presenting at a scientific meeting where slides are the norm, you should try to use slides in your own presentation. But if you're presenting in a situation where there is a wider range of choices, here are the basic options:

- Overhead transparencies ("foils")
- Slides
- Flip charts
- Chalkboard
- Handouts
- Three-dimensional objects
- Computer screen projection

Each of these types of visual aid has its own strengths and weaknesses, and must be evaluated with the following questions in mind: Is it easy to prepare? Can it be altered easily during the presentation? Will it allow you to control the audience's attention—or will it distract attention from what you're saying? Will it let you present at an appropriate speed? How much information can you convey with it? How large an audience can you use it with? How reliable is it—does it depend heavily on electronic or mechanical equipment? Can the audience keep it for future reference? How well does it work as a "cue card"? Although not based on any kind of scientific sampling, Table 19–1 gives you a general idea of how different visual aids compare when evaluated according to these criteria.

As can be seen, foils and handouts have the greatest number of good or excellent qualities, but they differ in their respective strengths and weaknesses.

Table 19-1 ❑ COMPARISON OF VISUAL AIDS FOR USE WITH ORAL PRESENTATIONS

	Foils	Slides	Flip Charts	Chalk- board	Handouts	3-D Objects	Computer Projections
Ease of preparation	good	poor	poor	exc.	good	good	fair
Ease of alteration	good	poor	poor	exc.	poor	poor	fair
Audience control	exc.	exc.	exc.	exc.	poor	exc.	exc.
Speed	exc.	exc.	exc.	poor	exc.	exc.	exc.
Amount of information	good	good	good	fair	exc.	fair	good
Audience size	exc.	exc.	fair	fair	exc.	fair	good
Reliability	good	fair	exc.	exc.	exc.	exc.	fair
Future reference	poor	poor	poor	poor	exc.	poor	poor
Cueing	exc.	exc.	exc.	poor	exc.	fair	good

Foils support the actual presentation better: you can show them whenever you want to, thus coordinating what you're saying with what the audience is looking at; and you can write on them. Handouts, by contrast, often distract the audience's attention from what you're saying. On the other hand, handouts provide material that the audience can keep for future reference. Each of the other media has certain advantages and disadvantages as well. Your choice, of course, should depend ultimately not on the general evaluation given above but on the particular requirements and purposes of your presentation.

If you think your audience may have trouble understanding your pronunciation—for example, if you are a nonnative speaker of English or if you are addressing a group of nonnative speakers—then it would be a good idea to write out your main points and keywords on your visual aids.

Keep the design of your visual aids as simple as possible (handouts can be more complicated). Include only details that you specifically discuss in your talk. For readability, use a mix of upper-case and lower-case lettering (all-upper-case lettering is hard to read). Make the lettering large enough so that people in the back of the room can easily read it. Test your visual aids out in an empty room long before your presentation, so that you can make any necessary changes.

7 *Prepare a suitable introduction.* It's essential that your listeners have enough background information to understand and appreciate your presentation. Are you addressing some problem? Make sure you define it so that your listeners know exactly what it is and can appreciate your proposed

solution. Are you taking sides on an issue and arguing for your point of view? If so, make sure your listeners know exactly what the issue is. In short, if you have any doubts about the audience's background knowledge, be sure to provide a basic orientation and to define important terms.

8 *Prepare a closing summary.* Listeners are typically very attentive at the beginning of a presentation, less attentive as it wears on, and then suddenly more attentive again as it comes to an end. In other words, they perk up at the end, hoping to catch a final summarizing comment or recommendation. This is a well-proven phenomenon, common to all of us: just think about the times you've listened to someone else's presentation and have come away remembering his or her final words best of all.

In preparing your oral presentation, therefore, you should plan to take advantage of this fact of human psychology—you should prepare a good, solid closing summary. Take this opportunity to repeat and thus reemphasize your most important "bottom line" conclusions and recommendations, along with the major reasons for them. Make these closing comments crisply and emphatically.

Example

You are a systems analyst employed by the Coronado Sugar Company (CSC). For a number of years the company has been using outside help in recording the amount of sugar beets received at its four processing plants. Since this service is quite costly, your supervisor has asked you to investigate the feasibility of using the company's own computer for this purpose. You have done so and are now ready to prepare a brief oral presentation of your findings. Accordingly, you go through the following eight steps:

1 *Audience analysis.* You expect the audience to include, in addition to your supervisor, the managers of the four processing plants and the company comptroller. All these people are familiar with the company's system of recording individual deliveries on weight tickets; they are also familiar with the daily and weekly summaries that are produced from these weight tickets. They do not, however, know how these summaries are put together (that's what the outside firm has been doing for $15,000 a year), and they do not know much about computers. Your supervisor and the company comptroller will be concerned mainly with how much money can be saved by adopting a new system. The four plant managers, on the other hand, will want to know mainly whether or not a new system will disrupt plant operations in any way.

2 *Purpose.* Your investigation has revealed that there are three in-house methods that would be substantially more cost-effective than using an outside firm and that any of these methods could be implemented without disrupting company operations. One of these methods would actually save the company more than $100,000 over 8 years, and you would like to see

it adopted. However, if you push this one method to the exclusion of the others, your presentation may sound too much like a sales pitch. So it might be a good idea to broaden your scope a bit and promote the idea that any of the three in-house methods would be better than the present system. This will give your listeners a range of options from which to choose. Surely at least one of the options should appeal to them.

Your primary purpose, then, will be simply to convince your listeners that the present system of compiling beet receiving data should be replaced by an in-house method. Once you've done that, it should be relatively easy to sell your own particular preference from among the three competing options.

3 *Supporting information.* Your supervisor (the decision maker in this case) and the comptroller will be most impressed, of course, with the cost figures, which favor the three in-house options over the present system; be sure to show both the short- and long-term cost advantages. The four plant managers will want to know how the three alternative methods work; in particular, they will want to be reassured that an in-house system will not disrupt their plant operations. All of your listeners will be pleased to hear that at least one of these in-house methods (the one you prefer) will provide a fringe benefit in the form of making extra equipment available for other uses during the off-season.

4 *Pattern of organization.* Your audience's main concern will clearly be how the four methods compare in terms of cost benefits and impact on plant operations. So the best way to present your supporting information would be a comparison-and-contrast pattern of organization.

5 *Outline*   I Problem statement
                        Need for compiling summaries
                        Present method
              II Alternative methods
                        Method 1
                            Costs
                            Advantages and disadvantages
                        Method 2
                            Costs
                            Advantages and disadvantages
                        Method 3
                            Costs
                            Advantages and disadvantages
             III Summary
                        All three alternative methods are more cost-effective
                            than the present system.
                        Method 2 is the most cost-effective in the long run.

6 *Visual aids.* Your audience will want to see a projected cost breakdown for each of the three alternative methods, as well as a summary graph comparing all four (Figure 19–1). The summary graph is particularly important, as it will allow you not only to emphasize your main point (that any of the three alternative methods would be cheaper than the present system) but also to show how Method 2 is ultimately the most cost-effective.

7 *Introduction.* You'll want to engage your audience's interest from the outset by explaining the full nature of the problem: CSC needs to compile regular summaries of beet receiving data for various purposes, but the present system of doing so is far too expensive. Don't get bogged down in too much background information: your listeners already know a lot about the present system, and much of what they don't know (for example, how the outside firm puts together the data) they don't really need to know. Also, this is supposed to be a *brief* presentation, so don't waste your audience's time telling long-winded jokes or anecdotes. Instead, get right to the point. After describing the problem, tell your listeners that you've investigated various possible solutions and have found three that deserve their consideration.

Your introduction might go something like this:

> As you all know, every fall we collect an enormous number of weight tickets from our four processing plants. In fact, we've been averaging about 100,000 of them over the past five years. From all of these tickets we have to compile regular summaries—daily summaries and weekly summaries of beets received for each plant, and then final summaries of delivered amounts for each grower. The daily summaries are particularly important because we need them to maintain smooth plant operations, keep to our shipping schedule, and allow for any sudden breakdowns. This means, of course, that the data have to be processed fast and accurately, and in the past we've always felt we needed the MFC people to do it. After all, they're the pros. And indeed, they've done a fine job for us. The problem is that their service is very expensive: $15,000, to be exact, for just the two months we use them.

> So last month Glenn asked me to look into the possibility of using our own company computer to do the job. I've now completed my investigation and am happy to report that there are indeed three different ways we could use our own computer to get the job done without spending nearly as much money or sacrificing any efficiency. One of these methods, in fact, could save us as much as $100,000 over the next eight years, without any adverse impact on plant operations. There are even some fringe benefits to be gained. . . .

8 *Summary.* You'll want to reiterate the fact that all three in-house methods are more cost-effective than the present system, and you'll want to use your summary chart (Figure 19-1) to emphasize this point. More specifically, you'll want to repeat your claim that Method 2 is the most cost-effective of all, noting that not only would it save CSC the most money in the long

Figure 19-1 ❑ COMPARISON OF PROCESSING METHODS: TOTAL COST FOR 8 YEARS

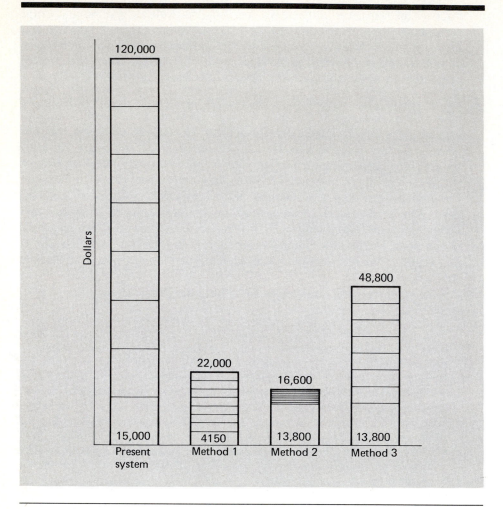

run but it would also allow the company to invest in equipment which could be used for other valuable purposes in the off-season. Be sure to make it clear to your listeners that these are not only your conclusions but also your strong recommendations:

> . . . and so I recommend strongly that we stop using MFC's services and create instead our own in-house program, using our own equipment and our own people. It would take us a little time to get it up and running, but we have the time and will start saving money the minute it's going. It will also increase our computing capability in all sorts of other ways. If you have any questions, I'd be glad to try to answer them for you.

Practice

Nothing is more helpful to the ultimate success of an oral presentation than practice. Not even the best of speakers can give a totally effective presentation without practicing it first. Practice allows you to spot the flaws in a presentation and eradicate them. It enables you to work on making smooth transitions from section to section, instead of awkward stops and starts. And practice gives you an idea of how long your presentation will take; if it's too long, you still have time to make changes so that you can ultimately deliver it at a tempo that's comfortable for you. All of these benefits promote greater self-confidence, which in turn leads to a more convincing, emphatic, effective style of delivery.

The best way to practice a talk is by rounding up a few friends and trying it out on them. Ask them to hear you all the way through, taking as many notes as possible but not raising any questions until you've finished. Then ask them for a complete "postmortem"; take note of spots where they had trouble following you, and immediately try out some other approach to see if it makes things clearer. Note also what the strong parts of your talk are: maybe you can use the techniques of the strong parts in other places.

If you'd prefer not to force your friends to listen to you stumble through a practice session, a good alternative would be to use a videotape recorder and then critique your own performance. This is particularly effective in allowing you to spot nervous mannerisms that you might not be conscious of while actually performing. An audiotape recorder, while not as useful as a videotape recorder, can also be effective, especially for the purpose of listening to yourself read from a manuscript (see point 6 below).

Here are some specific things to work on while practicing your oral presentation:

1 *Devise ways of reiterating your important points without being too repetitive.* Since your important points should all contribute to a single cumulative effect, it's a good idea to reiterate these points occasionally as you go along—especially in summary form at the end of your talk. However, since exact repetition of a point can become annoyingly monotonous the third or fourth time around, try to vary your wording.

2 *Create smooth transitions between sections.* Take note of places where the flow of your presentation seems to break down, and see if you can't insert a phrase or two to act as a bridge. (If you can't, there may be a fundamental flaw in the overall structure of your presentation; in that case, try to reorganize it.)

3 *Familiarize yourself with the equipment you'll be using.* It's embarrassing—and annoying to the audience—to waste precious time by fumbling around with slide projectors, television monitors, microphones, and other equipment when you're supposed to be giving your presentation. It's even worse if your presentation depends crucially on some piece of equipment

and you can't get it to operate at all. Therefore, *if you plan on using any equipment, check it out ahead of time and become familiar with it!* Learn how to use it so that it will help your presentation, not hinder it, and be prepared with some backup system just in case.

4 *Prepare yourself for questions.* Listeners may raise questions at any point in your presentation, and it is vitally important that you answer them satisfactorily. If you don't, your most precious asset as a speaker—your credibility—may be endangered. So be sure you know your topic *well.* To test your knowledge, have some friends listen to you while you practice and have them deliberately throw tough questions at you; if they succeed in stumping you, go do some research. It's not necessary that you have an answer for every conceivable question the audience might raise; there are questions for which *no one* has an answer (and you should certainly never try to fake an answer!). But if you have researched a topic well and have state-of-the-art knowledge about it, you'll be able to answer most questions and will be able to say confidently about the others, "We don't have an answer to that question yet."

5 *Develop your own speaking style.* Practice telling stories or jokes to a few friends at a time in an informal setting. Stand tall and face them as in an oral presentation; engage your listeners' interest. Use natural, animated gestures, and vary your intonation and rate of speech. In short, *be expressive! Let your enthusiasm show!* At the same time, take note of any distracting habits you might have: leaning against something, biting your nails, playing with a pencil, and so on. If you are aware of such habits, you can often take measures to keep them under control while you're "on stage."

6 *If you are going to be reading from a manuscript, work on giving it a lively intonation.* There is a strong tendency when reading aloud to adopt a monotonous style of delivery that is very boring for an audience to listen to. So practice varying your intonation as you read aloud. Raise your voice pitch for the more important words and allow it to drop for the less important ones. Pause occasionally at appropriate places. Put special emphasis on contrasting words, as in the following example:

> System 2986 is an experimental system with two processors which can work *independently* or work *together as coprocessors. One* processor is a commercially available single-board computer (Intel's ISBC 86/12A), which is also referred to as System *86. The other* processor is a custom-built microprogrammable special-purpose processor which is built around AMD 2900 series components and referred to as System *29.* The two processors can talk to each other over the Multibus using the Multibus protocols.

Delivery

As the time draws near for delivering your oral presentation, you will experience what all speakers experience: nervousness! If you have adequately prepared and practiced your presentation, of course, you will probably be less nervous than if you have not. No matter how well you prepare yourself, however, you will still be at least a little bit on edge.

The key to an effective delivery is to convert your nervousness into the kind of energy that injects liveliness, enthusiasm, and animation into your speech. If you try to either suppress your nervousness or let it dominate you, the result will be either a listless, boring, "laid back" style or a jittery, key-jingling, pacing-back-and-forth one—neither of which will appeal to your audience.

How can you control your nervousness? First of all, make sure you're properly prepared for your talk. This means getting all your visual aids and notes ready, and it also means getting your body and mind ready. Remember, *giving an oral presentation is a physical activity*. As with any physical activity, you should eat and drink properly beforehand and you should do some warm-up exercises. Exercising is especially effective in releasing tension. Also, make sure you're properly dressed and groomed. Your personal appearance is one of the most powerful "visual aids" you have. People will be looking at you throughout your talk, and they will be making judgments about your professional credibility partly on the basis of what they see. If you know you're looking your best, it will give you a boost in confidence.

Second, try to establish some contact with your listeners before you give your presentation. Converse with them, see what they're interested in, try to get to know them a little. This will help you think more about your audience and help prevent you from being paralyzingly self-conscious. And it will encourage you to use a more natural, conversational style of speaking.

Third, when you get up to give the presentation, walk purposefully and confidently to the front of the room. If the occasion calls for it, make appropriate acknowledgments. But don't feel you have to tell a prepared joke or story. If you enjoy telling jokes or stories and are good at it, fine—go ahead and tell one. If a spontaneous and harmless bit of humor comes to mind, don't hesitate to use it. Genuine humor is almost always effective in a public gathering. But don't force it. If you're not comfortable doing it, don't even try.

Finally, as you are actually giving your presentation, *concentrate your full attention on what you want to say*. Stick to your outline, make sure you cover all your main supporting points. Convince your listeners that the topic is important, and be enthusiastic about it. Show each of your visual aids long enough for the audience to understand and appreciate it, and then move on to the next one. Keep up the pace, don't dally. If you provide a steady stream of well-organized, interesting information, people will pay attention to you and give you their support. If you don't, they'll lose interest.

Encourage questions from the audience, but don't let questions disrupt your presentation. If you can't think of a short answer to a question, politely tell the questioner that you'll respond to it later, during the Question and Answer time. Above all, *do not show any antagonism toward a questioner*. It will make your listeners feel uncomfortable, and since you are in charge, they will probably hold it against you, even if the questioner is unfair or unpleasant. It's best to defuse any such hostility in public and try to deal with it in private.

❏ EXERCISE 19-1

A   Convert one of your written reports into a 5- or 10-minute oral presentation suitable for an audience of nonspecialists.

B   Convert the same report into a 5- or 10-minute oral presentation suitable for an audience of specialists.

C   Take a technical discussion section (300 to 500 words) from one of your reports and prepare it for oral reading to a large audience.

19.2 GIVING AN INFORMAL ORAL PRESENTATION

Occasionally you may be asked to give an oral presentation on short notice and may not have time to practice or prepare any visual aids. For example, as a member of a project team you may be asked at any time to give a spur-of-the-moment progress report. Or perhaps an orientation tour is taking place and you are asked to describe your laboratory. Perhaps your group leader has been asked to present a feasibility study but has been taken ill, and you are a last-minute replacement for her.

In such cases, the best thing to do is to concentrate on the first five steps in preparing any oral presentation, formal or informal: (1) audience analysis, (2) determination of primary purpose, (3) selection of supporting information, (4) selection of a pattern of development, and (5) outlining. Each of these steps is vitally important and should be performed as carefully and thoroughly as time permits. But *keep it simple*! Don't let yourself get bogged down in complicating details. Time does not allow it, and it will only result in disorganization if not outright confusion. Instead, keep your mind on your main points and how you can best and most clearly support them.

❏ EXERCISE 19-2

Take no more than 30 minutes to prepare a 10–minute informal oral presentation on a familiar topic. Then give it. (If you're a student, you might arrange to give the presentation to one of your teachers during office hours. If you're working for a company, try your supervisor or some colleagues or perhaps a technical writer or editor.)

REFERENCE

1   Peter M. Schiff, "Speech: Another Facet of Technical Communication," *Engineering Education*, November 1980, p. 181.

ADDITIONAL READING

Aristotle, *The Rhetoric of Aristotle*, translated by Lane Cooper (Englewood Cliffs, NJ: Prentice-Hall, 1932, 1960). Another good edition of this work is *The "Art" of Rhetoric*, translated by J. H. Freese (Cambridge, MA: Harvard University Press, 1926, 1967).

Joan Detz, *How to Write & Give a Speech* (New York: St. Martin's Press, 1984).

James G. Gray, Jr., *Strategies and Skills of Technical Presentations* (Westport, CT: Greenwood Press, 1986).

William E. McCarron, "Oral Briefing versus Technical Report: Two Approaches to Communication Problems," *Courses, Components, and Exercises in Technical Communication*, edited by D. W. Stevenson (Urbana, IL: National Council of Teachers of English, 1981), pp. 144–156.

Thomas M. Sawyer, "Preparing and Delivering an Oral Presentation," *Technical Communication*, First Quarter 1979, pp. 4–7.

20

Meetings and Negotiation

Meetings are an essential form of communication in every profession. They are also time-consuming: A recent survey found that the average business executive spends 21 workweeks a year in meetings—and this does not count the time spent traveling to and from those meetings. Some meetings work, others don't. In this chapter, we offer some advice on how to get the most out of this valuable but demanding type of communication.

20.1 ORGANIZING AND RUNNING MEETINGS

Let's begin with this basic question: When is it useful to have a meeting in the first place? Recent psychological studies have concluded that meetings are basically *political* instruments, that they generally work best to create a consensus on some issue where the possible solutions have already been carefully narrowed down.[1] Meetings usually do not work well for brainstorming or for other situations where the group is just beginning to think about a problem. If you are in the position of deciding whether to call a meeting, you should first ask yourself: "What do I hope to get out of it?" If the answer is "some ideas," you would probably do better to postpone any meeting and try to gather ideas in some other way—say, through one-on-one conversation.

If you decide to go ahead with a meeting, make sure you give yourself adequate time to prepare for it. First of all, decide who you want to invite to the meeting. Since the purpose of the meeting should be to reach some group agreement, include only those people, if possible, who will contribute to this goal. Do not include people who are likely to pursue their own agendas and disrupt yours. And do not include people who have nothing to say: they will only be dead weights.

Next, write up an agenda and send it to all the meeting participants. Tell them what you hope to accomplish. This will serve several purposes: (1) it will heighten the participants' expectations that this will be a useful meeting, giving them a more positive attitude; (2) it will encourage them to think about the topics beforehand and organize their thoughts, thus enabling them to make better contributions to the discussion; and (3) it will increase the chances of there being a clear focus to the meeting itself and a clear outcome.

Decide on an appropriate meeting time and place. If possible, the meeting should be held at a time when people are likely to be most alert. People vary in their basic metabolism, and a 5:00 P.M. meeting is not going to sit well with the "morning people" any more than a breakfast meeting will be welcomed by the "night owls." Probably the best compromise is to hold the meeting sometime in the middle of the day, say between 10:00 and 2:00. The room you choose should be conducive to a good meeting: comfortable, well-lit, quiet. It should be free of distractions: phone calls, garish artwork, maintenance workers, etc. Whatever you can do to put the participants into a positive mood will help make the meeting a constructive one.

Once the meeting begins, move quickly through the preliminaries and get down to business. Establish the "ground rules." For example, if you don't want people to smoke (a popular idea in some cultures but not in others), say so at the outset: it's easier to prevent smoking from occurring than to stop it in progress. Ask someone to take minutes. Then make an opening statement in which you state the purpose of the meeting and describe clearly whatever problems the meeting is intended to address. Set out the agenda, including any changes you may have made since notifying the participants. Then open the meeting up to discussion. The biggest problem in running a good meeting is keeping to the topic and thoroughly discussing it. Make sure the discussion stays on track and keeps moving.

Get everyone to participate as early in the meeting as possible: if you notice that someone is not saying anything, don't wait until the end of the meeting to solicit his or her thoughts. Keep the participation as balanced as possible: don't let any one person dominate, unless it's a designated speaker. Maintain a positive, constructive atmosphere. If someone makes a foolish comment, be gentle in your criticism. If someone launches a personal attack on anyone else, defuse it immediately. Try to reach a consensus on as many issues as you can. At the end of the meeting ask the person taking minutes to briefly summarize what's been agreed to.

20.2 PARTICIPATING IN MEETINGS

Most of your involvement with meetings is likely to be in the role of participant, not organizer. As a participant, there are a number of things you can do to maximize your effectiveness. First of all, *be a good listener*. Pay full attention to what the speaker is saying and try to understand everything that he or she is trying to communicate. Don't let your thoughts wander. Second, *be supportive*. If you agree with what the speaker is saying, show it by occasionally nodding your head. If you disagree, don't frown or scowl; wait for an opportunity to ask a polite question to make sure you haven't simply misunderstood. Then, if you want to express your disagreement openly, fine. But do so in a constructive way, so that you don't disrupt the meeting.

In a classic text on meeting behavior,[2] William M. Sattler and N. Edd Miller have identified nine participant roles that contribute to good meetings:

1 The Organizer, who keeps the discussion on track

2 The Clarifier, who points out misunderstandings and clarifies ideas

3 The Questioner, who detects gaps in the group's knowledge and tries to fill them

4 The Factual Contributor, who comes to the meeting well prepared and offers valuable information

5 The Energizer, who stimulates the group to keep going when things look difficult

6 The Idea Creator, who is able to synthesize new ideas and come up with imaginative solutions to problems

7 The Critical Tester, who evaluates ideas for validity and reasonableness

8 The Conciliator, who is good at resolving disagreements and finding a middle ground

9 The Helper of Others, who makes thoughtful efforts to promote group cohesiveness

No one can expect you to fill all nine of these roles at a single meeting, but if you can manage just three or four of them, you will be doing much toward making the meeting a success.

Finally, if you're in a meeting with a culturally diverse group of people, *try to adapt your behavior (though not necessarily your ideas) to what you perceive to be the majority norm*. In some cultures meetings are expected to be very polite affairs while in others they can be extremely noisy. Such differences of communication style can be found not only between countries but also within countries, in all segments of society. Even within a single company there may be a particular style in which meetings are run. If you are not sensitive to these patterns—for example, if you sit quietly in a meeting where people are expected to be talkative, or if you are loud and aggressive in a meeting where people are expected to be reserved—you can have a negative effect on the meeting and you can damage the personal image you project to other people. This does not necessarily mean that you should conform your *thinking* to everyone else's, only that you should conform your *behavior*.

20.3 TELECONFERENCING

An increasingly important type of meeting is the teleconference, which allows people in different geographic locations to "get together" via audio or video hookup. The major advantage of teleconferences over traditional face-to-face meetings, of course, is that they eliminate travel. This is an especially valuable feature at a time when communication needs are becoming more and more international. Teleconferences have also been found superior to traditional meetings for carrying out focused problem-solving activities. And they are no less effective than traditional meetings in the accuracy of the information transmitted or in the amount of time required to transmit that information.

But teleconferences have some disadvantages, too. First, even with a video component, they do not convey hand gestures, "body language," and other forms of nonverbal communication very well. Thus, they do not serve social purposes as well as traditional, face-to-face meetings do. Second, they are complicated to set up technically. Finally, they require special preparation and special presentation skills on the part of the participants. It is this last point that we will briefly address here.

If you are going to participate in a teleconference, keep these points in mind:

1 Any material you plan to present needs to be better prepared than it would be for an ordinary face-to-face meeting. For a video hookup, graphics should be sharper and clearer, with larger lettering. If you have a set presentation to make and you have access to a teleprompter, take advantage of it: have your script written out and installed on the teleprompter, then practice reading it aloud with a natural delivery, as is discussed in Chapter 19.

2 Eye contact is just as important with a remote video audience as it is in ordinary face-to-face conversation. To establish eye contact, look directly into the camera.

3 Since a TV camera or telephone line cannot transmit body language the way face-to-face conversation can, you have to convey enthusiasm and emphasis more with your voice. Therefore, put as much expressiveness into your voice as you can and vary your intonation. By all means, avoid a monotonous drone.

4 There is often a half-second delay in long-distance telecommunications. This can affect spontaneous jokes or other forms of humor that depend on timing. Also, if you are engaged in a dialogue or are fielding questions, with you and another speaker exchanging turns, you should try to make

each of your turns complete ones. You can do this by getting quickly to the point or, if that's impossible, by maintaining a steady flow of speech until you have made your point.

20.4 NEGOTIATING

One of the most important forms of professional communication is the negotiating of agreements between parties who have opposing views and interests. If you are a business executive charged with working out a contract with a labor union, you will be *negotiating* that contract. If you are a civil engineer seeking a variance from the local municipal government to help develop a new vacation resort, you may have to negotiate to get it. If you want to ask your boss for an unusual raise, you will almost certainly have to negotiate for it.

Rogerian Strategy

Although the settings and topics and personalities may differ, successful negotiations all seem to have certain underlying principles in common. Some of these principles constitute what is called Rogerian argumentation, after the psychotherapist Carl Rogers. According to Rogers, when people cling to an illogical or unreasonable position and refuse to even consider any alternatives, it is not necessarily because they believe in that position but because the alternatives are threatening to their self-image. For example, in the Westheimer Clinic case (Chapters 2–7), if you talked to the Westheimer doctors and suggested that they be more friendly and caring toward patients, they might feel that you were criticizing them not only as doctors but *as people*. Even though your advice might be supported by ample statistical evidence, they might refuse to consider it.

Rogers argues that the best way to break down this kind of psychological resistance is to open lines of communication with the other side, and to do this by establishing as much common ground as possible.[3] An important first step is to listen carefully to others and then show them that you understand their point of view. You might even paraphrase it back to them: "You feel that the patients are complaining unfairly. You think they don't fully appreciate the constraints you're laboring under and you feel that their demands are unreasonable." Try to state their side of the argument as well as you can, in a form that they accept. In this way, you succeed in starting the negotiations on a note of agreement, and you begin to break down their sense of threat.

Second, try to show your opponents that you share their aspirations and their moral ideals (honesty, integrity, fairness, etc.). If you and your opponents belong to the same culture, you can evoke this sense of sharedness most easily by making implicit references to that culture. A common and effective way of doing so is by telling a joke or by making a quip or some other type of "small talk." If you and your opponents belong to different cultures, it's more difficult to induce this sense of trust. In that case, you should look for some other kind of grouping that includes both you and your opponents (you belong to the same profession, can speak the same language, have similar political views, etc.). Even in cases where the two sides come from radically different cultures, as, for example, in diplomatic negotiations between Iran and Great Britain, there are usually enough points in common to establish some common ground.

Rogerian argumentation, rather than being a confrontational test of strength in which one side wins and the other loses, tries to create a cooperative atmosphere. By showing your opponents that you trust and respect them, you reduce their sense of threat. This encourages them to do likewise, and to consider *your* point of view. In short, by establishing a sympathetic dialogue with the other side, you can shake your opponents loose from a rigidly-held position and get them to consider various alternatives.

Principled Negotiation

Rogerian argumentation is not so much a step-by-step method as it is a general strategy. Its goal is to create an atmosphere in which meaningful negotiations can take place. For a more operational procedure incorporating the same basic strategy, we recommend the method of *principled negotiation* devised by Roger Fisher, William Ury, and their colleagues at the Harvard Negotiation Project.[4] This method is based on four principles:

1 Separate the people from the problem.

2 Focus on interests, not positions.

3 Invent options for mutual gain.

4 Insist on using objective criteria.

Separate the People from the Problem

There are two aspects to every form of negotiation: the substantive issue and the personal relationship. These two aspects tend to get entangled, but they should be kept separate. In many cases, the personal relationship is more important than the issue. If you lose sight of that, if you make no distinction between the substantive issue and the personal relationship, you may seriously damage the personal relationship; thus, you might win the immediate battle but lose in the long run.

For example, let's say you've just accepted a new job in another city and you're disappointed to find that your new employer is offering to pay only 80% of your moving costs. You see this as a personal insult and you become quite upset about it. You talk to your employer and tell him you would like to have more. When he denies your request in what strikes you as a cavalier manner, you see it as "proof" that he doesn't think very much of you. This disturbs you even more, and you can't help letting loose with some emotionally-fueled comments. Sensing how upset you are, he tries to calm you down by agreeing to pay the full amount. He adds, though, that he is doing this "just for your sake" and that he hopes he won't have to take such "unprecedented actions" in the future.

Can you claim victory in this case? Well, in a sense, yes: you've gained the additional 20% you were seeking. But you may also have tarnished your image in the eyes of your new employer: he may henceforth think of you as being quite self-centered and annoying. Was the 20% worth it? It could be that the 80% is a standard figure for this company and that you are not being slighted. Before jumping to conclusions about the symbolism of a substantive issue, it's best to ask a lot of questions. The people involved may have nothing to do with the problem.

Focus on Interests, Not Positions

Fisher and Ury note that "the basic problem in a negotiation lies not in conflicting positions but in the conflict between each side's needs, desires, concerns, and fears." If the two sides do not recognize this and if the positions they're taking are far apart, the negotiation can easily come to a standstill. Positions are often quite narrowly defined; thus, they have a way of solidifying themselves, of hardening into stubbornness. Interests, on the other hand, are broader. They can usually be satisfied in a number of ways. By focusing on interests instead of positions, negotiators can often find imaginative solutions to seemingly intractable problems.

For instance, in the moving costs example just cited, instead of seeing the conflict in terms of two positions (80% versus 100%), the two parties could look beneath the surface and realize that it is really a conflict of interests: your employer wants to stick to standard company practice and you want to save enough of your money to make a down payment on a small house. If you make these interests known, one of you may think of a way to accommodate both of them within a single position. For example, your employer might think of some other way to help you get your house (e.g., a company-sponsored special loan program or housing program) without violating the 80% policy.

Invent Options for Mutual Gain

As the example just discussed illustrates, focusing on interests can open up possibilities for inventing alternative solutions. Indeed, you should try to think of as many possibilities as you can; use brainstorming and other techniques described in Chapter 2. Remember that you are just inventing at this stage, not deciding. Don't restrict yourself to a "fixed pie" or a single solution. And don't consider just *your* interests, consider *theirs*, too: if you find an option that serves their interests as well as yours, you'll make it easy for them to come to an agreement with you.

Insist on Using Objective Criteria

Even after you've separated the people from the problem, focused on interests rather than positions, and invented options for mutual gain, you may still be faced with a situation in which you and the other party seriously disagree.

You want 100% of your moving costs covered; your employer wants to give you only 80%. What do you do then? Fisher, Ury, and others at the Harvard Negotiation Project have found that difficult negotiating is most likely to succeed when the two parties agree to decide the issue on the basis of independent standards, precedent, or other objective criteria. These criteria should be fair to both sides, and they should be based on reasoned principle.

In the moving costs example, if your employer decided to hold fast at 80% and you wanted to hold out for 100%, you might be able to resolve the impasse by establishing an independent, objective standard of some kind. For instance, you might get your employer to agree that "common practice among similar companies" should be the standard. If you then gather enough information from other companies to show that 100% is the norm, you will be in a strong position to win the full amount. The employer has already agreed to that standard, and so by accepting the higher figure, he will gain respect for being a man of his word. Of course, instead of accepting *your* objective criterion, he may propose one of his own (e.g., company policy). In that case, your only recourse would be to use reason and perhaps *other* objective criteria to challenge it.

❏ EXERCISE 20-1

Look for an opportunity to organize a meeting. Using the suggestions given in Section 20.1, write out a list of "things to do." Use it to run the meeting. Later, using the same list, analyze your performance.

❏ EXERCISE 20-2

Prepare a grid with Sattler and Miller's nine participant roles listed down the side and spaces for participant names across the top. Then go to a small meeting on campus or at your workplace, acting only as an observer. Write the name of each participant in the spaces on top. Whenever a participant says something, decide which of the nine categories it belongs to and put a check mark in the appropriate square. After the meeting, add up the "scores." Then, if possible, get together with the organizer of the meeting and ask him or her to evaluate, impressionistically, each participant's contributions to the meeting. Compare these evaluations with your own.

❏ EXERCISE 20-3

Do some brainstorming and try to invent some more "options for mutual gain" for the moving costs example.

❏ EXERCISE 20-4

Form a group with several colleagues or fellow students and develop a proposal about some local situation (for example, parking problems on campus or in your town), using all the concepts from this chapter.

REFERENCES

1 Daniel Goleman, "Why Meetings Sometimes Don't Work," *New York Times*, June 7, 1988, pp. 15, 24.
2 William M. Sattler and N. Edd Miller, *Discussion and Conference* (Englewood Cliffs, NJ: Prentice-Hall, 1966), pp. 338–340.
3 Carl R. Rogers, *On Becoming a Person* (Boston: Houghton, 1961).
4 Roger Fisher and William Ury, *Getting to Yes: Negotiating Agreement without Giving In* (Middlesex, England: Penguin, 1981).

ADDITIONAL READING

Michael Doyle and David Straus, *The New Interaction Method: How to Make Meetings Work* (New York: Jove, 1976).

Geoffrey Leech, *Principles of Politeness* (Harlow, England: Longman, 1983).

Daniel K. Rosetti and Theodore J. Surynt, "Video Teleconferencing and Performance," *Journal of Business Communication* 22(4):25–31 (Fall 1985).

Ron Zemke, "The Rediscovery of Video Teleconferencing," *Training: The Magazine of Human Resources Development* 23:28–43 (September 28, 1986).

PART FIVE

READABILITY

21

Readability: General Principles

As emphasized in earlier chapters, most readers of scientific or technical writing do not have as much time for reading as they would like to have and therefore must read selectively. This is especially true of managers, supervisors, executives, senior scientists, and other busy decision makers who often skim-read for main points and ideas. However, it is also true of professionals who often need to read more closely and slowly, for thorough understanding, and it is true of technicians, workers, and consumers who may need to read and follow operating instructions. These different types of readers are selective in different ways: the skim-reading decision maker may be looking for "bottom line" cost figures and performance data; the professional may be looking for the main thread of an argument; the technician, worker, or consumer may need to use operating instructions only as a checklist.

For such readers, writing is readable to the extent that it provides the information they need, located where they can quickly find it, in a form in which they can easily use it. This takes considerable effort on the writer's part; as the old adage has it, "It's easy to make things difficult, but difficult to make things easy." Nonetheless, it's worth the effort. If you can make your writing readable, you'll greatly increase its chances of being read and used; in short, you'll greatly increase its *effectiveness*. Conversely, if you don't make the effort, your readers may not either.

How can you make your writing more readable? Unfortunately, there is no simple formula to follow. (The so-called readability formulas are not designed to guide the writing process, and so prescriptions based on such formulas, such as "Use short sentences and short words," are not very reliable.) There are steps you can take, however, that should be of some help, and these are laid out in the next eight chapters. In this chapter, we make suggestions for selecting appropriate information and for making this information accessible to the reader. Then, in Chapters 22 to 28 we suggest a number of things you can do to make it easier for the reader to absorb details; the focus will be on sentences, phrases, and individual words.

Basic Principle
Put information that is new to the reader into a framework of information already known to the reader.

Communication involves the transmitting of information to someone who does not already possess that information, in other words, the transmitting of new information. When new information is absent from a message, the message does not communicate anything: it is noninformative. To most of us, the statement "The Pope is a Catholic" is noninformative: it tells us nothing we don't already know. In general, new information is a necessary ingredient of meaningful communication.

New information is not the only ingredient in successful communication, however. If statements consisted of nothing but new information, they would be incomprehensible. Try this, for example: "On big-wall Grade VIs, the second usually jumars all off-widths and overhangs." Got that—or did it go sailing right over your head? It's filled with technical rock-climbing jargon, and unless you're familiar with rock climbing, you probably found it totally incomprehensible. In general, for new information to be comprehensible to someone, it must be couched in a meaningful framework of information already known to that person, that is, a framework of given information. Given information is the background knowledge that we call upon in trying to make sense of new information.

Most of what we know about the world is stored in distant recesses of our mind, in what psychologists call long-term memory; it is not consciously on our mind. Thus, before this background knowledge can be put to use in interpreting new information, it must first (to use another psychological term) be activated, that is, brought to conscious awareness. Fortunately, our background knowledge exists not in isolated bits and pieces but rather in clusters and networks of associated concepts. Thus, mention of even a single word may trigger in our mind a whole host of related images, facts, beliefs, etc. Some of these related concepts may be useful in helping us interpret a particular piece of new information; others may not be.

Although there is almost no limit to the amount of knowledge we can store in long-term memory, there are very severe limits to the amount of knowledge we can maintain in active consciousness. We can think about only one thing at a time, as the saying goes. So it's important for you as a writer to activate the *right kind* of given information in the reader's mind—the kind of given information that will help most in understanding the new information. It should pertain directly to the topic of discussion so that it fits into the context already established and allows the reader to anticipate what's coming next. The word or words you use to refer to this given information should be familiar, concise, and loaded with imagery so as to maximize the amount of background knowledge you can bring to the reader's conscious awareness. Consider the following negative example and its rewritten version.

Negative Example

Modern cryptanalysis has attracted some of the most capable mathematical minds. In recent years the growing prospect that postal and diplomatic communication will soon be replaced by other forms of communication has furnished increased incentive for mathematicians and engineers to invent an unbreakable cipher. Messages will be able to be telephoned, delivered, and quickly typed on the other end. But unbreakable codes will be needed to protect senders against snoopers.

No doubt you can understand the basic meaning of this passage. But is it vividly memorable? Do you have a clear impression of what kind of unbreakable cipher or what kinds of snoopers the writer is talking about? If not, you may find this rewritten version more to your liking.

Revised Version

Modern cryptanalysis has attracted some of the most capable mathematical minds. In recent years the growing prospect that postal and diplomatic communication will soon be replaced by electronic communication has furnished increased incentive for mathematicians and engineers to invent an unbreakable cipher. Electronic devices will permit messages to be telephoned, delivered, and quickly typed on the other end. But unbreakable codes will be needed to protect senders against electronic snoopers.[1]

The major difference between the two versions is the use of the word *electronic* in the second one. This one word does not really add any new information to the passage; someone reading the first version, for example, could probably figure out that the phrase *other forms of communication* most likely includes electronic communication. But this word does bring to mind the kind of given information that's needed to fully understand what the writer is talking about. The word *electronic* is familiar and concise. Even more important, it activates in our minds a number of vivid images which we can associate with words like *communication*, *devices*, and *snoopers*, thus giving much greater meaning to these concepts. Notice, too, how the writer has used the word repeatedly so as to keep it in the reader's active consciousness and thus help maintain continuous focus on the topic of the passage on page 398.

As can be seen from this example, the use of key words can greatly enhance the coherence and memorability—and hence the readability—of a passage. Key words are most effective when they (1) trigger vivid imagery in the reader's mind, (2) are related in an obvious way to the topic of the passage, and (3) are related in an obvious way to the reader's purpose in reading the passage. If you select key words that satisfy these criteria, and if you put them in prominent positions in the text, you will be doing much to activate the right kind of given information in the reader's mind.

Modern cryptanalysis . . .

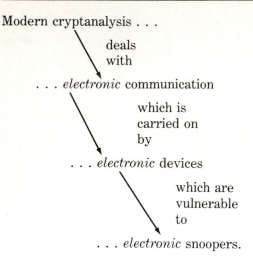

deals
with

. . . *electronic* communication

which is
carried on
by

. . . *electronic* devices

which are
vulnerable
to

. . . *electronic* snoopers.

What are the prominent positions in a text? Basically, these are the places that are visually prominent by virtue of having more white space around them and being more easily located: titles, headings, subheadings, captions, labels, paragraph topic statements, and sentence subjects. Some of these text features have already been or will be discussed (see Chapters 8, 9, and 22 especially); the remainder of this chapter and Chapters 24 and 25 will go into further detail.

21.1 ESTABLISH YOUR TOPIC AND PURPOSE

Make it clear what the main topic of the report or the section is. Then state your purpose explicitly, so that your readers can anticipate how you will be dealing with the topic. Readers of scientific and technical writing are typically purpose-directed and pressed for time. So, rather than reading word for word and cover to cover, they often prefer to merely "consult" a document, looking only for the information they need. When you define your topic and state your purpose, you make it easier for the reader to determine right away how to process the document: whether to read it closely, skim-read it, pass it on to someone else, or disregard it. A clear statement of topic and purpose allows the reader to form certain expectations about the rest of the text, specifically, how the topic is likely to be developed. It is a well-known fact that we process information most quickly and efficiently when it agrees with our preconceptions; this is why it's important to create the right preconceptions in the reader's mind in the first place.

Scientific and technical writing genres customarily have various features designed to announce the topic and set up initial expectations: Titles, Abstracts, Forewords, Summaries, Overviews, etc. Use these to full advantage by loading them with key words and main ideas instead of vague phrases. If you're writing a report dealing with some problematic issue—as is the case with most reports—be sure to include a well-written Problem Statement at the beginning. Engineering and other applied sciences are fundamentally problem-oriented, and so, as discussed in Chapter 5, a good problem statement usually has important orientational value.

21.2 USE KEY WORDS PROMINENTLY

Build sections and paragraphs around key words related to the main topic. If possible, make these key words visually prominent by using them in headings, subheadings, topic statements, and sentence subjects. Once you've established a conceptual framework at the beginning of your text, you can turn your attention to filling it in with appropriate details. To make sure that your discussion is a coherent one, you should strive to link these details as directly as possible to the main topic. The best way to do so is to establish a hierarchy of intermediate topics and subtopics for the various units and subunits of your text, with each being directly related to the immediately higher topic or subtopic. (If you've taken the time to outline your text before writing the draft, this should be a fairly straightforward matter.) Intermediate topics and subtopics should consist of appropriate key words, as discussed above.

This technique is illustrated in Figure 21-1, an excerpt from a scientific article. Notice how key words (circled) are used to weld the headings, subheadings, and topic sentences together into a hierarchy of topics and subtopics; notice, too, how the writer has given these key words prominence by putting them in the subject position of the sentences.

A well-structured discussion is highly functional in at least two respects. First, it builds on the basic framework established at the beginning of the text, allowing for easier interpretation and promoting greater coherence at the same time. As new information is progressively added to the initial framework, it is interpreted in terms of this framework and integrated into it. As such, this new information is transformed into given information and can then be used to help interpret succeeding pieces of new information. Second, as is discussed in Chapters 7, 9, 12, and elsewhere, a hierarchically structured text facilitates selective reading. Since the sections and subsections are arranged in a general-to-specific order, the reader can quite easily locate desired levels of detail—especially if the respective topics of these sections and subsections are made visually prominent through the use of headings and subheadings. (Paragraph-level topics can be made visually prominent by introducing them in topic statements at the beginnings of paragraphs, as discussed in Chapter 22.)

21.3 EXPLAIN IMPORTANT CONCEPTS WHEN WRITING FOR NONSPECIALIST READERS

When writing for nonspecialists, be sure to clarify the important technical concepts in your text by using examples, analogies, visual aids, or other

Figure 21-1 ❏ BUILDING SECTIONS AND PARAGRAPHS AROUND KEY WORDS RELATED TO THE MAIN TOPIC. THE KEY WORDS ARE CIRCLED
[From Ben Patrusky, "The Social Cell," *Mosaic* 12(4):9 (1981).]

Reach out and touch

Cell/cell junctions come in a variety of shapes and sizes.

When cells come together—and stay together—they do more than just touch. Contiguous cells actually form highly specialized, highly organized interconnective structures. And with the development of a technique called freeze fracture, a method for separating the inner face of a membrane from its outer face, scientists have managed to extract many details about the micro-architecture of these fine intercellular contacts or junctions.

One of the earliest known of the intercellular junctions is the revet- or weld-like desmosome. Adjacent cell membranes linked by way of desmosomes don't actually touch; the neighboring surfaces are separated by a space of about 200 angstroms. (One angstrom equals one hundred-millionth—10^{-3}—of a centimeter.) And the space is packed with dense, extracellular material, which serves as binder.

Desmosomes have been found primarily between the cells of sheetlike epithelial tissue, which covers the surface of the body and the internal surfaces of most organs. A variation of the desmosome, the half-desmosome, does not link cells to other cells but anchors individual cells to the underlying mesh of connective tissue, between the epithelium and the underlying organ. According to Norton B. Gilula of Rockefeller University, an authority on inter-cellular junctions, the function of desmosomes is now thought to be strictly that of adhesion—to keep tissues and organs intact despite mechanical stress. That belief is prompted in part by the fact that cells normally subjected to severe, repeated stress—heart muscle, for instance, or the cervix and uterine walls—possess unusually large number of desmosomes.

Tight junctions

Epithelial tissue is especially rich in another kind of distinctive intercellular connection called the tight junction, clearly characterized by Marilyn Farquhar and George Palade of Rockefeller Institute (now Rockefeller University) in 1967. Unlike the desmosome, with its 200-angstrom separation between cell surfaces, the tight junction, as the name implies, seems to represent a true contact between cells. No intervening space is evident. It's as if the outer surfaces of the two bilayers of neighboring cell membranes had melted into each other.

Found at the apex, or shoulder, of epithelial cells, tight junctions act as sealants, occlusions, or barriers that enable an organism or an organ to maintain an internal environment that is chemically distinct from its surroundings, explains the California Institute of Technology's Jean-Paul Revel, another leading figure in intercellular junction research. Tight junctions he says, abound in places where a sharp physical separation between two compartments—or from the outside world—is essential. It's no surprise, then, that tight junctions have been discovered in embryonic cells at the very early stages of development.

For all that, tight junctions may not be totally impermeable. Freeze-fracture studies by Philippa Claude and Daniel A. Goodenough of Harvard Medical School show that different kinds of tissue have structural differences in their tight junction that appear to correlate with varying degrees of leakiness: the bigger the junction, the less the leakage.

[Patrusky, Ben, "The Social Cell," MOSAIC. 12:4, July/August 1981, National Science Foundation, p. 9] *forms of verbal or visual illustration.* Research by information theorists in the past few decades suggests that communication proceeds best when there is a fairly even balance between given information and new information. This is what you should strive for in your own writing. To do so, you must have some idea of who your readers are and what sort of background knowledge they have; as illustrated by the rock-climbing example, "givenness" and newness are partly functions of the knowledge a reader brings to a text—and this can vary from reader to reader and text to text. To a rock climber, the sentence about big-wall Grade VIs would be perfectly comprehensible; to anyone else, it wouldn't be. Thus, if for some reason you had to communicate that kind of technical information to a nonspecialist reader, you would have to insert some background information more familiar to the reader to provide a proper framework for interpreting the new information. In so doing, you would be creating a better balance between given and new.

In technical writing, it frequently happens that the writer feels it necessary to introduce key concepts that may be unfamiliar to the reader. Sometimes these key concepts even occupy topic roles: topics of paragraphs, topics of sections, perhaps even the topic of the entire text. In general, it's important to define such concepts, not necessarily with a formal definition but rather with some kind of *illustration.* How is the concept used? What is it similar to? What does it look like? If technical terminology is used, what is a nontechnical way of saying more or less the same thing? Not only will answering such questions with the reader's needs in mind help the reader understand that particular concept but, more important—especially if the concept is a topical one—it will enrich and sharpen the reader's interpretation of the text as a whole. It will provide some of the given information that a specialist reader would automatically and implicitly associate with that particular concept but which a nonspecialist reader would not.

There are several ways to illustrate and explain unfamiliar concepts for the nonspecialist reader. *Visual aids,* of course, should be used whenever the concept is suited to visual representation (see Chapters 8 and 9). Often, however, a concept is too abstract to be represented visually. In such cases, specific *examples* of the concept are usually the most powerful means you can use to help the nonspecialist reader. Research by cognitive psychologists indicates that readers confronted with an unfamiliar abstract concept will often try to construct a concrete example they can relate to; in some cases, they will actually build a scenario with themselves as the principal actor acting out the concept. As a writer, you can sometimes save the reader this effort by providing an example, or even a scenario, yourself. This will also prevent the reader from constructing a misleading example or scenario. *Analogies* help explain an unfamiliar concept by showing that it is similar in certain ways to a familiar concept; they are useful in situations where the concept is so unfamiliar that you simply cannot think of any ordinary examples of it. *Paraphrases,* on the other hand, are useful in precisely the opposite situation: where the concept is familiar to the reader but only if restated in more recognizable terms. Paraphrases have a distinct advantage over examples and analogies in that they usually take up less space; sometimes even a one-word paraphrase will accomplish the purpose. *Definitions*, of course, are a familiar way of explicating new concepts. They can be combined with some of the techniques mentioned above to form *extended definitions*.

Here is an example of an extended definition, explaining what the technical term *Remrak coefficient* means:

The Remrak Coefficient

In the production of powdered detergents, spray drying is the technique used to evaporate the solvent from the liquid reaction mixture and physically form the finished powder product. In spray drying, the liquid is sprayed into the top of a tall tower and allowed to fall freely to the bottom of the tower, where it is removed as a dry powder. The solvent evaporates during the course of the fall. Particles dried in this fashion have an unusual

Analogy   shape, *like that of a saddle (or a Pringle's potato chip)*, and consequently fall through the air in an unusual manner. Rather

Paraphrase   than falling in a vertical path, the particles fall in a helical *(spiral)* path. The shape of the helical path is described by the Remrak

Definition   coefficient, *which is the ratio of the diameter of the helix to the height required for one passage of the particle around the perimeter of the helix.* The coefficient, which is a function of drying conditions,

Paraphrase   is sought to be maximized, *so that the length of flight of the particle is made much greater than the actual height of the spray-drying tower.* [Italics added]

The writer of this passage has obviously gone to considerable lengths to help us understand what the Remrak coefficient is. Among other things, most of the standard forms of verbal illustration have been used: analogy, paraphrase, simple definition, and extended definition. Has it been worth the trouble? Yes, if we didn't already know the concept and were interested in finding out; no, if we already knew or weren't interested in knowing. In general, verbal and visual illustrations are powerful devices, but they work only under the following conditions:

1 The concept is not already familiar to the reader.

2 The information used to illustrate the concept *is* familiar to the reader.

3 The concept being illustrated is an important one in that particular context.

4 The information used to illustrate the concept focuses on features of that concept that are relevant to that particular context.

Do take advantage of the power of illustrations in explaining unfamiliar technical terms to nonspecialist readers. When used correctly, illustrations can clarify things in an instant.

21.4 USE STANDARD TERMINOLOGY WHEN WRITING FOR SPECIALIST READERS

When writing for specialists, on the other hand, do not overexplain. That is, do not exemplify, define, illustrate, paraphrase, or otherwise explain concepts the reader is likely to already be familiar with. Instead, simply refer to such concepts with the standard terminology of the field. Part of what it means to be a specialist in a given field is to know the standard technical terminology of that field. Technical terms permit efficient and precise communication among specialists who know the concepts that such terms refer to. They should be used for that purpose, and used freely, even if they appear to be incomprehensible jargon to an outsider. When used among specialists, standard technical terms are not only comprehensible but are often "information rich" in the sense that they may trigger a host of associated concepts in the reader's memory. These associated concepts then become part of the given information in the message. Thus, in the jargon-laden writing of specialists for other specialists, there is usually more than enough given information. Adding more given information in the form of examples, analogies, etc., would only produce a disproportionate and inefficient given-to-new ratio for that type of reader.

What do you do, though, if you are writing to a *mixed* audience of specialists and nonspecialists? This is always a very challenging situation, but there are a few things you can do. First, you might divide and conquer: produce two separate pieces of writing, or a single piece with two parts to it, so that each group of readers can be addressed with appropriate terminology. Alternatively, you might stick to a single text but briefly define the technical terms as you go along. The least objectionable way of doing this, usually, is to insert a short, familiar paraphrase immediately after each technical term; in the Remrak coefficient example, for instance, notice how the writer has inserted the paraphrase *(spiral)* after the less familiar term *helical.*

21.5 STRUCTURE YOUR TEXT TO EMPHASIZE IMPORTANT INFORMATION

Structure the different parts of the text so as to give greatest prominence to the information you expect the reader to pay most attention to. For main ideas, use a hierarchical (general-to-particular) structure; for details, use a listing (coordinate) structure. As discussed above, a hierarchical text structure allows the reader to move quickly through a text, seeing what the main ideas are, how they're linked together, and what kind of detailed support they have. Many readers, especially busy decision makers, habitually read this way. Thus, if you are writing for that type of reader, you should try to organize and present your information in a highly hierarchical pattern, with many levels of subordination.

On the other hand, if you are writing for a reader who will be focusing more on details, try to use a more coordinate structure—that is, with the details arranged in lists. A listlike structure, whether it's formatted as a list or not, draws the reader's attention to all the items making up the list. Instead of one statement being subordinated to another, as in a hierarchical structure, the statements in a list are all on the same level and thus share equal prominence. Perhaps the most obvious examples of this phenomenon are lists of instructions, which are expected to be read and followed step by step. The same phenomenon can also be seen in carefully reasoned arguments and explanations, which are often cast in the form of a listlike sequence of cause-and-effect statements. Chronological sequences, too, as found in descriptions of test procedures or in progress reports, are often presented as lists. (As Chapter 23 emphasizes, remember to use parallelism for all lists, formatted or not.)

21.6 CONSTRUCT WELL-DESIGNED PARAGRAPHS

Make sure that each paragraph has a good topic statement and a clear pattern of organization. The paragraph is a basic and highly functional unit of discourse in scientific and technical writing. By definition, a paragraph is a group of sentences focusing on one main idea. If you use a topic statement to capture the main idea and a clear pattern of organization to develop it, you make it easy for the reader to either read the paragraph in detail or read it selectively. The topic statement should generally be presented within the first two sentences of the paragraph, and it should contain one or more key words for readers to focus their attention on. The pattern of organization you select for the remaining sentences in the paragraph should (1) be consistent with expectations likely to be raised by the topic statement, (2) be appropriate to the subject matter, and, most important, (3) be appropriate to the anticipated use of the paragraph by the reader (see, for example, Section 21.5). If you adhere to these principles with all your paragraphs, you will greatly enhance the overall readability of your writing.

For a thorough discussion and review of the principles of paragraph writing, see Chapter 22.

21.7 FIELD-TEST YOUR WRITING

Field-test your manuscript with its intended users or with representative substitutes. Up to this point, you've had to make guesses about whether or not you're providing your readers with a proper mix of given information and new information for their purposes. Your decisions about what kind of terminology to use, what kind of structure to use, when to use verbal or visual illustrations, and so on, have been made on the basis of guesswork about your readers' background knowledge and the reasons they will have for reading your writing. Educated guesswork, perhaps, but guesswork nonetheless.

This is why field testing with actual users (or representative users) is an essential part of making any manuscript maximally useful. (See Chapter 7.) Field testing allows you to see whether the assumptions you have made about your readers are accurate. This is so important that you should not put it off until the final editing stage (though you might want to do a second round of field testing at that time). As soon as you've finished writing a good first or second draft, try it out on a few intended users. Have them read it as if it were the final draft submitted for actual use. Tell them to mark it up, raise questions about it, criticize it. Talk to them about it, ask them for their comments. Does it leave anything out? Does it mislead them? Does it raise unanswered questions? If they're using it for reference purposes, can they easily find what they need? If they're skimming it for main points, can they easily locate and understand them?

If your manuscript contains any instructions or any other material designed to be acted on, have your readers read and try to follow these instructions while you observe. Take note of places where they have to pause and ponder. Take note also of places where they go astray. Do not interfere with their efforts—let the manuscript speak for itself. Later, meet with your readers. Ask them where they had trouble. Did something confuse them? Did something seem to be missing? Would they prefer to have the material written in some other way?

This is also a good stage at which to consult experts, to make sure that nothing you have written is substantively wrong. If you're writing a research proposal or article, for example, you might want to show your draft to other researchers in that area, so as to guard against the possibility that you've overlooked something important or misrepresented someone else's research. If you're writing a progress report for a group project, this would be a good time to show it to other members of the team.

In all of this testing, both with intended users and with experts, take note of *all* trouble spots—even those that do not seem to pertain to inadequate selection or structuring of information. Some trouble spots will probably be due to sentence-level problems: problems of phrasing, wordiness, diction, sentence structure, lack of emphasis, etc. These are addressed in Chapters 24, 25, 26, and 27.

21.8 TAKE CULTURAL DIFFERENCES INTO ACCOUNT

All of the foregoing suggestions assume that you are writing for a Western audience. Western readers welcome explicit purpose statements, clearly-articulated main points, full explanations, etc. Other cultures, however, may have different norms and different expectations. For example, Japanese writers normally prefer to be less direct in their assertions. Instead of stating their main points and recommendations early on, they like to lay the ground slowly and state their important points at the end or not at all. Rather than making all the steps of an argument explicit, they usually prefer only to hint at the argument and let readers make the connections on their own. And the Japanese generally are not willing to sacrifice this politeness for small gains in readability. For example, although a short, to-the-point letter may be very readable in the sense of being quickly understood, it may also be seen as curt and thus somewhat impolite by a Japanese.

Readability is a function not only of basic psychological mechanisms but also of culture-dictated norms. If you try to meet the cultural expectations of your reader, he or she will find your writing more readable than otherwise. Specifically, if you are writing to a Japanese audience, you should try not to be too direct, too assertive, or too explicit. Don't overexplain. On the other hand, if you are Japanese and are writing to an American audience, you should try to adjust in the opposite direction and be more direct and explicit than you normally are.

❑ EXERCISE 21-1

A Take a highly technical term or concept from your field and define it in standard terminology so that another specialist in the same field can quickly and easily understand it. Then check your definition against that found in a technical dictionary.

B Take the same term or concept you chose in A and write an extended definition of it (using at least three of the techniques described in this chapter) so that a nonspecialist can understand it. Then test your definition by having a nonspecialist read it; if he or she has any trouble understanding it, make appropriate modifications.

❑ EXERCISE 21-2

Take a full-length sample of your own writing and field-test it with one or more intended users (or representative substitutes). Then revise it according to the guidelines laid down in this chapter.

REFERENCES

1 Julie Wei, "Pure Mathematics: Problems and Prospects in Number Theory," *The Research News 30*(3):20 (1979).

ADDITIONAL READING

R. C. Anderson, R. E. Reynolds, D. L. Schallert, and E. T. Goetz, "Frameworks for Comprehending Discourse," *American Educational Research Journal 14*:367–381 (1977).

Marshall Atlas, "The User Edit: Making Manuals Easier to Use," *IEEE Transactions on Professional Communication PC24*(1):28–29 (1981).

F. C. Bartlett, *Remembering* (London: Cambridge University Press, 1932).

Thomas N. Huckin, "A Cognitive Approach to Readability," in *New Essays in Technical and Scientific Communication* , edited by Paul V. Anderson, R. John Brockmann, and Carolyn R. Miller (Farmingdale, NY: Baywood, 1983), pp. 90–108.

Koreo Kinosita, "Differences between Japanese and Western Styles of Scientific and Technical Communication." Paper delivered at the Second International LSP Conference, Eindhoven Institute of Technology, Eindhoven, Netherlands, August 1988.

Walter Kintsch and Douglas Vipond, "Reading Comprehension and Readability in Educational Practice and Psychological Theory," in *Perspectives in Memory Research*, edited by L. G. Nilsson (Hillsdale, NJ: Erlbaum, 1979), pp. 329–365.

James R. Miller and Walter Kintsch, "Knowledge-Based Aspects of Prose Comprehension and Readability," *Text 1*(3):215–232 (1981).

J. M. Royer and G. W. Cable, "Illustrations, Analogies, and Facilitative Transfer in Prose Learning," *Journal of Educational Psychology 68*:205–209 (1976).

D. E. Rumelhart and A. Ortony, "The Representation of Knowledge in Memory," in *Schooling and the Acquisition of Knowledge* (Hillsdale, NJ: Erlbaum, 1977).

D. Schallert, "Improving Memory for Prose: The Relationship between Depth of Processing and Context," *Journal of Verbal Learning and Verbal Behavior 15*:621–632 (1976).

M. N. K. Schwarz and A. Flammer, "Text Structure and Title: Effects on Comprehension and Recall," *Journal of Verbal Learning and Verbal Behavior 20*:61–66 (1981).

Jack Selzer, "What Constitutes a 'Readable' Technical Style?" in *New Essays in Technical and Scientific Communication* , edited by Paul V. Anderson, R. John Brockmann, and Carolyn R. Miller (Farmingdale, NY: Baywood, 1983), pp. 71–89.

G. J. Spilich, G. T. Vesonder, H. L. Chiesi, and J. F. Voss, "Text Processing of Domain-Related Information for Individuals with High and Low Domain Knowledge," *Journal of Verbal Learning and Verbal Behavior 18*:275–290 (1979).

A. B. Tenenbaum, "Task-Dependent Effects of Organization and Context upon Comprehension of Prose," *Journal of Educational Psychology 69*:528–536 (1977).

Perry Thorndyke and Barbara Hayes-Roth, "The Use of Schemata in the Acquisition and Transfer of Knowledge," *Cognitive Psychology 11*:82–106 (1979).

22

Writing Paragraphs

As emphasized in Chapter 21, readability is largely determined by how closely the text fulfills the reader's expectations. This is a dynamic process, not a static one. At each step of the way, your writing should set up certain expectations in the reader's mind and then promptly satisfy those expectations. This is true at all levels of a text: sentences, paragraphs, sections, entire documents. In this chapter, we discuss the writing of paragraphs.

Unlike teachers, who are expected to read every word of their students' papers, technical and scientific professionals often read *selectively*. Faced with "the information explosion," they simply don't have time to digest thoroughly every piece of writing they encounter. Instead, they often scan a document, looking for particular pieces of information, or skim-read it, trying just to get a general impression of its message.

As a writer, you should do everything possible to ease such a reader's burden. In particular, you should make your writing *easy to skim-read*. One of the best ways to do this is to write good paragraphs. What makes for a good paragraph in scientific and technical writing? First of all, a good paragraph has *unity:* it focuses on a single idea or theme. Paragraph unity helps a busy reader skim-read a document because he or she can bounce from one paragraph to the next and quickly build an impression of the whole. Second, a good paragraph has *coherence*: one sentence leads to the next in some kind of logical sequence. Paragraph coherence supports paragraph unity and makes it easier for the reader to quickly see how the ideas in a paragraph are related. Third, a good paragraph has *adequate supporting content*: it has an appropriate selection and number of details to support the main idea of the paragraph. Even the fastest skimmer may slow down occasionally and digest a paragraph in detail, and in such cases you'd better provide the details that he or she needs! Fourth, a good paragraph fits in with the overall structure of points being made in the text as a whole. And finally, a good paragraph contributes to the creation of appropriate emphasis in the text as a whole.

There are two principal tools you can use to invest your paragraphs with the qualities just described: (1) a good topic statement, (2) an appropriate pattern of organization, and (3) appropriate functions and emphasis.

22.1 WRITE A GOOD TOPIC STATEMENT

The topic of a paragraph is its main idea or theme—what the paragraph is about. In formal scientific and technical writing, the most common way of presenting the topic of a paragraph is the so-called deductive pattern: the opening sentence introduces the topic and may even indicate how it will be developed. In some cases, a second sentence is used to refine the topic, to summarize it, or to shift the direction of development. Busy readers generally want to know right away what the topic is. They will use whatever cues they can to quickly generate expectations about the paragraph as a whole. This strategy serves two purposes: (1) it allows readers to guess what's coming and thus digest it more easily and (2) it allows them to avoid reading the paragraph altogether if the subject matter holds no interest for them.

For special purposes you can use an inductive pattern of development, in which you delay the topic statement until the end or near the end of a paragraph. This pattern is appropriate only for situations where you expect the reader to slow down and digest every word of the paragraph. For example, if you are laying out the logic of an important point in step-by-step fashion and want your reader to follow it each step of the way, you might use an inductive pattern. Or if you have to write a "bad news" letter and want to soften the blow before delivering the news, you might consider this pattern.

In most cases, though, you can help your readers best by providing a good topic statement right at the beginning of the paragraph. It does not have to be confined to a single sentence: often a topic statement is extended over the first two sentences of a paragraph. It should, however, always contain one or more key words directly related to the topic, and it should be as complete a statement of the main idea as possible without getting into too much detail and making the sentence(s) excessively long. In addition, if possible, it should suggest how the topic will be developed (by comparison and contrast, by cause-and-effect analysis, etc.).

Here is an example of an effective topic statement.

Unlike gasohol-powered cars, the fuel cell alternative is virtually pollution-free. A methanol fuel cell system works through chemical reactions that leave the air clean. A fuel processor breaks the methanol down into carbon dioxide and hydrogen; the hydrogen is then pumped to the cell itself, where it combines with oxygen to form water. Current is then produced when the electrons traded between molecules in this reaction travel through an external circuit. The net products are carbon dioxide, water, and electricity. By contrast, when gasohol is burned in an internal combustion engine, it produces the same nitrous oxides that gasoline does.[1] [Italics added.]

This topic statement is a good one because it tells the reader immediately what the theme of the paragraph is (fuel cell cars don't pollute) and because it's consistent with how the rest of the paragraph is developed (as a cause-and-effect description of how the fuel cell process works). Notice how the writer has used the key term *fuel cell* in the most important position in the sentence, the main-clause subject position, thus establishing it as the paragraph topic—what the paragraph is about.

For an example of what *not* to do, here is a paragraph from a student report on whether to use an argon recovery process or a hydrogen recovery process in a proposed ammonia plant.

Negative Example

Utility costs for the argon process are 75% greater than for the proposed hydrogen process. Initial capital cost is $5.4 million, roughly three times the hydrogen process cost. However, annual income from the sale of argon, increased ammonia production, and reduced natural gas requirements elsewhere in the plant is 160% higher than that generated by the hydrogen process. Present-worth analysis shows that the argon process is the better investment. The present worth of the argon process is $10.25 million. The present worth of the hydrogen process is $4.14 million.

Most readers will quickly conclude, on the basis of the first two sentences of this paragraph, that the argon process is more costly than the hydrogen one and should therefore not be chosen. But this is just the opposite of what the writer wants them to understand! For later on, buried near the bottom of the paragraph, the report states that "the argon process is the better investment." This statement is actually the topic statement of the paragraph. By "burying" it, the writer is running a serious risk of having the readers completely overlook it.

A few simple changes can easily remedy the situation: (1) promoting the topic statement to initial position in the paragraph, (2) combining the next two sentences and subordinating them to the next one, and (3) adding a few words for emphasis:

Revised Version

The argon process is clearly a better investment than the hydrogen process. Although it has higher utility costs (by 75%) and a higher initial capital cost (by 300%), it generates annual income—from the sale of argon, from increased ammonia production, and from reduced natural gas requirements elsewhere in the plant—that is 160% greater than that generated by the hydrogen process. Present-worth analysis shows that the argon process is valued at $10.25 million. The hydrogen process, by contrast, is valued at only $4.14 million.

Notice how much more readable the rewritten version is. The topic statement serves to establish the main point and also to suggest how the rest of the paragraph will be developed (as a comparison-and-contrast pattern). The key term *argon process* (or its pronoun equivalent *it*) is used repeatedly in sentence-subject or clause-subject positions, thus keeping the reader's mind focused on it; all reference to the hydrogen process, by contrast, is deliberately subordinated.

The basic principle behind a well-written topic statement (except for inductive paragraphs) is this: by the time a reader has finished reading the first sentence or two of a paragraph, he or she should be able to predict what the rest of the paragraph will be generally about and how it will probably be developed. Suppose, for example, you were reading an article on transport across membranes in a popular-science magazine and began with this paragraph opening:

> The human body is made up of millions and billions of cells, each of which contains, among other substances, millions and billions of protein molecules . . . .[2]

How do you think the rest of the paragraph will go? What will it be about? Do you expect it to elaborate on other substances, on cells, on the human body generally? Probably not. Instead, there seems to be a narrowing down of focus to the term *protein molecules*; this is probably what the paragraph is about. How will this topic be developed? Well, the pattern of development used in this opening sentence is one of classification-division. Perhaps that pattern will be continued. Or maybe the writer has used some other pattern— a general-to-particular ordering of details, say, or a comparison of protein molecules with other kinds of molecules. Maybe the writer has used two patterns together. Of all the possibilities, though, you'd probably expect the classification-division pattern to be continued. It's a general fact about human nature that once we perceive a pattern in something, we expect it to continue— unless, of course, it's explicitly broken. Let's see what happens with this paragraph:

> . . . An average cell contains hundreds of different kinds of proteins, and all of the cells of the human body contain, among them, as many as 100,000 different kinds of proteins. These proteins can perform millions of different functions, a versatility which is largely responsible for the phenomenon called "life."[2]

The writer has indeed continued with the classification-division pattern set up in the first sentence. Notice, too, how the pattern proceeds from general to particular; the grammatical subjects of the three sentences show this progression quite clearly: *The human body. . . . An average cell. . . . These proteins. . . .*

Though paragraphs sometimes exist in isolation, they are usually linked to other paragraphs, forming larger conceptual units. In such cases, either the topic and pattern set up in one paragraph may be carried on to the next or the break between paragraphs can be used to switch to a new topic and/or a new pattern. In any event, it is usually desirable to maintain some kind of continuity when moving from one paragraph to the next. This is most often done by incorporating one or more key words from one paragraph into the first sentence of the next. In the membrane transport article, for example, the first sentence of the next paragraph begins as follows:

> The proteins derive their versatility from their structure—they are made up of chains of molecules of amino acids, substances of which there are 20 different ones in the human organism . . . .[2]

Notice how the writer has picked up on the key words *The proteins* and *versatility* from the last sentence of the first paragraph and used it as a transition to the subject matter of the second paragraph.

And what *is* the subject matter of this new paragraph? If the topic statement (above) is a good one, you should be able to predict with some assurance what it is. First of all, two new words appear prominently in this statement: *structure* and *amino acids*. We might suppose, therefore, that the theme of this paragraph is the structure of amino acids. Furthermore, since the writer has begun discussing amino acids by saying that there are 20 different types of them, we might expect the rest of the paragraph to be a discussion of the structural variety of amino acids, perhaps according to a classification-division pattern, a comparison-contrast pattern, or a general-to-particular ordering of details. Let's see how it actually does continue:

> . . . An average protein molecule consists of about 500 molecules of amino acids of different kinds (seldom all 20), arranged in some particular sequence. A sequence of 500 amino acids composed of all the 20 different ones would have as many as 1×10^{60} (1 followed by 60 zeros) possible arrangements, each arrangement having particular chemical properties and therefore chemical capabilities. From these few facts alone, we can easily appreciate how important the study of the amino acids is to our understanding of proteins, of the cell, and of life.[2]

As you can see, the paragraph as a whole *does* satisfy the expectations raised by the topic statement. It *is* about the structural variety of amino acids, and it *does* follow a general-to-particular pattern of development.

❑ EXERCISE 22-1

For each of the following paragraphs, circle an appropriate topic statement from among the three possibilities given. Be prepared to defend your choices.

A   i   The quantity of coal left in the earth is impressive.
      ii Coal is a more viable source of energy than petroleum.
     iii The mining of coal entails a number of difficulties.

There are known to be 198 billion tons of coal at a depth of less than 1000 feet and lying in beds at least 3.5 feet deep for bituminous coal and at least 10 feet thick for beds of lower grade coal. An equal quantity of coal at the same depth is identified as "undiscovered reserves." In addition, there are even larger quantities of less available coal resources, amounting to 1 trillion, 400 billion tons. At 35 million Btu's per ton, coal can provide a great deal of energy for many years to come.[3]

B i   Many molecular biologists believe that the discovery of movable genetic elements will help solve several long-standing mysteries.
   ii In eucaryotes, the genes coding for protein production do not exist as one continuous stretch of DNA.
   iii There are approximately 50,000 genes in the human body.

It seems to go a long way, for instance, toward explaining how the human body is able to synthesize a million and more different molecular antibody species, each tailor-made to grapple with a specific antigen. Movable elements may help answer the age-old question of differentiation: how a fertilized egg divides and ultimately becomes, in the course of embryonic development, many different kinds of tissue cells. Jumping genes may also provide a mechanism for satisfying scientists who have been arguing that point mutations alone were far from enough to account for the story of evolution.[4]

❏ EXERCISE 22-2

Each of the following sentences has been taken from an original text where it serves as an effective topic statement for a well-written paragraph. See if you can guess roughly how each paragraph is developed beyond the topic statement: what the key words are, what general strategy the writer uses.

A The first modular home to be tested by government engineers for durability exceeded the criteria for the National Bureau of Standards.[5]

B At the time of its explosion, Mount St. Helens was probably the most closely watched volcano in the world.[6]

C The production of an important heavy chemical, nitric acid (HNO_3), requires large quantities of ammonia.[7]

D Until 1922, no one knew how a signal crosses the junction between one nerve cell and another.[8]

E The basic property of gyroscopic action is that the gyroscope stays spinning in exactly the same direction in space over both short and long periods of time.[9]

22.2 DEVELOP A CLEAR PATTERN OF ORGANIZATION

Once you have written a satisfactory topic statement, you'll want to follow it up with a number of supporting statements. These statements should follow a consistent pattern of organization, one that flows naturally or even predictably from the topic statement. That way, you'll satisfy the reader's expectations and allow him or her to process the paragraph as a unified whole.

Some of the most commonly used patterns of organization in scientific and technical writing are chronological description, cause-and-effect analysis, comparison and contrast, listing, and general-to-particular ordering of details. Each of these patterns has certain characteristic features, and by using these features you can make it easier for the reader to perceive which pattern you're using.

Chronological Description

The use of a time frame to tie sentences together is a well-known pattern of organization which you have no doubt used many times in your writing. It is commonly used, for example, to either describe or prescribe a step-by-step procedure: *First connect the vacuum tube . . . then return the plate . . . finally, close the contact key . . .* It is used to recount a sequence of past events, as when you want to bring a reader up to date—for example, in a progress report or in the Review of Literature section of a journal article.

The most characteristic features of chronological description are:

Time adverbs and phrases	in 1980, last week, at 10:15, first, second, finally, soon after the project began
Verb tense sequencing	Originally we *wanted* to. . . . More recently we *have attempted* to. . . . Now we *are trying* to. . . . In the future we *shall try* to. . . .
Grammatical parallelism	*Mount* the grating near the end. . . . *Locate* a rider on the scale. . . . *Adjust* the grating. . . . *Read* the distances on the scale. . . .

Not all of these features are likely to be found in any one type of chronological description. For example, descriptions of standard procedures (e.g., test procedures, experimental procedures, assembly instructions) strongly favor the use of parallelism over the other features. Descriptions of past events, on the other hand, tend to rely more on time adverbs and phrases and on different verb tenses.

Below is a well-written paragraph using chronological ordering as its basic pattern of organization:

Total U.S. research and development spending *is projected* to reach a current-dollar level of $66.7 billion *in 1981*. This is an increase of 10% over *the 1980 projected level* and nearly double the amount spent on these activities *in 1975*. Even in constant dollars, U.S. R&D spending *in 1981 is expected to follow* the growth trend of the *past five years, when* R&D funding *grew* at an average annual rate of better than 3%. That growth *resulted* in large part from increased emphasis on searching for means to resolve energy and environmental problems and a resurgence of defense R&D activity. *Between 1975 and 1978*, the last year for which survey data are available, energy *accounted for* one-third of the R&D spending increase while, at the same time, amounting to 10% of the national R&D effort.[10] [Italics added.]

This paragraph contains a great deal of information, and it would probably be quite confusing were it not so well structured. Notice how the topic statement provides a clear overview of the paragraph as a whole: it tells you not only what the main theme is (that U.S. R&D spending is continuing to follow a significant growth trend) but also how this theme will be developed (by chronological order, featuring the years 1981, 1980, and 1975). The remainder of the paragraph is then devoted to fulfilling these expectations. Notice how the writer not only has repeated the key dates but also has taken care to use correct verb tense forms when referring to them.

Cause-and-Effect Analysis

This pattern of organization is commonly used in scientific and technical writing for a number of purposes, including (1) making a logical argument, (2) explaining a process, (3) explaining why something happened the way it did, and (4) predicting some future sequence of events. Whenever you are explaining causes and effects, it is usually best to explain them in straight chronological order: causes *before* effects. That way, you can minimize the number of "traffic signals" you need. Even when the description is clearly one of causes followed by effects, however, your readers will appreciate it if you occasionally insert such signals. The characteristic signals of cause-and-effect analysis include:

Connective words and phrases	therefore, thus, consequently, accordingly, as a result, so
Subordinate clauses	since, because (of), due to
Causative verbs	causes, results in, gives rise to, affects, requires, produces
Conditional constructions	when; where; given; if, then

In addition, when causes and effects are described in chronological order, the features associated with chronological description can be used.

A good example of cause-and-effect patterning is this paragraph from a physics text explaining what surface tension is:

One of the most important properties of a liquid is that its surface behaves like an elastic covering that is continually trying to decrease its area. A *result* of this tendency for the surface to contract is the formation of liquids into droplets as spherical as possible considering the constraint of the ever-present gravity force. Surface tension arises *because* the elastic attractive forces between molecules inside a liquid are symmetrical; molecules situated near the surface are attracted from the inside but not the outside. The surface molecules experience a net inward force; and *consequently*, moving a surface molecule out of the surface *requires* energy. The energy E required to remove all surface molecules out of range of the forces of the remaining liquid is proportional to the surface area; *therefore*,

$$E = \sigma A$$

where σ, the proportionality factor, is called the surface tension,

$$\sigma = E/A$$

and is measured in joules/m². [Italics added][11]

Notice how, in addition to using signal words such as *consequently* and *therefore*, the writer has linked the sentences together in a steplike sequence. This has been accomplished mainly through the following technique: after introducing and discussing a new term in one sentence, the writer then uses it in the next sentence as part of the framework for introducing and discussing the next new term:

Surface tension . . . *molecules situated near the surface* . . . .

The surface molecule . . . *requires energy.*

The energy E required . . . is *proportional* to . . .

the *proportionality* factor . . . is measured in joules/m².

If this kind of step-by-step linking is done properly, it reduces the need to insert many signal words and markers of subordination. (See Chapters 24 and 25 for further discussion of this technique.)

Comparison and Contrast

As a technical professional, you will often find it necessary to compare two or more things that are similar in some ways but different in others. This is particularly common in business and industry, where one is constantly wrestling with cost-benefit trade-offs and other choices that must be made from among various alternatives under various constraints.

In writing a comparison-and-contrast paragraph, *try to avoid jumping back and forth from one alternative to another.* Suppose, for example, that you are comparing Products X and Y and are using criteria A, B, C, and D to compare them; suppose further that the first three criteria favor X and the fourth favors Y. In such a case, you should present the comparison in terms of these two criteria groupings: first A through C, then D. This will make it easier for the reader to see that Product X wins out over Product Y in three of the four criteria. And that's what the reader will most likely want to know—

the proverbial bottom line. (If you think the reader will be more interested in the details of the comparison, provide a table.)

Characteristic features of comparison-and-contrast paragraphs include:

Connective words and phrases	however, on the other hand, conversely, similarly, likewise, in contrast to
Comparative constructions	more than, -er, than, less than, as . . . as, rather than, is different from
Verb tense differences	Program X *will* be easy to implement, whereas Program Y *would* entail a number of complications . . .
Subordinate clauses	while, whereas, but
Parallelism	Model X is reliable and efficient, whereas Model Y is unreliable and relatively inefficient . . .

One final principle of comparison-and-contrast writing is this: *phrase your descriptions in such a way that the reader can immediately see how they bear on the argument you are making.* In other words, don't routinely make "neutral" statements of comparison. For example, instead of saying, "Item X weighs 3.2 pounds, and Item Y weighs 2.7 pounds," consider the possibility of saying, "Item X weighs 3.2 pounds, whereas Item Y weighs only 2.7 pounds." In many cases, using subordinators and descriptors like *whereas* and *only* can serve to emphasize the point you are making. This makes it easier for the reader to follow your argument. Even the same quantity can be depicted in different ways, depending on the words used to describe it. For example, if you say that a certain product costs "less than" $10, you are casting it in a more favorable light than if you say it costs "almost" $10, even though the price may be exactly the same in both cases. Such slanting helps the reader see your point of view, and it is not unethical, since you are not saying anything that's not true.

Below is a model comparison-and-contrast paragraph. Note how it uses some of the features listed above.

A one-million-fold increase in speed characterizes the development of machine computation over the past thirty years. The increase results from improvements in computer hardware. In the 1940's ENIAC, an early electronic computer, filled a room with its banks of vacuum tubes and miles of wiring. Today one can hold in the hand a computing device costing about $200 that is twenty times faster than ENIAC, has more components and a larger memory, is thousands of times more reliable, costs 1/10,000 the price, and consumes the power of a light bulb rather than that of a locomotive.[12]

Listing

Scientific and technical writing presents frequent opportunities to put information in the form of lists. If you are describing an experiment, for example, you will probably want to make a list of the equipment used. If you are writing a progress report, you may want to make a list of things already done and another list of things still to do. If you're writing a report recommending the development or purchase of some new product, you may want to list its outstanding features or the reasons you're recommending it.

Lists may be either formatted or unformatted. Formatted lists are set off from the rest of the paragraph by means of indentation and/or numbering or lettering. Unformatted lists do not have such visual cues. In both cases, *all items in a list should be cast in parallel grammatical form.* This principle is especially important in the case of unformatted lists, since it can be quite difficult otherwise for the reader to detect the presence of the list. (Parallelism is important not only in lists but also in comparative constructions, in descriptions of procedures, and in other rhetorical patterns; see Chapter 23.) The following is an example:

> In addition to coal and nuclear energy, a wide variety of other power sources are frequently discussed in the news and in scientific literature; unfortunately, most are not yet ready for practical use. Geothermal energy is one of the more practical of proposed new sources. It is already in use in Italy, Iceland, and northern California but is not yet meeting all expectations for it. Solar energy seems an elegant idea because it is inexhaustible and adds no net heat or carbon dioxide to the global environment. Yet present methods of exploiting it make solar energy hopelessly inadequate as a major power source in the next few decades. Sophisticated windmills to generate electricity are also under study by some. Biomass conversion is also getting under way. Some of these sources of energy, which we now generally regard as esoteric, may well prove themselves and make a substantial contribution over the long run if their costs can be brought within reason.

The different power sources discussed in this paragraph constitute an unformatted list:

> Geothermal energy . . .
>
> Solar energy . . .
>
> Sophisticated windmills . . .
>
> Biomass conversion . . .

Notice how the effective use of parallelism enables the reader to easily locate the four items making up the list, even though no visual formatting is used.

A second important principle to follow when constructing lists is this: *If the items in a list are not equally important, they should usually be arranged in descending order of importance.* A list, by definition, is a set of items all of which have something in common and yet are independent of each other. Thus, in principle, these items can be arranged or rearranged any way at all. You can take advantage of this freedom by arranging them with the most important item in the most prominent position, namely, on top.

An alternative to this pattern of descending order of importance is to put the most important item last. If you adopt this strategy, be sure to signal the fact that the last item, not the first, is the most important one. Public speakers sometimes do this by using the old cliché "Last but not least . . ." Writers would do better to use a less stale expression like "Finally, and most important . . ."

Finally, *every list should be introduced by an informative lead-in*. This lead-in should tip the reader off as to what kind of list is coming and perhaps how many items it contains. In the example just cited, notice how the phrase "a wide variety of other power sources" serves this purpose. Not only do you expect to find a list of power sources; you expect to find at least four or five such sources on the list.

General-to-Particular Ordering of Details

A final rhetorical pattern commonly used in scientific and technical paragraphs is the ordering of supporting details from the more general to the more particular. Each sentence in this pattern focuses on a smaller frame of reference than the sentence before it. Here is an example:

> Magnetic bearings have been developed for aerospace applications, but only recently has their practicality been demonstrated as the heart of energy storage systems. The breakthrough is partly due to the recent development of stronger permanent magnets, such as those made from rare-earth cobalt compounds. Only ten pounds of such magnets could support two tons of rotor. Although the free suspension of a weight with permanent magnets is an unstable condition, an electromagnet servo loop has been used successfully to stabilize the rotor position.[13]

Notice how the topic of the first sentence (and of the paragraph as a whole), *magnetic bearings*, is developed in the second sentence by the writer's focusing on one component of such bearings, the *permanent magnets*. This subtopic is then discussed in even greater detail in the third sentence (*Only ten pounds of such magnets . . .*). Finally, the writer focuses on a subcomponent of these magnets, the *electromagnet servo loop*.

Other Patterns

The patterns of organization described above are not the only ones commonly used in scientific and technical writing. Others—such as classification-division, exemplification, extended definition, and analogy, to mention but a few—are also used. (An example of paragraph development by classification-division was given earlier in this chapter.)

In addition, two or more patterns can often be used together. For example, you might want to give an *extended definition* of a process by describing it in terms of *causes and effects* or in terms of an *analogy* to some more familiar process. You might want to explain something by providing a *list* of *examples*. Or you might want to *divide* some subject into different classes and then *compare and contrast* these classes. Regardless of what kind of combination you elect to use, the important point is to establish at least one pattern of organization that runs through most if not all of the paragraph and thus provides a structural backbone for that paragraph.

❑ EXERCISE 22-3

Each of the following paragraphs follows a particular pattern of organization. Identify the pattern, and point out as many of its features as you can.

A   Flywheels appear to be the ideal energy storage element in solar electric or wind power systems. They can smooth the load on the generators by providing the energy to generate electricity when the sun is not shining or the wind is not blowing. And they can also provide power to an electrical load during periods when demand exceeds supply, such as during motor start-ups. Indeed, once spinning, flywheels can deliver energy rapidly for variable load conditions, which makes them especially useful in industrial and agricultural applications.[13]

Pattern: _____

Features: 1 _____

　　　　　 2 _____

　　　　　 3 _____

B   *The Vapor-Compression Refrigeration Cycle.* A simple vapor-compression refrigeration cycle is shown in Figure 22-1. The refrigerant enters the compressor as a slightly superheated vapor at a low pressure. It then leaves the compressor and enters the condenser as a vapor at some elevated pressure. There the refrigerant is condensed as a result of heat transfer to cooling water or to the surroundings. The refrigerant then leaves the condenser as a high-pressure liquid. The pressure of the liquid is decreased as it flows through the expansion valve, and, as a result, some of the liquid flashes into vapor. The remaining liquid, now at a low pressure, is vaporized in the evaporator as a result of heat transfer from the refrigerated space. This vapor then enters the compressor.[14]

Pattern: _____

Features: 1 _____

　　　　　 2 _____

　　　　　 3 _____

　　　　　 4 _____

Figure 22-1 ❑ SCHEMATIC DIAGRAM OF A SIMPLE REFRIGERATION CYCLE

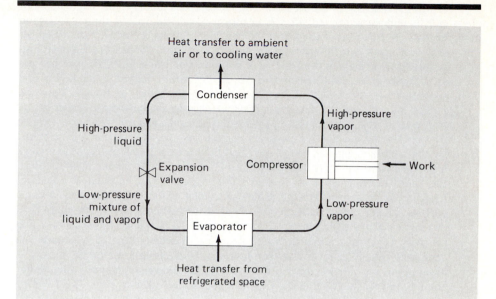

C   One of the newest branches of number theory is *analytic number theory*. A vast and intricate subject, it is largely a creation of the twentieth century. It has been called the science of approximation, for it is concerned mainly with determining the order (relative size) of the errors made when a calculation is approximate rather than exact. Its techniques have had an important impact on many departments of applied mathematics, including statistical mechanics and the kinetic theory of gases, where exact results are sometimes impossible to attain.[15]

Pattern: _____

Features: 1 _____

2 _____

❑ EXERCISE 22-4

The following passage is taken from a popular-science journal article, where it was divided into seven paragraphs. Using your knowledge of topic statements and patterns of paragraph organization, see if you can determine where the six paragraph divisions should be.

We generally think of volcanoes as cone-shaped mountains that belch lava and smoke from craters at their summits. The essential element of a volcano, however, is not its aboveground structure but the underground conduit that brings molten rock to the surface. The molten rock is known as magma as long as it is under the ground; after it erupts it is called lava. As the lava flows or explodes from a vent at the top of the conduit, it starts to build a volcanic edifice that may grow into a Vesuvius, a Mauna Loa, or a Mount St. Helens. Few people have witnessed the birth of a volcano, but Dionisio Pulido was present on February 20, 1943, when the Mexican volcano Paricutin made its first appearance in a cornfield on his small farm in the valley of Rancho Tepacua. Pulido, his wife and son, and a neighbor watched as smoke and ash began to rise from a small hole in the middle of the field. The smoke and ash were accompanied by rumbles, hisses, and hot particles of rock that set fire to nearby pine trees. By the next day, ash and rock debris had built a cone 10 meters high; in a week it had grown to 140 meters. By the time Paricutin finally quieted down in 1952, the cone towered 410 meters above the original level of Pulido's cornfield. Paricutin is an unusually large cinder cone, one of four basic types of volcanic edifices. Cinder cones, made up entirely of lava fragments, rarely rise more than 300 meters above their surroundings. They are the simplest type of volcano and are very common in western North America. A second type of volcano is the composite cone built up of alternate layers of lava and ash. Mount St. Helens is a composite cone. So are many other spectacular volcanic peaks, including Mount Rainier and Mount Hood in the Cascades and Japan's Mount Fuji. The summits of many composite cones tower 1800 to 2400 meters above their bases. Shield volcanoes are built almost entirely of lava flows. They are broad, gently sloping structures that resemble a convex shield laid on the ground face up. Many of the world's largest mountains are shield volcanoes; clusters of them form the Hawaiian Islands. Mauna Loa, on the island of Hawaii, is the world's largest active volcano. Its base is more than 4500 meters below the surface of the ocean, and its crest is 4170 meters above the sea level—a total height of more than 8500 meters. Lava domes, the fourth type of volcano, are built up of thick, pasty lava that tends to pile up in a rough knob rather than to flow outward as it emerges from its vent. The main factors that determine whether lava will erupt in a smooth flow or a violent explosion are its chemical composition and the characteristics of the gases that are dissolved in it. Magma is a mixture of several oxides, mainly silicon dioxide. The rock can be principally basalt, andesite, dacite, or rhyolite, which are distinguished one from the other, among other things, by the fact that each has a larger proportion of silicon dioxide than the one before. Basaltic magmas usually erupt in a highly fluid state, and their dissolved gases escape easily, so that the lava flows freely. The more viscous rhyolites and dacites usually erupt explosively. Andesites may erupt in explosions or in flows that are not as smooth as those of basaltic lava.[16]

22.3 GIVE EACH PARAGRAPH A FUNCTION WITHIN THE LARGER CONTEXT

The previous sections have focused on the structure of individual paragraphs. This section will focus on the function of a paragraph within a larger piece of text: as part of either the whole text or a related group of several paragraphs. In this sense, an individual paragraph needs to be structured so that it is a part of a series of paragraphs with a larger, overall purpose. Thus, an individual paragraph will need to fit into the larger context, complement some points, and emphasize others—all depending on the needs of the larger context. Various modes of emphasis within sentences are treated in Chapter 26, including the use of formatting to highlight especially important points. We will concentrate here on the function of a paragraph, given the larger context. For instance, consider the following 2 paragraphs from the opening of Section 1.6 in this book.

1.6

THE SPECIFIC SKILLS NEEDED

So far, we have argued that scientific and technical writing is important, time-consuming, and often collaborative; is computerized and becoming more so; will be relying even more on nonprose elements; and has important international analogues and legal implications. Let us now consider what communication skills are needed to support the needs of technical professionals in the near future.

What are the specific communication skills most needed by technical professionals? In terms of particular types of communication, the most common written forms are memos, letters, and short reports, followed by step-by-step procedures and proposals to clients. The most common oral forms are one-to-one talks, telephone calls, and small group or committee meetings. Frequency, however, is not necessarily the best measure of importance. Even infrequent forms of communication can be extremely important. For example, at one research laboratory we know of, each staff scientist is required to give a 10-minute oral presentation to upper management once a year summarizing the work he or she has done during that year. It may be an infrequent form of communication, but you can imagine how important it is to these scientists! (In fact, some of them practice for weeks ahead of time.)

While the second paragraph illustrates the characteristics of paragraph construction described earlier in this chapter, the first paragraph does not. It does not really have a topic sentence: the first sentence does not introduce material which the second sentence talks about. Further, it is only two sentences long, and the second sentence does not really add development information. This paragraph is really functioning as a transition paragraph between the two main logical parts of Chapter 1. When a reader encounters this paragraph, the first part of the chapter (through Section 1.5) has sketched out several features of technical communication in the real world, and the second part of the chapter is going to outline many of the skills that a professional needs to compete successfully on the job. This paragraph reflects that two-part structure: the first sentence sums up what has been covered so far, and the second sentence introduces what is going to be covered.

In this sense, the first paragraph of Section 1.6 illustrates that paragraphs can be rhetorical units as well as logical units—that is, that paragraphs can exist to serve functional purposes such as transitions, to create better visual effects, or to conform to the paragraph conventions of a particular institution or form of writing. Since we have seen how a paragraph can function as a transition, let us briefly consider how a paragraph serves other functional purposes.

If you consider the paragraphs of the following example from Chapter 9, you will notice that the paragraphs could appear in the different forms presented below, the formatted form and the unformatted form.

Formatted Version of Part of Figure 9-13

1.1   Damages Due to a Combination of Temperature and Moisture

The most damaging environment that an LCD is likely to experience is produced by the combination of high temperature and high humidity. Depending on how the display is constructed, three failure modes can occur under this condition: failure of the polarizers, failure of the adhesive on the polarizer and reflector, and, finally, failure of the liquid crystal mixture.

1.1.1   Failure of the Polarizers

The most common failure occurs with the polarizers. Regular polarizing material will begin to degrade after a short period of time (4 days) when subjected to a temperature of 50°C and 95% relative humidity. Once moisture penetrates the plastic backing of the polarizing material, it affects the polarizing properties. The digit area of the display will start to lose contrast, gradually turning from black to brown until it finally disappears completely.

1.1.1.1   Preventing Failure of the Polarizers

To provide protection against the effects of high temperature and humidity, the use of a polarizing material known as "K-sheet," introduced by Polaroid Corporation, is recommended. This material has polarization properties which are relatively unaffected by moisture.

1.1.2   Failure of the Adhesive

Moisture can also attack the adhesive on the polarizer and reflector and cause it to begin peeling away from the glass. If the alignment is lost, dark patches will appear in the viewing area like those as already described.

1.1.2.1   Preventing Failure of the Adhesive

To prevent this kind of failure, the use of a truly hermetic seal is recommended to protect the adhesive.

1.1.3   Failure of the Liquid Crystal Mixture

If any of the components in the liquid crystal mixture (such as Schiff-base compounds) are attacked by moisture, the liquid crystal molecules will start to break down, thereby lowering the clearing point of the mixture and the effective upper operating temperature of the display. Due to the ionic by-products of the chemical reactions, the electrical currents of the display will also show a very significant increase.

1.1.3.1   Preventing Failure of the Liquid Crystal Mixture

The procedure described in part 1.1.2.1 should be adopted.

Unformatted Version of Part of Figure 9-13

The most damaging environment that an LCD is likely to experience is produced by the combination of high temperature and high humidity. Depending on how the display is constructed, three failure modes can occur under this condition: failure of the polarizers, failure of the adhesive on the polarizer and reflector, and, finally, failure of the liquid crystal mixture. The most common failure occurs with the polarizers. Regular polarizing material will begin to degrade after a short period of time (4 days) when subjected to a temperature of 50°C and 95% relative humidity. Once moisture penetrates the plastic backing of the polarizing material, it affects the polarizing properties. The digit area of the display will start to lose contrast, gradually turning from black to brown until it finally disappears completely. To provide protection against the effects of high temperature and humidity, the use of a polarizing material known as "K-sheet," introduced by Polaroid Corporation, is recommended. This material has polarization properties which are relatively unaffected by moisture. Moisture can also attack the adhesive on the polarizer and reflector and cause it to begin peeling away from the glass. If the alignment is lost, dark patches will appear in the viewing area like those as already described. To prevent this kind of failure, the use of a truly hermetic seal is recommended to protect the adhesive. If any of the components in the liquid crystal mixture (such as Schiff-base compounds) are attacked by moisture, the liquid crystal molecules will start to break down, thereby lowering the clearing point of the mixture and the effective upper operating temperature of the display. Due to the ionic by-products of the chemical reactions, the electrical currents of the display will also show a very significant increase. The procedure described above should be adopted.

In the formatted form, the supporting detail for the claims in Section 1.1 are presented in the separate little paragraphs (which are also numbered and indented to indicated their logical relationship to the material in Section 1.1). This type of paragraphing is common in technical reports and other forms of technical writing produced for very harried readers. In the unformatted form, all of the material in the paragraph supports the claim made in the topic sentence and narrowed in the second sentence by the phrase "three failure modes can occur under this condition." This type of paragraph is common in some fields which typically have long paragraphs of supporting detail. Thus, the form of the paragraph—its length and the amount of supporting detail— is rhetorical to the extent that it reflects the conventions of its type. There are other examples of such conventionalized paragraphs in the types of writing presented in this book, including the formatting of paragraphs in job letters and resumes described in Chapter 10.

❑ EXERCISE 22-5

The following passages (extracted from longer reports) were submitted as is by students. As you'll discover, they require very careful reading in order to be understood. Revise them—by rearranging or rewriting sentences, or perhaps by dividing them into two or more paragraphs—so that they can be more quickly and easily understood.

A   Advantages and Disadvantages

The balance system is the simplest and cheapest system. However, it has drawn numerous complaints from users in California. The nozzles are heavy, hard to hold in place, and hard to remove due to the necessary airtight seal. The major problem has been that gasoline has been sucked out of the customer's tank back into the storage tank. This occurred when customers attempted to top off their tank and in the process overrode the automatic shutoff. The brands of the nozzles (OPW) were tested by the ARB in actual stations by inserting a small vial in the vapor return hose and checking for liquid in the vial after filling the vehicle. In the ARB tests the nozzles did not fail. However, in the MS test the nozzle failed. A 3-gallon container fitted with a standard auto filler pipe was used. The container had a mechanism to allow the inside pressure to be increased, and a 22.5% failure rate was reported. In one trial the container was filled to 15 gallons. The vacuum and assisted systems do not permit this overfilling, and no force is required to hold the nozzle. The prices of the three systems vary significantly. The balance system costs the amount of the assisted system, and the vacuum system is the most expensive.

B Technical Discussion

The most important part of this modification program is the addition of a larger engine to the Corsair. The 18–cylinder Pratt & Whitney R-2800-CB-13 and three-bladed propeller that are the standard fit for an FG-1D are rated at 2500 hp, though there are cases where R-2800s have supplied up to 3300 hp (Greenmayer, 1969) in modified form. One possible alternative to the R-2800, a Curtiss-Wright R-3350-24 Cyclone 18, has been substituted for an R-2800 in an F8F-2 Bearcat. Rated at 2800 hp, this particular example has been pushed to 3800 hp (Shelton, 1971). Another alternative is the 28-cylinder Pratt & Whitney R-4360-4 Wasp Major. These were raced in the late 1940s on F2G Corsairs with four-bladed propellers. These engines were rated at 3000 hp, though in races they produced up to 4000 hp (Cleland, 1949). After some careful checking, I have chosen to recommend fitting a Pratt & Whitney R-4360-B-20, with a four-bladed propeller, to your Corsair. While rated at 3500 hp, I expect it to produce over 4000 hp after modification. One particular advantage in using the R-4360-B-20 is that it won't increase the aircraft's frontal area, being the same diameter as an R-2800. In contrast, an R-3350 would increase the frontal area by 11% because of its larger diameter. However, in fitting this larger engine, there will be a weight penalty of 1130 pounds. To keep the aircraft's center of gravity from moving too far forward, the firewall will have to be moved back at least a foot.

REFERENCES

1 Joel Gurin, "Chemical Cars," *Science 80*, September/October 1980, p. 96.
2 Julie Wei, "Membranes as Mediators in Amino Acid Transport," *The Research News 31*(11–12):3 (1980).
3 Adapted from Blanchard Hiatt, "Coal Technology for Energy Goals," *The Research News 24*(12):3 (1974).
4 Henry Simmons, "DNA Topology: Knots No Sailor Ever Knew," and Ben Patrusky, "Gene Segments on the Move," *Mosaic 12*(1):9, 41–42 (1981).
5 Karl G. Pearson, *Industrialized Housing* (Ann Arbor: University of Michigan, Institute of Science and Technology, 1972), p. 39.

6   Henry Lansford, "Vulcan's Chimneys: Subduction-Zone Volcanism," *Mosaic 12* (2):46 (1981).

7   J. Waser, K. Trueblood, and C. Knobler, *Chem One* (New York: McGraw-Hill, 1976), p. 589.

8   Edward Edelson, "The Neuropeptide Explosion," *Mosaic 12*(3):15 (1981).

9   Henry Hooper and Peter Gwynne, *Physics and the Physical Perspective* (New York: Harper & Row, 1977), p. 207.

10 *Mosaic 11*(5):cover 4 (1980).

11 Hooper and Gwynne, pp. 223–224.

12 Blanchard Hiatt, "Big Computing, Tiny Computers," *The Research News 30*(4):3 (1979).

13 Alan Millner, "Flywheels for Energy Storage," *Technology Review*, November 1979, p. 33.

14 Richard Sonntag and Gordon Van Wylen, *Introduction to Thermodynamics: Classical and Statistical* (New York: Wiley, 1971), pp. 9–10.

15 Julie Wei, "Pure Mathematics: Problems and Prospects in Number Theory," *The Research News 30*(3):31 (1979).

16 Adapted from Lansford, p. 50.

ADDITIONAL READINGS

Frank D'Angelo, "The Topic Sentence Revisited," *College Composition and Communication 37*(4):431–441 (December 1986).

Rick Eden and Ruth Mitchell, "Paragraphing for the Reader," *College Composition and Communication 37*(4): 416–430 (December 1986).

R. M. Gagne and V. K. Wiegand, "Effects of a Superordinate Context on Learning and Retention of Facts," *Journal of Educational Psychology 61*:406–409 (1970).

Marcel A. Just and Patricia A. Carpenter, *The Psychology of Reading and Language Comprehension* (Boston: Allyn and Bacon, 1987).

David Kieras, "The Role of Major Referents and Sentence Topics in the Construction of Passage Macrostructure," *Discourse Processes 4*:1–15 (1981).

David Kieras, "Initial Mention as a Signal to Thematic Content in Technical Passages," *Memory and Cognition 8*:345–353 (1980).

Bonnie J. F. Meyer, *The Organization of Prose and its Effects on Memory* (Amsterdam: North-Holland, 1975).

23

Using Parallelism

Writing consists of both form and content. When the form *reflects* the content, readability is enhanced: readers can use formal features to guide them in their efforts to interpret the content. This is especially true of readers who are reading selectively. By using various formal cues, such readers can skim or scan a text for particular pieces of information. But it is true of other readers as well. As a writer, you should always try to maximize the readability of your writing by matching form to content.

One major way of using form to signal content is to use parallelism in the writing of lists. Using parallelism means putting each item of a list in the same grammatical form. For example, if you write out a list of "Things to Do" that looks like this:

Pick up mail

Call Hazlitt

Work on tailgate project

Ask RJ for $$$

you're using parallelism because you've put each item on the list in the same grammatical form (imperative verb plus one or more nouns). If you write the list like this, on the other hand:

Mail needs to be picked up

Call Hazlitt

Tailgate project

$$$ from RJ—don't forget!

you're not using parallelism.

In cases like these, where the presence of a list is marked by formatting, parallelism simply adds a touch of elegance to the presentation. In other cases, however, parallelism plays a much more important role: it points out a list that would otherwise be difficult to identify.

Before we go further, however, perhaps we should clarify what we mean by the word *list*. In our terms, a list is any set of two or more items having a coordinate relationship. This definition includes not only obvious cases such as the one described above but also others that may not be so apparent. For example, the following sentence contains a list since two different but related functions (*metabolite transport* and *metabolite signaling*) are mentioned (the word *both* is a signal of this):

> Evidence that the receptor sites on the membrane can serve both for metabolite transport and for metabolite signaling has raised many questions as to the mechanisms of membrane information transfer.

This sentence also contains a two-item list:

> A pollution-free inertial-electric system has greater energy efficiency than an internal combustion engine.

Here the items forming a list are the two systems that are being compared, and the word *than* is a signal of this.

23.1 TYPES OF LISTS

Lists are especially common in scientific/technical writing. We find lists of experimental apparatus, lists of instructions, lists of task objectives, and so on. Some lists are *fully formatted* as such, with alphanumeric sequencing and vertical alignment used as formal features to make the list stand out. Here is an example from a company memo:

REFS: (a) Government invoice #79–1018A

      (b) Bonded Stores invoice #31285

      (c) DECAS letter 5/20/79

      (d) Bonded Stores invoice #44590

Some lists are *partially formatted,* in the sense that they use either alpha-numeric sequencing or vertical alignment but not both, as in this example:

> Factors taken into consideration include the following: (1) size of lot, (2) parking requirements, (3) need for elevator, (4) cost per gross square foot, and (5) expected annual return per gross square foot.

Some lists are completely *unformatted,* having neither alphanumeric sequencing nor vertical alignment:

> Compared to standard bipolar types, VMOS transistors offer higher input impedance, faster switching times, wider operating range, and smaller chip area.

In all cases, however, whether a list is fully formatted, partially formatted, or unformatted, *all of the items making up the list should be cast in the same grammatical form.*

Putting the items of a list in parallel grammatical form allows a busy reader to easily perceive the nature of these items and of the list as a whole. It is especially important to use this principle with unformatted lists. In the VMOS example above, notice how easy it is to glance at the sentence and see what the four criteria of comparison are: . . . *input impedance . . . switching times . . . operating range . . . chip area.* By contrast, notice how difficult it is to perceive the nature of the list, and the items making up the list, in the following badly written example:

Negative Example

The TFC engineers and I found the selection of Hybrid Analog Transmission using FDM (Frequency Division Multiplexer) to be highly reliable, improved security of communications; complete ground isolation, free from cross-talk, sparking, short circuit loading; RFI (radio frequency interference), EMI (electric magnetic field), and EMP (electric magnetic polarization) immunity.

What are the major items constituting the list in this example? What role are they playing? You may be able to figure out the answers if you work at it, but you shouldn't *have to* work at it. The writer should have used parallelism to make the content clearer.

As a general habit, you should use parallelism whenever you are constructing any kind of list, even when the nature of the list is already clear. Often this is a simple process, involving only minor adjustments. Suppose, for example, you have written a draft of a report about wastewater processing and have included this statement:

Negative Example

The principal processes are:

1. Coagulating and flocculation

2. Removing the solids

3. Nitrogen-removal

4. Disinfection

It would be an easy matter to recast this list in a more grammatically parallel form:

Revised Version
The principal processes are:

1. Coagulation and flocculation

2. Removal of solids

3. Removal of nitrogen

4. Disinfection

Instead of mixing verb forms (*-ing*) and noun forms (*-al* and *-tion*) as in the first version, the second version uses all noun forms. This results in a more sharply defined list; it looks more like the work of a careful, precise technical professional.

Sometimes it is useful to include prepositions and other small words in the making of parallelism. Consider this example:

Negative Example
The purpose of this report is to present evidence that the contracting officer acted within the range of his authority and his final opinion was binding.

Here the two items in the embedded list are both grammatical clauses: *the contracting officer acted within the range of his authority* and *his final opinion was binding.* They have parallel form. But the parallelism is not as strong as it should be. The reader could easily overlook this weak parallelism in favor of the stronger (but misleading) parallelism of the two *his*'s. To prevent a possible misreading, the writer should use a second *that*:

Revised Version
The purpose of this report is to present evidence *that* the contracting officer acted within the range of his authority and *that* his final opinion was binding.

The proper use of parallelism can do far more than just enhance your professional image; it can also help prevent misinterpretations of your writing. To see what can go wrong when parallelism is *not* used, consider the following example:

Negative Example
This filter has two important functions: to reject impulse noise signals and passing low frequency command signals without amplitude or phase distortion.

This excerpt could easily be misread (especially by a busy nonspecialist reader) as meaning that two things are rejected: the impulse noise signals and the passing of low frequency command signals. Notice that *reject* and *passing* can be combined to produce such a misreading. Even those readers who can figure out what the writer is trying to say may have to slow down their reading just to make sure of what's being rejected. Such ambiguity is completely unnecessary. By putting the two key verbs in parallel form, the writer could have created a much clearer version:

Revised Version
This filter has two important functions: *rejecting* impulse noise signals and *passing* low frequency command signals without amplitude or phase distortion.

Other parallel forms would work equally well: *to reject . . . to pass;* or *the rejecting of . . . the passing of.*

☐ EXERCISE 23-1

Each of the following contains a list that lacks parallelism. See if you can devise an improved version.

A The building is 140 feet in length, 78 feet wide, and has a height of four stories.

B Test results indicate that soil 1 is likely to settle, since high plasticity is equivalent to highly compressible.

C The widest employment of DDT is in the control of insects of public importance (as a mosquito larvicide, a spray for malaria eradication, and to control typhus by dust application).

D The reason for this is that most small businesses have a lower budget for their managers than do government or industrial managers.

E In this particular case the most important variables are the following:

   (1) pressure and temperature of the boiler

   (2) what type of fuel is required

   (3) the amount of oxygen

   (4) fuel temperature

F As a result of the above problem, this report addresses the following tasks:

   a. To redesign the mix for the concrete slab

   b. An evaluation of the compressive strength with the use of test cylinders for various designs

   c. Determining both the theoretical and actual material costs per cubic yard

G   We recommend the purchase of the New Orleans heat exchanger because (i) it can be obtained for $25,000, a savings of 80% over the new cost, (ii) production could be increased by 28% if needed, 8% higher than requested, and (iii) it will help recoup losses incurred in our benzene plant.

H   Advantages of this system are

1.   Automatically controlled

2.   Lower operating cost

Disadvantages are

1.   May cause slugging of liquid refrigerant to compressors

2.   Complex automatic controls

3.   Substantial replumbing of existing system is required

I   This buoy terminal has three components that rotate as a single unit:

(a)   The Rotating Mooring Bollard allows floating mooring lines to lie on the lee side when weather conditions become bad.

(b)   Products are carried by a Rotating Cargo Manifold from the terminal-to-tanker hoses to the multiproduct distribution unit in the center of the buoy.

(c)   A Rotating Balance Arm not only maintains the buoys on an even keel but also provides an accommodation ladder.

23.2 MISLEADING PARALLELISM

Given the fact that readers use parallelism as an aid in perceiving and interpreting lists, it is important that you *not* use parallelism in situations where lists are *not* involved. Otherwise, readers are apt to be misled and may misanalyze the passage *as* a list, which in turn can lead to comic interpretations or to unnoticed misinterpretations. Here is an example:

Negative Example
Richard Clarke, Senior Systems Programmer, asked me to develop a magnetic tape management system, to reside permanently on the computer, to give better control, and to coordinate the numerous magnetic tapes.

The parallel use of infinitives in this sentence makes it appear that a four-item list is being presented:

Richard Clarke, Senior Systems Programmer, asked me:

■   to develop a magnetic tape management system

■   to reside permanently on the computer

- to give better control

- to coordinate the numerous magnetic tapes

This would mean, among other things, that it is the writer, not the system, that will "reside permanently on the computer, give better control," etc.— surely a strange state of affairs! Of course, most readers would probably figure out what the writer means, but this requires extra time and effort on the reader's part—precisely what good writing avoids.

In cases like this, the writer should break up the misleading parallelism so that only those items that really are part of the list appear in parallel form. In this particular case, only the last three items belong together; the first item should be kept separate. If the last three items were written as relative clauses, say, instead of as infinitives, they would be distinguished from the first item:

Richard Clarke, Senior Systems Programmer, asked me to develop a magnetic tape management system:

- which will reside permanently on the computer

- which will give better control

- which will coordinate the numerous magnetic tapes.

This can be simplified by factoring out the relative pronoun and auxiliary verb (*which will*) and making them part of the lead-in:

Revised Version
Richard Clarke, Senior Systems Programmer, asked me to develop a magnetic tape management system which will *reside permanently on the computer*, *give better control*, and *coordinate the numerous magnetic tapes.*

Compare this version to the original one and see what a difference there is in readability.

❑ EXERCISE 23-2

Each of the following passages contains misleading parallelism. Correct each passage by making appropriate grammatical changes.

A My present occupation is repairing typewriters, printing machines, and duplicating machines.

B You learn many reasons why our product failed by reading, observing, talking, and listening to our salespeople.

C This technology consists of three methods of scrambling, which are the coupling of light source and optical fiber in low-order mode, splicing, bending and tightening the fiber near its connectors, and installing the scrambler into the existing fiber optic in intervals of 1 km along the route.

23.3 PARALLELISM IN PARAGRAPHS

The most important use of parallelism is in helping the reader detect unformatted lists in paragraphs and larger units of text. Two or more sentences may be related coordinately within a paragraph and yet not be marked by any kind of formatting. Likewise, two or more paragraphs may have a coordinate relationship that is not marked by formatting. In such cases, parallelism is often the only formal cue the reader can use to detect the coordinate relationship.

Below is an example of an embedded, unformatted list within a paragraph. The list is not easy to detect, however, because the writer failed to use parallelism to mark it.

Negative Example

All-Savers Certificates will not benefit all investors. Investors exceeding a deposit of $7931 ($15,861 joint return) would have an after-tax yield far lower than alternative investments such as money market funds or Treasury bills. Alternative investments would also yield better after-tax yields and no penalty if the certificate were redeemed within the 1-year maturity period.

A common strategy used by skimreaders is to first read the topic sentence of a paragraph and then read the beginning words of the following sentences to see what kind of support these sentences provide for the topic sentence. In the case above, a person using this technique could easily be misled into perceiving the paragraph as having a general-to-particular structure. The first sentence is clearly the topic sentence. The second sentence supports it by describing a class of investors who would benefit more from alternative investments than from All-Savers Certificates. The third sentence then appears to provide additional details about alternative investments. The diagram below illustrates this structure:

All-Savers Certificates will not benefit all *investors*.

Investors exceeding a deposit of $7931 . . . would have an after-tax yield far lower than *alternative investments* . . .

Alternative investments would also yield better after-tax yields and no penalty . . .

Notice how the chain of repeated words links one sentence to the next, each one appearing to be subordinate to the one above it. In actuality, however, the writer means to say that there are not one but *two* classes of investors who will not benefit from All-Savers Certificates: those who exceed a deposit of $7931 and those who redeem their certificates before 1 year has elapsed. The purpose of the paragraph (as we were informed by the writer) is to describe these two classes. The second and third sentences, therefore, constitute a two-item list, with each of the sentences describing one of the two classes. If the writer had cast these two sentences in parallel form, it would be much easier for readers to see this listlike structure:

> All-Savers Certificates will not benefit all investors. Investors exceeding a deposit of $7931 ($15,861 joint return) would have an after-tax yield far lower than alternative investments such as money market funds or Treasury bills. Investors redeeming the certificate within the 1–year maturity period would have to pay a penalty and would also have a lower after-tax yield than with alternative investments.

The diagram below illustrates why this revision has a more transparent listlike structure:

> All-Savers Certificates will not benefit all *investors*.
>
> *Investors exceeding a deposit of $7931* . . . would have an after-tax yield far lower than alternative investments . . .
>
> *Investors redeeming the certificate within the 1-year maturity period* would have to pay a penalty and would also have a lower after-tax yield . . .

The use of parallelism enables the reader to see very quickly that the second and third sentences constitute a two-item list.

❏ EXERCISE 23-3

Each of the following passages contains information that can be presented in the form of one or more lists. Reconstruct each passage accordingly, using appropriate parallelism.

A   In order to meet the job requirements, it became clear that a microcomputer would be required to do this type of work. Microcomputers are very compact and portable. They are easily programmed and can be used for a wide range of data processing. These computers are normally quite easy to operate and capable of storing large amounts of data both inside and outside of the computer itself. The most attractive part of using a microcomputer is that it is very inexpensive in comparison to other large-scale systems.

B   Reuse of treated water is most applicable where large amounts of water are used and the wastes are not too contaminated. Industrial wastes may be heavily contaminated and therefore may not be very suitable for reuse. The location of the treatment plant and the possible transport of the renovated water are also important factors. The treatment process works most efficiently and economically when dealing with a steady flow of wastewater. A very important point is whether the wastewater will be reused only once or whether it will be reused many times. Multiple recycling results in a buildup of certain dissolved materials, especially inorganic ions, that may make demineralization necessary. Most reuses do not lead to a high degree of recycling.

C   The new design meets all of the important criteria. The new design uses the same sulfur dioxide scrubbing process used in the present scrubbing system. Therefore, the new design gives the same sulfur dioxide removal rate as the present system. The new design includes a regeneration loop for the absorption reactant, thereby cutting the absorption reactant consumption considerably. The new design also terminates with a solid waste product. Thus, a very manageable waste product is produced.

23.4 PARALLELISM BEYOND THE PARAGRAPH

Parallelism can be used to reveal listlike structures in any piece of text, even in units larger than paragraphs. Here is an example (with sentence numbers added):

> Corn, the largest and most important crop in the United States, is an annual in the grass family (1). Modern varieties of corn are descendants of a wild relative called teosinte that was domesticated thousands of years ago by farmers living in what is now Central America (2).
>
> Like its relatives in the family of grasses that produce cereal grains—wheat, sorghum, barley, rye, oats—corn produces male and female flowers (3). But unlike most of its relatives, corn does not normally pollinate itself (4). Reproduction is accomplished by joining pollen and silks, usually from different plants (5).
>
> The male flower, known as the tassel, forms at the top of the stalk, which under normal conditions is 6 to 8 feet tall in July (6). Within the tassel are anthers producing grains of pollen (7). A healthy tassel sheds millions of yellow pollen grains (8).
>
> The female flowers form in the elbows where leaves and stalk join and in early July are ears no larger than thumbs (9). Each ear is surrounded by a husk from which a cluster of long, soft silks emerge (10).
>
> Normally in early July tassels release pollen and winds carry the pollen grains to the silks, where they germinate (11). But the drought and heat have weakened corn plants and altered their normal development (12). Hundreds of millions of plants simply did not germinate, according to scientists monitoring the crop in South Dakota, Iowa, Missouri, Indiana, and other states across the Corn Belt (13).
>
> Hundred of millions of others have been stunted by lack of moisture (14). The young plants, unable to secure enough moisture, sharply reduced the rate at which they were converting carbon dioxide from the atmosphere in the presence of sunlight into carbohydrates, organic compounds, and amino acids (15). With less energy available to form stalks, leaves, and other structures, much of the crop is no more than 4 feet tall (16).[1]

Because of the parallelism between sentences 3 and 4, sentences 6 and 9, and sentences 13 and 14, it is easy to see the sets of contrasts this author is laying out:

3. Like its relatives . . . corn produces male and female flowers.
4. Unlike most of its relatives, corn does not normally pollinate itself.

6. The male flower . . . forms at the top of the stalk . . .
9. The female flowers form in the elbows . . .

13. Hundreds of millions of plants simply did not germinate . . .
14. Hundreds of millions of others have been stunted by lack of moisture.

These contrasts form the heart of this text. Corn does not normally pollinate itself . . . male flowers and female flowers form separately . . . this year millions of male flowers and millions of female flowers failed to form properly. And two of the three pairings span two paragraphs, helping to tie the text together.

Whenever you are describing lists or pairs of any kind in any size of text, use grammatical parallelism. By using form to reflect content, you make it easier for the reader to see both the structure and the meaning of your writing.

❑ EXERCISE 23-4

You are the director of a health and exercise center and need to write a series of information sheets for potential customers outlining the forms of exercise and combinations of exercises appropriate for different types of people. You need information sheets for (1) high school athletes trying to get into shape before football and track start, (2) office workers trying to improve their cardio-vascular systems and to get in better overall shape, (3) members of a community dance program trying to get in shape for weekly 2–hour practice sessions, and (4) members of a local chapter of Weight Watchers trying to lose weight through a combination of dieting and exercise. The following table[2] contains information about the benefits of different forms of exercise.

Write a three- to five-paragraph interpretation of this table for each information sheet in which you evaluate the different activities for each purpose. Be sure to use parallelism where appropriate.

Activity	Calories Consumed in 20 Min	Value in Improving Heart and Lungs	Value in Improving Suppleness of Joints	Value in Improving Muscle Power
Bicycling	220	Excellent	Fair	Good
Dancing	160	Fair	Good	Minimal
Easy walking	60	Minimal	Minimal	Minimal
Fast jogging	210	Excellent	Fair	Fair
Football	180	Good	Fair	Good

Activity	Calories Consumed in 20 Min	Value in Improving Heart and Lungs	Value in Improving Suppleness of Joints	Value in Improving Muscle Power
Golf	90	Minimal	Fair	Minimal
Gymnastics	140	Fair	Excellent	Fair
Ice skating	160	Good	Good	Fair
Skiing	160	Good	Good	Fair
Swimming	240	Excellent	Excellent	Excellent
Tennis	160	Good	Good	Fair

REFERENCES

1 Adapted from Keith Schneider, "Drought Hits Corn at Critical Stage," *New York Times*, July 12, 1988, p. 21.
2 Adapted from J. Kunz and A. Finkel (eds.), *The American Medical Association Family Medical Guide*, rev. ed. (New York: Random House, 1987), p. 17. Used with permission.

24

Maintaining Focus

When readers say that sentences are hard to read or "unreadable" or don't "flow," they usually mean one or both of the following: (1) that individual sentences are hard to understand and/or (2) that a series of sentences does not seem to have clear connections from sentence to sentence. These problems with the readability of a text are important because, if a reader cannot understand the individual sentences and how they connect, the reader is not going to be able to deal with the text in any significant way.

As an example of the differences in reading ease which can occur, consider the following versions of a passage and decide which is easier to read.

Version A
The 5-year plan does not indicate a clearly defined commitment to long-range environmental research. For instance, the development of techniques rather than the identification and definition of important long-range issues is the subject of the plan where it does address long-range research.

Version B
The 5-year plan does not indicate a clearly defined commitment to long-range environmental research. For instance, where the plan does address long-range research, it discusses the development of techniques rather than the identification of important long-range issues.

For most people, Version B is significantly easier and more pleasant to read, that is, more readable. This greater readability depends partly on issues of cohesion and partly on issues of sentence structure, two sets of issues which are discussed in this and the next chapter.

Our discussion of cohesion and sentence structure relies critically on the concept of the noun phrase. The *noun phrase* is one of the most important parts of speech in scientific and technical writing. It can be defined as *any noun, any noun-plus-modifier combination, or any pronoun that can function as the subject of a sentence*. While noun phrases can be found throughout a sentence, you can determine whether a unit is a noun phrase by testing to see if it can serve as the complete subject of a sentence. Some examples of noun phrases used as subjects are *tables, water, we, a potential buyer, the growing demand for energy*, and *strict limitations on the weight that can be mailed*. Notice that each of these noun phrases can serve as the subject of a sentence:

Tables usually have four legs.

Water can be dangerous.

We have an emergency.

A potential buyer has arrived.

The growing demand for energy is obvious to all home owners.

Strict limitations on the weight that can be mailed have been established by the post office.

By contrast, a singular countable noun, such as *table*, is not a noun phrase because it cannot function by itself as the subject of a sentence. We cannot say

Table usually has four legs. [an ungrammatical sentence]

Instead, we would have to say one of the following:

A table usually has four legs.

The table has four legs.

John's table has four legs.

Noun phrases are important on the grammatical level because they account for most of the principal parts of a sentence, as illustrated below. The four italicized units are noun phrases, and each underlined unit is a separate noun phrase.

<u>*We*</u> are sending   <u>*your company*</u>   <u>*the new TK-140 model*</u>

on   <u>*the assumption that you need an all-purpose sorter.*</u>

Noun phrases are important on the functional, or communicative, level because they carry the main information of the sentence. In the example just cited, for instance, notice how much information is contained in the italicized noun phrases—almost all of it.

24.1 OPTIMAL ORDERING OF NOUN PHRASES

In English, noun phrases are expected to occur in certain orderings according to grammatical and functional criteria. These criteria will be discussed in this and the next chapter, beginning with the most important criterion.

Readability Criterion 1: Put Given Information before New Information

As with all languages, English sentences typically contain a mixture of *given* information and *new* information. That is, some noun phrases in a sentence refer to concepts or objects that have already been discussed or that are presumed to be understood from the context; this is *given information*. Other noun phrases refer to concepts or objects that have not yet been discussed and are not presumed to be understood from the context; this is *new information*. In Version A above, when the reader has finished reading sentence 1, the information contained in that sentence is *given* and the additional information which is being added by sentence 2 is *new*.

When people read, they constantly try to fit the incoming information into a context created by what they already know, the given information. Therefore, the optimal ordering of noun phrases within a sentence is *given* before *new*, because the given information can serve as a kind of "hook" or "glue" to attach the new information into the context established by the prior sentence(s).

Let us use this view of things to suggest one reason why Version A above was harder to understand than Version B. Once you had read the first sentence of either version, you had heard about *the 5–year plan, a clearly defined commitment*, and *long-range environmental research*. As you are reading the second sentence of each version, you are trying to fit the information you are seeing in the second sentence into the context created by the first. You are trying to see what the second sentence has to do with the three noun phrases you saw in the first sentence, and you are trying to do this within the processing limits of your brain.

In Version A, once you started reading the main clause of the second sentence (which begins after the *For instance*), you had to read 19 words into the main clause—all the way to *the plan*—before you saw that the second sentence was adding information about *the plan* from the first sentence. This places a great load on your short-term memory, the part of your brain which processes incoming words and connects them to the context established by the prior sentences. Unfortunately, short-term memory has serious limitations. As described by George Miller, a famous psychologist, short-term memory can usually hold only 7 plus or minus 2 pieces of information at one time before it gets overloaded. For the sake of this discussion, if we say that a piece of information is roughly equated to a word, then short-term memory has roughly a 9–word limit before it starts to have trouble remembering what it has already seen in an incoming sentence and thus to have trouble connecting the incoming sentence to the context created by the first sentence. To return to the specific case at hand, Version A makes you read 19 words into the main clause of the second sentence before you see how it connects to the first sentence (it adds information about *the plan*). This stresses the limits of short-term memory (here roughly defined as 9 words) and thus makes the second sentence hard to process and hard to understand.

On the other hand, the second sentence of Version B is much easier to read partly because it announces right away that it is talking about *the plan*. This makes it very easy to see how it relates to the first sentence (it adds information about the given information, *the plan*). This placement of given information early in the sentence, ideally in the subject position, makes it easier for readers to process incoming sentences. Thus, the suggestion for all sentences within paragraphs (or contexts) is to put given information "up front" before new information. This is not done in Version A but clearly is done in Version B, as shown below.

Version A
The 5-year plan does not indicate a clearly defined commitment to long-range environmental research. For instance, the development of techniques rather than the identification and definition of important long-range issues is the subject of *the plan* where *it* does address *long-range research*.

Version B
The 5-year plan does not indicate a clearly defined commitment to long-range environmental research. For instance, where *the plan* does address *long-range research*, *it* discusses the development of techniques rather than the identification of important long-range issues.

24.2 TYPES OF GIVEN OR REPEATED INFORMATION

If you are going to put given information first in a sentence, you need to be able to recognize given information when it occurs, that is, to recognize the different ways that given information can appear, to recognize the different ways that noun phrases can be repeated. There are five main ways of repeating information: repeating the full form of a noun phrase; repeating a shortened form; repeating with a pronoun (such as *it* or *they*) or with another substitute (such as *this*, *that*, *these*, or *those*); repeating with a synonymous noun phrase (another noun phrase which means the same thing); and repeating with an "associated" noun phrase. The first three of these are usually easily recognized as given, but the last two do not seem given to many readers, especially to nonspecialist readers.

Full-Form Noun Phrase Repetition

A writer may refer to something by repeating the word or entire phrase—the full form—which originally expressed the idea. This is an appropriate option when the writer needs the full form to *clarify meaning* or to *establish special emphasis*. For instance, consider the following example:

Negative Example
The flue dampers shall be tied with the inlet dampers of the fans located at the exit side of the electrostatic precipitators. *These dampers* will open automatically in the event of any failure in the electrostatic precipitator system.

Here it is not clear what *these dampers* refers to. It could refer to the flue dampers, to the inlet dampers, or to both sets of dampers. To avoid this confusion, the author should have repeated the full noun phrase of the original reference as follows:

Rewritten Version
The flue dampers shall be tied with the inlet dampers of the fans located at the exit side of the electrostatic precipitators. *The flue dampers* will open automatically in the event of any failure in the electrostatic precipitator system.

A writer can also repeat the full form to establish special emphasis. For instance, the writer of the following example wanted to emphasize that it is the proposed system which meets the criteria, not the current system:

The proposed system meets the criteria specified by the design team:

1. *The proposed system* is compact enough to fit into a 4′ × 5′ area adjacent to the conveyor line.

2. *The proposed system* is economically competitive with the manual system.

This sort of repetition for emphasis should not be used too frequently or it will clutter your writing and finally destroy emphasis.

Short-Form Repetition

If writers do not need full-form repetition for clarity or emphasis, they should use a shorter form which will usually consist of a pronoun or the main word and one or a few modifiers. This short-form repetition is appropriate because it eliminates awkward full-form repetition and still provides a clear reference to the original noun phrase. For instance, in the following example, the shortened form *these standards* repeats and refers back to *the same safety standards that gasoline cars must meet*.

Electric cars must be able to meet *the same safety standards that gasoline cars must meet. These standards* are derived from an established crash test, in which the car is propelled against a solid wall at 30 MPH.

Notice that there is only one set of standards in sentence 1 and thus only one possible referent for *these standards* in sentence 2.

Short-Form Repetition Using an Acronym

One short-form repetition which appears frequently in some kinds of documents is the acronym/abbreviation. An acronym is a word formed from the first letter(s) of each word in a name such as *COMP* for *C*ommittee *o*n *M*anagement *P*ractices or a word such as *radar* from *r*adio *d*etecting *a*nd *r*anging).

> Light Emitting Diode (LED) displays have been used in many electrical devices for more than 20 years. Now the LED is facing serious competition from the Liquid Crystal Display (LCD). In comparing the LCD and LED, it is clear that the LCD is more appropriate for use in our equipment.

Short-Form Repetition Using Increasing Compression

Sometimes a noun phrase will first appear as a long unit and later be progressively compressed, as illustrated below.

> *the increased radiation dosages that would be experienced at sea level*
> *the radiation dosages experienced at sea level*
> *the radiation dose at sea level*
> *sea-level radiation*

This progressive compression is very useful for readers who are not very familiar with the content, for they are introduced to the final complex noun phrase in a gentle way. The phrase first appears in a definitional mode, with a noun phrase followed by a full unit introduced by *who, which,* or *that.* This is the most expanded and gentle introduction of a complex new noun phrase. After the readers have seen this version, they see progressively compressed forms, each relying on the familiarity with the concept created in earlier stages.

The most compressed type of repetition involves the *noun compound*, a noun phrase consisting of two or more nouns functioning together as a single unit. This is illustrated by the italicized noun compounds below.

> A *management decision* must consider *worker productivity*.

In all noun compounds, the rightmost noun is the main noun and all nouns preceding it serve as qualifiers in some way. Thus, the full meaning of a noun compound can often be expressed by simply "unwinding" it from right to left—that is, reversing the order of the nouns—and inserting appropriate connecting words. For example, in the sentence above, a *management decision* is a *decision* made by *management*, and *worker productivity* is the *productivity* of the *worker*.

However, the exact relationship between qualifiers or between one or more qualifiers and the main noun is not always clear—at least not to a nonspecialist. Take, for example, the noun compound *wall stresses*. Does it mean the stresses *on* a wall (from outside) or the stresses *produced by* a wall or the stresses *inside* a wall? Unless you're a construction engineer, you might not know for sure. (It's the last.) Or how about *mission suitability*: Does that mean suitability (of something) *for* a mission or suitability *of* a mission (for something)? Unless you're an aerospace specialist, you probably are not sure which. (It's the former.)

Thus, if you use a noun compound, only use one that is clear and free of ambiguity so that your intended meaning will be understood by your entire audience.

Sometimes in the process of compressing a long unit into a noun compound, you can textually define the meaning of the noun compound by the context in which it appears. For instance, consider the following sentence.

> A *management decision* must consider *worker productivity* and the *production air filtration system*.

Here, the *production air filtration system* is either a *system for the filtration of production air* (air in the area in which production occurs) or a *system for air filtration which is being produced by the company*. The correct interpretation will be determined by the context—that is, by whether you are talking about filtering the air in the production area or talking about the production of an air filtration system. For instance, in the following passage, the only meaning possible is *system for the filtration of production air*.

> Management is considering ways of improving both the efficiency of our production line and the health of our workers. Two areas of concern have been suggested: the productivity of individual workers and the adequacy of the new system for filtering noxious gases from the air in the production areas. (The new filtration system has greatly improved the quality of the air in the quality control areas, but has not substantially improved the production air.) A study of these concerns has convinced us that a *management decision* must consider both *worker productivity* and the *production air filtration system*

Noun compounds are especially useful for presenting the given information at the beginning of the sentence since they are short and in that part of the sentence there is a premium on keeping things short. The usefulness of the noun compound is illustrated below where the relatively long *backlog of active and waiting requests* is repeated by the much shorter *backlog size*:

> The smooth flow of requests through the hospital's Correspondence Unit is marred by a number of operational problems. First, the Correspondence Unit has an average worker productivity of 59.1%, well below the suggested 65% level. Second, this low worker productivity has caused the *backlog of active and waiting requests* to rise above 2000. This large *backlog size* causes excessive processing delays.

Repetition Using Pronouns or Other Substitutions

Pronouns such as *they* and *it* or substitutions such as *this*, *that*, *these*, and *those* all allow a writer to repeat a previous noun phrase in a concise way.

However, it must be clear which noun phrase is being repeated, or the substitute will only lead to confusion. For instance, consider the following passage, where *one* clearly means *choice* and *it* clearly means *the best choice*.

> A survey of the field was made to determine possible choices for the new analysis technique. I was assigned to review the choices, select the best *one*, submit *it* to a feasibility test, and, if *it* seemed feasible, to identify initial development and implementation concerns.

It versus This

If you are a nonnative speaker of English, do not make the common mistake of freely substituting *it* for *this* or vice versa. If the repeated unit is the whole idea of a sentence or clause, you must use *this, that, these,* or *those* (usually *this*):

> Computers can solve problems at the rate of 100,000,000 steps per second. *This* is one reason why they have become so popular.

Here, the repeated idea is the entire first sentence. On the other hand, *it* is used to repeat a single noun or noun phrase:

> I was assigned to evaluate *the proposed analysis technique* and to compare *it* to other potential techniques.

A single noun or noun phrase can also be repeated by *this, that, these,* or *those*:

> A simple measure of the rate of increase of the output of the scientific community can be obtained by looking at *the Royal Society Catalogue of Scientific Literature*. *This* eventually covered all the scientific literature in all subject fields in the 19th century. *It* occupies about the same shelf space as last year's output in chemistry alone.[1]

In this example, both italicized words (*this* and *it*) repeat the single noun phrase *the Royal Society Catalogue of Scientific Literature*. When is one used rather than the other? Usually, *this* is used to refer back to a noun phrase or idea just introduced as the topic. In the example above, the topic and subject of the first sentence is *a simple measure* (and its modifiers). By the end of sentence 1, a new topic has been introduced, *the Royal Society Catalogue of Scientific Literature*, which then becomes the topic and subject of the next two sentences. Notice that *this* would not usually be used for a second repetition of the original noun phrase and would not usually be used at all if the first repetition were an *it*.

❑ EXERCISE 24-1

The following passages contain many full-form noun phrases. Try to replace as many of these as possible with shorter forms—short-form noun phrases, pronouns, or other substitutions. Check your choices against those of your classmates.

A   The main conveyor line in Building 9 is used to transport three of our products. The main conveyor line in Building 9 has three stations, which are located at various positions along the main conveyor line in Building 9. Each station of the three stations located along the main conveyor line of Building 9 has a particular worker who is assigned to one of our three products. Each time the worker's one out of three assigned products passes by, it is the worker's responsibility to lift his one out of the three assigned products off the main conveyor line and to place the product in the appropriate chute, which transfers the product to another conveyor line.

B   Digital Systems Division (DSD) is currently involved in the design and implementation of an advanced Digital Signal Processor Unit (DSPU). The advanced Digital Signal Processor Unit (DSPU) will be used in applications such as runway roughness measurements and highly accurate spectroscopy experiments. The Digital Signal Processor Unit (DSPU) design was completed last October by the Digital Systems Division (DSD) design team and is now being implemented. The Digital Systems Division (DSD) design team expects a prototype of the Digital Signal Processor Unit (DSPU) design to be completed by April. However, the Digital Systems Division (DSD) design team needs to know the type of logic component best suited for use in the advanced Digital Signal Processor Unit (DSPU) and thus asked me to investigate the possibilities and to recommend the type of logic component best suited for use in the advanced Digital Signal Processor Unit (DSPU). This report presents the results of my investigation and the proposals based on the results of my investigation.

Repetition Using Synonymous Terms

A frequent but potentially misleading type of repetition involves the use of a synonymous term, a different noun phrase meaning the same thing as an original noun phrase. For instance, consider the following:

> Because requesters desire medical information as soon as possible, hospital management feels that *requests should be processed within two weeks*. Since *this standard* is far from being met, management undertook a study of request processing procedures.

In this example, the writer has implicitly equated *requests should be processed within two weeks* and *this standard*. If the reader understands that *processing within two weeks* is a standard, and in this context the reader probably does, then the writer has chosen a good repeated form.

However, a writer cannot freely use synonymous terms if the reader does not know enough to see that the terms mean the same thing. For instance, consider the following:

> It is the combination of *an inorganic polymer* with a short organic (methyl) group in the structure that accounts for many of the unique properties of PDMS. The *siloxane backbone* is extremely flexible and gives many useful properties.[2]

Unless you have a good background in silicon chemistry, you will probably not see that the two italicized phrases are synonyms. The compound under discussion is polydimethylsiloxane (PDMS), which has an inorganic polymer backbone consisting of alternating silicon and oxygen atoms with methyl groups (whose names end in *-ane*) attached to the silicon.

$$\begin{array}{ccc} CH_3 & CH_3 & \\ | & | & \\ -O-Si-O-Si-O- & & \\ | & | & \\ CH_3 & CH_3 & \end{array}$$

Thus, the backbone is *sil*(icon) + *ox*(ygen) + *ane*, i.e., *siloxane*.

If you were writing this passage to a group of silicon chemists who would already knew these things, then the use of the synonymous term (*siloxane*) is probably acceptable. However, if you were writing to nonchemists or even to chemists who did not specialize in this particular area of silicon chemistry, you might want to provide some extra information before using the synonymous term *siloxane backbone*.

In closing, please note the following.

Warnings

1 If you want to use synonymous repetitions, you must be very careful that the repeated form is one the reader will easily understand, especially if it is several sentences away from the original full form of the noun phrase.

2 If you need to repeat a noun phrase several times and feel an overwhelming urge for variety in your repetitions, try to use short-form repetition combined with at most one synonymous form, rather than using several synonymous forms. Readers get easily confused, and the apparently shifting terminology of synonymous terms is likely to create confusion.

Repetition Using Associated Noun Phrases

Sometimes a writer "repeats" a noun phrase by representing it at a more general or more particular level. That is, a writer can "call up" or refresh the idea of an original noun phrase by using another noun phrase closely associated with it. For instance, consider the following passage, in which the first mention of a noun phrase is a general term, *the thermal properties of glassy materials at low temperatures*. The next times the idea of *thermal properties* appears, it is "called up" by more specific or particular noun phrases. In sentence 2, *the thermal conductivity* (and *it*) calls up the idea of *thermal properties* since thermal conductivity is a thermal property. In sentence 3, *the specific heat below 4K* (and *it*) "calls up" the idea of thermal conductivity because *specific heat below 4K* is also a thermal property. Finally, the idea of *thermal properties* is called up or "repeated" in sentence 4 by *the thermal behavior*, since *thermal properties* are an obvious cause of *thermal behavior*.

> *The thermal properties of glassy materials at low temperatures* are still not completely understood. *The thermal conductivity* has a plateau which is usually in the range 5 to 10K, and below this temperature *it* has a temperature dependence which varies approximately at T. *The specific heat below 4K* is much larger than that which would be expected from the Debye theory, and *it* often has an additional term which is proportional to T. Some progress has been made towards understanding *the thermal behavior* by assuming . . . that there is an additional system of low-lying two-level states.[3]

In addition to using a general-to-particular or particular-to-general "repetition," a writer sometimes "repeats" a noun phrase by using the whole-part or part-whole relationship. If the part is very closely associated with the whole in the reader's mind, naming the part will recall the idea of the whole, and vice versa. Thus, if the original noun phrase names the whole, naming the part could be considered a repetition of the whole. However, this form of repetition will be effective only if the audience already clearly knows the relationship between the whole and the part. As illustrated below, *a manufacturing company* in sentence 1 is repeated or recalled by the mention of its parts in the next sentences.

> It is difficult to coordinate and direct *a large manufacturing company* so successfully that the company makes a profit and keeps its employees happy. *Sales, manufacturing, and design engineering* are all at odds with the executive group. *The sales department* wants low selling prices to make the goods easier to market, *the design engineering group* tends to overdesign the product to insure certainty of operation, while *the manufacturing group* tends to make frequent purchases of new equipment and to stockpile materials in order to maintain uninterrupted plant operations. Now it happens that *the executive arm of the firm* is also interested in low selling prices, quality design, and continuous plant operation, but not to the extent of running the business into bankruptcy. Consequently, *the management's* simultaneous desire to set prices which are certain to return a profit, to limit the quality of the design, and to keep a tight rein on equipment expenditures and inventories makes every problem solution a compromise with respect to the interests involved.[4]

Obviously, this type of repetition works only if the audience is aware of the relationship between the whole and its parts.

Another type of "repetition" sometimes used by writers involves the use of a noun phrase strongly related to or associated with the original noun phrase. The association will usually not be as strong as a part-whole or whole-part association, but it will have to be significant if it is to create a sense of repetition for the reader. Some common powerful associations in our culture are (1) cause and effect (e.g., *running in a long race* and *being tired*), (2) an item and a quality or characteristic of the item (e.g., *water* and *wetness*), and (3) an item and some other item sharing its class or type (e.g., *a university* and *a college* where both belong to the class of *institutions of higher learning*). To see how this type of connection might "feel" in a passage, decide how hard or easy this passage is for you:

> Another report of tests made on the concrete roof of a second building confirms the findings of earlier tests. An intense 1 and ½-hour fire subjected the roof to extreme heat and sudden load changes. The only damage visible to the 3 and ½-inch roof was minor cracking of the concrete. To confirm this observation, a full-scale loading test was ordered. About a month after the fire, and before any repairs were made, a uniform load of 40 pounds per square feet was placed on one of the damaged panels. The maximum deflection was 0.1 inch and disappeared on removal of the load. The theoretical deflection was calculated to be 0.027 inch, whereas the actual deflection was 0.025 inch. The structural strength of the building was thus shown not to have been weakened by the fire.[5]

This passage is quite interesting from several points of view. First, many non-civil engineers find it hard or very hard to understand, though a civil engineer usually finds it quite easy. What causes the difference in difficulty for the two audiences? The level of background knowledge that is required to see the "repetition" and connections between some of the associated noun phrases. In particular, in sentence 1, the general idea of a *test* is introduced. In sentence 2, this idea is "repeated" at the particular level of a specific test, *an intense 1 and ½-hour fire*. This idea is then recalled by the subject noun phrase of sentence 3, *the only damage visible to the 3 and ½ inch roof*. This does not directly repeat *fire*, the main noun in the subject of sentence 2, though it does repeat *roof*. However, the main sense of "repetition" or cohesion in sentence 3 probably arises from the strong association we have between *an intense 1 and ½-hour fire* and the resulting *damage* usually following such a fire.

This same way of "repeating" given information appears again in sentences 5 to 7, where an original noun phrase is recalled by using others associated with it. Here the original noun phrase in sentence 5, *a uniform load of 40 pounds per square foot*, is recalled by the three types of deflection in sentences 6 and 7 associated with loading tests: *the maximum deflection*, *the theoretical deflection*, and *the actual deflection*. Finally, the subject of the last sentence, *the structural strength of the building*, recalls the ideas of deflection, load, and damage: the association here is that *structural strength* is the underlying cause of the small *actual deflection*.

As was mentioned earlier, non-civil-engineering readers often find this passage hard to read. This is because, for these readers, the associated noun phrases just discussed are not seen to be repetitions of earlier noun phrases, and so it is hard to see the connections between the sentences. This response points out a very important lesson:

Warning

If you are using a form of repetition which is not a full form, short form, or pronoun or other substitution, then you should examine the form carefully to see if your audience could easily understand the repeated nature of the synonym or association you are using. If you are not sure that the audience would connect the original mention and repetition, then you should probably find a short-form reference to make the connection.

❑ EXERCISE 24-2

Identify the various forms of repetition in the following passages. If you find forms which do not involve direct repetition of some key word, be prepared to explain what type of repetition it is.

A   Air-traffic radar of the 1950's could readily determine the direction of incoming and outgoing aircraft, but not their height above the ground. The planes might be a mile apart in altitude or on the verge of colliding— but air-traffic controllers on the ground had no way of obtaining this information from current radar data alone, or without some previous radar history on both planes before they entered the same air space "capsule." Recognizing this obvious deficiency, the Airways Modernization Board (the forerunner of the FAA) decided to investigate the radar "height finder" as a likely solution to the problem.

   Any new air-traffic radar would have to be able to resolve targets to within 1000 feet at 50 miles and be able to determine whether two aircraft, approaching each other from opposite ends of the sky, might pose a threat to each other. This meant that the radar had to measure the altitude of each aircraft before it entered the range of another and to allow for any distortions in the target's return signals due to blending in the atmosphere.

   Looked at another way, the problem was to hold steady a lever 50 miles long—the length over which the radar beam had to be effective to do its job. This problem brought with it the sub-problems of wind, temperature, and humidity variations, and even the light, imperceptible seismic motions of the ground itself. In addition to these structural problems, the major problems facing the designers were "ground clutter"— the bane of all radar—and, unexpectedly, system noise.[6]

B   Organizational behavior is directly concerned with the understanding, prediction, and control of human behavior in organizations. It represents the *behavioral* approach to management, not the whole of management. Other recognized approaches to management include the process, quantitative, systems, and contingency approaches. In other words, organizational behavior does not intend to portray the whole of management. The charge that old wine (applied or industrial psychology) has merely been poured into a new bottle (organizational behavior) has also proved to be groundless. Although it is certainly true that the behavioral sciences make a significant contribution to both the theoretical and the research foundations of organizational behavior, it is equally true that applied or industrial psychology should not be equated with organizational behavior. For example, organization structure and management processes (decision making and control) play an integral, direct role in organizational behavior but have at most an indirect role in applied or industrial psychology. The same is true of many important dynamics and applications of organizational behavior. Although there will probably never be total agreement on the exact meaning or domain of organizational behavior—which is not necessarily bad, because it makes the field more exciting—there is little doubt that organizational behavior has come into its own as a field of study, research, and application.[7]

C   Now even some of the sickest heart patients with clogged coronary arteries can be considered for heart-saving balloon therapy, researchers say. That's important news for up to 60,000 older "high risk" patients with multiple coronary artery plugs, blocked grafts from previous bypass surgery, or massive congestive heart failure. Many have such weak hearts that they were considered too weak for balloon angioplasty. In that procedure, a tiny balloon is snaked through an artery and inflated to break up a clot. This works best for patients under 65 who have only a single blockage, according to just-issued guidelines to doctors.

    Now there's hope for high-risk hearts: a new-generation heart-lung machine that so far has helped 30 patients through balloon therapy—many of whom had almost no chance for survival otherwise, says Dr. James O'Toole, Director of Cardiology at Shadyside Hospital in Pittsburgh. Unlike conventional machines that require rib-spreading surgery, the Bard Cardiopulmonary Support system (CPS) requires only a small incision in the leg. Once it's connected, it takes over all the functions of the heart and lungs, so patients remain stable and doctors can take the time to do a more thorough job clearing arteries.[8]

REFERENCES

1   D. J. Urquart, *Times Literary Supplement* (London), May 7, 1970.

2   Adapted from P. G. Pape, "Silicones: Unique Chemical for Petroleum Processing." Paper delivered at the 56th Annual Fall Technical Conference and Exhibition of the Society of Petroleum Engineers of AIME, October 1981.

3   Quoted by John Swales, *Aspects of Article Introductions*, Aston ESP Research Reports No. 1 (Birmingham, England: University of Aston, Language Studies Unit, 1981).

4   Adapted from R. R. Raney, "Why Cultural Education for the Engineer?" Quoted in J. G. Young, "Employment Negotiations," *Placement Manual, Fall 1981* (Ann Arbor: University of Michigan, College of Engineering, 1981), p. 50.

5   Adapted from American Society of Civil Engineers, *Design of Cylindrical Concrete Shell Roofs*, ASCE Manual of Engineering Practice No. 31 (1951), p. 11.

6   Adapted from E. A. Torrero, "The Big Pan of 3–D Radar," *IEEE Spectrum 13* (10):51 (1976)

7   Fred Luthans, *Organizational Behavior*, 4th ed (New York: McGraw-Hill, 1985) pp. 7–8.

8   Linda Wasmer Smith, "Machines That Mend," *American Health*, November 1988, pp. 16, 18.

ADDITIONAL READING

Wallace Chafe, "Cognitive Constraints on Information Flow," in *Coherence and Grounding in Discourse*, edited by Russell S. Tomlin (Amsterdam: John Benjamins, 1987).

Herbert Clark and Susan Haviland, "Comprehension and the Given-New Contract," in *Discourse Production and Comprehension*, edited by R. Freedle (Norwood, NJ: Ablex, 1977).

Barbara Hayes-Roth and Perry W. Thorndyke, "Integration of Knowledge from Text," *Journal of Verbal Learning and Verbal Behavior 18*:91–108 (1979).

Michael P. Jordan, "Some Associated Nominals in Technical Writing," *Journal of Technical Writing and Communication 11*(3):251–264 (1981).

David Kieras, "The Role of Major Referents and Sentence Topics in the Construction of Passage Macrostructure," *Discourse Processes 4*:1–5 (1981).

Charles Li (ed.), *Subject and Topic* (New York: Academic Press, 1976).

Leslie A. Olsen and Rod Johnson, "A Discourse-Based Approach to the Assessment of Readability," *Linguistics and Education 1*:3 (Winter 1989).

C. A. Perfetti and S. R. Goldman, "Thematization and Sentence Retrieval," *Journal of Verbal Learning and Verbal Behavior 13*:7–79 (1974).

25

Creating Flow between Sentences

25.1 OPTIMAL ORDERING OF NOUN PHRASES II

Chapter 24 has discussed sentence readability in terms of repetition and cohesion and has introduced the first of three "criteria" for writing sentences that are as readable as possible. This chapter will continue the discussion about sentence-level readability begun in Chapter 24 and add two additional considerations or criteria.

Readability Criterion 2: Put Topical Information in Subject Position

Often, more than one noun phrase in a sentence carries given information. In that case, which of these noun phrases should be the subject? The ideal choice is the noun phrase which the paragraph is really talking about at that point. Consider this example:

> Not all investors will benefit from All-Savers Certificates. Investors exceeding a deposit of $7931 ($15,861 joint return) would have an after-tax yield far lower than with alternative investments, such as money market funds or Treasury bills. Alternative investments would also yield better after-tax yields and no penalty if the certificate were redeemed within the 1-year maturity period.

The grammatical subject of a sentence is what the sentence is apparently talking about. Thus, the first sentence is talking about *not all investors* and the second sentence is talking about a class of investors who will not benefit: *investors exceeding a deposit of $7931 ($15,861 joint return)*. The last sentence is apparently talking about *alternative investments* (which is the subject of that sentence) and has three noun phrases which contain given information: *alternative investments, after-tax yields,* and *the certificate.* But what is the *real* topic of this paragraph? The *different kinds of investors (who will not benefit)* . The word *investors* appears in the subject of the topic sentence, in the subject of the second sentence, and as the unspecified actor in the third sentence: *were redeemed (by investors).* Ideally, then, we should try to insert the word *investors* as the subject of the third sentence, if it is at all possible. Indeed it is:

> Not all investors will benefit from All-Savers Certificates. *Investors* exceeding a deposit of $7931 ($15,861 joint return) would have an after-tax yield far lower than with alternative investments, such as money market funds or Treasury bills. *Investors* redeeming their certificates within the 1–year maturity period would also have a lower after-tax yield and would pay a penalty besides.

Not only does this rewritten version keep the focus on the topic of the paragraph—and thus contribute to paragraph unity—it also establishes parallelism between the second and third sentences. This makes it much clearer to the reader that the passage is talking about two different classes of investors: those who exceed a deposit of $7931 ($15,861 joint return) and those who redeem their certificates early. In contrast, the original version links the third sentence to the second through the use of *alternative investments* as the subject of sentence 3. Thus, the third sentence seems to have shifted the topic from *investors* to *alternative investments* and might easily be misunderstood as simply modifying or adding detail to the second sentence rather than as adding a whole new class of investors.

Readability Criterion 3: Put "Light" before "Heavy" Noun Phrases

As can be seen in the examples at the beginning of Chapter 24, noun phrases vary considerably in length, complexity, preciseness, etc. If we use the word *light* to describe noun phrases which are short and simple and the word *heavy* for noun phrases which are long and complex, then the preferred ordering is light noun phrases before heavy noun phrases. For instance, consider the following passage:

> We have received and acted upon requests for equipment from several branch offices. We have sent the research, development, and testing office in Chicago a gas analyzer.

The second sentence of this passage has a very heavy indirect object (*the research, development, and testing office in Chicago*) and a light direct object (*a gas analyzer*). Because the noun phrases have been ordered heavy-before-light, the sentence is awkward and somewhat difficult to read. A more readable and thus better version would order the noun phrases light-before-heavy as follows. Notice that in moving the heavy noun phrase to the end, we have to insert the word *to*.

We have sent *a gas analyzer*

to *the research, development, and testing office in Chicago.*

Sentences are more readable if they are ordered light-before-heavy because of a limitation on the capacity of the mind to process information, as discussed in Chapter 24. This limitation is that the mind can usually process a maximum of 9 (7 plus or minus 2) separate items of information at a time before it begins to get overloaded. This means that as the mind reads through a sentence, it must try to make sense out of the sentence as quickly as possible. Ideally, the mind should see the entire structure of the sentence—and certainly the major units of subject and verb—all within the first 9 items of information—roughly equated to 9 words. This cannot happen when the main units of a sentence span much more than 9 words, and they often do.

However, even if you can't put *all* of the main units into the first 9 words, you can try to put in as many of them as possible and to get in at least the subject and verb. One way to do this is to put light units before heavy ones. Let us consider our heavy-before-light sample sentence again:

1		2	3
We		have	sent
SUBJECT		VERB	

4	5	6	7	8	9	10	11	12	13	14
the	research,	development,	and	testing	office	in	Chicago	a	gas	analyzer.

In this version of the sentence, the reader sees only the subject, the verb, and part of the next unit (2+ "chunks") before reaching the ninth word. However, in the light-before-heavy version shown below, the reader sees the subject, verb, the next unit (*a gas analyzer*) and part of the *to* unit (3+ "chunks") before reaching the ninth word.

1		2	3	4	5	6
We		have	sent	*a*	*gas*	*analyzer*
SUBJECT		VERB				

7	8	9	10	11	12	13	14	15
to	*the*	*research,*	*development,*	*and*	*testing*	*office*	*in*	*Chicago.*

Thus, by the ninth word in the light-before-heavy version, the reader has a fuller sense of the sentence structure and a better chance of understanding the sentence with a minimum amount of effort.

From this discussion, it should be obvious that a heavy noun phrase early in the sentence generally causes more trouble for a reader than a heavy noun phrase late in the sentence. In a sentence with a heavy subject noun phrase, the reader may not even get through the subject before reaching the ninth word. For instance, consider the following:

> The idea of designing and producing an economical AM/ FM receiver that is both affordable for the average consumer and profitable for the company was introduced.

Here, all of the sentence before *was introduced* is the subject. This is a 23-word subject and poses serious comprehension problems for the reader. The subject's main noun, *idea*, must be held in memory while the reader processes 21 other words before reaching the verb, *was introduced*. Note how much easier the sentence is to read if it is rewritten as follows:

> It was suggested that we design and produce an economical AM/FM receiver that will be both affordable for the average consumer and profitable for the company.

In this rewritten version, the reader sees the complete subject and verb by the ninth word, giving a fuller sense of the sentence's structure early on and thus a better chance of understanding the sentence with a minimum amount of effort. Producing this fuller sense of structure is the main reason writers try to order their noun phrases light-before-heavy.

One other situation should be considered before leaving this discussion of light-before-heavy ordering. That situation is the processing and "counting" of long introductory units, as in the following sentence:

> At the September and October meetings of all company engineers, it was suggested that we design and produce an economical AM/FM receiver that will be both affordable for the average consumer and profitable for the company.

From what has been said so far, we would expect this example to be harder to read than the previous one: it begins with a 10–word unit, *at the September and October meetings of all company engineers*, which comes before the subject (*it*) and verb (*was suggested*) in the main clause. However, it doesn't seem to be any harder to read. This is because when readers process a sentence, they first process introductory units and then start over again with the main clause. Thus, in ordering units light-before-heavy, a writer should count from 1 again at the beginning of each clause or introductory unit, as illustrated below.

1	2	3		5		7	9	10	1	2

At the September and October meetings of all company engineers, it was

    3   4

suggested that we design and produce an economical AM/FM receiver that will be both affordable for the average consumer and profitable for the company.

Then the writer should order the units light-before-heavy first within the introductory unit and within the main clause. For ease of reading, the writer might also try to keep the introductory unit as short as possible.

Sometimes a long subject is followed by a very short unit after the verb:

That we are in the midst of a pollution crisis   is obvious.

              SUBJECT   VERB

Although such a construction may be used to convey topical information or emphasis, when it is not used for such purposes, it will probably sound strange to most people, perhaps because it violates the criterion of light-before-heavy. In such cases, a writer might shift the heavy subject noun phrase to the end of the sentence and replace it with the pronoun *it* (this last step satisfies the requirement that all English sentences have a subject noun phrase):

It is obvious that we are in the midst of an energy crisis.

Of course, if the unit after the verb is quite long and complex itself, then the need to shift the subject is not so strong:

That we are in the midst of a pollution crisis is

              SUBJECT

obvious to everyone in the country except the big politicians.

This sentence is more balanced than the first version, does not violate our light-before-heavy guideline, and sounds better to most people.

The three situations we have discussed can be summarized as follows (where S stands for *subject* and V for *verb*):

S____V_____.   preferred

S_____ V_____.   acceptable if subject is not too long

S_____ V____.   not preferred except for special effects

25.2 WAYS OF SATISFYING THESE CRITERIA

In most cases, sentences can be constructed so that at least the first two criteria, if not all three, can be satisfied. Sometimes this procedure is straightforward. However, at other times writers may find themselves boxed in, having written a sentence that satisfies the light-heavy criterion, say, but not the other two (that is, given information before new and topical information in the subject position). The purpose of this section is to describe several of the most common ways that good writers get themselves out of such a situation by shifting around the noun phrases.

We often have a sentence such as the following, consisting of a subject noun phrase, an active verb, and some other noun phrase:

<u>Christopher Sholes</u> invented   <u>the typewriter.</u>
      SUBJECT

We can create a passive form of such a sentence by (1) interchanging the two noun phrases, (2) inserting an appropriate form of the verb *to be*, and (3) inserting the preposition *by*:

<u>The typewriter</u>   was invented by   <u>Christopher Sholes.</u>
      SUBJECT

Although these two sentences have the same meaning, they differ in word order and thus should be used in different contexts. For example, consider the following two passages.

> A typewriter is a machine that prints alphabetic characters, numbers, and other symbols. It was invented by Christopher Sholes and made the life of the secretary easier.

> Christopher Sholes was a nineteenth-century inventor who made the life of the secretary much easier. He invented the typewriter, a machine that prints alphabetic characters, numbers, and other symbols.

The second sentence of each begins with a unit saying that Christopher Sholes invented the typewriter, but this unit differs in the ordering of the two noun phrases: *The typewriter/it* and *Christopher Sholes/he*. In the first passage, the *typewriter (it)* is the subject of the second sentence because that is the given information from the first sentence. In the second passage, *Christopher Sholes (He)* is the subject of the second sentence because that is the given information.

Notice that it is the choice of the noun phrase that determines the form of the verb. If the subject noun phrase requires an active verb, then an active verb (*invented*) is used. However, if the subject noun phrase requires a passive verb (*was invented by*), then that is used. One should not always prefer active to passive verbs, as recommended by some style books. That could introduce processing trouble for the reader by moving a noun phrase out of the subject which the reader needs to tie the sentence into the context. For instance, consider the following.

Negative Example

A typewriter is a machine that prints alphabetic characters, numbers, and other symbols. Christopher Sholes invented it and made the life of the secretary easier.

In this negative example, *typewriter* is the given information in sentence 2 and the topical information (what the sentence is really about). However, this is masked by having Christopher Sholes as the subject when *typewriter* or *it* is needed to tie the second sentence to the first. Let the subject determine the verb, not the opposite.

Some sentences consist of two noun phrases which are connected by a form of the verb *to be* and which are interchangeable without any change of meaning:

Air pollution is one major form of pollution.
One major form of pollution is air pollution.

Here again the choice of which form to use depends mainly on given-new considerations. If *air pollution* is given information and we want to add new information about it (namely, that it's a major form of pollution), then the first version is preferred:

There are several things that the average person should know about air pollution. *It* [*air pollution*] *is one major form of pollution*. It exists in all large cities and most small ones. And it is a secondary cause of death among the sick and elderly.

However, if we have already been talking about different forms of pollution and now want to introduce air pollution as one such form, then *major forms of pollution* is given information and *air pollution* represents new information and should come after the verb:

There are several forms of pollution with which the average person should be familiar. *One major form of pollution is air pollution*. Other major forms of pollution are water pollution and noise pollution.

Sometimes you face a situation where you have a heavy-before-light unit somewhere toward the end of the sentence. Here again, you shift the two noun phrases and do whatever else is needed to make a more readable sentence. Consider this example:

We are sending   *your branch office in Longview*   *a copy.*

Suppose this sentence were to occur in a context where a report is being discussed:

> We have finished our final report on the new product line. We are sending your branch office in Longview a copy.

In this case, *a copy* (of the report) would not constitute much new information, since reports are customarily made in copies, but the location where one of the copies is to be sent probably does constitute new information. Furthermore, the location is a longer, more complex (i.e., heavier) element in this case. Given these considerations, a preferred version of the sentence would shift the heavier and newer information to the end of the sentence (and here you have to add *to* before *your branch office in Longview):*

> We have finished our final report on the new product line.
>
> We are sending   *a copy*   to   *your branch office in Longview.*

25.3 A PROCEDURE FOR PRODUCING MORE READABLE TEXTS

The purpose of this section is to summarize the material presented in Chapters 24 and 25 and to outline a procedure for applying it. The two chapters provide a number of criteria and suggestions for producing functional and readable sentences. By applying the criteria given—in the order they are given—to each sentence as it appears in a paragraph or text, a writer should consistently produce more readable prose.

A summary of these criteria and a procedure for applying them are given in the flowchart in Figure 25-1. To illustrate the use of this chart, let us consider this paragraph:

> The sound reproduction from a radio can be quite inferior to that from a record or tape. One of the causes of radio's inferior sound reproduction, and in some environments perhaps the major cause, is multipath distortion. This type of distortion occurs when there are tall structures or buildings in the receiver's environment. Most of the radio signal is reflected and diffracted by these tall structures before it reaches the radio's antenna. When the signal is received by the radio, the radio interprets it as distorted and can only produce an inferior sound from it. In contrast, the sound on a record or tape suffers none of this distortion.

To evaluate this paragraph and revise it where necessary, let us follow the procedure outlined in Figure 25-1. Step 1 asks us to read the paragraph, and then step 2 asks us to decide if the topic sentence (the first sentence of the passage) is adequate. The answer is "yes" here because the topic sentence introduces the main idea, that *sound reproduction from a radio can be quite*

Figure 25-1 ❏ FLOW CHART FOR EDITING SENTENCES IN PARAGRAPHS

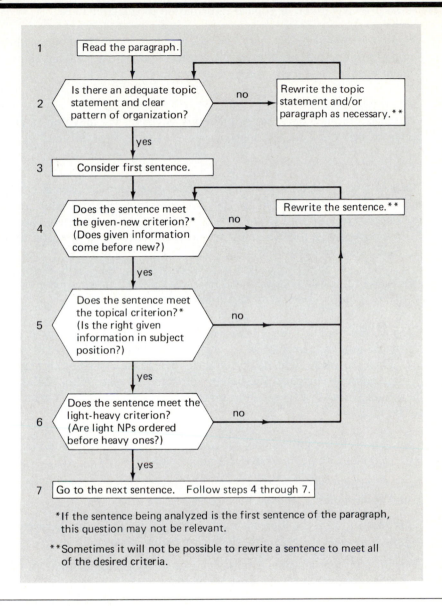

1 Read the paragraph.

2 Is there an adequate topic statement and clear pattern of organization? →no→ Rewrite the topic statement and/or paragraph as necessary.**

yes

3 Consider first sentence.

4 Does the sentence meet the given-new criterion?* (Does given information come before new?) →no→ Rewrite the sentence.**

yes

5 Does the sentence meet the topical criterion?* (Is the right given information in subject position?) →no→

yes

6 Does the sentence meet the light-heavy criterion? (Are light NPs ordered before heavy ones?) →no→

yes

7 Go to the next sentence.   Follow steps 4 through 7.

*If the sentence being analyzed is the first sentence of the paragraph, this question may not be relevant.

**Sometimes it will not be possible to rewrite a sentence to meet all of the desired criteria.

inferior to that from a record or tape, and the rest of the paragraph goes on to describe why this is so.

First Sentence

We then consider whether or not the first sentence meets the criteria outlined in steps 4 to 6.

Steps 4 and 5   It does not violate given-new or topical criteria since there is no previous given or topical information to consider here.

Step 6   It meets light-heavy criteria because the subject noun phrase (*the sound reproduction from a radio*) is not significantly heavier than the object noun phrase (*that from a record or tape*).

Second Sentence
Since the first sentence does not require any revision, we move on to step 7 and decide whether the second sentence meets the criteria outlined in steps 4 to 6.

Steps 4 and 5   It does not really violate given-new or topical criteria, although it would be desirable to get the given and topical information (*radio's inferior sound reproduction*) closer to the beginning of the sentence. We could rewrite the sentence as follows and it would then meet given-new and topical criteria.

One cause of radio's inferior sound reproduction, and in some environments perhaps the major cause, is multipath distortion.

Step 6   Revised sentence 2 does not meet light-heavy criteria at all. It has a 15-word subject noun phrase (everything before *is*) and only two words after *is*. Thus, to meet light-heavy criteria, we must rewrite the sentence again. Four possible revisions are listed below and all meet all of the criteria in steps 4 to 6. Since all of these revisions are quite similar, we shall arbitrarily choose the first.

1   One cause of this inferior reproduction is multipath distortion; in some environments, this is perhaps the major cause.

2   Radio's inferior reproduction is partly caused by multipath distortion; in some environments, this is perhaps the major cause.

3   This inferior reproduction is partly caused by multipath distortion; in some environments, this is perhaps the major cause.

4   This inferior reproduction is partly caused by multipath distortion, which is perhaps the major cause in some environments.

Third Sentence
Since the second sentence does not require any more revision, we move on to step 7 and consider whether the third sentence meets the criteria outlined in steps 4 to 6.

Steps 4 and 5   It meets given-new and topical criteria because the first noun phrase (*this type of distortion*) is given and topical information. However, since we have moved *multipath distortion* away from the end of sentence 2, we may wish to repeat *multipath distortion* at the beginning of sentence 3 as indicated. This new version meets given-new and topical criteria.

> Multipath distortion occurs when there are tall structures or buildings in the receiver's environment.

Step 6 It also meets light-heavy criteria because the subject noun phrase (now *multipath distortion*) is lighter than the other noun phrase (*tall structures or buildings in the receiver's environment*).

Fourth Sentence

Since sentence 3 does not require any more revision, we move on to decide whether sentence 4 meets the criteria outlined in steps 4 to 6.

Steps 4 and 5 It violates given-new and topical criteria, since the given and topical information (*these tall structures*) occurs after significant new information (*most of the radio signal is reflected and diffracted*). We could revise the sentence to meet given-new and topical criteria as follows:

> These tall structures reflect and diffract most of the radio signal before it reaches the radio's antenna.

Step 6 The rewritten sentence also meets light-heavy criteria.

Fifth Sentence

Since sentence 4 does not need any more revision, we move on to consider sentence 5.

Step 4 It does not violate given-new criteria in the main clause, since the previous context has been talking about both *the signal* and *the radio*.

Step 5 The sentence does violate the topical criterion, since the previous sentence introduced the *radio signal* and this is the logical topic of this sentence as well. We could revise the sentence to meet the topical criterion as follows:

> These tall structures reflect and diffract most of *the radio signal* before *it* reaches the radio's antenna. When *the signal* is received by the radio, *it* is seen as distorted and produces an inferior sound.

If we rewrite sentence 5 with *the radio* as the subject, it would violate topical criteria:

Negative Example

> These tall structures reflect and diffract most of *the radio signal* before *it* reaches the radio's antenna. When *the radio* receives the signal, *the radio* sees it as distorted and produces an inferior sound from it.

While *the radio* is mentioned in sentence 1 and thus is given information, sentence 4 introduces the topic of *the radio signal*, not *the radio*. Thus, to satisfy the topical criterion set up in sentence 4, sentence 5 should place *the radio signal* (or *the signal*) in the subject position:

> These tall structures reflect and diffract most of *the radio signal* before *it* reaches the radio's antenna. When *the signal* is received by the radio, *it* is seen as distorted and produces an inferior sound.

Step 6 Sentence 5 now also meets light-heavy criteria.

Last Sentence

Since sentence 5 does not require any more revision, we move on to consider sentence 6, which meets all of the criteria in steps 4 to 6. The revised paragraph follows. (Note that this is not the only possible revision.)

> The sound reproduction from a radio can be quite inferior to that from a record or tape. One cause of this inferior reproduction is multipath distortion; in some environments, this is perhaps the major cause. Multipath distortion occurs when there are tall structures or buildings in the receiver's environment. These tall structures reflect and diffract most of the radio signal before it reaches the radio's antenna. When the signal is received by the radio, it is seen as distorted and produces an inferior sound. In contrast, the sound on a record or tape suffers none of this distortion.

Once we have arrived at this reordered version, we can combine and cut the individual sentences to produce a more concise version of the paragraph:

> The sound reproduction from a radio can be quite inferior to that from a record or tape. One cause of this is multipath distortion; in some environments it is perhaps the major cause. This distortion occurs when there are tall structures in the receiver's environment which reflect and diffract the signal before it reaches the radio's antenna. The signal is thus distorted and can only be used to produce an inferior sound. In contrast, the sound on a record or tape suffers none of this distortion.

Note that such cutting makes sense only after the sentences are structured in a logical and appropriate order. Procedures for effectively combining and cutting sentences are discussed in Chapter 26.

❏ EXERCISE 25-1

Where necessary, reorder the noun phrases in the following passages to create more readable sentences.

A   The labor movement has existed in the United States for about 180 years. In the past 40 years, labor unions have become an integral part of the workplace. Over 19 million workers belong to labor unions. The wage rates, working hours, and working conditions of both unionized and nonunionized industries are influenced and determined by organized labor. Clearly, unions are permanent and powerful institutions in America.

B *Advanced Component Technology for the Ford Pinto* A number of systems, devices, and technologies to work with in meeting your specified requirements are available. One such device is the aneroid control. This device is used to limit the fuel while the turbocharger is building up manifold pressure during acceleration. The pressure-reducing valve is another such device. It reduces the pressure of fuel entering the carburetor. We have also developed other devices that we think will be useful in the future.

C Several recent developments have encouraged a reevaluation of wind-powered vessels for the American merchant fleet. The sharp rise in the cost of energy required to drive a fuel-powered ship is the most significant of these developments. A second development is the improvement of technology associated with wind-powered ships.

D *General Appraisal of the EPA Plan* The plan prepared by the Environmental Protection Agency (EPA) lacks several essential characteristics. It does not clearly outline program priorities nor does it relate priorities to overall program goals. The planning process is vague, and no guidelines are offered for future updates of the plan. It is difficult to see a rationale for the major thrusts suggested in the budget. For example, the plan offers no basis for the dominant expenditure on developing control technology over the 5-year period covered by the plan.

E *Accomplishments of the Massachusetts Institute of Technology's Sea Grant Program* The sea grant program estimated the environmental effects which would be associated with oil exploration on the Georges Bank, off the New England coast. Among other matters, we reviewed what is now known about the biological effect of oil, particularly its toxicity. We analyzed how long a Georges Bank oil spill would be likely to stay on the bank and where it would be likely to go. Available statistics on oil spills were reviewed, and the amounts of oil which might be spilled in New England under a number of developmental hypotheses were estimated. Also the effect of offshore drilling and of possible oil spills on the fishing industry were estimated. Finally, we made a preliminary analysis of the loss in regional income which would be associated with oil spills as a result of cleanup costs.

F In the 1950s, fundamental changes in finance began to occur. The analytical methods and techniques traditional to economics began to be applied to problems in finance, and the resulting transformation has been significant. A change in the focus of the literature accompanied this evolution. The focus of the literature on normative questions such as "What should investment, financing, or dividend policies be?" began to shift to positive theories addressing questions such as "What are the effects of alternative investment, financing, or dividend policies on the value of the firm?" This shift in research emphasis was necessary to provide the scientific basis for the formation and analysis of corporate policy decisions.[1]

REFERENCES

1 Adapted from M. Jensen and C. Smith (eds), *The Modern Theory of Corporate Finance* (New York: McGraw-Hill, 1984), pp. 2–3.

ADDITIONAL READING

Mary Ann Eiler, "Thematic Distribution as a Heuristic for Written Discourse Function," in *Functional Approaches to Writing Research Perspectives*, edited by Barbara Couture (Norwood, NJ: Ablex, 1986), pp. 49–68.

Lester Faigley and Stephen P. Witte, "Topical Focus in Technical Writing," in *New Essays in Technical and Scientific Communication*, edited by Paul V. Anderson, R. John Brockmann, and Carolyn R. Miller (Farmingdale, NY: Baywood, 1983), pp. 59–68.

C. R. Fletcher, "Markedness and Topic Continuity in Discourse Processing," *Journal of Verbal Learning and Verbal Behavior*, 23:489–493 (1984).

Talmy Givon, "Topic Continuity in Discourse: An Introduction," in *Topic Continuity in Discourse: A Quantitative Cross-Linguistic Study*, edited by Talmy Givon (Amsterdam: John Benjamins, 1983).

M. A. K. Halliday and Ruqaiya Hasan, *Cohesion in English* (London: Longman, 1976).

Barbara Hayes-Roth and Perry W. Thorndyke, "Integration of Knowledge from Text," *Journal of Verbal Learning and Verbal Behavior* 18:91–108 (1979).

Ronald Langacker, "Movement Rules in Functional Perspective," *Language 50*:629–664 (1974).

Charles Li (ed.), *Subject and Topic* (New York: Academic Press, 1976).

C. A. Perfetti and S. R. Goldman, "Thematization and Sentence Retrieval," *Journal of Verbal Learning and Verbal Behavior 13*:7–79 (1974).

26

Editing for Emphasis

Although some readers may prefer to skim-read, others have to read more thoroughly, concentrating on details. For these readers, there is a danger of getting lost in the details, of overlooking main points and "not seeing the forest for the trees." When this happens, the reader fails to see how the details fit into the larger picture and is unable to evaluate their importance; ultimately, the reader bogs down and is forced to start over.

When readers get bogged down like this, it's often the writer's fault. Many writers make little effort to organize details in a coherent, unified way, preferring instead to have the reader do all the work. But this invites the kind of failure just described. Many readers are pressed for time, or tired, or have other things on their mind, or lack the writer's background knowledge. Still others have poor reading techniques and cannot understand poor writing, no matter how hard they work at it. In general, readers are at the mercy of the writer; they depend on the writer to present details so that they reinforce main points in a unified fashion.

This is somewhat similar to the demands made on a speaker in a serious face-to-face conversation. Conversation is an intense form of communication in which the speaker is acutely aware of the listener and vice versa. Because of this close speaker-listener relationship, conversations are governed by certain unwritten rules: say what you mean, get to the point, be honest, etc.[1] If the speaker violates any of these rules, the conversation will begin to break down unless the listener rescues it with a corrective comment, such as "I don't see what you're driving at" or "What's your point?" Most listeners are simply intolerant of irrelevant details; they either intervene or break the conversation off if the speaker strays too far from the topic. This forces the speaker to make every detail relevant—and relevant in a fairly obvious way.

469

Of course, good conversationalists are aware of the unwritten rules, and they employ various techniques to make it clear to the listener that they are observing these rules. For one thing, they use emphatic intonation, physical gestures, inverted sentence structure, intensifiers, or other devices to signal important words—key words, topical words, words carrying new information. Conversely, they use none of these devices for the less important words—those that carry redundant information—and they omit empty, meaningless words that serve no communicative purpose at all. In general, both by giving prominence to important words and by subordinating or omitting unimportant ones, good conversationalists emphasize those aspects of a detailed discussion that link the discussion to the main point or purpose of the conversation. As a result of this emphasis, the listener not only absorbs those details but also sees just how they support the main point.

Writers should do the same kinds of things as good conversationalists. They may not be in as close touch with their audience, they may not have the possibility of immediate feedback, and they cannot use intonation and gestures in their writing. But writers do have at their disposal many devices for helping the reader see how the details support the main points. And readers do need this kind of help. Thus, it is as appropriate and necessary for writers to use emphasis as it is for conversationalists.

This chapter describes several ways in which you can build emphasis into your writing: (1) combine closely related sentences, (2) use signal words, (3) be concise, and (4) use physical highlighting. In most circumstances, following these stylistic guidelines should help make your ideas more accessible to more readers. Bear in mind, though, that guidelines are merely suggestions, not rules; they should not be considered infallible. Stylistic choices involve not only readability but also sociocultural acceptability, that is, conformity to the ways in which groups of people prefer to express themselves—and sometimes these two goals are somewhat in conflict. For example, in a study of two *Fortune 500* companies, Robert L. Brown, Jr. and Carl G. Herndl found that the most respected writers in these companies sometimes used wordy expressions like *undertake an examination* instead of more concise alternatives like *examine*.[2] Apparently it was common, accepted practice in that particular corporate culture for employees to express themselves that way. The more concise *examine* might be more readable, but *undertake an examination* would help identify the writer as a member of the group. In general, when making stylistic choices, do not follow guidelines blindly. Expressing an idea in a certain way may have not only some benefits but also some undesirable side effects. Expressing the same idea in a different way may have fewer (or more) benefits but also fewer (or more) undesirable side effects. Keep this trade-off relationship in mind and strive for optimization, not perfection.

26.1 COMBINE CLOSELY RELATED SENTENCES

Combine closely related sentences unless there is a compelling reason not to (such as maintaining independent steps in a list of instructions or avoiding extreme sentence length); put main ideas in main clauses. Many inexperienced writers have a tendency to use nothing but short, simple sentences. Such sentences are easier to write and, in general, easier to read than long, complex ones. However, constant use of short, simple sentences produces a choppy, irritating style of writing that fails to give sufficient emphasis to important ideas. In any piece of writing, some ideas are going to be more important than others and some ideas are going to be more complex than others. In writing sentences to capture such ideas, you should try to let the *form* of the sentence reflect the *content* of the sentence. If an idea is simple and straightforward, a simple sentence may be the best way to present it. If an idea is complex, a complex sentence may be needed to capture it. Main ideas are usually best presented in main clauses. Secondary or modifying ideas often go best in modifying clauses or phrases.

In writing a first draft, you may find it easiest to stick mainly to simple sentences. But as you edit your writing, you may notice that some of these simple sentences represent closely related ideas, with one modifying another. In such cases, you should try to combine these sentences, so that the form of your writing reflects the content of your ideas. These stylistic decisions are *strategic*: they can be made properly only within the context of an entire paragraph or text. For example, consider the following excerpt from a student's response to Situation 13–1, Assignment 1 [sentence numbers have been added]:

Draft Version

We have established a Communications Systems (CS) Planning Team [1]. This team includes user representatives from each of the 12 departments in the Forever-Seal Division [2]. We hold regular meetings to assess state-of-the-art communications equipment [3]. We discuss how new products are being used by other companies [4]. At these meetings, the user representatives describe their future needs [5]. They also work out their requirements for the new communication network [6].

In reviewing his draft paragraph, this student writer noticed that he had produced a string of simple sentences that were not grouped together in any particular way. The six sentences appeared to represent six independent ideas. As the writer of this text, he knew that some of the ideas in it were closely related—specifically, 1–2, 3–4, and 5–6. Since he wanted his readers to be able to see this relatedness as quickly and easily as possible, he decided to revise the paragraph by combining these pairs of sentences.

He began with sentences 1 and 2. The first sentence contained the main idea of the paragraph; the second sentence simply added some detail to it. In other words, the second sentence was logically subordinate to the first. So he kept sentence 1 in the form of a main clause and converted sentence 2 into a modifying clause:

We have established a Communications Systems (CS) Planning Team which includes user representatives from each of the 12 departments in the Forever-Seal Division.

Next, he turned his attention to sentences 3–4. The writer felt that these two sentences were also closely related, but as equals, not as one subordinate to the other. So he combined these sentences using a coordinating conjunction (*and*):

> We hold regular meetings to assess state-of-the-art communications equipment and to discuss how new products are being used by other companies.

Finally, he inspected sentences 5 and 6. Here too there was a logically coordinate relationship between the two ideas. So, as with sentences 3–4, he used the connective *and* to combine these two sentences:

> At these meetings, the user representatives describe their future needs and work out their requirements for the new communication network.

The result of these sentence-combining operations was a revised paragraph that would enable readers to see more quickly the logical relationships among the writer's ideas. Instead of six simple and independent ideas, the paragraph now depicted three sets of more complex ideas. Readers would now be able to see the architecture of the writer's thinking more easily.

Revised Version

We have established a Communications Systems (CS) Planning Team which includes user representatives from each of the 12 departments in the Forever-Seal Division. We hold regular meetings to assess state-of-the-art communications equipment and to discuss how new products are being used by other companies. At these meetings, the user representatives describe their future needs and work out their requirements for the new communication network.

In general, combining sentences is often a good way to create emphasis in your writing. By making it easy for your readers to see the relatedness of ideas, you make it easier for them to absorb these ideas. You can also show explicitly that one idea is logically subordinate to another by putting the more important idea in the main clause of the sentence and the less important idea in a subordinate clause. For example, suppose you wanted to combine the two sentences in italics in the following paragraph:

Negative Example

Electric cars must be able to meet the same safety standards that gasoline cars must meet as set up by the Department of Transportation. *These standards are derived from an established crash test. In the crash test, the car is propelled against a solid wall at 30 mph.* The data obtained from the crash test are analyzed for fuel spillage, fuel system integrity, windshield retention, and zone intrusion.

In combining the two italicized sentences, we could subordinate the more detailed second sentence to the more general first one:

These standards are derived from an established crash test, in which the car is propelled against a solid wall at 30 mph.

Alternatively, we could maintain prominence on the details and subordinate instead the idea that the crash test is an established one:

These standards are derived from propelling the car against a solid wall at 30 mph, which is an established crash test.

Clearly, the first option is more appropriate in this context: the fact that the crash test is established underscores the main idea stated in the topic sentence.

Revised Version

Electric cars must be able to meet the same safety standards that gasoline cars must meet as set up by the Department of Transportation. These standards are derived from an established crash test, in which the car is propelled against a solid wall at 30 mph. The data obtained from the crash test are analyzed for fuel spillage, fuel system integrity, windshield retention, and zone intrusion.

There are times when it is best *not* to combine sentences. For example, if you are giving a list of instructions and want to emphasize the independent steps that the user should take to carry out the instructions, you might want to state these steps in independent sentences. For instance, consider the following instructions for replacing a brake line in an automobile:

1   Disconnect the union nuts at both ends.

2   Unclip the line from the chassis.

3   Pull the line out.

4   Install the new line in the chassis clips.

5   Moisten the ends in brake fluid.

6   Tighten the union nuts.

You could leave this set of instructions as is, in the form of a formatted list. Or you could combine some of the steps (2 with 3, 5 with 6) to create a more realistic four-step sequence of disconnect-remove-install-reconnect, as is done in this excerpt from a repair manual:

To replace a brake line, disconnect the union nuts at both ends. Unclip the line from the chassis and pull it out. Install the new line in the chassis clips. Moisten the ends in brake fluid, then tighten the union nuts.[3]

To combine sentences beyond this, however, would be a mistake because it would destroy the emphasis we want to maintain on certain individual steps. For example, if we were to combine sentences 2 and 3 in the repair manual version, this would be the result:

Negative Example

To replace a brake line, disconnect the union nuts at both ends. Unclip the line from the chassis, pull it out, and install the new line in the chassis clips. Moisten the ends in brake fluid, then tighten the union nuts.

By lumping together the remove and install steps like this (*Unclip the line from the chassis, pull it out, and install the new line in the chassis clips*), we would be creating an imbalance in the sequence: no mechanic would consider this to be a single step, as the form of the description implies.

It's also best not to combine sentences when the result would be too long a sentence. Suppose, for example, you have been writing a proposal for a computer-aided design system and have included this paragraph in your summary:

> The proposed system is required to alleviate the increase in demand. The system will do that by removing the burden of data entry from the present system, CADDS. This is accomplished by using the microcomputer as a stand-alone data entry system. The microcomputer has all of the graphics and software capabilities required to implement this concept.

As it stands, this paragraph is a nicely written one, with an adequate topic statement, a clear general-to-specific pattern of development, and properly constructed sentences satisfying the given-new, topical, and light-heavy criteria discussed in Chapters 24 and 25. The result is a highly readable paragraph with appropriate emphasis on the main ideas and key words. If you were to combine the four sentences into one, on the other hand, much of this emphasis would be destroyed:

Negative Example

> The proposed system is required to alleviate the increase in demand by using the microcomputer as a stand-alone entry system with all the necessary graphics and software capabilities to remove the burden of data entry from the present system, CADDS.

This is a more economical version, since it contains 16 fewer words than the original. But is it more readable? Obviously not. In fact, it's a perfect example of the kind of incomprehensible gobbledygook that so many readers of technical writing complain about. This example illustrates the following point: Do not combine sentences just for the sake of combining; do it only when it serves a purpose.

❏ EXERCISE 26-1

The following paragraphs lack flow and emphasis because the sentences are short and choppy. Break up this choppiness by combining the most closely related sentences. Be sure to put main ideas in main clauses.

A The Cleveland Engine Plant will begin making diesel engines in 1992. The diesel engines will be produced along with the gas engines already in production. However, there is only one area for machining pistons. Both gas pistons and diesel pistons must be machined there.

B   I recommend using the cart system to reduce the costly existing production layout for machining gas pistons and diesel pistons. The cart system is the least expensive of the three systems. Also, the cart system meets the desired production rate for all probable product mixes. The cart system is the most compact of the three systems and fits in the existing piston machining area. The cart system layout should be used in the piston machining area.

C   The maximum speeds of the electric cars are much lower than those of the gas cars. The maximum speeds of the electric cars are barely fast enough to maintain freeway speeds. This causes problems with utilization of electric cars on freeways, and it almost limits their use to residential streets. The maximum speed must be increased to give electric cars a better chance for utilization.

26.2 USE SIGNAL WORDS

Emphasis is lost when the reader must figure out how the details and main points are related; emphasis is created when this relationship between main points and details is established in an obvious way. So, make the connection between your main points and the supporting details *obvious*. If this means taking sides in an issue, so be it. In many situations, readers will expect you to take a stand and make clear recommendations. This does not mean that you should abandon your objectivity. It means that once you have used your objectivity to assess the details and reach a certain conclusion, you should then make every effort to present that conclusion and write toward it. Present the details in such a way that the reader can easily see what your conclusion is and how the details support that conclusion.

How can you do this? In addition to a number of techniques already discussed—such as using good topic statements, putting topical words in the subject position, and combining sentences—you can use appropriate signal words and phrases at the beginning of sentences: signals like *specifically, in general, conversely, furthermore, on the other hand, however, therefore* and *since, because, although, despite*. (Many of these are discussed in connection with the paragraph patterns described in Chapter 22. Users of *Technical Writing and Professional Communication for Nonnative Speakers of English* will find additional helpful material in Chapter 34.) Longer phrases like *for this purpose, with this in mind, before taking any further steps*, and *in order to determine the market for this product* can also help guide the reader.

To see the effect of using signal words and phrases, consider first this negative example:

Negative Example
The more-complete-expansion cycle is more efficient than the conventional diesel cycle. It uses the same four strokes but a relatively smaller combustion chamber. The intake valve closes when the piston is at some point A between TDC and

BDC in its intake stroke. This draws in less air-fuel mixture. The compression, power, and exhaust strokes are identical to those of the diesel cycle. The increase in volume of the air-fuel mixture due to the piston moving to BDC is the same. The temperature and pressure of the gases after the power stroke is over are lower. More of their energy has been utilized. This increases the efficiency of the cycle. The smaller combustion chamber is necessary. The temperature and pressure at TDC in the compression stroke must be high enough to cause combustion of the mixture. The mixture must be squeezed into a smaller volume to get the same temperature and pressure.

It's pretty hard to make sense of this, isn't it? Certainly there's little emphasis on main points. However, notice what happens when we insert a few signals:

Revised Version

The more-complete-expansion cycle is more efficient than the conventional diesel cycle. It uses the same four strokes but a relatively smaller combustion chamber. *In its intake stroke*, the intake valve closes when the piston is at some point A between TDC and BDC, *thus* drawing in less air-fuel mixture. The compression, power, and exhaust strokes are identical to those of the diesel cycle. *However, since there was less mixture to start with and since* the increase in volume due to the piston moving to BDC is the same, the temperature and pressure of the gases after the power stroke is over are lower. More of their energy has been utilized, *thus* increasing the efficiency of the cycle. The smaller combustion chamber is necessary *because* the temperature and pressure at TDC in the compression stroke must be high enough to cause combustion of the mixture. *Since there is less mixture in the chamber*, the mixture must be squeezed into a smaller volume to get the same temperature and pressure.

Clearly, this rewritten version is far more comprehensible than the original. The signal words and phrases help us see the logic of the paragraph—how the sentences work together to explain how the more-complete-expansion cycle is more efficient than the diesel cycle.

Another way to highlight details and main points is to use descriptors like *significantly, completely, easily, obviously, only, much,* and *many.* Nonspecialists often need the kind of guidance that these descriptive signal words provide. Stubbornly "letting the facts speak for themselves" may work with specialist readers—who understand the full import of those facts—but it often does not work with nonspecialist readers. For example, if you say "The present worth of the hydrogen process is $4.14 million," you are assuming that the reader will know how to evaluate that figure. Is $4.14 million a lot for such a process or isn't it? Many readers wouldn't know. But if you say "The present worth of the hydrogen process is only $4.14 million," you are helping the reader to arrive at an interpretation. The word *only* may be small but it can have a big effect on interpretation. Of course, powerful descriptors like these should be used only when the occasion calls for them, and they should never be used inaccurately: as a careful professional, you do not want to be guilty of exaggeration.

❏ EXERCISE 26-2

Pretend that you are the fleet manager of a taxicab company. Your profit margin was much lower than expected last year because your costs soared while revenues remained about the same. In light of this problem, several months ago the company president had you purchase and field-test several new front-wheel-drive automobiles. You kept careful records of each car's characteristics and performance, as summarized in the following table:

CHARACTERISTICS AND PERFORMANCE OF NEW CARS TESTED

Car	Fuel Cost per Mile	Turning Circle (Feet)	EPA Volume Index	Cost per Mile	Purchase Price
Old fleet average	10.0¢	41.6	150	4.6¢	$ 7,800
Buick Skylark	8.8¢	33.5	135	4.3¢	$ 9,100
Plymouth Horizon	8.4¢	32.8	121	4.8¢	$ 8,200
Volkswagen Dasher	7.9¢	33.2	129	4.3¢	$10,100

*The EPA volume index is a measure of relative roominess and cargo capacity.

Write the Summary section of a short informal report to the President describing these test results and recommending that the company replace its present fleet with one of these new models. Be sure to use signal words and intensifiers to emphasize your point of view.

❏ EXERCISE 26-3

The following writing samples fail to give proper emphasis to important points. Correct this deficiency by providing topic statements or summarizing statements, by combining sentences, by inserting intensifiers or signal words, by eliminating unnecessary words, and/or by making other appropriate changes.

A   The fuel cost for the electric car is much lower than that for the gas car. When the price of battery replacement is added, the fuel cost rises considerably. For an average distance of 10,000 miles/year, the total fuel cost of the electric car is about 33% greater than that of the gas car. This amount may seem to be significant, but gas prices are rising faster than electricity prices. Also, the price of battery replacement should go down slightly as batteries improve. The total fuel cost poses a problem for utilization at the present time, but it could be alleviated in the near future.

B   *Labor Cost*   The same number of workers is not used for each layout, but each layout uses six machine operators. Four extra workers are needed to operate the forklifts in the process layout. The workers make $10 per hour. Based on a 40-hour workweek, 50 weeks a year, the extra workers cost $20,000 per worker per year. No extra workers are needed in the product layout, but its production rate is well below that needed to meet forecasted demand. More worker-hours will be needed to meet production. This would require a second shift or overtime to be used. Overtime is less costly than running a second shift to meet required demand (see Appendix B [not included] for the calculations). There is no extra labor cost in the cart system since it is fully automated. The labor costs associated with each layout are given below. All three include the labor cost of the six machine operators.

Layout	Labor Cost
Product layout	$228,000
Cart system	120,000
Process layout	200,000

26.3 BE CONCISE

Be concise. Although the more important words and phrases of a text should be emphasized, the less important ones should be subordinated—or perhaps eliminated altogether. Unnecessary words and phrases will only dilute the emphasis you've carefully built up through the use of combined sentences and signal words. A wordy style can blur your message, making it hard for readers to see your main points and to be convinced by them. In fact, wordy language is more likely than not to turn readers against you. Recall the Macintosh survey cited in Chapter 1 of the 182 company executives, senior scientists, project leaders, and other prominent technical people who were asked to evaluate the writing they read. All 182 complained about "generally foggy language." More specific complaints were "failure to connect information to the point at issue" (169 respondents), "wordiness" (164), and "lack of stressing important points" (163).

Inexperienced writers sometimes think that they must use a wordy, pretentious style to create a professional image or to enhance their image as experts in their field. Actually, what evidence there is suggests just the opposite: pretentious, wordy language is less likely to promote one's credibility as an expert than is concise, direct, simple language. For example, in a survey[4] taken at a conference of the British Ecological Society, 74 scientists were asked to read two versions of the same information; one version (attributed to a person named "Brown") was noticeably wordier than the other (by a person named "Smith"):

Brown's Version
In the first experiment of the series using mice it was discovered that total removal of the adrenal glands effects reduction of aggressiveness and that aggressiveness in adrenalectomised mice is restorable to the level of intact mice

by treatment with corticosterone. These results point to the indispensability of the adrenals for the full expression of aggression. Nevertheless, since adrenalectomy is followed by an increase in the release of adrenocorticotrophic hormone (ACTH), and since ACTH has been reported (P. Brain, 1972) to decrease the aggressiveness of intact mice, it is possible that the effects of adrenalectomy on aggressiveness are a function of the concurrent increased levels of ACTH. However, high levels of ACTH, in addition to causing increases in glucocorticoids (which possibly accounts for the depression of aggression in intact mice by ACTH), also result in decreased androgen levels. In view of the fact that animals with low androgen levels are characterized by decreased aggressiveness the possibility exists that adrenalectomy, rather than affecting aggression directly, has the effect of reducing aggressiveness by producing an ACTH-mediated condition of decreased androgen levels.

Smith's Version

The first experiment in our series with mice showed that total removal of the adrenal glands reduces aggressiveness. Moreover, when treated with corticosterone, mice that had their adrenals taken out became as aggressive as intact animals again. These findings suggest that the adrenals are necessary for animals to show full aggressiveness.

But removal of the adrenals raises the levels of adrenocorticotrophic hormone (ACTH), and P. Brain found that ACTH lowers the aggressiveness of intact mice. Thus the reduction of aggressiveness after this operation might be due to the higher levels of ACTH which accompany it.

However, high levels of ACTH have two effects. First, the levels of glucocorticoids rise, which might account for P. Brain's results. Second, the levels of androgen fall. Since animals with low levels of androgen are less aggressive, it is possible that removal of the adrenals reduces aggressiveness only indirectly: by raising the levels of ACTH it causes androgen levels to drop.

Not surprisingly, the vast majority of the scientists (88%) found Smith's version easier to read. They also found Smith's style "more appropriate" (58% versus 8% for Brown, with 33% finding both styles equally appropriate). Most significantly, however, those who were willing to judge the authors' probable scientific competence on the basis of their writing styles clearly favored Smith, as can be seen in Table 26-1. A larger sample of 338 other respondents showed similar results.

Smith's style is more concise—155 words versus 179 for Brown's—and was definitely preferred by the scientists, to the point of seeing Smith as probably a better scientist. What is it, aside from the better formatting, that makes Smith's style so appealing? First, it's not so "noun-heavy": it has a higher percentage of less wordy verbs and adjectives. For example, instead of saying *effects reduction of*, Smith simply says *reduces*. Instead of *point to the indispensability of the adrenals*, Smith has *suggest that the adrenals are necessary*. Instead of *producing . . . a condition of decreased androgen levels*, Smith has *causes androgen levels to drop*.

Second, the Smith style has simpler sentence structure, with fewer and shorter phrases before the sentence subject. This means that the reader

Table 26-1 ❑ RESPONSES FROM 74 SCIENTISTS JUDGING THE SMITH AND BROWN ABSTRACTS

	Yes, Smith (%)	Yes, Brown (%)	No Difference (%)
Does one author give the impression of being more competent?	31	12	57
Does one author seem to have a better organized mind?	73	3	23
Does one author seem to be more objective?	24	8	68
Does one author inspire more confidence in what is being said?	44	15	41
Does one author seem to have a more dynamic personality?	46	7	47

SOURCE: *Bulletin of the British Ecological Society* 9(3):8 (1978).

reaches the main verb of the sentence sooner, making it easier to process the sentence as a whole (see Chapter 25 for a detailed explanation). In fact, the average number of words preceding the main verb in Smith's abstract is only 6.8 versus 17.6 for Brown's. In the Brown version, the third sentence (beginning with *Nevertheless*) has 32 words before the main verb, and the next sentence has 26!

Third, the Smith style avoids unnecessary technical terms in favor of more commonplace equivalents, even when more words are required. In place of *adrenalectomised mice*, for example, Smith has *mice that had their adrenals taken out*; instead of *are a function of*, Smith has *are due to*. Finally, Smith's style uses more pronouns and demonstratives (*this, that, these, those*): *their* in sentence 2, *these* in sentence 3, *this* and *it* in sentence 5, and *it* in the last part of sentence 9. By contrast, Brown's style has only the one demonstrative *These* leading off sentence 2. Pronouns and demonstratives, in general, help make a text more cohesive—provided, of course, that the reader sees what they refer to.

This last point deserves some discussion. Technical people, managers, and other white-collar workers sometimes use full noun phrases repeatedly to avoid being "imprecise." They've heard of cases, perhaps, where a single misinterpretation of a pronoun by a single reader has led to some accident or mishap, which in turn has led to the writer's company being sued for damages. Therefore they tend to avoid pronouns and demonstratives altogether, pre-ferring instead to repeat full noun phrases over and over. This strategy is certainly a safe one, and indeed it should be used in appropriate circumstances (such as when writing operating instructions for a potentially hazardous machine or when writing a legally binding contract). There are many circum-stances, however, where such caution is uncalled for, and where in fact it simply disrupts the coherence of the text. Consider this example:

Negative Example

In order to keep from delaying the construction phase of the Parkway Office Building, the Geotechnical Division needs to know the loads that will be placed upon the footings. I have investigated the proposed use of the structure and various flooring systems to determine the loads that will be placed upon the footings. This report presents the loads on the footings and explains how these loads were derived.

There is no reason to *describe* the loads every time they're *referred* to! Pronouns and demonstratives can be used instead without real risk of misinterpretation, and the result will be a more coherent and concise text:

Revised Version

In order to keep from delaying the construction phase of the Parkway Office Building, the Geotechnical Division needs to know the loads that will be placed upon the footings. I have investigated the proposed use of the structure and various flooring systems to determine these loads. This report presents the loads and explains how they were derived.

In general, when you have to refer repeatedly to some object or concept that has first been introduced with a long noun phrase, you can usually use a shortened version of the phrase and a demonstrative or definite article without much, if any, risk of ambiguity. Here is a typical example:

Damper position indicators shall be provided on all damper units. *These indicators* shall be located a minimum of 1'-0" from the breeching to allow for installation of thermal insulation.

It would also be safe to substitute *The indicators* in this context. (Note, however, that the demonstrative *These* by itself could be misinterpreted as referring to the *damper units*.) It would not be necessary to repeat the full noun phrase *damper position indicators*.

Sometimes, however, it is necessary to repeat full noun phrases to ensure proper interpretation. Consider this example:

Negative Example

The two dampers installed in the flues shall be tied with the inlet dampers of the fans located at the exit side of the electrostatic precipitators. *These dampers* will open automatically in the event of any failure in the electrostatic precipitator system. Contractor shall provide proper instrument and control for the actuation of *these dampers.*

Now, what does *these dampers* refer to? The flue dampers? The inlet dampers? Or both? In cases like this, the writer has little choice but to repeat the full noun phrase. In fact, it is often important in such cases for the writer to do so *with exactly the same wording*, especially if the paragraph is long and complex. For instance, see if you can keep track of the different orders involved in the following:

Negative Example

On July 8, 1989, I was given the assignment of determining a method to reduce the problems associated with *lost unit down orders* and *special orders*. The misplacement of *special orders* in the shipping and receiving area has caused many customers to become unsatisfied with our services. This situation has also caused our partsmen to spend many hours each day locating *lost orders*. These problems have arisen from the increase in the number of *special orders* waiting to be picked up in our Shipping and Receiving Department. Currently, there are between 200 and 300 *orders* awaiting customer pickup. *These parts* are stored on shelves, on the Shipping and Receiving room floor, and outside of the building. The Shipping and Receiving Department has no records at this time concerning which *special orders* have arrived. To correct this situation a system is needed which reduces employee work. This system must also minimize the number of *lost orders* at a low cost.

Do you know for sure what *lost orders* refers to in the third sentence? (Is it just *lost special orders* or is it *all* lost orders?) How about the *orders* in sentence 5? Or *These parts* in sentence 6? We didn't know how to interpret these terms when we first read the passage, and we still don't.

❏ EXERCISE 26-4

The verb phrases in column A are representative of those that characterize a wordy, noun-heavy style of writing. Use column B to provide one-word equivalents. (The first two have been done for you.)

A	B
effect a reduction in	reduce
accomplish a modification of	modify
put emphasis on	_____
come to the conclusion that	_____
provide with information	_____
increase by a factor of two	_____
give an explanation of	_____
have a deleterious effect upon	_____
create an improvement in	_____
do an analysis of	_____
make a recommendation that	_____
conduct an investigation of	_____

❑ EXERCISE 26-5

Often there is a slight difference of meaning or emphasis between a verb phrase like *effect a reduction in* and a one-word equivalent like *reduce*. Sometimes an expert writer will deliberately use the longer form in order to capture this slight difference. For example, the distinguished essayist E. B. White begins one of his letters with this sentence: "I think it might be useful to stop viewing fences for a moment and take a close look at Esquire magazine's new way of doing business."[5] Presumably White chose to use *take a close look at* instead of the shorter *look closely at* because he wanted to give the idea a little more emphasis.

Review the verb phrases in Exercise 26-4 and, for each, think of a situation where the longer version might be more effective than the shorter one.

❑ EXERCISE 26-6

The writing samples given below can all be made much more concise. Rewrite them, eliminating all unnecessary words.

A At this point in time we are engaged in a reevaluation of the budget.

B The design that is recommended for adoption would accomplish the removal of particulates with the desired degree of efficiency.

C According to our estimates, the cost of the system would be on the order of approximately $1.2 million.

D It is the purpose of this report to provide you with information about the current status of the no-smoking experiment.

E In our judgment, the optimal course of action would be to maintain continued use of the cycle of first testing and then revising.

F My telephone number is 221-3600 if you would like to contact me about any questions you might have pertaining to the contents of this report.

G At our meeting of 9/4/89 we had a discussion about the fact that people have been making numerous complaints concerning vertical movement in the Hudson wing. The problem at hand was the excessive wait time people have been experiencing while waiting for the Hudson elevator.

26.4 USE PHYSICAL HIGHLIGHTING

You can emphasize important words and other units by marking or highlighting them with bold or italic type, larger or smaller type, or formatting such as indentation or lists. However, you want to be sure that your use of such devices is logical and signals some important difference between the marked text and the normal text: that the form you use for your text reflects and reinforces its meaning. For instance, the headings in this chapter are bold and larger than the normal text because they are especially important: they summarize key concepts and provide high-level organization. Italics are used to indicate quoted phrases from the sample sentences or paragraphs, or to provide an expanded summary of the key concepts in the headings. Formatting in the form of indentation and smaller size is used to mark sample passages and to set them off from the discussion in the normal text. Finally, a type of indented list is used in the exercises to set off the individual items. All of these are examples of marking special features of the text with special forms.

There are two problems associated with using such forms of marking:

1 Deciding when something is important enough to be highlighted

2 Deciding how you will highlight it (what form of marking you will use)

The first problem occurs because there is sometimes a tendency to overdo the highlighting once you get started. For example, you may see that you have made lots of things bold and created lots of indented lists. This is problematic because if everything is bold or indented, it is hard to see the few things that are especially important. Thus, you should think carefully about what things are most important and only use highlighting to emphasize those points.

The second problem can occur if you have not decided before you write what uses you will have for each type of highlighting. This problem can be solved either (1) by deciding before you write how you will use each type of marking or (2) by writing first, then deciding what kinds of things you need to highlight and how to best allocate your highlighting devices, and then editing your text to create logical and consistent highlighting.

REFERENCES

1 H. P. Grice, "Logic and Conversation," in *Syntax and Semantics 3: Speech Acts*, edited by P. Cole and J. Morgan (New York: Academic Press, 1975), pp. 41–58.

2 Robert L. Brown, Jr., and Carl G. Herndl, "An Ethnographic Study of Corporate Writing: Job Status as Reflected in Written Text," in *Functional Approaches to Writing: Research Perspectives*, edited by Barbara Couture (Norwood, NJ: Ablex, 1986), pp. 11–28.

3 *Volkswagen Service-Repair Handbook* (Los Angeles: Clymer, 1971), p. 235.

4 Christopher Turk, "Do You Write Impressively?" *Bulletin of the British Ecological Society* 9(3):5–10 (1978).

5 E. B. White, "Letter to the Editor of the Ellsworth (Maine) *American* — North Brooklyn, January 1, 1976," in *Letters of E. B. White*, edited by Dorothy Lobrano Guth (New York: Harper & Row, 1976), p. 657.

ADDITIONAL READING

Richard A. Lanham, *Style: An Anti-Textbook* (New Haven and London: Yale University Press, 1974).

Jack Selzer, "What Constitutes a 'Readable' Technical Style?" in *New Essays in Technical and Scientific Communication: Theory, Research, and Criticism*, edited by P. Anderson, C. Miller, and J. Brockmann (Farmingdale, NY: Baywood, 1982).

Joseph Williams, *Style: Ten Lessons in Clarity and Grace*, 3rd ed. (Glenview, IL: Scott, Foresman, 1989).

27

Choosing Appropriate Words

Words are the "nuts and bolts" of communication. Words convey both basic meanings and shades of meaning. English is especially powerful in its ability to capture shades of meaning. Thanks to its hybrid development, English has perhaps the richest vocabulary of any language on earth. Over the past 15 centuries, it has drawn on Latin, Greek, German, French, and many other languages to develop a vocabulary which is now estimated to surpass one million words. As a result, for any basic concept there are usually different ways of referring to it, each carrying a slightly different nuance of meaning or a slightly different tone. The words you select reveal much about how you see the subject and about how you see yourself; they project a "voice." If you choose your words carefully, you will greatly enhance your communication. If you do not, you may ruin it. Even if you pay careful attention to your document format, your paragraph writing, your sentence structure, etc., you may alienate or confuse your reader if you do not pay equally careful attention to your choice of words.

There are many ways to talk about words, but we will concentrate on three that we feel are most important for technical and professional communication: *accuracy, comprehensibility*, and *tone*.

First of all, the words you use should be accurate. They should depict what it is you are talking about. If you refer to a three-dimensional polymer as a *thermoplast* rather than a *thermoset*, you cannot blame your reader for being confused. You should be accurate in your use of nontechnical words as well. For example, if you use a masculine term to refer to a group of men and women, you are being inaccurate (and sexist, too!).

Second, the words you use should be comprehensible to your reader or listener. Communication simply does not occur if your audience fails to understand what you're saying. If you are addressing a nonspecialist audience, you should try to define technical terms and avoid jargon. If you are addressing an international audience, you should avoid unusual words and idiomatic expressions.

Third, the words you use should convey the right tone. It's not only *what* you say that counts, but *how* you say it. People will judge your message—and you—partly on the "voice" that you project. In a job letter, your words should convey enthusiasm. In a letter of complaint, they should suggest a little anger or frustration. In an invitation to a company picnic, they should display some lightheartedness and maybe even a little humor.

27.1 ACCURACY

Since words are the prime carriers of meaning, it is essential that you choose words that accurately reflect your meaning. If you don't, communication can break down. To a great extent, the training you receive in school is training in *vocabulary*. Whether you study chemical engineering, biophysics, or medicine, much of your time is spent learning new words and definitions. It is part of your professional responsibility to use those words accurately. Indeed, you probably take a lot of pride in your knowledge of technical language.

The Need for Precision

Accuracy is such an important dimension of word choice that we are constantly modifying the language to this end. As the world around us changes, we make changes in our language so that we may talk accurately about it. In particular, we create new words or we modify the meanings of old words, and in so doing we increase both the accuracy and the efficiency with which we use language. For example, *download* has become a standard technical term in computer technology because it is more accurate than *transfer* and more efficient than *transfer from a larger computer to a smaller computer*. We should take delight in the fact that our language is flexible enough to let us create new words or new meanings when the situation calls for it.

It's good to remember, though, that new technical terms should be created to serve a particular purpose, to refer precisely and efficiently to some new technical concept. If you use them in situations where a more traditional, nontechnical term would serve just as well, your audience may think you are just using jargon to show off your knowledge. For example, if you were talking to a group of nonspecialists and said something like "I *interfaced* with the sales rep," your listeners might well wonder why you didn't just say "I *talked* with the sales rep." Don't use technical language just for the sake of using technical language; use it for its intended purpose.

Accuracy in word choice includes not only technical words but nontechnical words as well. If you mean *defective*, you should not say *deficient*. If you mean *imply*, you should not say *infer*. Different words almost always differ in meaning, even if only slightly. Knowing how to use words correctly will help make your communication more precise and will also enhance your credibility as a careful, well-educated professional. Thus, if you think you may be confusing one word with another, you should check your dictionary.

Avoiding Sexism

Accuracy in word choice includes accuracy and fair treatment in referring to groups of people. A major issue with this kind of accuracy today concerns the treatment of women. In the past, whenever writers or speakers used singular forms to refer generically to groups of men and women, they would normally use a masculine form: *If a student turns in an assignment late, he will be penalized for it*. Today, however, we have come to realize that this kind of linguistic bias can be seen as reinforcing an analogous social bias. And in a democratic society, such bias is unacceptable. If not all the students are male, we should not use language that suggests they are.

There are a number of ways to avoid the generic pronoun *he*. Each has certain strengths and weaknesses. Here are the main ones:

1   Avoid using a pronoun: *Any student turning in an assignment late will be penalized for it*. This is often the most elegant solution to the problem, but it requires more creativity and effort.

2   Pluralize the sentence: *If students turn in an assignment late, they will be penalized for it*. This is simple and easy to do, but it lacks the vividness that a singular form provides.

3   Use *he or she* : *If a student turns in . . . , he or she will be penalized for it*. This retains the power of the singular, and it is easy to do. But too many of these can create verbal clutter and make you look pedantic.

4   Use *they* as a "singular": *If a student . . . , they will be penalized for it*. This is becoming quite common in spoken English, but it is technically ungrammatical. You should avoid it in written English.

5   Use *he* in one paragraph and *she* in another. This retains the power of the singular, but using *she* as a generic form may confuse or irritate some readers.

There are many other words that some people might consider sexist: *workman, foreman, chairman*, etc. Here is a short list, with some suggested substitutes:

workman   →   worker

foreman   →   supervisor

chairman   →   chairperson, chair

manpower	→	workforce, personnel
stewardess	→	flight attendant
gal Friday	→	assistant

❑ EXERCISE 27-1

Convert the following expressions into nonsexist equivalents:

policeman	salesman
man-hours	man-made
old wives' tale	the common man
errand boy	lady doctor
male nurse	coed

"Correctness"

It is important to use words accurately and correctly. Keep in mind, however, that "correctness" is not a permanent, fixed condition. The correct (i.e. accurate) use of a word can change over time and from place to place. Words are like living organisms: they adapt to their environment. What's correct in a certain place or at a certain time may not be correct in some other place or some other time. Just think of how the word *elevator* is used in the United States to describe the same contraption that is called a *lift* in England. *Elevator* is correct in the United States but incorrect in England; *lift* is correct in England but incorrect in the United States. There is no absolute sense in which either form is "inherently" correct.

To get a sense of how language changes over time, look at these four versions of the same few lines from the Bible:[1]

> (9th century) Eornustlīce þā sē Hǣlend ācenned wæs on Iūdēiscre Bethleem, on þæs cyninges dagum Herodes, þā cōmon þā tungolwītegan fram ēastdǣle tō Hierusalem, and cwǣdon, "Hwǣr ys sē Iūdēa Cyning þe ācenned ys?"

> (14th c.) Therfor whanne Jhesus was borun in Bethleem of Juda, in the daies of King Eroude, lo! astromyenes camen fro the eest to Jerusalem, and seiden, Where is he, that is borun King of Jewis?

> (17th c.) Now when Jesus was borne in Bethlehem of Judea, in the dayes of Herod the king, behold, there came Wise men from the East to Hierusalem, Saying, Where is he that is borne King of the Jewes?

> (20th c.) Jesus was born at Bethlehem in Judaea during the reign of Herod. After his birth astrologers from the east arrived in Jerusalem, asking, 'Where is the child who is born to be king of the Jews?'

Aside from all the other differences, notice how differently these four writers refer to the visitors from the east: *tungolwītegan* > *astromyenes* ("astronomers") > *Wise men* > *astrologers*. Today most of us would not call astrologers either "wise men" or "astronomers." Yet we would not accuse those early Bible translators of being "incorrect."

Correctness is a function of how the people you are communicating with use language. If everybody in your office calls your computer printer *Pokey*, then that is the correct way to refer to it *in that environment*. Note, however, that if you were to use it with other people, it would be confusing to them and therefore incorrect. Likewise, consider the word *hopefully* as in "Hopefully I'll finish this project by tomorrow." This word is said to be "illogical" by many self-anointed authorities on language, yet most people use it freely and effectively. Who is right? If the vast majority of people in a certain society use a word in a certain way, then that is the correct way to use that word within that society.

Writing tends to be more formal and more tradition-bound than speech. Thus, some words that are acceptable in speech may be less acceptable in writing. For example, the word *data* is traditionally the plural form of *datum*, and many writers insist on using it that way. In speech, however, *data* is being used more and more as a singular: *This data is really impressive.* In such cases, it's likely that the speaker is thinking of data as an undifferentiated mass of information rather than as a collection of data points. Although using *data* as a singular may seem incorrect to a traditionalist, the fact that people are using it more and more in speech means that it will probably gain acceptance in writing, too. In cases like this, as in all questions of usage, it is generally best to do what you see and hear most other educated people doing, unless you or your readers have a preference for traditional forms.

Dictionaries can provide some help. They generally provide good descriptions of current usage by educated people, and for this reason they are very useful guides. But dictionaries are not perfect. Some are outdated, and many are not in touch with special environments where words are being used for special purposes. To be correct in your use of English words, you need to pay close attention to how other people in your environment are using them, especially people who have the power to judge you. And when you "step out of" your customary environment—for example, if you write a letter to a client—be aware that you may run into other standards of word usage.

27.2 COMPREHENSIBILITY

Technical Terms

Technical terminology is an essential part of any technical or professional field. How could biochemists do their work without terms like *glycogen*, *proteolytic cleavage*, and *mitochondrial adenosinetriphosphatase complex*? How could financiers get by without terms like *amortization*, *depreciation*, and *dividend*? If words like these did not exist, somebody would have to invent them! Technical terms are efficient and useful simply because they are shorter than the corresponding full description or definition. This means that they can be used repeatedly with a minimum of clutter, which allows them to be used as *names*, which in turn makes it easier for you to talk and think about the concepts they represent. Technical terms can be used with other terms to produce complex technical terms. Sometimes too they participate in entire systems or taxonomies of related terms. Much of what it means to know a field is to know the vocabulary of that field. You should be proud of your knowledge of technical terms, and when you're communicating with your fellow specialists, you should use technical language freely.

However, a good part of your life as a technical professional involves communicating with nonspecialists, people who do not share your knowledge of certain technical concepts. If you use a lot of technical language with *them*, you will only confuse them. If they think you are doing it deliberately, just to show off your knowledge and intimidate them, they may resent you for it. That hardly makes for good communication! Those who really understand a subject well—like the astronomer Carl Sagan and the biologist Isaac Asimov—are able to talk about it without resorting constantly to highly technical language. If you are unable to talk about your subject in a fairly nontechnical way, it could be that you do not know the subject as broadly and as deeply as you should.

When addressing nonspecialists, try to use only those technical terms that are strictly necessary. And be sure to define those terms. In general, the technical terms that are considered "necessary" in such cases are *key* terms which serve major roles in the communication. Sometimes they are what the communication is about; sometimes they refer to objects or concepts that are part of a step-by-step procedure. Usually these key technical terms do not have simple paraphrases, and so you should always try to define or explain them, either by using language that you're sure the reader will understand or by making reference to some visual aid. For example, if you were writing assembly instructions for a 10–speed bicycle, you would have to make sure that the reader knew key technical terms like *front derailleur*, *seat stay*, and *rear sprocket*. The easiest way to "define" them would be to present a sketch of the bicycle and label those particular parts, as is done in Figure 27-1.

In cases where a technical term has a simple nontechnical equivalent, you may be able to insert the nontechnical term in parentheses right after it. Notice how this is done for the technical term *1-bit* in the following example: "The original Macintosh II with an expanded video card used 8-bit color images and required significantly more storage than the 1-bit (black and white) images on the Macintosh Plus and SE." (For further discussion of definitions, see Chapter 4.)

International Audiences

Our world is becoming more and more of a "global village"; communication with international audiences is becoming more and more common. There are millions of technical professionals and managers around the world who speak

Figure 27-1 ❑ IDENTIFICATION AND LOCATION OF THE BICYCLE PARTS
(Used with permission of the Murray Ohio Mfg. Company)

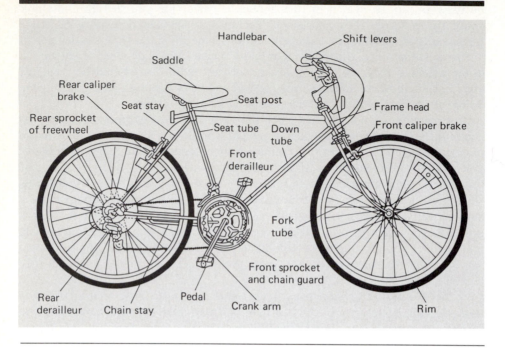

English as a second language, including many in English-speaking countries such as the United States and England. They are playing an increasingly important role in every field of engineering, science, and business. Although many of these people speak excellent English, many others do not. Even those who do speak excellent English sometimes do not have as large a vocabulary as their native-speaking peers. To increase the chances that your writing or speaking will be comprehensible to international audiences, there are certain things you should do when choosing words.

First, *avoid idiomatic expressions.* Expressions like *talk shop, shoot the breeze,* and *put our heads together* are difficult for nonnative speakers to understand. Although these expressions consist of multiple words, they have only a single meaning; and that meaning is not a product of the individual word meanings put together. *Shoot the breeze,* for example, has nothing to do with either shooting or wind. Also, they typically occur in speech, not in formal writing. Since most nonnative speakers learn English by reading it rather than hearing it, they often are not exposed to these kinds of expressions. Finally, they often are not listed in dictionaries; and if they are listed, it's difficult to find them. Many nonnative speakers depend on a dictionary, which won't help them when it comes to decoding or learning idiomatic expressions.

Many international professionals enjoy learning new idiomatic expressions. If you're willing to spend time teaching them, fine; your efforts will no doubt be appreciated. But if you want to have rapid communication, you would be wise to simply avoid such expressions. Instead of *talk shop*, say *talk about business* or *talk about work*. Instead of saying *shoot the breeze*, say something like *relax and talk*. And above all, don't use sports metaphors like *They pulled an end run on us.*

Second, **avoid uncommon words.** Nonnative speakers may know many common English words, but they often do not know less common ones. Try not to use uncommon words like *blemish, surmise, tenet,* and *remiss* unless the context makes their meaning clear. Also, although nonnative speakers can be counted on to know many technical terms in their field, they may not know many technical terms from other fields.

Avoiding idiomatic expressions and uncommon words is also a good idea when working with interpreters or translators. Interpreters and translators can be expected to have an excellent command of English, of course. But if they have learned their English mainly through books, they are likely to be deficient in their knowledge of idiomatic expressions. And unless they have spent many years reading English extensively, they may not know as many uncommon words as you would expect.

❏ EXERCISE 27-2

The following sentences contain idiomatic expressions or uncommon words. Revise them so that they will be more comprehensible to an international audience.

A   If we put our heads together, I'm sure we can find a solution to this problem.

B   Lee has a lot on the ball; he'll make a good officer.

C   We would be remiss if we didn't send at least a letter of condolence.

D   To get my report finished on time, I'll have to burn the midnight oil.

E   Looks like we're back to square one.

F   My résumé is attached for your perusal.

G   Do you think we can get a handle on this?

H   The test results warrant an immediate response.

 I   There's so much red tape involved, it may take us months to get this project off the ground.

 J   I was about to take their offer but then decided I'd better sleep on it.

27.3 TONE

In addition to its basic meaning, a word often conveys a certain tone, a certain attitude about your relationship with your readers or listeners. Partly this has to do with degree of formality. By using more formal language, you show respect and politeness, but you also create distance between yourself and your audience. By using very informal language, you show less respect and less politeness, but also less social distance. Thus there is a tension between respect and friendliness. If you want to be both friendly *and* respectful, you should choose your words very carefully.

For example, let's say you've written a draft of an important report and you want to have your boss critique it. Do you ask her to *examine* it, *check* it, or *eyeball* it? The first is formal, the second informal, and the third even more informal. If you have a formal relationship with your boss, *examine* would probably be the most appropriate choice; the other two (especially the third) might seem a little disrespectful. On the other hand, if you have a more relaxed relationship with your boss, you could probably use *check* or *eyeball*; the more formal *examine* might not seem friendly enough. Thus, how you choose your words can reveal how you see the relationship between yourself and your reader.

In general, words that came originally from Latin and French, such as *acquire, develop,* and *velocity,* are longer and more formal than words that came from German or Anglo-Saxon, such as *get, build,* and *speed.* Whenever you suspect that the tone of a word "is not quite right," consult a thesaurus. It will give you a list of other possibilities to choose from. Do not, however, use a word you're not wholly familiar with. When you use a word, your readers will assume that you know what it means. Thus, if the word carries unwanted connotations, you will be held accountable.

Words are the basic carriers of meaning. You should have a large enough vocabulary so that you can express your ideas clearly and maintain good relations with the people with whom you interact. If you want to improve your vocabulary, the best way is simply to read as much as you can from a wide variety of materials, both technical and nontechnical. There are also many vocabulary-building books and computer programs on the market. (If you are a nonnative speaker of English, we recommend that you study and apply the principles laid out in Chapter 36.)

❑ EXERCISE 27-3

In each of the following sets of near-synonyms, arrange the words according to degree of formality. What patterns do you notice across these sets? Be prepared to discuss other differences of usage within each set.

A   discard, throw away, reject, abandon, get rid of

B   gripe, complain, air a grievance, grumble, register a complaint

C   assist, aid, help, lend a hand, succor

D   heavy, ponderous, hefty, weighty, a load

E   sidekick, friend, acquaintance, pal, confidant

REFERENCES

1 See A. G. Rigg (ed.), *The English Language: A Historical Reader* (New York: Appleton-Century-Crofts, 1968), and the references therein.

ADDITIONAL READING

Don Bush, "Correctness vs. Communication," *Technical Communication*, Third Quarter 1986, pp. 133–135.
M. Stanley Whitley, *"Hopefully*: A shibboleth in the English Adverb system," *American Speech*, 1983, pp. 126–149.

28

Proofreading

The last step in preparing any piece of writing—be it a letter, a memo, a formal report, or whatever—is to proofread it for errors. Proofreading doesn't take much time, and it can really pay off in terms of conveying both accurate information and an image of credibility. Don't forget that readers will often judge you by the quality of your writing. If your writing is full of typographical errors, misspellings, and other kinds of easily noticeable mistakes, your readers may well judge you to be not only a careless writer but a careless person in general. This may lead them to question the quality of your technical work: the accuracy of your measurements, the preciseness of your calculations, the soundness of your judgment. In short, your professional image can be damaged if you allow your writing to circulate without first checking it over for errors.

Even more serious than errors of form are errors of substance: miscalculations, misrepresentations of data, misleading claims. Errors of this type may easily go unnoticed by your readers and thus may genuinely misinform them, sometimes to the point of having very serious consequences. Needless to say, such errors of substance must be eradicated from any piece of writing.

Given the fact that there are several different kinds of errors to look out for, it makes sense to divide the proofreading process into different phases so that you can "sweep" the manuscript, concentrating on one type of error at a time. We recommend the following sequence of steps: first, proofread for substantive errors; second, proofread for sequencing errors; finally, proofread for spelling and grammar errors. Once you've finished proofreading—especially for substantive errors—you might want to get someone else to help you proofread for sequencing and spelling and grammar errors or to check your proofreading. If you do rely on outside help, you have several options: You can hire someone to proofread for you, you can get a friend or colleague to proofread for you, or you can use one of the computer-based spelling checkers and/or style and grammar checkers. These computer based tools and some potential problems with them are discussed in Chapter 6.

Substantive Errors

Are your calculations correct? Are your data correct? Have you plotted your curves accurately? Have you got the right dates, times, places, job titles, model numbers? Remember, this kind of information may be of critical importance to your readers—so make sure it's right!

Sequencing Errors

There may be one or more numerical or alphabetical sequences in your manuscript: page numbering, references, illustrations, footnotes, report sections, lists, figures, steps in a procedure, etc. Make sure these sequences are all correct and complete.

Misspellings

For many writers, proofreading is particularly fruitful in checking spelling. The English spelling system is notoriously irregular, causing writers no end of trouble. Yet many readers—including many readers in supervisory positions—attach great importance to spelling, seeing it as a sign of how careful and even of how well educated the writer is. So don't take any chances—correct all spelling errors!

Most spelling errors occur with ordinary words that can be found in any pocket-size dictionary: words like *foreword*, *receive*, and *vacuum*. So, as you proofread for spelling, keep a small dictionary near you and use it for any words you're not sure of. If you're using a word processor with a text editor, it may have a built-in dictionary and spelling program; you can save considerable time by using that. Another way to save time, of course, is to ask someone who's known to be a good speller to proofread your manuscript for you.

Although any of these tactics may work for you, try not to become overly dependent on them; after all, you may someday have to write a report in a hurry and may not have a text editor or friend available to help you out, or the time to look up every other word in a dictionary. Also, no electronic text editor or ordinary dictionary is likely to be of much help when it comes to technical terms or proper names. And electronic text editors are notoriously bad at detecting misused homophones (see below). So try to learn the correct spelling of words as you encounter them, especially those words that you use often in your writing. Practice writing them down, so that you can see them in your own handwriting as well as in print. Make a list of words that give you particular trouble, and then test yourself on them from time to time.

❏ EXERCISE 28-1

Below is a list of some of the most frequently misspelled words in scientific and technical writing. Study the list; then have someone dictate the words to you while you write them down. Repeat the exercise until you've spelled all the words correctly.

accidentally	environment	occasionally
accommodate	exaggerate	occurred
achieved	exceed	omit
acoustic	excel	omitted
address	exhaust	parallel
alignment	existence	personnel
allot	feasibility	precedent
allotted	February	probably
analyze	fluoride	quantity
apparent	foreign	receive
appropriate	foreword	recommend
argument	government	refer
auxiliary	irrelevant	reservoir
basically	judgment	retrieve
beneficial	knowledgeable	schedule
calendar	laboratory	seize
ceiling	likelihood	separate
commission	lose	therefore
definitely	maintenance	truly
dependent	manageable	unnecessary
desirable	mathematics	vacuum
develop	misspell	
development	necessary	
different		

Many writers have trouble with homophones—words that have the same pronunciation but are spelled differently and have different meanings. Probably the most frequently misused homophones in English are *its* and *it's*. Are you sure which is which? (Answer: *its* is a possessive form; *it's* = *it is*.) It's especially important to learn the difference between homophones, because if you're using a spelling checker on your computer, it will not be able to catch errors involving homophones. That is to say, if a writer uses *it's* where *its* is appropriate, most computer-based spelling checkers will fail to notice the error because they don't take usage into account; they look only at the form of the word and see if that form is found in the dictionary. So, if you are using a word processor to check your spelling, don't expect it to pick up spelling mistakes when homophones are involved.

Here are some other homophones that often cause writers trouble:

affect	(The new policy will not affect us.)
effect	(We decided to effect a change.) (The extra dosage had no effect.)
all ready	(They are all ready to begin.)
already	(She has already made her decision.)
complement	(The illustrations nicely complement the written text of the report.)
compliment	(Linda's supervisor complimented her for having devised such a workable system.)
foreword	(Many technical reports begin with a foreword.)
forward	(We look forward to your visit.)
lead	(That idea went over like a lead balloon.)
led	(past tense of the verb *to lead*: Our company has led the way in this field.)
plain	(Use plain English whenever possible.)
plane	(The two lines formed a plane angle.)
principal	(The principal feature of this approach is its simplicity.)
principle	(The basic principle in this design is the conversion of heat energy to work.)
stationary	(A land-based missile system would provide a stationary target for the enemy.)
stationery	(Please don't use company stationery except for official business.)
their	(We would like to try some of their products.)
there	(There are too many terminals already in place there.)
they're	(What they're trying to do is capture the market.)
weather	(We'll launch at 700 hours, provided the weather's OK.)
whether	(Whether we succeed or not probably depends on our marketing strategy.)
who's	(Who's there?)
whose	(Whose jacket is this?)
your	(Do you have your calculator handy?)
you're	(When you're ready to begin, give the signal.)

❏ EXERCISE 28-2

A Keep a list of any words you misspell and review them from time to time. After you've accumulated 20 of them, have someone test you as in Exercise 28-1.

B   Proofread the letter of application for employment shown in Figure 28-1 for misspellings and other errors.

Figure 28-1 ☐ LETTER OF APPLICATION FOR EMPLOYMENT
(To Be Proofread for Misspellings and Other Errors)

225 Nugent Hall
Brookfield Polytechnic Institute
Ames, NY 12181
(555) 270-4391

Febuary 25, 1990

Mr. Bill Dahlen
Sperry Univac Semiconductor Divisions
U2X26, P.O. Box 3525
St. Paul, MN 55165

Dear Mr. Dahlem:

In response to your advertisement in IEEE Spectrum (August 1990), I am writing to apply for the position of Bi-polar Developement Engineer. After considering your challenging requirements; I believe that my experience and educational background would enable to make significant contributions to your expanding division. My backround includes—

Employment by the Electron Physisc Laboratory, Brookfeld Polytechnic Institute. I participated in designing an innovative computer modeling process of a MOS submicron devise. I was responsible for designing the numerical analysis subroutines of the computer program.

Completion of a senior desing project in designing and fabricating bi-polar device chips using state-of-the art technics. I was also responsable for testing and characterizing the device chips.

Completion of several graduate courses in semi-conductor devices, intergrated circiuts and digital logic circuit design.

I will recieve my B.S.E.E. from BPI in August 1990 and would appreciate being considered for permanant employment starting therafter. I will call on you next Wed. during your office hours to set up an interview at your convience. If you desire additional information, I can be reached at (555) 270-4381 during the mornings.

Thank you for your kind consideration.

Yours truely,

Perry C. Culbert

PART SIX

REVIEW OF GRAMMAR, STYLE, AND VOCABULARY BUILDING

29

Indefinite Articles

Articles (*a*, *an*, *the*) are an important part of the English language. If you're not sure how to use them, you should study this chapter and the next.

In order to understand the use of articles, you have to keep in mind the following basic fact about common nouns. Common nouns can refer to objects or concepts at different levels of generality: the same noun that refers to an entire category of objects in one sentence may refer only to a particular, unique object in another. For example, consider the word *window* in a sentence like "Every room in this building has a window." The writer of this sentence is probably thinking not of any particular window, but of windows in general. On the other hand, in a sentence like "The window in Room 303 is broken," the writer is referring to a particular window, one that has a unique identity.

This chapter explains the use of indefinite articles (*a* or *an*). In general, if you are using a common noun to refer to a single, "countable" example of a general class of things, you should use an indefinite article with it.

29.1 UNCOUNTABLE NOUNS

Many English nouns represent objects or concepts that do not have a clear shape or form: *water, rice, gravity, magnetism, information, engineering,* etc. Such nouns are called "uncountable" because they refer to things that we don't usually think of in terms of units. We do not say "5 waters" or "8 rices" or "15 magnetisms." In fact, we do not normally use the plural form at all for such nouns.[1]

Some common types of uncountable nouns are:

1   Those representing a physical mass without definite shape: *sugar, sand, salt, rice, flour,* any liquid or gas, etc.

2   Those representing an abstract concept: *gravity, information,*[2] *curiosity, satisfaction, magnetism,* etc.

3   Those representing a continuous process: *photosynthesis, pollution, osmosis, combustion,* etc.

4   Those representing a field of study: *mathematics, chemistry, business, engineering,* etc.

Rule 1
Whenever you use an uncountable noun to refer to a general type of substance, concept, process, etc., do not use an article with it.

Examples
Magnetism is the force that causes iron to be attracted to a magnet.
My daughter says she wants to study *business.*

We can represent the above rule with the flowchart shown as Figure 29-1.[3]

29.2 COUNTABLE NOUNS

In contrast to uncountable nouns, many nouns in English represent concepts or objects that do have a clear form. These are the "countable" nouns. Some examples of countable nouns are *book, automobile, molecule, computer,* and *microscope.* Unlike uncountable nouns, words like these occur:

1   In the plural form (*books, automobiles, molecules*)

2   With numbers (*four computers, nine microscopes*)

3   With quantifiers such as *several, many,* and *few* (*several books, many microscopes, few molecules*)

Included among the countable nouns of English are words like *idea, concept, theory,* and *hypothesis.* Although nouns like these do not have a physical form like *book, automobile, molecule,* etc., they do have an abstract form. Therefore, these more abstract nouns can occur:

1   In the plural (*ideas, concepts, theories*)

2   With numbers (*three hypotheses, two concepts, four theories*)

3   With quantifiers like *several, few,* and *many* (*several ideas, few theories, many hypotheses*)

Figure 29-1 ❏ PARTIAL FLOWCHART FOR CHOOSING THE CORRECT ARTICLE (STEP 1)

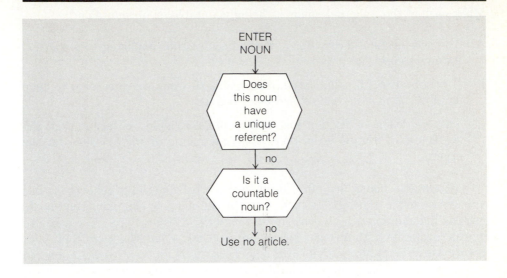

❏ EXERCISE 29-1

Mark each of these nouns as being countable or uncountable:

rocket	program
budget	economics
inflation	satellite
taxes	snow
carbon dioxide	ecology
wheel	molecule of carbon dioxide
friction	radar
gravity	combustion
atom	motor
computer	injury
aluminum	book
weather	meter
machine	sodium nitrite
information	photosynthesis

❑ EXERCISE 29-2

Mark each of the italicized nouns as being countable or uncountable:

A   The acceleration of *gravity*, g, is the *acceleration* imparted to a *body* by its own *weight*.[4]

B   If *friction* could be eliminated, no force at all would be necessary to keep an *object* in motion, once it had been started.[5]

C   The *coefficient* of friction just before *motion* begins is larger than the coefficient of friction when there is actual sliding of one *surface* over the other.[6]

Countable nouns are often used to refer to something general, as our window example showed. In such cases, if the noun is singular, you should use the indefinite article: *a book, an automobile, a theory*, etc. (Use *a* before words beginning with a consonant sound: *a meter, a handle, a unit*, etc.; *an* before words beginning with a vowel sound: *an atom, an hour, an uncontrolled reaction*, etc.). If the noun is plural, you should use no article at all: *books, automobiles, theories*, etc.[7]

Rule 2

Whenever you use a singular countable noun to refer to a general type of substance, concept, process, etc., use *a* or *an* with it.

Example

A betatron is a device used to accelerate electrons. The electrons travel around a circular path in a vacuum tube known as a "doughnut." The force of acceleration is supplied by a magnet.[8]

This example is taken from an introductory physics textbook. The author is trying to tell his readers what a betatron is, in general; he is not talking about any particular betatron. To indicate this level of generality, he uses the indefinite article. He also uses indefinite articles for the other singular, countable nouns (*device, path, vacuum tube*, etc.).

Notice that the first sentence in this example is a definition conforming to the standard pattern:

$$\text{Term} = \text{Class} + \text{Distinguishing Features}$$

In this case, the term being defined is *betatron*; it belongs to the class of devices; it is distinguished from other members of this class by the fact that it is used to accelerate electrons. All definitions, of course, are general statements: they refer not to particular things but to classes of things. This is why we seldom find definite articles used in definitions. (One exception: the term being defined may have the generic definite article, e.g., "The betatron is a device used to accelerate electrons." See Chapter 30 for discussion of this use of the definite article.)

Figure 29-2 ❑ PARTIAL FLOWCHART FOR CHOOSING THE CORRECT ARTICLE (STEP 2)

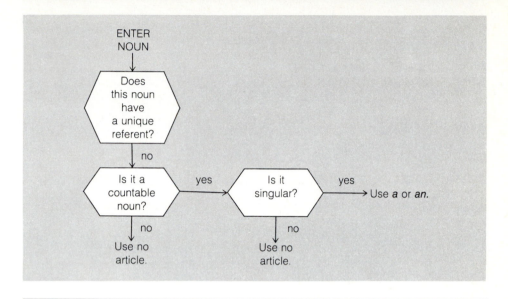

Other Examples of Countable but Nonspecific Nouns
An ideal machine is one that has no friction.
Bill hopes to buy *a computer* someday.
An effective way of learning mathematics is by working on *problems*.
Valence electrons are those located in the outermost shell of *an atom*.

Adding Rule 2 to our previous flowchart, we have the diagram in Figure 29-2:

To see how this flowchart works, let's take the sentence *Bill hopes to buy ____ computer someday*. Do we use *a, an,* or no article at all before the noun *computer*?

By answering the questions in the flowchart and following the arrows, we arrive at the correct answer, *Bill hopes to buy a̲ computer someday*.

❑ EXERCISE 29-3

In the following passages, insert indefinite articles (*a* or *an*) where necessary. (Use Figure 29-2 as an aid.)

A **Storing Electric Power**
In ____ space the sun is always shining, so there is no problem with storing ____ electricity. On earth, however, the sun shines only half the time in ____ good weather and not at all in bad. Thus, ____ electricity must be stored for ____ sunless periods. Currently this is done with ____

Figure 29-3 ❑ USE OF PARTIAL FLOWCHART FOR CHOOSING THE CORRECT ARTICLE

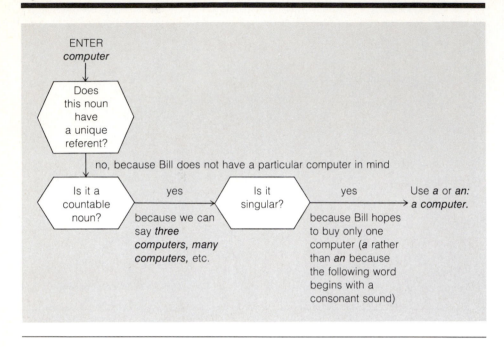

lead-acid storage batteries, similar to those used in ____ automobiles. ____ day's electricity for ____ average single-family house can be stored in ____ batteries occupying the space of ____ closet; ____ row of such "closets" in the basement stores ____ power for ____ sunless periods.[9]

B   Power Requirements in an Average Home
____ average single-family residence (____ four-person, 1500–square-foot, non-air-conditioned house) uses about ____ 700 kilowatt-hours per month, the equivalent of ____ 1-kilowatt generator running continuously. Because this average house needs ____ 1 kilowatt of ____ average power, it requires ____ 5 kilowatts of peak power. This, in turn, would require about ____ 500 square feet of ____ solar cells at ____ present efficiency levels under ____ optimum conditions. The Energy Research and Development Administration (ERDA) says 1500 square feet would now be required in ____ northern city like Boston.[10]

29.3 COUNTERS

Although uncountable nouns themselves cannot be counted, they can usually be modified by certain types of phrases—called counters—that can be counted.

One example of a counter is *cup of* _____. It can be used to modify an uncountable noun like *rice* to create a countable noun phrase, *cup of rice*. Thus, although you would not normally say "a rice" or "two rices" or "three rices," you could say

> a *cup of* rice
>
> two *cups of* rice
>
> three *cups of* rice
>
> etc.

Some other common counters are *liters of, quarts of, ounces of, grams of,* etc.

❑ EXERCISE 29-4

A list of uncountable nouns follows. Insert appropriate counters or measure phrases so that you change these uncountable nouns to countable units. The first two examples have been done for you.

Uncountable	Countable	
oxygen	3 liters of oxygen	
	3 moles of oxygen	
	3 molecules of oxygen	
electric power	3 watts of _____	electric power
current	_____	current
force	_____	force
work	_____	work
pressure	_____	pressure
temperature	_____	temperature
random-access memory	_____	random-access memory
pollution	_____	pollution
chemistry	_____	chemistry
electrolysis	_____	electrolysis
gasoline	_____	gasoline
combustion	_____	combustion
friction	_____	friction
electric potential	_____	electric potential

29.4 TWO-WAY NOUNS

Many nouns in technical English can refer to either countable or uncountable concepts, depending on the writer's focus. Such "two-way" nouns all have one thing in common: when they are used in the uncountable sense they refer to a general concept, but when they are used in the countable sense they refer to a specification of that general concept. Consider the word *metal*: it is commonly used in an uncountable sense, as in "knives are usually made of *metal*." However, it can also be used as a countable noun, to mean a *type* of metal. For example, the sentence "Brass is an alloy composed of two *metals*, zinc and copper" means the same thing as "Brass is an alloy composed of two types of metal, zinc and copper." Some other two-way nouns that usually have the meaning *type of x* in the countable sense (where *x* is the name of a class) are *fuel, acid, soil, material*, and *gasoline*.

Note, for instance, the following:

1 Most American automobiles run on *gasoline*. (Here *gasoline* is general, uncountable.)

Our local service station usually carries several *gasolines*: unleaded regular, unleaded premium, and unleaded silver premium. (Here *gasolines* is specific, countable. Note that we could have said, Our local service station usually carries several *types of gasoline*: unleaded regular, etc.)

2 Railcars transporting *acid* must be specially designed. (Here *acid* is general, uncountable.)

Electrolytic action in an automobile battery produces an *acid* that can lead to corrosion of the terminals. (Here *acid* is specific, countable. We could have said, Electrolytic action in an automobile battery produces a *type of acid* that can lead to corrosion of the terminals.)

In many other cases, however, the specification of a two-way noun does not mean *type of x* but instead means *amount of x*. Note this example:

The two circuits have different *resistances*.

Here *resistance*, a word commonly used as an uncountable noun referring to *general* opposition against electric current, refers specifically to the *amount* of opposition. The above sentence means, in effect, "The two circuits have different *amounts of resistance*." Some other two-way nouns that usually mean *amount of x* in the countable sense are *mass, velocity, force, pressure, power*, and *acceleration*.

Sometimes the specification of a two-way noun means, roughly, a *complete process of x*. For example, the word *distillation* normally is used to refer to the general process of distillation and is thus used as an uncountable noun. Sometimes, however, the writer may want to focus on one or more specific applications of the process, as in this sentence:

At the refinery, the crude oil undergoes two complete processes of distillation.

In such a case, the sentence could be shortened as follows:

At the refinery, the crude oil undergoes two distillations.

There are still other possible interpretations that can be attached to the specification of a two-way noun; it is advisable to exercise caution in your own writing and not to perform a specification that you are not sure of.

❑ EXERCISE 29-5

Mark each of the italicized nouns as countable or uncountable. The first two examples have already been done.

A 1   u   c
     Water expands when it is heated in a *container*.

           c   u
  2 *Molecules* occupy *space*.

  3 Effective business *communication* is the *lifeblood* of every *organization*.

  4 Chemistry *students* study the *concept* that *matter* is made up of *particles*.

  5 In the 1950s, fundamental *changes* in *finance* began to occur.

B 1 For many *people*, *fitness* and *strength* are more important than good *looks*.

  2 *Velocity* and *speed* are both the *ratio* of a length to a time.[11]

  3 Management *decisions* that increase *earnings* but do not affect cash flows represent wasted *effort*.

  4 In cancer *patients*, radiation *therapy* usually depresses white blood *cells*.

  5 In the *year* 1642, Sir Isaac Newton (the famous *physicist*) was born and Galileo died.

C 1 *Mechanics* is the branch of physics and engineering which deals with the interrelations of *force*, matter, and motion.[12]

  2 Many *businesses* spend large sums of *money* on *advertising* to generate new *business*, while neglecting their regular customers.

  3 *Mass*, as used in *mechanics*, refers to that *property* of matter which in everyday *language* is described by the word inertia.[13]

  4 The newton is a *force* which imparts to a *mass* of one kilogram an acceleration of one meter per *second*, per second.[14]

  5 *Pressure* is expressed in lbs/ft^2, newtons/m^2, or *dynes*/cm^2.[15]

29.5 INFORMAL USAGES

There are certain informal usages that you might hear spoken but will rarely see in writing. It is important to avoid such usages in your writing, since they are quite informal.

For example, in the informal situation of the chemistry laboratory or small discussion group, student chemists sometimes talk about normally uncountable nouns, such as the names of elements and molecules, as if they were countable nouns. A student might read aloud this equation, for instance,

$$Ag_2O + H_2O_2 = 2 Ag + O_2 + H_2O$$

as "one silver oxide plus one hydrogen peroxide yields two silver(s) plus an oxygen plus a water." Similarly,

$$Cu + 2 H_2SO_4 = CuSO_4 + SO_2 + 2 H_2O$$

could be read aloud as "one copper plus two sulfuric acid(s) yields one copper sulfate plus one sulfur dioxide plus two water(s)." Although this type of language is frequently heard in informal settings, it would be highly unsuitable for formal purposes. A more formal reading of the same formula would be "one atom of copper plus two molecules of sulfuric acid, etc."

REFERENCES

1 See the discussion of two-way nouns and informal usages (Sections 29.4 and 29.5) for some exceptions.

2 The word *information* deserves special attention, since the corresponding noun in many other languages (e.g., French, German, Spanish) is countable.

3 The flowchart should be used as follows: Taking the example "Magnetism is the force that causes iron to be attracted to a magnet," let us consider the noun (properly speaking, the noun phrase) *magnetism*. We first ask of it, "Does it have a unique referent?" The answer to this question is no (since magnetism in a general, not particular, sense is meant here), and so we follow the arrow from the "no" cell to the next question, "Is it a countable noun?" The answer here is no, and so we follow the "no" arrow to the final instruction, "Use no article." The other flow charts in this chapter are used in similar fashion.

4 F. W. Sears, *Mechanics, Heat, and Sound* (Reading, MA: Addison-Wesley, 1950), p. 283.

5 Adapted from Sears, p. 20.

6 Sears, p. 24.

7 Certain verbs, especially verbs of volition, can allow a specific interpretation for a following noun with an indefinite article. For example, the sentence "John wants to marry a Norwegian" can be interpreted as meaning that John has a specific person in mind that he wants to marry, who is of Norwegian nationality (as opposed to an alternative reading according to which one of John's requirements for any future spouse is that she be of Norwegian nationality). This specific

reading, focusing only on the nationality of John's intended spouse, should not be confused with cases of unique reference (as embodied, for example, in "John wants to marry the Norwegian"), where complete identity (of John's intended) is implied. For further discussion of this distinction between specificity and uniqueness, see J. Lyons, *Semantics* (Cambridge, England: Cambridge University Press, 1977), pp. 187–192.

8   C. H. Bachman, *Physics for the Non-Scientist*, privately published, 1968, p. 101.

9   Adapted from D. Morris, "Solar Cells Find Their Niche in Everyday Life on Earth," *Smithsonian*, October 1977, p. 40.

10   Adapted from Morris, p. 41.

11   Sears, p. 51.

12   Sears, p. 1.

13   Sears, p. 72.

14   Adapted from Sears, p. 77.

15   Sears, p. 300.

30

The Definite Article

Recall the example given at the beginning of the preceding chapter, involving the word *window*. We pointed out that—in contrast to a sentence such as "Every room in this building has a window," where the writer is thinking of windows in a general sense—a sentence such as "The window in Room 303 is broken" refers to a particular window, one that has a unique identity. Notice that this distinction between generality and particularity is marked by the use of the indefinite article *(a window)* in the first sentence and the definite article *(the window)* in the second.

In general, the definite article *the* is used to show that what the noun (or noun phrase) refers to is *unique*; that is, it has a *unique referent*. Unlike the indefinite article, the definite article may occur with any type of noun—singular or plural, countable or uncountable.

There are many different circumstances under which a noun may be said to have a unique referent:

1 The noun may have a *special adjective* as a modifier (Section 30.1).

2 The noun may be a *special noun* referring to some unique time or place in our common existence (Section 30.2).

3 The noun may be *generic*, that is, it may refer to an entire species or type of something (Section 30.3).

4 The noun may have the same referent as some *previously mentioned* noun in the present context (Section 30.4).

5 The noun may have a *following modifier* that restricts it to a unique referent (Section 30.5).

514

6   The noun may have a unique referent by virtue of *shared knowledge* between writer and reader (Section 30.6).

7   The noun may have a unique referent by *implication* (Section 30.7).

30.1 SPECIAL ADJECTIVES

Certain adjectives modify a noun so that it has a unique referent in almost any context. In other words, such a noun is assumed to refer to a particular person, place, or thing.

One such type of highly restrictive adjective is the *superlative* adjective. Superlative adjectives are words like *tallest, fastest, heaviest, least important,* and *most useful.* In any given context, there is normally:

Only one tallest building (tree, person, etc.)

Only one fastest car (plane, runner, etc.)

Only one heaviest machine (person, book, etc.)

and so on. Thus, each of these nouns has a unique referent and must take the definite article: *the tallest building, the fastest car, the heaviest machine,* etc.

Similarly, ordinal adjectives—*first, last, second, fifth, nth,* etc.—each represent a unique position in an ordering: given any ordered set, there can be only one first position, only one last position, only one second position, and so on. Consequently, whenever a noun or noun phrase is modified by such an adjective, it almost always requires the definite article: *the first attempt, the last investigation, the second stage, the 19th century.*[1]

Some other adjectives by their very nature also restrict nouns to a unique referent and therefore normally occur with the definite article. Examples of such adjectives are *only, sole, exact, current,* and *present.* Three of these adjectives can be found in the following report about the *Viking* mission to Mars.

The *Viking* lander's surface sampler is *the only means* for getting small amounts of Martian soil and then delivering them to the three analytical instruments located deep inside the lander. Without these samples of the Mars surface, *the sole use* for these instruments would be to analyze wind-blown dust that might accumulate in them over a long time—a very unattractive alternative. Thus, on the second day, when the surface sampler failed during its initial operation, a team of experts at the *Viking* mission control center immediately tried to help the situation.

The exact nature of the problem became evident during tests of a model of the lander, known as the Science Test Lander (STL), at the Jet Propulsion Laboratory in Pasadena, California.[2]

30.2 SPECIAL NOUNS

Certain nouns are commonly used with the definite article to refer to periods of time or to certain physical features of our world: *the past, the present, the future, the 1940s, the early 1980s, the sky, the earth, the sun, the moon, the ground.* Such nouns take the definite article because each refers to some unique aspect of our common existence.

There are also many proper nouns that take the definite article. (Proper nouns are names, usually of persons or places: *Einstein, Mao Ze-dong, New York, Tokyo, Iran, Vietnam,* etc.; as can be seen from this sample list, they often occur without any article.) In cases where the "head noun" (usually the rightmost one) of the name is derived from a common noun, the definite article is often used: *the United States, the Twin Cities, the Soviet Union, the Merritt Parkway, the Ventura Expressway, the Mississippi River, the Indian Ocean, the Rocky Mountains.* Words such as *states, cities, union, parkway, expressway, river, ocean,* and *mountains* exist as common English nouns as well as in names such as these listed here.

One should be careful, however, for there are many exceptions to this pattern. Consider, for example, the following proper nouns: *Washington State, Salt Lake City, Lydecker Street, Fifth Avenue, Cripple Creek, Walden Pond, Bunker Hill, Lookout Mountain.* Each of these head nouns is derived from a common English noun *(state, city, street, avenue, creek, pond, hill,* and *mountain)*, and yet *no* definite article is used. How can we account for such a difference?

In general, the distinction between these two classes of nouns seems to be based on relative *size:*

A group of states or cities	is usually larger than a single state or city.
A parkway or an expressway	is usually larger than a street or an avenue.
A river or an ocean	is usually larger than a creek or a pond.
A chain of mountains	is usually larger than a single mountain or hill.

Since large geographical objects like rivers, oceans, and mountain ranges tend to be uniquely identifiable for many people, the definite article is usually attached to their names. On the other hand, smaller objects like creeks, ponds, and hills are not so uniquely identifiable. Thus, one says <u>the</u> *Atlantic Ocean* (with the definite article) but *Walden Pond* (without one). There is only one Atlantic Ocean, but there may be many Walden Ponds.

❑ EXERCISE 30-1

In the following passage, find all the examples of the special adjectives and nouns just discussed in Sections 30.1 and 30.2. Be prepared to classify the various *definite article plus noun* and *indefinite article plus noun* units and to explain your classification.

An Introduction to 3-D Radar

Air-traffic radar of the 1950s could easily determine the direction of incoming and outgoing aircraft, but not their height above the ground. The planes might be a mile apart in altitude or about to collide—but air-traffic controllers on the ground had no way of obtaining this information from current radar data alone, or without some previous radar history on both planes before they entered the same radar zone. Recognizing this obvious deficiency, the Airways Modernization Board (the forerunner of the FAA) decided to investigate the radar "height finder" as a likely solution to the problem.

The new air-traffic radar would have to resolve targets to within 1000 feet at 50 miles and be able to determine whether two aircraft, approaching each other from opposite ends of the sky, might pose a threat to each other. This meant that the radar had to measure the altitude of each aircraft before it entered the range of another and to allow for any distortions in the target's return signals due to blending in the atmosphere.

Looked at another way, the problem was to hold steady a lever 50 miles long—the length over which the radar beam had to be effective to do its job. Designers had to contend with wind, temperature, and humidity variations, and even the slight motions of the ground itself.

In addition to these structural problems, the major problems facing the designers were "ground clutter"—the enemy of all radar—and, unexpectedly, system noise. However, an unexpected benefit of the system was the height finder's unique ability to track targets in the rain.[3]

❏ EXERCISE 30-2

Insert *the*, *a*, or *an* where appropriate in the following passage and be prepared to justify your choice.

The Big-Dish Radio Telescope

"One of ＿＿ most challenging engineering problems of ＿＿ century" . . . "Dwarfing anything constructed in ＿＿ past for the study of the universe" . . . " ＿＿ largest movable land-based structure ever constructed in ＿＿ world."

These were only three of the descriptors enthusiastically applied to the U.S. Navy's attempt in ＿＿ late 1950's to build ＿＿ huge radio telescope with ＿＿ 600–foot-diameter antenna that would be fully steerable. The structure, which would have been twice as large as any fully steerable telescope built since, would have stood more than 750 feet above ＿＿ ground on ＿＿ 7-acre foundation near Sugar Grove, West Virginia. Unfortunately for astronomers, however, in 1961–1962 original cost estimates of $52 million were reevaluated and reestimated to be between $200 and $300 million for ＿＿ future. As a result, the so-called Big Dish was abandoned. Now all that remains of that dream to see more of ＿＿ sky and farther into ＿＿ space than ever before is the concrete foundation for the telescope tracks and the pintle bearing.[4]

30.3 GENERICS

The definite article is sometimes used generically to indicate that a countable noun or noun phrase refers to an entire *type* of something. In such usage, the noun or noun phrase is always in the singular. Here is an example:

> For simplicity and efficiency, *the Hawker Siddeley Harrier* is one of the best present-day VTOL aircraft. This plane uses the concept of "vectored thrust," where four rotating exhaust nozzles are used to deflect the exhaust from vertically down to directly behind.[5]

The author of this passage is referring not to a single plane but to a single *type* of plane. Since there is only one such type, the reference is unique. Some other examples are *the Ford Pinto, the IBM 3600, the Polaroid One-Step,* and *the Honda CVCC.*

It is possible, of course, to use a noun or noun phrase generically without using the definite article. One can say *Hawker Siddeley Harriers* or *a Hawker Siddeley Harrier* to refer to the entire type, just as one can say *the Hawker Siddeley Harrier.* In all three cases, the meaning is essentially the same, though it derives from different sources. With the indefinite forms (plural or singular), individual planes are being referred to as *representative* of the type; with the definite form, the type as a whole is being directly referred to.

30.4 PREVIOUS MENTION

Often a writer will refer more than once to the same object or concept. In such cases, the definite article is used with each mention after the first. (In some cases a pronoun or demonstrative can be substituted; see Chapter 24.) Consider the following example:

> Soil physicists have characterized the drying of *a soil* in three stages. They are:
>
> The wet stage, where the evaporation is solely determined by the meteorological conditions;
>
> An intermediate, or drying, stage, where *the soil* occurs in the wet stage early in the day but then dries off because there is not a sufficient amount of water in *the soil* to meet the evaporation rate; and
>
> The dry stage, where evaporation is solely determined by the molecular transfer properties of water within *the soil*.
>
> There is a striking change in the evaporation rate as *the soil* dries during the transition from the wet stage to the drying stage.[6]

In this example, the first mention of the noun *soil* does not have a unique referent. The author is referring to any *soil* and so marks the noun with the indefinite article: *a soil* (line 1). After this, however, the author refers repeatedly to the same soil and so uses the definite article with each of these later mentions: . . . *the* soil . . . *the soil* . . . *the soil.*

Not all cases of "previous mention" *the* involve repetitions of the exact same noun. In many cases, *the* is used when the repetition involves a synonymous noun or associated nominal (see Chapter 24). In the following paragraph, for example, the three noun phrases in italics vary somewhat in form but nonetheless have the same referent.

> The simplest approach to passive space heating is through direct gain of solar radiation by means of *a south-facing expanse of glass.* This approach works best when *the south window area* is double-glazed and when the building has considerable thermal mass in the form of concrete floors and masonry walls insulated on the outside. What results is, in effect, a live-in solar collector thermal storage unit. If *the south-facing window area* is vertical, seasonal temperature control is basically automatic. . . .[7]

Here, as in the previous example, the first mention of the noun phrase *south-facing expanse of glass* does not have a specific, unique referent. Hence, the indefinite article is used: *a south-facing expanse of glass.* Then, the author uses this *expanse of glass* as a model and refers to it twice. The first time it is called the *south window area;* the second time it is called the *south-facing window area.* In form, these are both slight variants of the original *south-facing expanse of glass*, but they both refer uniquely to that particular *expanse of glass.* Thus, the definite article must be used: . . . *the south window area* . . . *the south-facing window area.*

30.5 MODIFIERS FOLLOWING THE NOUN

Often a writer will designate a referent as unique by attaching one or more modifying phrases to a noun. For example, consider the noun *growth.* By itself, this noun does not have a unique referent, but if the writer adds appropriate modifiers to it, its meaning can be narrowed to a unique referent:

> *The growth of the American garment industry in the mid-19th century* was made possible by Isaac Singer's improvement of the sewing machine.

Now the word *growth* refers to a specific growth and thus requires the definite article: *the growth.*

This use of the definite article seems to occur most often when a noun is followed immediately by an *of* phrase:

theory	*the* theory *of* relativity
construction	*the* construction *of* Aswan Dam
principles	*the* principles *of* thermodynamics

cost   *the* cost *of* producing nuclear energy

invention   *the* invention *of* the electric light bulb

It also occurs with other types of following modifiers, including relative clauses:

> At *the time the Mariner-9 spacecraft began its Martian orbit, in November 1971,* an intense, planet-wide dust storm was in progress. *The infrared spectroscopy experiment which was carried on the spacecraft* obtained information on the thermal structure of the atmosphere, both during and after the storm.[8]

In the first sentence, the noun *time* refers not to just any time but rather to a single, specific time, namely, the time when *Mariner-9* began its Martian orbit (November 1971). Because of this unique reference, the definite article is used: *the time*. Similarly, in sentence 2, the noun *experiment* is modified so that it refers not just to any experiment but to a specific one, namely, the one that was carried out on the spacecraft and used infrared spectroscopy.

❏ EXERCISE 30-3

Find all the examples of *the* used for generics, previous mention, or a following modifier (the *the* discussed in Sections 30.3 to 30.5). Be prepared to classify the various *definite article plus noun* and *indefinite article plus noun* units and to explain your classification.

An Introduction to the Physics and Physiology of Acceleration

In the days of the frail, canvas-covered aircraft which could not take stresses easily tolerated by the human body, acceleration was not much of a problem. Today, aircraft of much stronger construction travel at sonic and supersonic speeds and thus can impose tremendous forces for appreciable periods of time on the now relatively frail human occupant. Since aviation medicine has as yet little understanding of these important forces, a study of the fundamental principles involved in the physiology of acceleration is needed. Such a study should proceed through the following stages:

1   The history of acceleration and its relation to aviation medicine should be described.

2   The physiological effects and the clinical response to such forces should be understood by the flight surgeon.

3   The conventional terminology for discussing these forces and their effects must be established.[9]

❏ EXERCISE 30-4

Insert *the*, *a*, or *an* where appropriate in the following passages and be prepared to justify your choice.

A   Awareness of Technological Revolutions

It is easy to be aware of ＿＿ revolution brought about by ＿＿ internal combustion engine. ＿＿ effects of ＿＿ revolution are part of our world. Most of us own ＿＿ car and know something about ＿＿ pistons, ＿＿ cylinders, and ＿＿ horsepower. It is also easy to be aware of ＿＿ revolution brought about by ＿＿ electronic tube. Most of us own ＿＿ radios and ＿＿ television sets and know something about waves and ＿＿ electrical interference. Unfortunately, it is harder to be aware of ＿＿ chemical revolution because its products are hidden. Yet this revolution is important, because it is ＿＿ basis for the other revolutions. ＿＿ automobile could not run without fuel, which is a chemical. ＿＿ car or ＿＿ electronic tube could not be built without ＿＿ metal, fabric, adhesive, glass, paint, and plastics, which are ＿＿ chemicals or ＿＿ results of chemical processes. If we are to understand our world, we must be aware of ＿＿ chemical revolution.

B   The Dynamics of Rotation Applied to the Centrifuge

＿＿ classic treatment of rotational physics usually found in ＿＿ engineering textbook and in ＿＿ advanced text on gyrodynamics is ordinarily quite rigorous in its mathematical treatment of ＿＿ subject. This paper is intended to present ＿＿ subject in such manner, first, as to relieve ＿＿ physicians and medical officers of ＿＿ labor required to gain ＿＿ rigorous insight into ＿＿ subject and, second, to eliminate ＿＿ nonessentials associated with ＿＿ classic developments, which are not pertinent to ＿＿ work at hand.

To this end, ＿＿ treatment used in this paper will be ＿＿ largely intuitive, extensively graphical one, and ＿＿ use of mathematics beyond algebra will be studiously avoided.[10]

30.6 SHARED KNOWLEDGE

Often the writer and intended readers share certain knowledge of the world because they belong to the same culture. (*Culture*, in this case, can be defined broadly or narrowly, depending on whether the writer is addressing a broad, general audience or a narrow, specialized one.) Thus, the writer will sometimes use a noun or noun phrase that has only one referent in the real world, believing that the readers already know about this uniqueness of reference, and he or she will accordingly use the definite article to mark the noun or noun phrase. Consider the following passage, for example, which begins a report presented to a group of aerospace scientists and engineers at a NASA symposium:

> This paper presents the preliminary results from *the Goddard–University of New Hampshire cosmic-ray experiment* during *the recent Pioneer-11 encounter with Jupiter*. Before continuing, however, I would like to say a few words about

> *the other Goddard experiment* on *Pioneer-10, the flux-gate magnetometer experiment.* This experiment performed flawlessly throughout the encounter, and it observed a maximum magnetic field strength of 1.2 gauss. . . .[11]

Now it happens to be a fact that each of the italicized noun phrases in this passage has a unique real-world referent; there was only one Goddard–University of New Hampshire cosmic-ray experiment on *Pioneer-11*, only one *Pioneer-11* encounter with Jupiter, etc. Of course, not everybody knows these facts, but in the author's immediate culture—that is, in the circle of fellow scientists and engineers, the ones being addressed in this case—these facts seem to be *common knowledge:* everyone in the group knows that each of the noun phrases in question has a unique referent. Consequently, even at first mention, the author marks these noun phrases with the definite article.

If this writer had been addressing a broader, more general audience of readers, the indefinite article might have been chosen instead:

> This paper presents the preliminary results from *a Goddard–University of New Hampshire cosmic-ray experiment* during *a recent Pioneer-11 encounter with Jupiter.* Before continuing, however, I would like to say a few words about *another Goddard experiment* on *Pioneer-10, a flux-gate magnetometer experiment.* This experiment performed flawlessly throughout the encounter, and it observed a maximum magnetic field strength of 1.2 gauss. . . .

In deciding whether to use the definite or indefinite article, therefore, you sometimes have to consider *two* things: (1) whether or not the noun or noun phrase has a unique referent and (2) whether or not your reader *shares* this knowledge with you. If you do not know who your readers are likely to be or if there is a great variety in the backgrounds of your readers—as is often the case with technical reports, for example—you cannot depend on the sharing of specialized knowledge. In such cases you may want to use the indefinite article.

30.7 IMPLIED UNIQUENESS

A less frequent use of the definite article occurs when the writer wants to imply that a noun has a unique referent even though the reader may not know of this uniqueness. Consider the following passage from a Volkswagen repair manual:

> ### Removing and Installing Regulator
> The regulator is mounted in the engine compartment, next to the battery. Always disconnect *the battery ground strap* before removing the regulator. . . .[12]

The manual had previously discussed the regulator, the engine compartment, and the battery, but it had not yet said anything about a "battery ground strap." Yet even a reader who knows nothing about cars would infer from the writer's use of the definite article that there is *only one* battery ground strap involved.

Here is another example of implied uniqueness (from the same repair manual):

Valve Train

The camshaft is gear-driven off the crankshaft and runs in three split-shell bearings. A Woodruff key positions *the crankshaft gear* while *the camshaft gear* is riveted in place. Solid lifters, pushrods, and adjustable rocker arms make up *the valve-operating linkage*.[13]

This passage is part of the manual's introductory description of the Volkswagen engine. There has been no previous mention in this manual of any *camshaft*, *crankshaft gear*, *camshaft gear*, or *valve-operating linkage*. Yet even a completely unspecialized reader would be able to infer from the writer's use of the definite article that there is only one of each of these items in a Volkswagen engine. In contrast, notice that the writer refers to *non*unique items by using either the indefinite article (*a Woodruff key*) or the plural with no article (*three split-shell bearings, solid lifters, pushrods, adjustable rocker arms*).[14]

30.8 A FLOWCHART FOR ARTICLES

Reviewing the main points of this chapter and the preceding one, we find:

1 The indefinite article (*a* or *an*) is used with a singular countable noun having a nonunique referent (see Chapter 29, Figure 29-2).

2 The definite article (*the*) is used with *any* noun (excluding most proper names) that has a unique referent.

3 A noun can be judged to have a unique referent for any one or more of the following reasons:
   a The noun may have a *special adjective* as a modifier (Section 30.1).
   b The noun may be a *special noun* (Section 30.2).
   c The noun may be *generic* (Section 30.3).
   d The noun may refer to some *previously mentioned* referent (Section 30.4).
   e The noun may have a *following modifier* (Section 30.5).
   f The noun may refer to *shared knowledge* (Section 30.6).
   g The noun may have a unique referent by *implication* (Section 30.7).

If you are undecided as to which article to use, try using Figure 30-1. To see how this flowchart works, let's consider some of the noun phrases in the following introductory paragraph from a report by the U.S. Energy Research and Development Administration:

Figure 30-1 ❑ FLOW CHART FOR CHOOSING THE CORRECT ARTICLE.

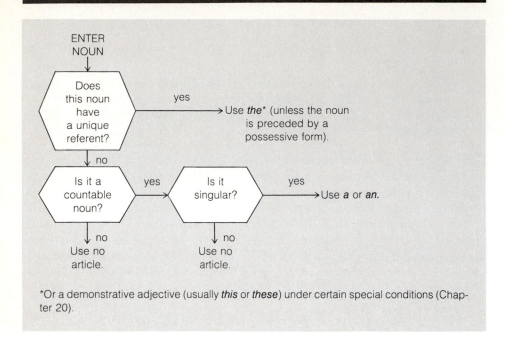

*Or a demonstrative adjective (usually *this* or *these*) under certain special conditions (Chapter 20).

> *The continuing depletion of domestic fossil fuels* may be a signal of one of the most significant long-term issues facing the United States as it enters its third century. The impacts of decisions made today concerning our remaining natural resources will persist for generations, beyond the lifetime of today's children. *Solar energy* will be *an important part of these decisions.*[15]

In the first sentence, the head noun of the italicized noun phrase is *depletion*; the other words of the phrase are its modifiers. If we were the authors of this paragraph and were not sure about which article to use, we could begin by asking ourselves the first question on the flow diagram in Figure 30-2 (in the upper left-hand cell): "Does this noun have a unique referent?" The answer is yes (1) because *depletion* has *a following modifier*, which identifies a particular kind of depletion, and (2) because this kind of depletion has been brought to public attention in recent years and is thus *shared knowledge*. Thus, we follow the "yes" arrow and arrive at the correct instruction: "Use *the*."

Next, consider the noun phrase *solar energy* in the last sentence. We begin, "Does it have a unique referent?" The answer is no because *solar energy* is being used here to refer to solar energy in general, not to any particular form of it. Second, "Is it a countable noun?" The word *energy* is being used here in its usual uncountable sense, so we answer no to this

Figure 30-2 ❑ USE OF FLOW CHART FOR CHOOSING THE CORRECT ARTICLE.

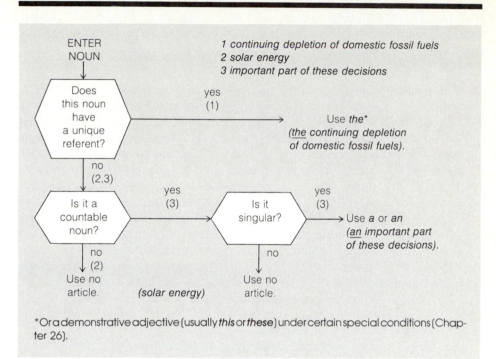

ENTER NOUN

1 continuing depletion of domestic fossil fuels
2 solar energy
3 important part of these decisions

Does this noun have a unique referent?

yes (1) → Use *the** (*the* continuing depletion of domestic fossil fuels).

no (2,3)

Is it a countable noun?

yes (3) → Is it singular?

yes (3) → Use *a* or *an* (*an* important part of these decisions).

no (2)

Use no article. (*solar energy*)

no

Use no article.

*Or a demonstrative adjective (usually *this* or *these*) under certain special conditions (Chapter 26).

question and arrive at "Use no article." Again, this result agrees with the choice made by the ERDA authors.

Finally, let's analyze the noun phrase *an important part of these decisions*. The head noun in this case is *part*, and so we begin by asking the first question, "Does this noun *(part)* have a unique referent?" The answer is no, even though this noun has a following modifier *(of these decisions)*; the writer wants to imply that there may be *other* important parts of these decisions, aside from solar energy. We therefore follow the "no" arrow and arrive at the next question, "Is it a countable noun?" The answer is yes, since we customarily think of decision making as involving the consideration of many different factors, or "parts." Therefore we follow the "yes" arrow to the next question, "Is it singular?" The answer is obviously yes again, which yields the final instruction: "Use *a* or *an*." (Since the first word following this article—*important*—begins with a vowel sound, the proper choice here is *an*.)

A word of caution in using the flowchart: be flexible and be prepared for occasional exceptions. No natural language can be completely precise or rule-governed, and many choices may depend on cultural knowledge or on subtle word meanings that happen to be unknown to you.

❑ EXERCISE 30-5

Now look at the following passages and determine, using Figure 30-1, which article to use for each blank space where appropriate. Be prepared to explain and justify your choices.

A   The economy was going strong just before ____ big financial crash of 1929. Production specialists had achieved ____ amazing results. What went wrong? With maximized production, would not ____ "invisible hand" of laissez-faire economics take care of ____ rest? Obviously, it did not. Most economic analyses include one or more of ____ following as ____ major causes of ____ depression: (1) an expanding of ____ business inventories and ____ accumulation of ____ large stocks of ____ new durables in ____ consumers' hands, (2) ____ consumer resistance to ____ rising prices and ____ increasing business costs, (3) ____ end of ____ upward acceleration effect and ____ resulting decline in ____ investment spending, (4) ____ accumulation of large amounts of ____ new productive capacity and new technological developments, (5) ____ reduction of promising large-scale investment outlets and ____ exhaustion of excess bank reserves, and (6) ____ weakening of ____ confidence and expectations.[16]

B   Standard ESSA Wind Observations
By far ____ most complete wind data available in ____ United States reside in ____ records of ____ U.S. Weather Bureau (ESSA) stations, which number more than 1000. Many of ____ stations have been in operation for one or more decades, and ESSA has provided ____ excellent record-keeping center at Asheville, North Carolina, so that data can be easily retrieved. ____ typical "surface" wind observation is usually obtained from ____ slow-moving cup anemometer and vane assembly, mounted in ____ well-exposed location about 30 feet above ____ ground or ____ building structure. ____ vast majority of ____ sites are at airports, and urban observations are comparatively rare. ____ standard observation procedure is to note ____ indicated wind speed and direction for ____ brief period once each hour and record them to 10° (formerly 16 compass points) and ____ nearest mile per hour. Normally ____ contacting device is associated with ____ anemometer, from which one can obtain ____ time required for one mile of wind flow to pass ____ instrument. Observations called PIBALS or RABALS are taken at many of ____ stations, using either visual or electronic tracking of ____ rising balloon several times per day to determine ____ variation of wind with height. ____ data are processed to reflect ____ wind at 1000-foot intervals above sea level.[17]

C   Manned Flight: A Success Story
On December 17, 1903, Orville and Wilbur Wright successfully achieved ____ sustained flight in ____ power-driven aircraft. The first flight that day lasted only 12 seconds over ____ distance of 37 meters (120 feet). This

is about ____ length of ____ space shuttle *Orbiter*. The fourth flight of ____ day (and ____ longest flight) traveled 260 meters (852 feet) in ____ 59 seconds. The initial notification of ____ event to ____ world was ____ telegram to ____ Wrights' father.

Sixty-six years later, ____ human being first stepped on ____ lunar surface. An estimated 500 million people throughout ____ world saw the event on ____ television or listened to it on ____ radio as it happened. This was surely ____ historic event.

Historic events are spectacular. The space program, however, has always been much more than a television spectacular. Today, ____ space transportation is working in ____ many ways for us all, and we have come to expect this.

____ whole new era of transportation came into being in ____ 1980s with ____ development of the space shuttle. As ____ transportation system to Earth orbit, it offers ____ workhorse capabilities of such earthbound carriers as trucks, ships, and ____ airlines. It is as vital to ____ nation's future in ____ space as the more conventional carrier of today is to ____ country's economic life and well-being.[18]

D   The Edsel: A Modern Antisuccess Story

Not since Ford's introduction of ____ Model A, 30 years earlier, had so much attention accompanied ____ arrival of ____ new car. When ____ same company formally introduced ____ Edsel in September 1957, it had already invested ____ quarter of ____ billion dollars in development, production, and distribution, which, according to one account of ____ day, made ____ Edsel ____ costliest consumer product in ____ history. Ford counted on selling at least 2,000,000 Edsels ____ first year.

However, a little over 2 years and 2 months later, Ford had sold only ____ 100,000 Edsels and ____ automaker permanently discontinued production. ____ total loss to ____ company reached $350 million, according to some estimates. ____ loss was equivalent to that which would be incurred if ____ Ford were simply to give away 110,000 models of its comparably priced car, ____ Mercury.

Conventional wisdom has held that ____ rapid decline of the Edsel was due to ____ company's dependence on ____ results of ____ public-opinion polls and motivational research. According to ____ view, ____ results dictated ____ way the Edsel was promoted, ____ way it was named, and also the way ____ car was designed. It is further argued that such efforts will always fail, for when ____ car-buying public perceives itself pursued in ____ overly calculated manner, it will always turn away in favor of ____ more spontaneously attentive competitor.

When conceived in ____ late 1940s, ____ idea that eventually led to the Edsel was one of putting on ____ market ____ new and completely different medium-priced car. It would be designed to keep ____ prosperous owners of ____ Fords, intent on trading in their symbols of low-income earnings, in ____ Ford family.

_____ Edsel design was certainly different. A novel radiator grill, in _____ shape of _____ horse collar, was set vertically in _____ center of _____ conventionally low, wide front end. _____ unique rear end was marked by widespread horizontal wings. Another feature, within _____ driving compartment, was a group of _____ automatic-transmission push buttons on the center of _____ steering wheel. _____ push buttons controlled what was then _____ most powerful engine (345 horsepower) for _____ automobile at _____ time of its introduction.

The Edsel's failure has been attributed to _____ rather long delay between conception and market introduction. Forced by _____ Korean War in 1950 to postpone the car's development, Ford came out with the Edsel precisely at _____ time the car-buying public was moving decisively toward _____ smaller, less powerful compacts. Moreover, within 2 years, _____ stock market would nose-dive, marking _____ beginning of _____ recession, and _____ automobile industry would end its season with _____ second largest number of unsold cars in history.[19]

E The Role of Standard Reference Materials in Environmental Engineering

_____ provisions of _____ Clean Air Act Amendments of 1970 and _____ Federal Water Pollution Control Act of 1972 include _____ requirements to limit _____ emission of _____ pollutants from various points of _____ discharge, such as _____ automobile tailpipe and _____ wastewater effluent. It is fairly obvious that such measures are required if we are to maintain our air and water in _____ sufficiently clean state for protection of _____ public health and welfare. What is not obvious is _____ exact extent to which it is necessary to limit _____ discharges in order to reach _____ sufficient purity. As long as _____ costs rise exponentially with degree of _____ purification, we may expect only enough public pressure to control discharges for adequate protection and no more. _____ economic fact underlines the critical necessity for _____ development of _____ valid environmental measurement system.

There are two fundamental elements to such _____ system. First, we must establish as accurately as possible _____ dose-response relationships so that _____ most proper ambient air and water quality standards can be set. Second, we must establish _____ relationship between _____ pollutant concentrations at _____ point of discharge and at _____ point of human contact. Basic to _____ measurement system are requirements that:

1   _____ health effect can be measured with sufficient accuracy.

2   _____ accurate model is available.

3   _____ pollutant measurements at ambient and source concentrations are internally consistent.

_____ part of _____ measurement system with which we at _____ National Bureau of Standards have been primarily concerned is _____ third of these. In the discussion that follows, it is presumed that ____ other two requirements are met.[20]

REFERENCES

1   It should be pointed out that ordinal adjectives do occasionally occur with the indefinite article: "There are several problems that need to be resolved before we can make progress on this project. One problem is that. . . . *A second problem* has to do with the auxiliary generator. . . ." In such cases, we would argue, the adjective is being used not in a truly ordinal sense but rather in an enumerative one.

2   Adapted from C. R. Spitzer, "Unlimbering Viking's Scoops," *IEEE Spectrum* *13*(10):92 (1976).

3   Adapted from E. A. Torrero, "The Big Pan of 3-D Radar," *IEEE Spectrum* *13*(10):51 (1976).

4   Adapted from M. F. Wolff, "The Navy's Big Dish," *IEEE Spectrum 13*(10):89 (1976).

5   T. A. Talay, *Introduction to the Aerodynamics of Flight*, NASA-SP-367 (Washington, DC: NASA, 1975), p. 149.

6   T. J. Schmigge, "Measurement of Soil Moisture Utilizing the Diurnal Range of Surface Temperature," in *Significant Accomplishments in Science and Technology: Goddard Space Center* (Washington, DC: NASA, 1975), p. 2.

7   *Pacific Regional Solar Heating Handbook* (Los Alamos, NM: Los Alamos Scientific Laboratory, Solar Energy Group, University of California, 1976), p. 82.

8   B. J. Conrath, "Dissipation of the Martian Dust Storm of 1971," in *Significant Accomplishments in Science and Technology: Goddard Space Center* (Washington, DC: NASA, 1975), p. 135.

9   NATO: Advisory Group for Aerospace Research and Development, Biodynamics Committee, "An Introduction to the Physics and Physiology of Acceleration," in *Principles of Biodynamics: As Applied to Manned Aerospace Flight, Section A, Prolonged Acceleration: Linear and Radial* (Paris, 1967), p. 1.

10   NATO: Advisory Group for Aerospace Research and Development, Biodynamics Committee, "The Dynamics of Rotation Applied to Centrifuges," in *Principles of Biodynamics: As Applied to Manned Aerospace Flight, Section A, Prolonged Acceleration: Linear and Radial* (Paris, 1967), p. 10.

11   B. J. Teegarden, "Late Results from the Pioneer-11 Flyby," in *Significant Accomplishments in Science and Technology: Goddard Space Center* (Washington, DC: NASA, 1975), p. 157.

12   *Volkswagen of America Official Service Manual* (Cambridge, Mass.: Robert Bentley, 1974), p. 4–17.

13   *Volkswagen Manual*, p. 5–4.

14   In cases of implied uniqueness, the writer's implication that there is only one referent will be accepted by the reader only to the extent that it conforms to that reader's experience in that particular culture. For example, "I took a stroll in the park yesterday and sat down for a while next to the person" sounds distinctly odd unless you have already identified which person you mean. On the other hand, "I took a stroll in the park yesterday and sat down for a while next to the bandstand

(drinking fountain, softball field, etc.)" sounds all right in U.S. culture because many parks contain only one bandstand (drinking fountain, softball field, etc.).

15 *Solar Energy in America's Future: A Preliminary Assessment*, 2d ed. (Washington, DC: Energy Research and Development Administration, Division of Solar Energy, 1977), p. 2.

16 Adapted from F. Luthans, *Organizational Behavior*, 4th ed. (New York: McGraw-Hill, 1985), p. 9.

17 Adapted from I. A. Singer and M. E. Smith, "The Adequacy of Existing Meteorological Data for Evaluating Structural Problems," in *Wind Loads on Buildings and Structures*. (Washington, DC: U.S. Department of Commerce, National Bureau of Standards, 1970), p. 23.

18 Adapted from "A New Era in Space," in *Space Shuttle* (Washington, DC: NASA, 1976), p. v.

19 Adapted from D. Christianson, "The Edsel: A Modern Antisuccess Story," *IEEE Spectrum 14*(11):94 (1977).

20 *Health, Environmental Effects, and Control Technology of Energy Use* Report No. 600/7-76-002 (Washington, DC: U.S. Environmental Protection Agency, 1976), p. 58.

31

Verbs

Although nouns and noun phrases are the principal carriers of information in scientific and technical English, verbs are the principal means by which these units of information are tied together in a sentence. Thus it is important that you have full command of the English verb system and of the most frequently used verbs in scientific and technical English.

This chapter addresses the most common areas of difficulty for nonnative speakers with regard to the use of verbs: (1) major tense distinctions, (2) the use of the progressive (*-ing*) form of the verb, (3) subject-verb number agreement, and (4) grammatical irregularities of particular verbs.

31.1 MAJOR TENSE DISTINCTIONS IN SCIENTIFIC AND TECHNICAL ENGLISH

Of the 12 traditional verb tenses in English, only 5 are used with any frequency in scientific and technical writing: simple present, simple past, present perfect, future (*will*), and present progressive. Each of these has particular meaning(s) associated with it and so should be learned thoroughly and used carefully. (*Will*, which is not a tense in the strict sense, is treated in Chapter 32.)

The Simple Present Tense

In formal scientific and technical English, the simple present is used primarily to express "timeless" generalizations—that is, general statements which do not specify any particular time frame. For example, the statement "Water boils at 212°F" is not restricted to a particular time frame and so is properly cast in the present tense (*boils*). Similarly, this recommendation in a company report—"I *recommend* that we continue to use the lime reactant-agent in our desulfurization process"—is given in the simple present tense because it represents a general statement that holds true not only at the time of writing but also indefinitely into the future.

Be sure to take full advantage of the simple present tense in your own writing. It is the most useful verb tense in scientific English, predominating in almost every type of writing situation except those explicitly set in the past (e.g., historical reviews, laboratory writeups) or in the future (e.g., proposals, speculations, recommendations). Even in these circumstances, the simple present tense can be used occasionally in an important contrastive way to indicate a generalization that is not restricted to the past or to the future. Such generalizations often represent evaluative judgments or interpretations on the writer's part and thus are crucially important to good writing. Unfortunately, many technical people often fail to make interpretive statements in situations where such statements are called for, thinking that their role as an objective scientist or engineer requires them to report only the facts. They consequently tend not to use the simple present tense as often as they should, preferring instead to use long sequences of past-tense forms. This sort of thinking is misguided, however. Scientists and engineers and other professionals are trained and employed precisely to make educated judgments and interpretations, not simply to report facts. Therefore, look for appropriate opportunities to make generalizations, and when the opportunity arises, use the present tense.

The Simple Past Tense

In contrast to the present tense, the simple past tense specifies a particular event or condition which occurred or existed at some time in the past but which no longer occurs or exists. For example, if you carried out a procedure as part of an experiment and are now reporting on it, you would probably use the simple past tense in your description:

> Commercial cholesterol *was recrystallized* three times from 95% ethanol. Radioactive cholesterol monohydrate *was prepared* by mixing 5 g of the recrystallized cholesterol. . . .

The immediate results of carrying out an experimental procedure are also usually reported in the simple past tense:

> NMR studies quantitatively *confirmed* the monohydrate nature of the crystals. TLC studies *indicated* the absence of any impurities.

Similarly, if you are writing a report in response to some request, you might begin by referring, in the past tense, to that request:

> On 15 December 1982, you *asked* us to help you find ways to lower your energy costs.

You might then continue with a description of the actions you undertook, also in the past tense.

Accordingly, we *visited* your plant and inspected the complete heating system.

In short, the simple past tense is used when simply reporting facts.

The Present Perfect Tense

Nonnative speakers often confuse this tense and the simple past tense, thinking that the two are more or less interchangeable. This is wrong, however: each has its own meaning and range of uses. Basically, the simple past tense is used for completed actions, whereas the present perfect is used for actions which were begun in the past but which are still going on. In the Review of Literature section of a scientific report, for example, single isolated studies are usually referred to in the simple past tense (Adler *reported* that . . .), whereas multiple studies, suggesting an ongoing sequence of studies, are usually referred to in the present perfect (Adler, Pierce et al., and other researchers *have reported* that . . .).

Similarly, the present perfect is used to report on actions that were carried out in the past but are still producing effects in the present. Consider this example taken from the opening paragraph of a company memo:

> Our Patent Department lawyers *have asked* this research group to provide them with data for a patent infringement suit. A competitor *has* recently *marketed* a new tower packing which they claim has superior pressure drop characteristics compared with our product.

In the first sentence, the use of the present perfect tense indicates that even though the lawyers may have asked only once, their request is still in effect and the research group is still expected to act on it. In the second sentence, the use of the present perfect not only reports on a past action (the marketing of a new product by a competitor) but also implies that this product is currently being actively promoted by the competitor. The choice of the present perfect tense in both cases thus serves to emphasize the immediate, ongoing nature of the threat posed by the competitor's past actions.

❏ EXERCISE 31-1

In each of the following passages, fill in the blank spaces with the most appropriate form of the verb given in parentheses. Be prepared to justify each of your choices in accordance with the guidelines provided above. The first one has already been done for you.

A   (beginning of a memo in response to the memo discussed above)

On September 14, 1982, you _____*requested*_____[1] a comparison
(request, requested, have requested)

of our patented ⅜″ raschings rings with a competitor's packing to support a patent infringement suit. Specifically, you _____*requested*_____ (2)
(request, requested, have requested)

that data on the packing factor and pressure-drop characteristics be compared. Accordingly, we _____*have carried out*_____ (3) your re-
(carry out, carried out, have carried out)

quest, using a column of the competitor's packing in our research facility. The purpose of this report _____*is*_____ (4) to provide our lawyers with
(is, was, has been)

the requested data for the patent infringement suit.

The results _____*show*_____ (5) that the packing characteristics
(show, showed, have shown)

of the competitor's packing differ significantly from those of our packing. Our packing factor _____*is*_____ (6) 1000; the competitor's packing factor
(is, was, has been)

_____*was*_____ (7) determined to be 681.
(is, was, has been)

Justification:

1   Simple past tense because the request was made in the past and has been carried out

2   Simple past tense for the same reason as 1

3   Present perfect because it relates past actions (measuring and comparing) to present purpose

4   Simple present because the purpose of the report is not restricted to a particular time period but applies whenever anyone reads the report

5   Simple present because this is a generalization that is not restricted to a particular time frame

6   Simple present because this is a general fact about the packing factor of the company's product that applies at all times

7   Simple past because it refers to the action of measuring the competitor's packing factor, which was done in the past

B   Before the *Voyager I* encounter with Saturn we _____ far
(have, had, have had)

less information about its satellites than we _____ about Jupiter's
(do, did)

before 1979. This _____ due to Saturn's greater distance from
(is, was, has been)

Earth and the sun, the smaller diameter of its satellites, and their closeness to Saturn. Only Titan _____ comparable in size to
(is, was, has been)

the Galilean satellites. The icy satellites _____ intermedi-
(are, were, have been)

ate in size between the largest asteroids and the moon. *Voyager I* _____ the first close look at bodies in this size range.[1]
(provides, provided, has provided)

C (the introductory paragraph of a company memo)
At our meeting on October 15, 1989, I _____ asked to in-
                                        (am, was, have been)
vestigate the noise-related problems that one of our production supervisors, D. P. Stonear, _____ having on the job lately. As you
                         (is, was, has been)
may recall, Mr. Stonear _____ that he
                          (complains, complained, has complained)
_____ having increasing difficulty understanding people on the
(is, was, has been)
telephone in his small office located in the middle of the wheel-stamping
room. He _____ that his office _____ too
          (believes, believed, has believed)   (is, was, has been)
noisy and _____ continually apprehensive about answering the
           (is, was, has been)
phone. Most important, his nervousness over this communication problem
_____ to be undermining his effectiveness as a super-
(seems, seemed, has seemed)
visor. The purpose of this report _____ to determine the cause
                                    (is, was, has been)
of Mr. Stonear's communication problem and to suggest corrective measures.

D (the introductory paragraph of a science journal article)
Recently great efforts _____ made to develop geochemical
                         (are, were, have been)
techniques to estimate environmental parameters such as temperature
and salinity. While the oxygen isotope method for paleotempera-
tures _____ employed extensively, efforts to refine methods for
       (is, was, has been)
paleosalinities _____ less successful. Most paleosalinity methods
                 (are, were, have been)
_____ based on Goldschmidt's classical observations on
(are, were, have been)
the occurrence and distribution of trace elements in different sedimentary environments. For example, Degens et al. (1957)
_____ marine from freshwater shales on
(differentiate, differentiated, have differentiated)
the basis of spectrochemical analysis for boron, gallium, and rubi-
dium. Similarly, Potter (1963) _____ that
                                 (demonstrate, demonstrated, has demonstrated)
each of the elements boron, chromium, copper, nickel, and vanadium
_____ more abundant in marine than in freshwater argillaceous
(is, was, has been)
sediments.[2]

31.2 THE USE OF THE PROGRESSIVE (-ING) FORM OF THE VERB

Another common area of difficulty for nonnative speakers is the use of the
progressive (-ing) form of verbs. This form should be reserved for situations

where you are describing an event in the process of occurring—that is, an incomplete, ongoing event. It is a particularly useful form in progress reports, in letters, and in introductions to technical reports. Here is an example:

> In this day and age, the computer *is finding* more applications than were ever conceived possible. Computers *are* now *controlling* heating in buildings, traffic signals in cities, and guidance control in commercial and military jets. The feasibility of using an onboard computer to control fuel flow in personal transportation has also been a point of interest because a substantial amount of fuel can be saved with the proper equipment controlling fuel allocation.
>
> Global Design Corporation *is* currently *developing* a computer-controlled vehicle that uses both an internal combustion engine and lead-acid batteries as power sources.

This passage appears at the beginning of a technical report about electric hybrid vehicles. By using the progressive form of the present tense, the writer focuses attention on the here and now, on the state of the art in computer-controlled vehicles. This emphasis on actions that are not yet complete makes the description more dynamic and attention-getting than it would otherwise be.

In using the progressive form of the verb, however, you must keep in mind an important restriction: generally only "event verbs" can take the progressive form, not stative verbs. This makes sense since the progressive form, as mentioned above, is used to describe events. Notice, for example, that the three verbs in the progressive form in the above example are all event verbs: *find*, *control*, and *develop*. The only main verb—*be*, in sentence 3—is a stative verb and thus does not appear in the progressive form. It would sound very much like a nonnative speaker error to read, *"The feasibility of using an onboard computer . . . *is also being* a point of interest. . . ."

Below is a list of all the stative verbs found among the 200 most common verbs in scientific and technical English.

Verbs That Do Not Occur in the Progressive Form

afford	correspond	represent
appear	differ	result in
be	exist	satisfy
believe	involve	seem
concern	know	suppose
consist of	mean	understand
constitute	need	yield
contain	possess	

Note, however, that although stative verbs cannot be used in the progressive aspect, they *can* be used in short-form (*-ing*) relative clauses, as

---

* Indicates that the following sentence is ungrammatical.

described in Chapter 33. For example, the sentence *"The procedure is consisting of five main steps" is ungrammatical because *consist of* is a stative verb and cannot be used in the progressive aspect; however, a sentence like "This is a procedure consisting of five main steps" would be perfectly correct because the *-ing* form in this case is not the progressive but rather part of a short-form relative clause.

31.3 SUBJECT-VERB NUMBER AGREEMENT

A basic rule of English grammar is that the subject noun phrase and the main verb of a sentence or clause must agree in number—that is, they must both be singular or both be plural. No singular-plural mixes are allowed. Although your native language may not have such a rule, you should try to observe it carefully in English. It not only makes you look bad to have such grammatical errors in your writing; it can also create difficulties for the reader trying to comprehend your meaning. For example, if you wrote *"The other method for determining K use the Arrhenius Equation," it would not be clear to the reader whether you meant one method or more than one method because you've put the subject noun phrase (*method*) in singular form but the main verb (*use*) in plural form. In such a case, the reader will simply be forced to guess your meaning—and he or she may guess wrong.

If you're using a grammatically complex subject noun phrase, be sure to use the head noun in determining subject-verb number agreement. Otherwise, you may commit errors like these:

*The above procedures in the dark stage is based on tracer studies.

*A lot of research programs has already been carried out.

In the first sentence, *procedures* is the head noun in subject position, and so the main verb should be in the plural; the fact that there is an intervening noun in singular form (*stage*) has no bearing on the form of the verb because this intervening noun is not the head noun. In the second sentence, *research programs* is the head noun in subject position, and so the main verb here, too, should be in the plural. Do not be misled by the singular noun *lot*: it is simply part of the quantifier phrase *a lot of*, equivalent in meaning and in use to *many*.

Another common mistake in subject-verb agreement involves the word *each*, e.g., *"Each of the components have a different function." In this type of sentence, *each* is the head noun of the subject phrase; since it is singular, the main verb should also be singular: "*Each* of the components *has* a different function." On the other hand, if the sentence were constructed like this—"The components each have a different function"—*have* would be the correct form of the verb because the head noun is now *components*, not *each*.

---

* Indicates that the following sentence is ungrammatical.

31.4 GRAMMATICAL IRREGULARITIES OF PARTICULAR VERBS

Many mistakes made by nonnative speakers in the use of verbs derive from the irregular features those verbs have. We saw an example of this in Section 31.2, where it was pointed out that a small set of English verbs—stative verbs—do not occur in the progressive. In this section, we shall discuss several other irregularities associated with particular verbs: (1) verbs that do not occur in the passive form, (2) verbs taking *-ing* or unmarked complements, and (3) verbs with irregular forms of inflection.

Verbs That Do Not Occur in the Passive Form

A passive sentence (as discussed in Chapter 25) is one in which the recipient of the action appears in subject noun phrase position, not in its normal direct object position. For example, instead of saying, "We then measured the reaction rate," we could put the direct object phrase, *reaction rate*, in subject position and say, "The reaction rate was then measured (by us)." The passive is a particularly common and useful form for describing experimental procedures, chemical processes, cause-and-effect relationships, etc.

The problem is that a fairly large number of English verbs do not take direct objects to begin with; these are the so-called intransitive verbs, such as *exist, fall, differ*, and *occur*. Using the passive form with such verbs is therefore completely ungrammatical, as in *"This contamination is occurred frequently." Quite a few other verbs exist, too, which logically could be used in the passive form but which virtually never are, at least not in scientific and technical English: *yield, suffer, possess*, etc. Below is a complete list, from among the 200 most frequently used verbs in scientific and technical English, of those verbs that do not occur in the passive form.

appear	fall	possess
arise	flow	proceed
be	function	remain
become	get	result in
begin	go	rise
come	happen	seem
consist of	have	suffer
correspond	lead	tend
depend on	let	travel
differ	lie	undergo
enable	live	work
exist	occur	yield

* Indicates that the following sentence is ungrammatical.

Verbs Taking *-ing* or Unmarked Complements

Most complement-taking verbs in scientific and technical English take *to* or *that* complements:

> Recently you *asked us to* undertake an investigation of. . .

> Our findings *indicate that.* . .

Thus, many nonnative speakers tend to rely exclusively on these two types of complements. This strategy works in most cases, but not in all, for there is a small set of verbs which take *-ing* or unmarked complements. If you fail to observe these exceptions, you'll make mistakes like *"We recommend to choose the first option" and *"This modification should make the system to function better." *Recommend* takes an *-ing* (or *that*) complement, and *make* takes an unmarked complement; therefore the correct versions of these two sentences would be "We recommend choosing the first option" and "This modification should make the system function better."

Below is a list of frequently used verbs in scientific and technical English which take *-ing* or unmarked complements. Commit them to memory so that you do not repeatedly make mistakes with them.

Verbs taking *-ing* or *to* complements	*try, start*
Verbs taking *-ing* or *that* complements	*suggest, recommend, emphasize*
Verbs taking *-ing* complements only	*discuss, consider, resist, reject, consist of, insist on, depend on, result in*
Verbs taking unmarked complements only	*make, let*

NOTE: This last set of verbs also requires a noun or noun phrase between the verb and its complement, as in "This modification should make *the system* function better." If you do not observe this rule, the result will be ungrammatical: *"This modification should make function better the system."

Verbs with Irregular Forms of Inflection

Many of the most common verbs in English came originally from German, which typically changes the vowel of the verb stem to indicate past tense or participial forms rather than simply adding a suffix to the end of the word. Thus, whereas most English verbs simply add *-ed* or *-d* to the stem to form the past tense and participial forms (e.g., *demonstrate, demonstrated, have demonstrated*), many of our most common verbs make other changes instead (e.g., *give, gave, given*).

These irregular verbs are indeed irregular, and for that reason they cause all sorts of problems for nonnative speakers. However, since they are

---

* Indicates that the following sentence is ungrammatical.

among the most frequently used verbs in the language, you should make every effort to learn and use the correct forms. Below is a list of 50 of the most common ones.

Uninflected Form	Simple Past	Past Participle
arise	arose	arisen
bear	bore	borne
become	became	become
begin	began	begun
bend	bent	bent
break	broke	broken
bring	brought	brought
build	built	built
catch	caught	caught
choose	chose	chosen
come	came	come
cut	cut	cut
do	did	done
draw	drew	drawn
fall	fell	fallen
feed	fed	fed
find	found	found
get	got	gotten or got
give	gave	given
go	went	gone
grow	grew	grown
have	had	had
hold	held	held
keep	kept	kept
know	knew	known
lead	led	led
leave	left	left
let	let	let
lie	lay	lain
lose	lost	lost
make	made	made
mean	meant	meant
put	put	put
read [pron. "reed"]	read [pron. "red"]	read [pron. "red"]
rise	rose	risen
run	ran	run
say	said	said
see	saw	seen
send	sent	sent
set	set	set

Uninflected Form	Simple Past	Past Participle
show	showed	shown
shut	shut	shut
spend	spent	spent
spread	spread	spread
stand	stood	stood
sweep	swept	swept
take	took	taken
tear	tore	torn
think	thought	thought
write	wrote	written

REFERENCES

1   B. A. Smith et al., "Encounter with Saturn: *Voyager 1* Imaging Science Results," *Science 212*(4491):170 (1981).
2   N. A. Nugent and R. C. Fuller, "Sedimentary Phosphate Method for Estimating Paleosalinity," *Science 158*(3803):917 (1967).

ADDITIONAL READING

Practical

S. McKay, *Verbs for a Specific Purpose* (Englewood Cliffs, NJ: Prentice-Hall, 1981).

Theoretical

B. Comrie, *Aspect* (Cambridge, England: Cambridge University Press, 1975).

B. Gorayska, "The English Verb System," *Interlanguage Studies Bulletin 3*(2):234–249 (1978).

M. Joos, *The English Verb: Form and Meanings* (Madison: University of Wisconsin Press, 1964).

A. Ota, *Tense and Aspect of Present-Day American English* (Tokyo: Kenkyusha, 1963).

32

Modal Verbs

Technical professionals are often required to make careful, precise statements. One way of being precise is to use modal auxiliary verbs (or "modal verbs") appropriately. The most important modal verbs are *may, can, must, should, could, would, will,* and *might*. The difficulty in using these modal verbs correctly is that they all have two or more different meanings, depending on how they are being used. For example, the word *must* can mean that the statement in which it is used is logically necessary: "Since $2x + y = 11$ and $5y = 15$, x *must* equal 4." Or, it can mean that the statement in which it is used represents some kind of obligation: "We *must* thank Dr. Wilbur for her help." In all, these eight modal verbs can be used to express at least 16 different meanings.

Our purpose in this chapter is to make it as easy as possible for you to learn and use all 16 of these modal meanings. To do this, we shall divide them into three distinct groups, each of which constitutes a scale of degrees. These groups we shall label "obligation," "probability," and "ability."

32.1 OBLIGATION

Modal verbs are commonly used to indicate degree of obligation or prohibition, as in "We must thank Dr. Wilbur for her help." Here the obligation is a strong one, equivalent to "It is required that we thank Dr. Wilbur for her help." The writer or speaker could reduce the degree of obligation by using a weaker modal, such as *should* or *might* (or the informal *ought to*), depending on the situation. For example, if Dr. Wilbur has been a great help, then the writer would probably want to use the strong modal *must* to show it.

Figure 32-1 □ MODAL VERBS INDICATING DEGREES OF OBLIGATION

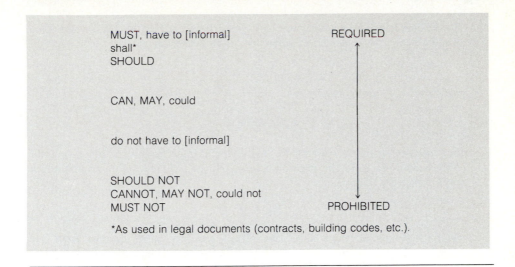

MUST, have to [informal]   REQUIRED
shall*
SHOULD

CAN, MAY, could

do not have to [informal]

SHOULD NOT
CANNOT, MAY NOT, could not
MUST NOT   PROHIBITED

*As used in legal documents (contracts, building codes, etc.).

The scale as a whole is given in Figure 32-1. Each modal verb is positioned on this scale according to the approximate degree of obligation or prohibition it imposes on the meaning of a sentence. *Must* is at the top end of the scale because it means strongly obligatory, whereas *must not* is at the opposite end because it means obligatory in the opposite (prohibitive) sense. The modals *can*, *may*, and *could* are located near the middle of the scale because their meaning lies midway between these two extremes; when you use one of these three modals, you are saying that the action of the main verb is neither required nor prohibited, but optional. Finer distinctions can be made, too. For example, the modal phrase *do not have to* also has an optional sense to it, but it differs from *can*, *may*, and *could* in that it means optional in a negative sense.

The most frequently used modals of obligation in scientific and technical English are *must*, *should*, *can*, and *may*; accordingly, we have written these with capital letters in Figure 32-1. The left-to-right ordering is also an indicator of frequency. For example, *can* is used more often than *may*, which in turn is used more often than *could*. It should be kept in mind, however, that frequency of use varies according to situation and context. In a situation requiring extensive use of the past tense, for instance, *could* (the past-tense form of *can*) is more likely to be used than *can*. In oral communication, *have to* might be heard more frequently than *must*.

In scientific and technical English the modals of obligation are most often used with statements describing procedural correctness. In the following extract, for example, the writer has used *may* to indicate that her statements describe legitimate options that one can take in explaining what an arithmetic progression is:

> To explain arithmetic congruence, let us start with an arithmetic progression. An arithmetic progression is a sequence of integers such as

$$3, 5, 7, 9, 11, 13, \ldots$$

> which we *may* also represent as

$$a_1, a_2, a_3, a_4, a_5, \ldots$$

> in which each term a_n (n stands for an integer) is generated by adding a common difference (here 2) to the preceding term. Hence we *may* define an arithmetic progression by the formula

$$a_n = qn + r$$

> where n stands for the rank of the term a_n and has values ranging from 1 to infinity, q for the common difference, and r for the remainder obtained by dividing the term by the common difference.[1]

This use of modals in descriptions of procedural correctness is particularly common in textbooks. In fact, most of the modal verbs used in science, mathematics, and engineering textbooks are modals of obligation used in describing procedures.

One other feature of the modals of obligation is that they are usually used in passive sentences with the agent not mentioned. In the extract just cited, for example, most writers would have written this sentence:

> Hence *we may define* an arithmetic progression by the formula

$$a_n = qn + r$$

like this:

> Hence an arithmetic progression *may be defined* by the formula

$$a_n = qn + r$$

After all, there is no real need to mention that it is "we" who are doing the defining; it is standard algebraic procedure, and anyone familiar with algebra could do the same. So, by putting the sentence in passive form, we can omit the reference to "we."

❑ EXERCISE 32-1

Below are two excerpts from a basic chemistry textbook. For each of the blank spaces, choose the modal of obligation that you think best fits the context.

A Errors and Significant Figures

All quantitative science is based on measurements, but measurements are usually afflicted by errors, which ____ be taken into account. Several strategies ____ be used to serve this purpose. . . .

An important aspect of the handling of data is that intermediate computations are required. This raises two questions. One concerns the relation of the uncertainty in the quantity of interest to the errors in the original measurements. In other words, how do the errors propagate through the equations that connect the quantity of interest with the data? The other question concerns the accuracy with which intermediate calculations ____ be carried out. To do justice to the data, the accuracy of the calculations ____ be sufficiently high that no new errors are introduced, but excessive accuracy is pointless and inefficient.[2]

B Chemical Equations

The reactions of chemical substances ____ be represented by chemical equations that show the formulas and relative numbers of the reactants and products involved. Equations ____ , by definition, be balanced: they ____ show the same number of atoms of a given kind on each side, and the net charge ____ be the same on each side. The condition, sometimes slavishly insisted upon, that all coefficients in an equation ____ be integral is not essential; coefficients ____ be fractional or ____ even have a common factor. . . .

Many equations are easy to balance, once the important reactants and products are known. For example, when dissolved in water, silver nitrate, $AgNO_3$, and sodium sulfate, Na_2SO_4, react to form solid silver sulfate, Ag_2SO_4, which precipitates from the solution. An unbalanced equation for the reaction is

$$AgNO_{3\ +}\ Na_2SO_4 \rightarrow Ag_2SO_4 + \ . \ . \ .$$

which ____ be balanced to

$$2\ AgNO_3 + Na_2SO_4 \rightarrow Ag_2SO_4 + 2\ NaNO_3$$

The arrow is often replaced by an equal sign for typographical reasons, but this ____ be done only when the equation is balanced.[3]

32.2 PROBABILITY

The modal verbs are also used to indicate degree of probability, as in the following sentence:

The x-ray count data were corrected for background but were not corrected for effects of absorption, so reported concentrations may be in error by up to ± 20 percent of the measured amount.[4]

Figure 32-2 ❑ MODAL VERBS INDICATING DEGREES OF PROBABILITY

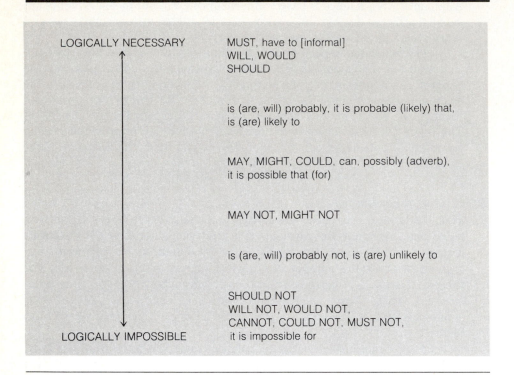

Here the writers are implying that an error of ± 20 percent is *possible*—that is, only moderately probable. If they had thought there was a high probability of error, they would likely have used a stronger modal, such as *should*, *will*, or *must* (or some adverbial form like *are probably*). The complete scale of choices, ranging from logically necessary to logically impossible, is given in Figure 32-2.

In scientific and technical English, the modals of probability are most often used in Conclusions, Summaries, Abstracts, etc.—wherever tentative generalizations are being made. They are also commonly used in reviews of literature, where the findings of previous investigations are being described. In all of these cases, *may*, *will*, and *would* are the most frequently used modals, followed by *might*, *must*, *can*, and *should*, in that order.

Figure 32-2 also includes a number of modal phrases that can be used to imply degrees of probability—phrases such as is *probably*, *is likely to*, *it is possible that*, and *it is impossible for*. Since most of these expressions occupy important positions on the scale and do not have exact modal verb equivalents, they should be learned and used as well.

Making Hypothetical Statements

From time to time you may find it necessary to make hypothetical or contrary-to-fact statements. For example, if you are describing the possible consequences of some action or theorizing about something or drawing an analogy to explain a difficult concept, you may want to make it clear to your reader that you are merely hypothesizing. In such cases, *would* is by far the most commonly used modal verb, followed by *might* and *could*.

Hypothetical statements are always made with regard to certain conditions. Consider the following sentence, for example:

> If the earth's magnetic field were reversed, the earth's surface would be exposed to a greater cosmic ray intensity than normal.[5]

Here the hypothetical statement (*the earth's surface would be exposed to a greater cosmic ray intensity than normal*) is directly preceded by a statement of conditions (*If the earth's magnetic field were reversed. . .*). Note that the verb in the *if* clause is in the simple subjunctive form, *were reversed*, in keeping with the simple conditional form of the main clause verb, *would be exposed*. It is important to maintain correct pairings of verb forms in such constructions and to not be confused by the similarity of subjunctive and past-tense forms. In this case, for example, *were reversed* looks like a past tense, yet has nothing to do with past time. (If the writer had wanted to refer to the past, he would have used the past subjunctive form, *had been reversed*, in the *if* clause and the past conditional, *would have been exposed*, in the hypothetical statement.) To make conditional statements that are less hypothetical (i.e., more likely to come true), use the present tense for the verb in the *if* clause:

> If the earth's magnetic field *is reversed*, the earth's surface *will be exposed*. . . .

Alternatively, you can use *should* in place of *if* and a simple nonfinite form of the verb:

> *Should* the earth's magnetic field *be reversed*, the earth's surface *will be exposed*. . . .

Although hypothetical statements are always made with regard to certain conditions, these conditions are not always stated explicitly in an *if* clause. For example, the meaning of the sentence given above is essentially retained in these versions without an explicit *if* clause:

> *It is possible for the earth's magnetic field to someday be reversed. In such an event*, the earth's surface would be exposed to a greater cosmic ray intensity than normal.

> *A reversal of the earth's magnetic field* would have the effect of exposing the earth's surface to a greater cosmic ray intensity than normal.

In the first instance, a full sentence is used to present the hypothetical condition and a connective phrase (*in such an event*) is then used to link this condition to the subsequent *would* statement. In the second instance, a noun phrase is used. The indefinite article preceding this noun phrase and the conditional *would* immediately following it make it clear to the reader that such a reversal of the earth's magnetic field is only a hypothetical possibility.

Once the conditions are established—whether by *if* clause, by full sentence, or by noun phrase—it is often possible to make a number of hypothetical statements without having to restate the full set of conditions each time. Continued use of the modal verb in these succeeding statements will make it clear to the reader that they are hypothetical statements, and the reader will then infer that the same conditions hold. Consider this passage, for example:

> If the earth's magnetic field were reversed, the earth's surface *would* be exposed to a greater cosmic ray intensity than normal. Some scientists argue that the consequent increase in radiation dosage at sea level *could* have a serious impact on animal life; they claim, in particular, that it *might* cause an increased mutation rate. Others argue, however, that this *would* be most unlikely, since the increased radiation dosages *would* be relatively small.

Here the basic condition is established in the first sentence, and it is then assumed to hold for the subsequent series of hypothetical statements. Because the writer has been careful to use the modals *would*, *could*, and *might* repeatedly and consistently, we as readers have no trouble understanding that the discussion as a whole is meant to be purely hypothetical.

❑ EXERCISE 32-2

Fill in the blanks with an appropriate modal verb or phrase.

A An employer plans to interview eight equally qualified people, including Robert, for possible employment. From these eight she will select two for job offers. Robert _____ be selected.

B Maria drove from Chicago to Detroit, a distance of 240 miles, in just under 4 hours. At some point, she _____ have been driving more than 60 mph.

C A new plastic underwent preliminary strength testing. The breaking load for each of the eight trials was found to be (in 1000 pounds per square inch) 8.0, 8.7, 7.2, 7.9, 7.8, 8.4, 7.8, and 8.1. If we subject this plastic to a load of 8.6 thousand lbs/in.^{2}, it _____ break.

D A certain transistor-manufacturing process has been known to have a defect rate in the 4–7% range. This means that as many as 70 transistors out of 1000 made by this process _____ be defective.

E If the number 4,294,967,297 is equal to the product of 641 times 6,700,417, it _____ be a prime number.

❏ EXERCISE 32-3

Write a one-paragraph description of what you would do if you won a $5 million lottery prize.

32.3 ABILITY

The final category of modal usage is that referring to ability, or capability. The dominant modal of this category is *can*, followed by its past-tense/ hypothetical variant *could*. The phrase *be able to* is also used quite frequently, combining with modals such as *should*, *may*, and *might* to indicate various degrees of ability.

Below is an example showing modals of ability being used in a scientific report about the possibility of constructing a gigantic "cosmic lift" in the earth's upper atmosphere. Notice how the writer artfully shifts back and forth between *can* and the more hypothetical *could*.

> The energy liberated by the lift *can* be used in different ways. First of all, it *can* be used for space flights. The energy developed over the upper section— above 36,000 km—*could* be fed to a power station supplying the lower section of the lift. This would result in an interesting situation. The expenditure of energy for ascending the "cable-way" to outer space *could* be reduced to a minimum. According to Artsutanov's calculations, a height of 144,000 km *can* be reached without wasting any energy at all. The amount of work obtained along the upper part of the path would equal that spent along the lower part. At heights greater than 144,000 km, operation of the cosmic lift would turn into pure profit. The lift would become a sort of power station.[6]

The complete scale of choices for the modal verbs and phrases of ability is given in Figure 32-3.

❏ EXERCISE 32-4

Below are two passages from chemistry and physics textbooks. For each of the blank spaces, choose the modal verb or phrase that you think best fits. (NOTE: This exercise tests your ability to use modals from all three of the categories discussed above.)

A   Significant Figures and Relative Precision
    In general, results of observations _____ be reported in such a way that the last digit given is the only one whose value _____ be in doubt. The digits that constitute the result, excluding leading zeros, are then termed significant figures. The number of significant figures is the same in 23.5 mg and 0.0235 g; this is reasonable, for the uncertainty in a result _____ not depend on the units in which the answer is expressed. Note that there is some uncertainty about the number of significant figures in a weight

Figure 32-3 ❑ MODALS INDICATING DEGREES OF ABILITY

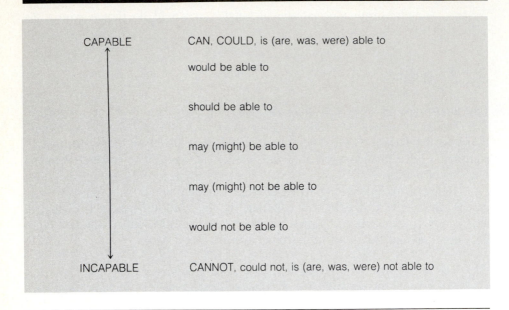

reported to be 2350 kg, because it is unclear whether the final zero is significant or not. The matter is clarified by reporting the weight either as 2.350×10^3 kg or as 2.35×10^3 kg, whichever applies.

The concept of significant figures does not apply to quantities known to be integers; for example, in a molecular formula, such as C_6H_{14}, the number of figures in each subscript does not imply that the composition is uncertain. We ＿＿ assume that there are exactly 6 mol carbon atoms per mole of compound, even though the formula ＿＿ to the unsophisticated seem to imply that this ratio is known only to low precision. Similar considerations apply to integral numbers of objects; the integers are regarded as infinitely precise.[7]

B Inhibition of Corrosion

Corrosion ＿＿ be curbed in a number of ways. The most obvious method is to cover the metal surface with a coating impervious to air and moisture. Because corrosion ＿＿ begin in even microscopic cracks, this coating ＿＿ itself be completely intact. It ＿＿ be a layer of paint, plastic, or some similar material, a coating of a strongly adherent oxide, or a layer of some metal that is itself resistant to corrosion.

An interesting phenomenon is *passivation*, in which a metal surface is made inactive by covering it with a very thin oxide layer. Iron ＿＿ , for example, be passivated by treatment with concentrated nitric acid or dichromate. Once in this state it does not react with acid or reduce cupric ions. Passivation is easily destroyed, however, and if the surface is scratched anywhere, the protection breaks down—not only at the location of the scratch but in a steadily increasing area with a fast-moving boundary.[8]

REFERENCES

1 J. Wei, "Pure Mathematics: Problems and Prospects in Number Theory," *The Research News* *30*(3):18 (1979).

2 J. Waser, K. N. Trueblood, and C. M. Knobler, *Chem One* (New York: McGraw-Hill, 1976), p. 33.

3 Waser, Trueblood, and Knobler, p. 41.

4 E. W. White and W. B. White, "Electron Microprobe and Optical Absorption Study of Colored Kyanites," *Science* *158*(3803):915 (1967).

5 Adapted from C. J. Waddington, "Paleomagnetic Field Reversals and Cosmic Radiation," *Science* *158*(3803):913 (1967).

6 J. D. Issacs et al., "Sky-Hook: Old Idea," *Science* *158*(3803):947 (1967).

7 Waser, Trueblood, and Knobler, p. 35.

8 Waser, Trueblood, and Knobler, p. 524.

33

Relative Clauses

A relative clause is a clause (or phrase) that is attached to a noun and serves to narrow the meaning of that noun. Relative clauses typically begin with a relative pronoun (*which*, *that*, *who*, etc.), though the relative pronoun is often omitted. Here are some examples:

> An exposure meter is a device that converts light energy into electrical energy.[1]

> The ignition system of an automobile contains several components which depend on the vehicle's 6– or 12–volt battery.[2]

> Any celestial body that emits light waves may be expected to emit radio waves because both wave forms are part of the same electromagnetic spectrum.[3]

> The abstracts for these papers are then printed out, together with an identifying number through which the article may be located.[4]

> A voltmeter is an instrument used for measuring the differences of potential between different points of an electric circuit.[5]

Sometimes two (or more) relative clauses overlap, that is, one relative clause may modify a noun which is itself part of another relative clause. Here is an example:

> Simple gears have teeth that are parallel to the axle or shaft on which they turn.[6]

Nonnative users of English sometimes have trouble constructing relative clauses and try to avoid them. This is understandable, since languages differ in the way they form relative clauses. If *you* have trouble making relative clauses in English, it's probably because your own native language does it differently.

The purpose of this chapter is to give you the basic rules for constructing grammatically correct and stylistically effective relative clauses in English.

33.1 GRAMMATICAL RULES

Positioning the Relative Clause

Rule 1
Put the relative clause directly after the noun it modifies.

Correct
The energy *which is consumed in overcoming friction* is not lost but is converted into heat.[7]

Incorrect
The energy is not lost *which is consumed in overcoming friction* but is converted into heat.

❑ EXERCISE 33-1

Check each of the following sentences for proper positioning of the relative clause. Make appropriate changes where necessary.

A   Any celestial body may be expected to emit radio waves that emits light waves because both wave forms are part of the same electromagnetic spectrum.[8]

B   The concepts were known previously which underlie these methods but could not be economically utilized.[9]

C   A turbine is a device that converts the kinetic energy of a moving fluid to the rotational energy of a spinning shaft.[10]

D   Today a jet engine develops 40,000 pounds of thrust that weighs 5000 pounds.[11]

E   With a half-life of 700 million years, the quantity of uranium-235 which was initially present on earth has dwindled greatly.[12]

F   A handicapped person may have no way of interacting with his community who feels he cannot trust a transit system to get him out of the house and back again.[13]

Exceptions to the rule are allowed in cases where the rule would produce either an extremely unbalanced sentence (Chapter 25) or a confusing one. Consider, for example, this sentence:

Exception
A popular large computer system can do computations in one minute *that it would take approximately 50 years to do manually.*[14]

Here the relative clause modifies the noun *computations* but is separated from it by the phrase *in one minute.* Following the rule in this case would have produced a potentially confusing result: "A popular large computer system can do computations that it would take approximately 50 years to do manually in one minute." In this version, it sounds like computations are being done manually in one minute!

Here is another example where violating the rule is called for:

Exception
Computer programs have been written *which analyze activities in order to calculate which are critical to the timely completion of a project and which can take longer than expected without delaying the completion date.*[15]

In this case the relative clause modifies the noun *computer programs* but is separated from it by the verb phrase *have been written.* If the writer had followed the rule, the result would have been an extremely unbalanced sentence with the verb stuck at the very end: "Computer programs which analyze activities in order to calculate which are critical to the timely completion of a project and which can take longer than expected without delaying the completion date have been written." As is discussed in Chapter 25, such a lack of balance would cause readers more frustration than would the violation of the relative clause rule.

Do keep in mind, however, that exceptions like those just described are indeed exceptions! They occur only rarely in formal written English. Therefore, it is a good idea not to break the rule unless very unusual circumstances prevail.

Heading the Relative Clause with a Relative Pronoun

Rule 2
Make sure there is a relative pronoun at the head of the relative clause (optionally preceded by a preposition).

Correct
A hygrometer is a device <u>which</u> *is used to measure the relative humidity in the atmosphere.*[16]

Incorrect
A hygrometer is a device *is used to measure the relative humidity in the atmosphere.*

Correct
The temperature *at which* *evaporation occurs* is called the saturation temperature.[17]

Incorrect
The temperature *evaporation occurs at* is called the saturation temperature.

❑ EXERCISE 33-2

Check each of the following sentences and see if you can spot grammatical errors involving relative clauses without relative pronouns. Make whatever corrections are necessary. (NOTE: Some of the sentences require two changes.)

A   Cryogenics is a term that refers to technology, processes, and research at very low temperatures.[18]

B   The flow of current through a wire conductor is equal to the number of electrons pass through its cross-sectional area in a particular unit of time.[19]

C   Thermodynamics is the science deals with heat and work and those properties of substances bear a relation to heat and work.[20]

D   A property can be defined as any quantity depends on the state of the system and is independent of the path by which the system arrived at the given state.[21]

E   A decade ago new electronic capabilities became available to increase greatly the kinds of information that could be gathered by remote means.[22]

F   For nonequilibrium processes, we are limited to a description of the system before the process occurs and after the process is completed and equilibrium is restored. We are not able to specify each state through which the system passes, nor the rate the process occurs at.[23]

G   This suggests that information-processing systems should be designed as man-machine systems in which each is assigned the part it is best suited for.[24]

H   All engines operate on the Carnot cycle between two given constant-temperature reservoirs have the same efficiency.[25]

I   A computer subroutine is essentially a program within a program: the main program transfers control to the subroutine, which accepts the necessary data, performs its function, and returns control back to the point in the main program from it came.[26]

Of course there are many cases where the relative pronoun can be omitted without making the relative clause ungrammatical. However, this can be done only with certain kinds of relative clauses and usually only if other changes are made at the same time (see Rules 5 to 7 below).

Avoiding a Duplicate Pronoun

Rule 3
Make sure you don't have a second pronoun duplicating the role of the relative pronoun.

Correct
The energy *which is consumed in overcoming friction* is not lost but is converted into heat.

Incorrect
The energy *which <u>it</u> is consumed in overcoming friction* is not lost but is converted into heat.

Many languages of the world (e.g., Arabic, Persian) differ from English in this regard: they do allow a "duplicate" pronoun to appear in the relative clause. This is because, in those languages, there is not really a "relative pronoun" as such. Instead, there is only a "relative marker." Let us explain.

A relative pronoun actually performs two functions. One, it serves as a *relative marker*, signaling the presence of a relative clause. Two, it serves as a *pronoun*, playing a grammatical role in the relative clause and sharing the same referent as the main-clause noun. This double function can be seen most easily in sentences like the following:

The punched card concept *which Hollerith devised in the 1880s* turned out to be a major step in the development of the computer.

Since the verb *devise* requires a direct object, we sense that there is a "hole" in this relative clause, after the verb. In effect, the relative pronoun (*which*) represents the "missing" direct object at the same time as it marks the relative clause. It is as if the sentence went through these "stages of construction":

1   The punched card concept *(Hollerith devised it in the 1880s)* turned out to be a major step in the development of the computer.

2   The punched card concept *which Hollerith devised in the 1880s* turned out to be a major step in the development of the computer.

Since relative pronouns in English serve not only as relative markers but also as pronouns, there is no need to fill the "hole" with another pronoun; in fact, it is grammatically wrong to do so.

Incorrect
The punched card concept which Hollerith devised *it* in the 1880s turned out to be a major step in the development of the computer.

❏ EXERCISE 33-3

Inspect each of the following sentences and cross out any pronouns that duplicate the role of the relative pronoun.

A   A pump is a device that it induces fluid flow through pipes, generally by means of a moving piston.[27]

B   Pulley systems provide a mechanical advantage which it is equal to the number of supporting ropes involved.[28]

C   Many different fuels and oxidizers have been tested, and much effort has gone into the development of fuels and oxidizers which they will give a higher thrust per unit mass rate of flow of reactants.[29]

D   Such a system would not provide the cooling power that people are now accustomed to it in air-conditioned cars.[30]

E   A difference between research and power reactors is in the kind of uranium that they use.[31]

F   The lowest weight of an element which it enters into chemical combination is called its atomic weight.[32]

G   Volta was the first experimenter in history who he was able to generate a continuous electric current.[33]

H   Dosimeters are small badges that people who work around nuclear radiation sources wear.[34]

33.2 STYLISTIC RULES

Once you have firm control of Rules 1 to 3 above, you can consider using additional rules for stylistic purposes. The rest of this chapter is devoted to four such rules, all of which serve to make sentences shorter, easier to pronounce, and often more emphatic. Unlike Rules 1 to 3, which must be followed for grammatical correctness, these next rules are strictly *optional*: if you do not feel comfortable using them, do not use them!

NOTE: Rules 4 to 7 apply only to true relative clauses, not to appositive (nonrestrictive) relative clauses, which are set off by commas.

Substituting *That* for *Which*

Rule 4
You may substitute *that* for *which* if it is not preceded by a preposition.

Correct

The energy $\left\{ \begin{matrix} which \\ that \end{matrix} \right\}$ *is consumed in overcoming friction* is not lost but is

converted into heat.

Incorrect
Multispectral sensing is a process *in that the earth's surface is surveyed from an airplane equipped with infrared and ultraviolet sensors.* (The correct form is *in which. . . .*)

Traditionally, the use of *that* as a relative pronoun was thought to be too informal for scientific and technical writing. Times have changed, however, and it is now fully accepted. Indeed, using *that* instead of *which* generally makes for a smoother, easier-to-pronounce sentence. It is also favored by many writers who feel it makes their writing less "stuffy" (i.e., less stiffly formal).

❏ EXERCISE 33-4

Convert *which* to *that* wherever possible in the following sentences.

A   Technology assessment is an opportunity for universities to carry on studies which go beyond the bounds of individual fields of study.[35]

B   The transit industry can benefit from other studies which focus on operational questions and scrutinize their cost-effectiveness.[36]

C   A thermodynamic system is defined as a quantity of matter of fixed mass and identity upon which attention is focused for study.[37]

D   An economizer is simply a heat exchanger in which heat is transferred from the products of combustion (just before they leave the steam generator) to the condensate.[38]

E   Even though automatic computing is much faster and cheaper than it has ever been, there are still problems which cannot be computed owing to their size and complexity.[39]

F   Energy which is useful is energy which can be transformed into another form.[40]

G   Combustion is a useful kind of chemical activity by which potential energy in chemicals turns to heat energy.[41]

H   A steam generator is a device in which the hot water which emerges under pressure from the reactor core transfers its thermal energy to a second water loop, turning the water in this secondary loop into the steam which will drive the turbine and electric generator.[42]

Omitting the Relative Pronoun and Auxiliary Verb

Rule 5
If the relative clause begins with a relative pronoun and some form of the auxiliary verb *to be* followed by a verb phrase, you may omit the relative pronoun and auxiliary.

Correct

The energy $\left\{ \begin{array}{c} \textit{that is} \\ \emptyset \end{array} \right\}$ *consumed in overcoming friction* is not lost but is

converted into heat.

In other words, you can omit any combination of relative pronoun and form of the verb *to be*,

which is	*who is*
which are	*that was*
that is	etc.
that are	

provided it is not immediately followed by a noun phrase or adjective phrase.

Incorrect
The technician *sick last year* has returned to work. (The correct form is . . .*who was sick last year*. . . .)

❏ EXERCISE 33-5

In the following sentences, reduce the relative clauses by omitting the relative pronoun and auxiliary verb:

A   In 1911 Ernest Rutherford showed that the atom was largely empty space and hence had to contain something that was smaller than the atom.[43]

B   "Water treatment" is a term that is traditionally applied to municipal treatment of water.[44]

C   In many power plants the air which is used for combustion is preheated in the air preheater by transferring heat from the stack gases as they are leaving the furnace.[45]

D   Simulation can thus offer a basis for decisions about a new system without the disruptions and the costs which are associated with trying a system which may fail.[46]

E   Cheapness of oil has been widely identified as a factor in the high standard of living that is enjoyed in many nations today.[47]

The fact that you *may* omit the relative pronoun and auxiliary does not mean that you always *should*. Good writers sometimes choose not to make such a reduction even though the opportunity presents itself. Why? It is mainly a question of whether or not you want to emphasize the word immediately following the auxiliary. If you do want to emphasize it, then you will probably want to retain the relative pronoun and auxiliary; this has the effect of giving greater prominence to the word following the auxiliary. If you do not want to emphasize it, you will probably want to omit the relative pronoun and auxiliary.

To see this, consider sentence E of Exercise 33-5 as it actually appeared in a well-written research article:

Cheapness of oil has been widely identified as a factor in the high standard of living enjoyed in many nations today.

The most important words and phrases of this sentence—those most likely to be stressed if one were to read the sentence aloud—are those given below in italics:

Cheapness of oil has been *widely identified* as a *factor* in the *high standard of living* enjoyed in *many nations today.*

Notice that there is a balanced spacing of these important words and phrases, so that they occur at rhythmic intervals. It is as if the sentence consisted of alternating peaks and valleys, with emphasis given only to the peaks. Now, this being the case, notice that the word *enjoyed* is located in between two peaks, i.e., in one of the valleys. This is appropriate, it seems to us, because the word *enjoyed* does not carry much importance in this sentence; in fact, it could even be omitted without really changing the meaning of the sentence: "Cheapness of oil has been widely identified as a factor in the high standard of living in many nations today."

In effect, the writer of this sentence deliberately deemphasized the word *enjoyed* by placing it in a valley. How was this effect achieved? By applying Rule 5. If Rule 5 had not been applied, this would have been the result:

Cheapness of oil has been widely identified as a factor in the high standard of living that is enjoyed in many nations today.

The presence of the relative pronoun and auxiliary verb (*that is*) in this version of the sentence would have moved *enjoyed* into a position where, as a result of the rhythm principle described above, it would more likely be treated as a peak than as a valley:

> *Cheapness of oil* has been *widely identified* as a *factor* in the *high standard of living* that is *enjoyed* in *many nations today*.

In other words, the writer would have incorrectly put emphasis on a word that does not deserve it.

❏ EXERCISE 33-6

In the following sentences, examine each relative clause headed by a relative pronoun and some form of the auxiliary verb *to be*. Is the next word an important one? If so, leave the relative clause as it is. If not, cross out the relative pronoun and auxiliary.

A   A steam-electric generating plant is a giant engine that is attached to an electrical generator.[48]

B   One type of refrigeration process involves passing the air through a throttle valve that is so designed and so located that there is a substantial drop in the pressure and temperature of the air.[49]

C   The high-pressure, high-temperature products of combustion expand as they flow through the nozzle, and as a result they leave the nozzle with high velocity. The momentum change that is associated with this increase in velocity gives rise to the forward thrust of the vehicle.[50]

D   The computer has made possible systems studies that were heretofore impractical.[51]

E   Decision simulations that are complex enough to be effective require a computer for making the computations that are involved.[52]

F   Much work has also been done on solid-propellant rockets. They have been very successfully used for jet-assisted takeoffs of airplanes, military missiles, and space vehicles. They are much simpler in both the basic equipment that is required for operation and the logistic problems that are involved in their use.[53]

G   As each electrical pulse reaches a synapse, chemicals which are called neurotransmitters come into play.[54]

H   After mining and milling, uranium ore is concentrated into an oxide form which is known as "yellow cake."[55]

I   We speak of substances as having different "phases." A phase is defined as a quantity of matter that is homogeneous throughout.[56]

J   The cloud chamber is a scientific device that is used to record the paths of subatomic particles, often in conjunction with collision experiments.[57]

Using the *-ing* Form

Rule 6
If the relative clause begins with a relative pronoun and a main verb (i.e., nonauxiliary verb), you may omit the relative pronoun and change the verb to its *-ing* form.

Correct

Persons $\left\{ \begin{array}{l} who\ do \\ doing \end{array} \right\}$ *research* find it very time-consuming and sometimes

impossible to make an adequate search of the literature $\left\{ \begin{array}{l} which\ bears\ on \\ bearing\ on \end{array} \right\}$

the subject they are investigating.[58]

This rule cannot be applied if the verb following the relative pronoun is an auxiliary verb, i.e., any form of the verb *to be*, any modal verb (*can, may, will, would,* etc.), or any form of the verb *to have* when it is used as an auxiliary. Here is an example:

A hologram is a photographic image $\left\{ \begin{array}{l} which\ is \\ *being \end{array} \right\}$ *produced without using*

any lens. (*The *-ing* form is incorrect.)

Sometimes the verb *to have* is used as a main verb, not an auxiliary, in which case the rule may be applied:

Correct

A pure substance is one $\left\{ \begin{array}{l} that\ has \\ having \end{array} \right\}$ *a homogeneous and invariable*

chemical composition.[59]

Incorrect
It is not possible to discuss experiments in particle physics without first considering the theoretical context *having spawned the experiments.* (The correct form is . . . *that has spawned.* . . .)

❏ EXERCISE 33-7

In the following sentences, reduce whatever relative clauses you can by omitting the relative pronoun and changing the verb to its *-ing* form. Make sure the verb is a main verb!

A   The model allows those who use it to compress real time, so that years of operation can be studied in a few minutes.[60]

B   Some of the liquid refrigerant that emerges from the condenser can also be used for air-conditioning.[61]

C   Today, marine engineers are responding to the ocean pollution problems that have now been widely acknowledged.[62]

D   Tankers that ply traditional shipping lanes in ocean waters can run aground while far out of sight of land.[63]

E   There is concern today over the potential dangers which attend the disposal of radioactive by-products produced in nuclear plants.[64]

F   The solution to a complex linear programming problem involves calculations which would be impractical without a computer.[65]

G   In a technical information-retrieval system which uses key words, the contents of each technical paper are identified by key words.[66]

H   All the things that can be done with rational numbers may be succinctly expressed in the form of the equation $ax + b = c$.[67]

I   The real and complex numbers may be subdivided into those that are algebraic and those that are not.[68]

J   The mission of the Pioneer Venus atmospheric probes was to determine at each level the amount of radiation which enters the atmosphere and rises from it.[69]

As in the case of Rule 5, good writers do not always apply Rule 6 whenever the opportunity presents itself. In fact, they avoid applying the rule more often than not. It depends mainly on how much emphasis or focus the writer wants to give the relative clause: a full relative clause tends to command more attention than an *-ing* type of relative clause.

   To see the difference, examine this excerpt from a report on coal technology, paying particular attention to the relative clause in italics:

> Coal differs from other familiar energy sources in its chemical makeup. Fuels like natural gas or gasoline consist of relatively small and uniform molecules *that contain carbon and hydrogen*. Natural gas is mainly methane (CH_4), and gasoline contains iso-octane (C_8H_{18}) and similar molecules. Heating-oil molecules are similar to gasoline molecules, but they have a longer chain of carbon atoms; lubricating oil and asphalt molecules are longer still. When these substances burn, their carbon and hydrogen atoms combine with oxygen to form carbon dioxide, water, and heat.
>
>   Coal, on the other hand, is not described by chemical formulas like the above. Coal contains varying amounts of carbon, volatile organic compounds, incombustible minerals, sulfur, nitrogen, and moisture. . . .[70]

The writer chose to use a full relative clause in this case because he wanted to emphasize the importance of chemical composition in differentiating between coal and other fuels; specifically, he wanted to focus attention on the fact that carbon and hydrogen characterize fuels such as natural gas and gasoline but do not characterize coal. If the writer had chosen to use the *-ing* form of relative clause instead of the full form, he would not have succeeded in putting as much emphasis on the idea that carbon and hydrogen are *defining features* of these noncoal fuels.

❏ EXERCISE 33-8

In each of the following paragraphs there are one or more relative clauses in italics. Look at each and decide whether or not it should be reduced to the *-ing* form.

A   The first class of numbers to be discovered was the natural numbers: 1, 2, 3, 4,. . . . They are familiar as the numbers used in counting. There is evidence that this class began very modestly. Some tribes still have only a vocabulary of three number words: "one," "two," and "lots." Mathematics, like any other language, had to enlarge its vocabulary to enable mathematicians to give full expression to their ideas. Thus, in the more advanced civilizations, the vocabulary of three terms was long extended to a vocabulary *that consisted of an infinite sequence of natural numbers*. But even this infinite vocabulary was eventually found to be insufficient for mathematical purposes, indeed even for everyday affairs, and the number system had to be extended repeatedly to include other classes of numbers.[71]

B   Particle physics is the study of the tiny fundamental constituents of matter. Another name for this science is high-energy physics, because to study the smallest parts of matter one must employ the highest levels of energy. High-energy devices are the synchrotrons and other particle accelerators *that impart energy and light-like speeds to beams of particles in a vacuum*. Such beams, which require vast amounts of electric power to create and control, are the probes or tools by which subatomic events are made to occur. What happens in such events, properly observed, tells physicists what the particles are like.[72]

C   In technical information retrieval using key words, the contents of each technical paper are identified by key words. For example, an article about the effect of radiation on metals in the Van Allen belt might be identified by "metals," "radiation," "Van Allen belt," and "space." In addition to the key-word identifiers, a short abstract is prepared and stored on secondary storage medium. A person *who wishes information on a topic* chooses a set of identifiers *which best describes the topic*. The computer is programmed to compare this set of identifiers with the list of all identifiers and to select papers *which have sets of identifiers which contain one or more identifiers common to the topic identifiers*. The abstracts for these papers are then printed out, together with an identifying number through which the article may be located.[73]

Omitting the Relative Pronoun

Rule 7
If the relative clause begins with a relative pronoun and another pronoun, you may omit the relative pronoun.

Correct
The free market system ~~that~~ *we are all familiar with* is increasingly under scrutiny.

❏ EXERCISE 33-9

In the following sentences, reduce whatever relative clauses you can by omitting the relative pronoun where appropriate.

A   At the conclusion of a cycle all the properties have the same value that they had at the beginning.[74]

B   A number of energy sources have serious potential and merit the research attention which they are getting, but most are not yet ready for practical use.[75]

C   The products of combustion from power plants are discharged to the atmosphere, and this constitutes one of the facets of the air pollution problem that we now face.[76]

D   Neuroscientists have in the last decade discovered that synapses take on a variety of characters, which are largely determined by the chemical neurotransmitters which they employ.[77]

E   Harbor safety requirements must be stiffer than ever in view of the widespread effects that a tanker accident may bring about.[78]

F   Mathematicians created larger and larger classes of numbers so as to have more room to move around in and to do all the things that they wanted and needed to do.[79]

❏ EXERCISE 33-10 (REVIEW EXERCISE)

In the following example there are a number of relative clauses in italics. Examine each and make stylistic changes (*that* for *which*, omission of relative pronoun and/or auxiliary verb, or change of verb to *-ing* form) where appropriate.

An investigation into the behavior of a system may be undertaken from either a microscopic or a macroscopic point of view. Let us briefly consider the problem *which we have* if we attempt to describe a system from a microscopic point of view. Consider a system *which consists of a one cubic inch volume of a monatomic gas at atmospheric pressure and temperature.* This volume contains approximately 10^{20} atoms. To describe the position of each atom, three velocity components must be specified.

Thus, to completely describe the behavior of this system from a microscopic point of view, it would be necessary to deal with at least 6×10^{20} equations. Even with a large digital computer, this is a quite hopeless computational task. However, there are two approaches to this problem *which reduce the number of equations and variables to a few which can be handled relatively easily in performing computations.* One of these approaches is the statistical approach, *in which*, on the basis of statistical considerations and probability theory, *we deal with "average" values for all particles which are under consideration.* This is usually done in connection with a model of the atom *which is under consideration.* This is the approach *which is used in the disciplines which are known as kinetic theory and statistical mechanics.*[80]

REFERENCES

1 *How Things Work* (New York: Bantam/Britannica, 1979), p. 56.
2 *How Things Work*, p. 186.
3 *How Things Work*, p. 147.
4 G. B. Davis, *Introduction to Electronic Computers*, 2d ed. (New York: McGraw-Hill, 1971), p. 42.
5 Adapted from *Webster's Seventh New Collegiate Dictionary* (Springfield, MA: Merriam, 1969), p. 997.
6 *How Things Work*, p. 19.
7 *How Things Work*, p. 13.
8 Adapted from *How Things Work*, p. 147.
9 Adapted from Davis, p. 39.
10 *How Things Work*, p. 24.
11 Adapted from B. Hiatt, "Heat into Work: The Second Law of Thermodynamics," *The Research News* 28(11/12):13 (1977).
12 B. Hiatt, "Nuclear Power, Nuclear Safety," *The Research News* 30(6):9 (1979).
13 Adapted from B. Hiatt, "Transportation and Energy," *The Research News* 30(1/2):11 (1979).
14 Adapted from Davis, p. 39.
15 Davis, p. 41.
16 *How Things Work*, p. 137.
17 R. E. Sonntag and G. J. Van Wylen, *Introduction to Thermodynamics: Classical and Statistical* (New York: Wiley, 1971), p. 4.
18 Sonntag and Van Wylen, p. 12.
19 Adapted from *How Things Work*, p. 89.
20 Adapted from Sonntag and Van Wylen, p. 17.
21 Adapted from Sonntag and Van Wylen, p. 20.
22 B. Hiatt, "Technology Assessment: Creating the Future," *The Research News* 28(9/10):4 (1977).
23 Sonntag and Van Wylen, p. 22.

24 Davis, p. 35.

25 Adapted from Sonntag and Van Wylen, p. 197.

26 Adapted from Davis, p. 200.

27 Adapted from *How Things Work*, p. 21.

28 Adapted from *How Things Work*, p. 17.

29 Adapted from Sonntag and Van Wylen, p. 16.

30 Adapted from Hiatt, "Heat into Work," p. 12.

31 Hiatt, "Nuclear Power, Nuclear Safety," p. 14.

32 Adapted from N. B. Reynolds and E. L. Manning (eds.), *Excursions in Science* (New York: McGraw-Hill, 1939), p. 278.

33 Adapted from W. Kaempffert, *Explorations in Science* (New York: Viking, 1953), p. 229.

34 Adapted from Hiatt, "Nuclear Power, Nuclear Safety," p. 7.

35 Hiatt, "Technology Assessment," p. 10.

36 Hiatt, "Transportation and Energy," p. 14.

37 Sonntag and Van Wylen, p. 17.

38 Sonntag and Van Wylen, p. 2.

39 B. Hiatt, "Big Computing, Tiny Computers," *The Research News* 30(4):5 (1979).

40 Hiatt, "Heat into Work," p. 3.

41 Adapted from Davis, p. 39.

42 Hiatt, "Nuclear Power, Nuclear Safety," p. 5.

43 B. Hiatt, "What's the Matter: Particle Physics at the University of Michigan," *The Research News* 26(3/4):3 (1975).

44 J. Wei, "Towards Cleaner Water: Studies in Toxic Substances and Other Contaminants," *The Research News* 29(8/9):6 (1978).

45 *How Things Work*, p. 137.

46 Davis, p. 41.

47 Adapted from Hiatt, "Transportation and Energy," p. 5.

48 Hiatt, "Heat into Work," p. 7.

49 Sonntag and Van Wylen, p. 13.

50 Adapted from Sonntag and Van Wylen, p. 16.

51 Davis, p. 41.

52 Davis, p. 43.

53 Adapted from Sonntag and Van Wylen, p. 16.

54 B. Hiatt, "Signals and Synapses: The Brain's Neurotransmitters," *The Research News* 30(10–12):4 (1979).

55 Hiatt, "Nuclear Power, Nuclear Safety," p. 8.

56 Adapted from Sonntag and Van Wylen, p. 20.

57 *How Things Work*, p. 152.

58 Davis, p. 42.

59 Sonntag and Van Wylen, p. 43.

60 Davis, p. 41.

61 Hiatt, "Heat into Work," p. 10.

62   B. Hiatt, "Oil on Troubled Waters," *The Research News* 26(2):12 (1975).

63   Hiatt, "Oil on Troubled Waters," p. 5.

64   B. Hiatt, "Coal Technology for Energy Goals," *The Research News* 24(12):3 (1974).

65   Adapted from Davis, p. 39.

66   Davis, p. 42.

67   J. Wei, "Pure Mathematics: Problems and Prospects in Number Theory," *The Research News* 30(3):10 (1979).

68   Wei, "Pure Mathematics," p. 13.

69   B. Hiatt, "Planetary Exploration," *The Research News*, 29(10–12):10 (1978).

70   Hiatt, "Coal Technology," p. 6.

71   Wei, "Pure Mathematics," p. 10.

72   Hiatt, "What's the Matter," p. 3.

73   Davis, p. 42.

74   Sonntag and Van Wylen, p. 23.

75   Hiatt, "Coal Technology," p. 3.

76   Sonntag and Van Wylen, p. 4.

77   Hiatt, "Signals and Synapses," p. 5.

78   Hiatt, "Oil on Troubled Waters," p. 5.

79   Wei, "Pure Mathematics," p. 10.

80   Sonntag and Van Wylen, p. 19.

34

Connectives

Good writing involves more than just well-chosen words and well-made sentences. The words and sentences must all fit together into a coherent whole, so that the reader can grasp the "big picture." Organizational patterns, parallelism, key words, and other devices will help give your writing this kind of coherence. But you should also try to use connective words and phrases like *therefore, however, although, on the other hand*, and *in theory*. Connective words help clarify the logic of your communication, letting your reader follow the flow of your reasoning from sentence to sentence.

The purpose of this chapter is to cover the more commonly used connectives in scientific, technical, and professional English. We have separated these connectives into different groups. In each group, the connectives mean basically the same thing—although there are slight differences in usage among them. As you gain in English language proficiency, you may want to learn these differences. In the following sections, the most commonly used connectives are presented in upper-case letters. Some connectives are generally used only in informal situations, and we have marked these accordingly. We have also provided for each group an example of usage.

Connectives as a whole are divided into two categories according to grammatical function: conjunctive adverbs and subordinating conjunctions. It is important not to confuse the two. Conjunctive adverbs (*thus, however, for example*, etc.) are normally used with independent clauses to show how that clause is logically related to some independent clause preceding it. Although they usually occur at the beginning of the clause, they can also be embedded within it. Subordinating conjunctions (*since, although, if*, etc.), on the other hand, are used only with a subordinate clause to express a logical relationship between that clause and an independent clause which usually follows it. Subordinating conjunctions cannot be used in sentences of only one clause:

Ungrammatical
Since dissolution of an organic compound into an organic solvent does not produce ions. Most solutions of organic compounds do not conduct electricity.

Grammatical
Since dissolution of an organic compound into an organic solvent does not produce ions, most solutions of organic compounds do not conduct electricity.

34.1 CONJUNCTIVE ADVERBS

THEREFORE, THUS, consequently, as a result, for this reason, so [informal]

Combustion catalysts consist of various shapes of basic material coated with a metallic compound. The variety of shapes and catalytic materials provides a multitude of catalysts for each application. ***Therefore,*** a good general rule to follow is to consult with a catalyst manufacturer on the most suitable catalytic equipment configuration.[1]

HOWEVER

At one time the atom was believed to be the ultimate unit in the subdivision of matter. Subsequently, ***however,*** it became known that the atom is composed of still smaller units.[2]

FOR EXAMPLE, for instance

Variation is a very important characteristic of data. ***For example***, if we are manufacturing bolts, excessive variation in the bolt diameter would imply a high percentage of defective product.[3]

FIRST, . . . SECOND(ly), . . . THIRD(ly), . . . FINALLY

A further elaboration of these facts falls naturally under two major headings. ***First,*** what are the quantitative relations between irradiation dosage and genetic damage? ***Second,*** to what dosages are people being exposed?[4]

IN ADDITION, FURTHERMORE, moreover, besides [informal]

All organic compounds, in principle, can be ranked according to their relative ability to undergo photochemical reactions characteristic of smog. For many, however, the data are not available. ***Furthermore,*** the ranking of a number of organic compounds on the basis of their rates of disappearance during photolysis would not necessarily be the same ranking of those same compounds on the basis of their ability to produce eye irritation.[5]

ON THE OTHER HAND	The increasing price of oil may cause a shift to coal in many countries. Worldwide, coal should last for at least 300 years. Oil, **on the other hand,** will be virtually depleted within 60 years.[6]
IN OTHER WORDS, that is	The chromosomes of higher organisms, including man, occur in pairs. **That is,** the genetic information found in any one chromosome is duplicated in another chromosome ordinarily identical in size and shape.[7]
OF COURSE, naturally [informal]	There are two types of kettles used for cooking varnish: the open kettle, which is heated over an open flame, and the newer totally enclosed kettle, which is set over or within a totally enclosed source of heat. **Naturally,** the open kettle allows vaporized material to be emitted to the atmosphere unless hooding and ventilation systems are provided.[8]
EVEN SO, NEVERTHELESS, nonetheless, still [informal]	Experiments on humans are essential; they have been the cornerstone of medical progress and will continue to remain so. **Even so,** many tests cannot be done on humans because of practicality or propriety; certain tests would be outrageously costly or unethical.[9]
as mentioned above (earlier, before), as shown in Figure ____	Conventional ultrasonic testing has its disadvantages: strong signals from one of the surfaces can mask the signal of a discontinuity near one of the surfaces. . . . [three paragraphs later] **As mentioned earlier,** conventional ultrasonic testing generally fails to clearly distinguish discontinuities close to the surface.[10]
IN GENERAL	The conversion of low-Btu gas to high-Btu gas entails the production of much CO_2, each molecule of which represents the waste of a carbon atom originally present in the coal. **In general,** for processes now under development, it appears that the production of low-Btu gas will have a thermal efficiency up to 20% greater than that of high-Btu gas.[11]
IN FACT, indeed, as a matter of fact	Physics is far from being a solitary pursuit. **In fact,** physicists today interact intensely with each other.[12]
IN PARTICULAR, SPECIFICALLY	It is the role of astronomy to ascertain the earth's probable future as determined by the

action of cosmic forces. ***Specifically,*** astronomy must seek to discover what the prospects are for the earth's continuing as a suitable abode for life, and study those events which could end the existence of mankind.[13]

actually, in actuality

Pythagoras was born in the year 569 B.C.(?) on the island of Samos, off the west coast of Asia Minor. It is claimed that he traveled extensively throughout Egypt, Babylon, and the Orient, where he supposedly learned of their mathematics, religions, and mysticism. ***Actually,*** all we know for certain is that he had a good knowledge of the beliefs and mathematical concepts of the people of Asia Minor.[14]

IN CONCLUSION, in summary, in conclusion, to summarize

[after a two-page discussion of the "Synthetic Aperture Focusing Technique" as a new method of ultrasonic testing of materials] ***In conclusion,*** the SAFT UT system represents a significant improvement in modern methods of nondestructive evaluation. If continued funding by the NRC allows further research to proceed. . . .[15]

in theory, in principle, theoretically

The problem with solar energy is that, although sunshine itself costs nothing, it takes ingenuity and money to capture it. ***Theoretically,*** 1 horsepower can be generated on about 1 square yard of land under the best conditions. But such are the unavoidable losses that not fewer than 4 and probably as many as 10 square yards are needed to generate 1 horsepower.[16]

in practice

If an experiment is repeated a large number of times, N, and event A is observed n times, the probability of A is

$$P(A) = \frac{n}{N}$$

In practice, the composition of the population is rarely known and hence the desired probabilities for various events are unknown. But we ignore this aspect of the problem, since our aim is only to create a mathematical model.[17]

INSTEAD, rather

The most striking feature of the Cray-1 supercomputer is its simplicity. It does not use any sophisticated devices such as magnetic bubble memory or supercooled conductors. ***Instead,*** it uses only three types of integrated circuits: memory, logic, and arithmetic.[18]

IN COMPARISON, BY CON-TRAST	Nuclear medicine differs from existing angiography procedures in that nuclear imaging is an emission effect caused by the emission of gamma rays from the decaying isotope. *In comparison,* angiography is a transmission effect caused by the transmission of x-rays through the patient, to be imaged on the opposite side.[19]
if so, in that case	The age of the engineer as a specialist may have reached its zenith and may now be starting to recede. *If so,* the development and implementation of computer-aided engineering may reestablish the importance of the engineer as a generalist.[20]
in most cases, generally, usually, for the most part	Large volumes of hydrocarbon gases are produced in modern refineries and petrochemical plants. *Generally,* these gases are collected and used as fuel or as raw material for further processing. Sudden or unexpected upsets in process units and scheduled shutdowns, however, can produce gas in excess of the capacity of the gas-recovery system.[21]

34.2 SUBORDINATING CONJUNCTIONS

ALTHOUGH, though, while, even though, despite the fact that	*Even though* automatic computing is faster and cheaper than it has ever been, there are still problems that cannot be computed owing to their size or complexity, at least not in a practical sense. *Though* a computer wades through a problem at 100 million steps per second, it may be in a losing race with the scientist or engineer applying his or her ingenuity to complex problems.[22]
SINCE, BECAUSE, owing to, the fact that	*Since* natural carbon contains approximately 1 percent C^{13} along with 98.9 percent C^{12}, the average atomic weight of carbon is about 12.011 amu.[23]
IF	The U.S. government considers computers to be strategic commodities with potential military applications. *If* an American firm wishes to export a computer, it must first obtain a license from the State Department (*if* the computer is intended for military use) or from the Commerce Department (*if* it is intended for civilian use).[24]

in order for (that)	For combustion of organic vapors and liquids, the concentrations of vapor and air must be within the limits of flammability. ***In order that*** a flame be self-sustaining, the mixture of air and combustibles must provide enough heat to maintain the combustion temperature.[25]
rather than, instead of [-ing]	The "ideal system concept" is a simple idea that anyone can understand. ***Rather than*** use the models of the present trouble-ridden system as a guide for developing a recommended system, this concept produces a model of the best or most ideal system as a guide.[26]

34.3 COMMON MISTAKES TO AVOID

Here are some of the most common mistakes made by nonnative speakers when using connectives:

Using *actually* to mean "currently, presently."   Speakers of Spanish, French, and other Romance languages tend to make this mistake, since similar forms in those languages (Sp. *actualmente*, Fr. *actuellement*, etc.) do mean "currently, presently." *Actually* is closer in meaning to *in fact*. The only real difference is that *actually* is normally used to play down or correct an earlier statement whereas *in fact* is used to upgrade or magnify one.

Using *after all* to mean "finally, in conclusion."   This phrase means roughly "regardless of what might be said about it." It is used to emphasize the fundamental essence of something, not to signal the end of a sequence or the end of a discussion.

Using *although* and *but* in the same sentence.   "Although I studied hard, but I only scored 56 on the exam" is incorrect. You can use either *although* or *but*, but not together.

Using *it* in idiomatic *as* phrases.   Phrases such as "as was mentioned above" and "as is shown in Figure 12" are idiomatic and should be used as fixed forms, without an *it* subject. "As it was mentioned above" is incorrect.

Using *comparing to (with)*.   The correct connective forms are *compared to* and *in comparison with*. You can use *comparing* as a gerundive participle followed immediately by a direct object: "Comparing these latest results to (with) our earlier ones, . . ."

Using *conclusively* to mean "in conclusion."   *Conclusively* is the simple adverbial form of the adjective *conclusive* (= "decisive, supported by solid evidence"). It is not a connective.

Using *on the contrary* to mean "on the other hand."   *On the contrary* is properly used to deny some earlier statement, whereas *on the other hand* is used simply to present an alternative to some earlier statement. *On the one hand* and *on the other hand* are often used together to present a pair of alternative statements.

Using *particularly* to mean "in particular."   *Particularly* means "especially"; it is an ordinary adverb, not a connective.

❑ EXERCISE 34-1

In each of the following passages, fill in the blanks with the most appropriate connective from among the choices given.

A   Studying how light (or other electromagnetic radiation) interacts with matter is an important and versatile tool of the chemist. _____,
(Still, Indeed, However)
much of our knowledge of chemical substances comes from their specific absorption or transmission of light.

Suppose you look at two solutions of the same substance, one a deeper color than the other. Your common sense tells you that the darker-colored solution is the more concentrated one. _____,
(In other words,
Consequently, In addition)
as the color of the solution deepens, you infer that its concentration also increases. This is an underlying principle of spectrophotometry: the intensity of color is a measure of the amount of a material in solution.

A second principle of spectrophotometry is that every substance absorbs or transmits certain wavelengths of radiant energy but not other wavelengths. _____, chlorophyll always absorbs
(Therefore, In general, For example)
red and violet light while transmitting the yellow, green and blue wavelengths. The transmitted and reflected wavelengths appear green—the color your eye "sees." The light energy absorbed or transmitted must match exactly the energy required to cause an electronic transition in the substance under consideration. Only certain wavelength photons satisfy this energy condition. _____, the absorption or trans-
(Thus, Nevertheless, Actually)
mission of specific wavelengths is characteristic for a substance, and a spectral analysis serves as a "fingerprint" of the compound.[27]

B   Before we can discuss the basic problem of the origin of our universe, we must ask ourselves whether such a discussion is necessary. Could it not be true that the universe has existed since eternity, changing slightly in one way or another in its minor features, but always remaining essentially the same as we know it today? The best way to answer this question is by collecting information about the probable age of various basic parts and features that characterize the present state of our universe.

_____, we may ask a physicist or chemist:
(In general, In practice, For example)
"How old are the atoms that form the material from which the universe
is built?" Only half a century ago such a question would not have made
much sense. _____, when the existence of natural radio-
(Instead, However, If so)
active elements was recognized, the situation became quite different. It
became evident that if the atoms of the radioactive elements had been
formed too far back in time, they would by now have decayed completely
and disappeared. _____, the observed relative abundances of
(Besides, Since, Thus)
various radioactive elements may give us some clue as to the time of their
origin.

 We notice that thorium and the common isotope of uranium
(U^{238}) are not much less abundant than the other heavy elements, such
as _____, bismuth, mercury, or gold.
(in comparison, furthermore, for example)
_____ the half-life periods of thorium and of common
(Since, Although, Of course)
uranium are 14 billion and 4.5 billion years, respectively, we must conclude
that these atoms were formed not more than a few billion years ago.
_____, the fissionable isotope of uranium (U^{235}) is
(So, On the other hand, However)
very rare, constituting only 0.7% of the main isotope. The half-life of U^{235}
is considerably shorter than that of U^{238}, being only about 0.9 billion years.
_____ the amount of fissionable uranium has been cut in
(Since, Therefore, While)
half every 0.9 billion years, it must have taken about seven such periods,
or about 6 billion years, to bring it down to its present rarity, if both
isotopes were originally present in comparable amounts. . . .[28]

❑ EXERCISE 34-2

Complete each of the following sentences with an appropriate line of development. (Do not worry about the factual accuracy of your answer; just provide
a continuation that fits the given connective.)

A Nuclear power plants do not produce chemical pollutants such as sulfur
   dioxide or PCBs. However, they _____
   _____
   _____.

B Synthetic fuels are difficult to manufacture. In fact, _____
   _____
   _____.

C White light appears to be devoid of any color. In actuality, though, _____
   _____
   _____.

D   More and more automobile engines are being built today that use fuel injection devices instead of carburetors. Nevertheless, _____

_____.

REFERENCES

1   "Control Techniques for Hydrocarbon and Organic Solvent Emissions from Stationary Sources," US DHEW. PHS. EHS (Publication No. AP-68). National Air Pollution Control Administration, (Washington, DC: March 1970), p. 3–5.

2   Lawrence Van Vlack, *Elements of Materials Science and Engineering*, 3d ed. (Reading, MA: Addison-Wesley, 1975), p. 29.

3   William Mendenhall, *Introduction to Probability and Statistics*, 2d ed. (Belmont, CA: Wadsworth, 1967), p. 29.

4   A. H. Sturtevant, "The Genetic Effects of High Energy Irradiation of Human Populations," in *Frontiers in Science*, edited by Edward Hutchings Jr. (New York: Basic Books, 1958), p. 71.

5   "Control Techniques," p. 3–23.

6   "Coal Technology for Energy Goals: No Fuel Like an Old Fuel," *The Research News 34* (12):3 (1974).

7   H. Eldon Sutton, *An Introduction to Human Genetics* (New York: Holt, Rinehart and Winston, 1965), p. 14.

8   "Control Techniques," pp. 4–21, 4–22.

9   Ray Barry, "Caution: Life May Be Hazardous to Your Health," *Michigan Technic 96* (4):10 (1978).

10   Nolan Van Gaalen, "Ultrasonic Synthetic Aperture Focusing as a Method of Nondestructive Evaluation," *Michigan Technic 96*(4):18 (1978).

11   "Coal Technology," p. 9.

12   "Can We Understand the Universe?" *The Research News 32*(6/7):8 (1981).

13   Albert G. Wilson, "Astronomy and Eschatology," in *Frontiers in Science*, p. 207.

14   Guy Zimmerman, "The Pythagoreans: More Than Just the Theorem," *Michigan Technic 97*(3):9 (1978).

15   Van Gaalen, p. 20.

16   Waldemar Kaempffert, *Explorations in Science* (New York: Viking, 1953), p. 190.

17   Mendenhall, p. 51.

18   Steve Hannah, "The United States: Maintaining a Firm Grip on the World Computer Market," *Michigan Technic 97*(2):19 (1978).

19   David Varner, "Nuclear Cardiac Imaging," *Michigan Technic 97*(2):20 (1978).

20   James J. Duderstadt, "Some Thoughts on CAD, CAM and Computer-Aided Engineering," *Michigan Technic 100*(2):7 (1981).

21   "Control Techniques," p. 4–4.

22   Blanchard Hiatt, "Big Computing, Tiny Computers," *The Research News 30*(4):5 (1979).

23   Van Vlack, p. 29.

24   Hannah, p. 18.

25   "Control Techniques," p. 3–2.

26   Gerald Nadler, *Work Systems Design: The IDEALS Concept* (Homewood, IL: Irwin, 1967), p. 23.

27   Alice S. Cohen (ed.), *Investigations in General Chemistry* (Ann Arbor: University of Michigan Department of Chemistry, 1978; revised by Nancy Konigsberg, 1979), p. 117.

28   George Gamow, *The Creation of the Universe* (New York: Viking, 1952), pp.6–20.

35

Noun Compounds

A noun compound is a noun phrase in which two or more nouns function together as a unit. In the following example, there are four compounds (italicized), each consisting of two nouns:

> The author provided an overview of *the research program* and described the *research objectives*, the potential *research opportunities*, and the impact of some past *research accomplishments*.

In all noun compounds, the rightmost noun is the head noun and all nouns preceding it serve as qualifiers. Thus, the full meaning of a noun compound can often be expressed by simply "unwinding" it from right to left and inserting the appropriate preposition(s). For example, a *research program* is a *program of research*. A *water purification system* is a *system for the purification of water*.

However, the exact relationship between qualifiers or between one or more qualifiers and the head noun is not always clear—at least not to a nonspecialist. Take, for example, the noun compound *wall stresses*: Does it mean the stresses *on* a wall (from outside) or the stresses *produced by* a wall or the stresses *inside* a wall? Unless you're a construction engineer, you might not know for sure. (It's the last.) Or consider *mission suitability*: Does it mean suitability (of something) *for* a mission or suitability *of* a mission (for something)? Unless you're an aerospace specialist, you probably are not sure which. (It's the former.)

35.1 CONSTRUCTING NOUN COMPOUNDS

As suggested above, the correct interpretation of noun compounds depends heavily on the reader's prior knowledge of the subject being discussed. For this reason, noun compounds work best in writing intended for specialists. Sometimes, more than 20% of the words in a specialized text are noun compounds. For nonspecialist readers, on the other hand, noun compounds should be used with caution. If you have any doubts about the reader's ability to correctly interpret a noun compound, do not use the compound form until you have first used a full-form noun phrase to express that same meaning. For example, if writing for nonspecialists, you might use the noun compound *sea-level radiation* only after first giving a full relative clause version and then gradually compressing it down:

> The increased radiation dosages that would be experienced at sea level . . .
> the radiation dosages experienced at sea level . . .
> the radiation dose at sea level . . .
> sea-level radiation.

Even when writing for specialists, however, be sure to observe the following precautions with noun compounds.

1 *Use only standard, well-established compounds*, ones that your readers will immediately recognize. Pay attention even to minor details such as the presence or absence of plural forms in the qualifier nouns. Generally, only the head noun of a noun compound can take the plural -*s* marking; qualifier nouns are usually unmarked for plural even though they may have a plural meaning. For example, an *electron beam* is a beam of electrons (plural), yet the word *electron* is not marked for plural when used as a qualifier noun in the compound form. Likewise, a *passenger ship* is a ship for many passengers, yet the word *passenger* appears in the singular form. There are many exceptions, though, where the plural form is used: *weapons system, materials science, parts shortage, emissions sampling*. When you learn a noun compound, take note of its standard form.

2 *Don't make compounds too long.* Three or four nouns in a row should be about maximum. If you find yourself constructing a noun compound longer than that, try to break it up into smaller groups by inserting prepositions. For example, instead of writing *the lift arm front bearing cup retainer*, you could write *the retainer for the front bearing cup of the lift arm*. This is a longer form but it is easier to understand, even for a specialist.

35.2 INTERPRETING NOUN COMPOUNDS

As a reader, you may sometimes be confronted with long noun compounds and may have trouble decoding them. In such cases, the best strategy is to "divide and conquer." This is basically a two-step procedure:

Step 1

Using your knowledge of the field, try to identify familiar smaller compounds within the longer one.

For example, suppose you were reading an article on aeronautics and came across the term *nozzle gas ejection ship attitude control system*. You might immediately recognize *ship attitude* and *control system* as familiar compounds used in aeronautics. You might also isolate *gas ejection* as a separate compound, either by having seen it before or by having seen such analogous forms as *fuel injection* or *rocket propulsion*. (If there are any hyphens in the term, they are usually there to indicate that the two words connected by the hyphen belong together as a unit; for example, some writers would probably hyphenate *gas-ejection*.) Thus, at this point, you would have four units to deal with instead of the original seven: [*nozzle*] [*gas ejection*] [*ship attitude*] [*control system*].

Step 2

Try to find plausible meaning relationships between subunits and combine subunits accordingly.

The meaning relationship between units of a noun compound in technical and scientific English is almost always one of the following: purpose, composition (material), principle of operation, mode of operation, shape, size, location, name of creator, or restricted reference. These relationships are all illustrated below, using the head noun *brakes*.

Type of Meaning Relationship	Example
purpose	*emergency brakes:* brakes designed for emergency use
composition	*carbon steel brakes:* brakes made of carbon steel
principle of operation	*air brakes:* brakes designed to operate according to the compressibility of air
mode of operation	*hand brakes:* brakes operated by hand
shape	*disk brakes*: brakes whose most important component is in the form of a disk
size	*6-inch brakes*: brakes with a diameter of 6 inches
location	*front brakes*: brakes located in the front
name of creator	*Ghirling brakes:* brakes designed and manufactured by the Ghirling Company
restricted reference	*car brakes*: the brakes of a car

If you keep in mind these nine types of semantic relationships as you look at the subunits of a long noun compound, you will probably be able to guess the intended meaning of the compound as a whole. For example, consider our earlier example: [*nozzle*] [*gas ejection*] [*ship attitude*] [*control system*]. Working from right to left, let's look at the first two units: *control system* and *ship attitude*. A plausible guess would be that they share a purpose relationship: the purpose of the control system is to control the ship attitude. Are there any alternative relationships to consider? Probably not. So let's bracket those two subunits together and move on to the other two: *gas ejection* and *nozzle*. Are these related in meaning? If so, how? Well, if you know what a *nozzle* is (a device used to spray liquid out the end of a hose) and you have some idea of what *gas ejection* might be, you can guess that these two terms share a mode-of-operation relationship: gas is ejected by means of a nozzle. At this point, then, we have combined subunits so as to have only two: [*nozzle gas ejection*] [*ship attitude control system*]. Can these two groups now be combined according to one of the types of relationships listed above? Yes, it appears that they share a principle-of-operation relationship: the ship attitude control system operates according to principles of fluid dynamics embodied in the ejecting of gas. This seems like a plausible interpretation, so we'd want to check the context and see if it fits.

❏ EXERCISE 35-1

For practice, convert the following full-form noun phrases into noun compounds. (The first one has already been done for you.)

Full-form Noun Phrase	Noun Compound
A A screw jack operated by a machine is	a machine screw jack
B Junctions made of polymer semiconductors are	_____
C A program of research in biomechanics is	_____
D The control of costs in building projects is	_____
E A device for jamming the fuses on warheads is	_____
F Tasks involving the handling of materials at low risk are	_____
G An award given for discovering a formula for succeeding at managing the feeding of people in large volumes is	_____

❏ EXERCISE 35-2

Reduce the length of the following writing samples by converting long noun phrases to noun compounds where appropriate.

A   At the Administrator's request in December 1989, a program to test exhaust emissions was initiated to gather data on approximately 20 passenger cars which get high mileage and which are equipped with a diesel engine.

B   The 360X diesel engine was selected for the purpose of determining the deterioration of its emissions over the life of the vehicle while the vehicle is in use.

C   In an attempt to predict the overheating of generators more accurately, I investigated alternative methods of monitoring based on the degradation of electrical insulation at elevated temperatures.

D   The Swales Equipment Company is seeking to reduce the high rate of recalls for maintenance of the 3-B pump.

E   Sterilization of media used for fermentation can be done by using the traditional method of treating batches or by using a continuous process.

❏ EXERCISE 35-3

If you can, try to unwind these noun compounds and present them as full-form noun phrases instead. This will require some educated guesses! (The first one has already been done for you.)

Noun Compound	Full-form Noun Phrase
A A coal liquefaction process is	a process by which coal is liquefied
B Roadside breath-testing surveys are	_____
C An auto entertainment center is	_____
D Toluene insolubles removal is	_____
E A land disposal system is	_____
F High latitude radio transmission is	_____
G Diffusion convection mass transfer resistance is	_____
H A chrome nickel steel forged connection rod assembly is	_____

35.3 USING NOUN COMPOUNDS TO PROMOTE COHESION

Noun compounds have several functions in scientific, technical, and professional English. They can be used as types of short-form reference, as names for concepts being introduced into a discussion, and as a means for more easily shifting noun phrases around in a sentence to follow the principles outlined in Chapters 24–25 (concisely putting old information in the subject position).

Noun Compounds Used as Short-Form References

The noun compound is related to the short-form relative clause, which is discussed in Chapter 33. It is the final stage of compression of a full-form reference using a relative clause. Examples of this use appear in earlier parts of this section on noun compounds.

Sometimes in the proces of compressing a long unit into a noun compound, you can textually define the meaning of the noun compound by the context in which it appears. For instance, consider the following sentence.

A *management decision* must consider *worker productivity* and the *production air filtration system*

A *production air filtration system* could be either a *system for the filtration of production air* (air in the area in which production occurs) or a *system for air filtration which is being produced by the company*. The correct interpretation will be determined by the context, that is, whether you are talking about filtering the air in the production area or talking about the production of an air filtration system.

Noun Compounds Used as Names

Sometimes a noun compound is used for the first mention of a concept. In such a situation, the compound functions as a name and is not a shortened form of some other naming unit. For instance, in the following passage notice that all of the italicized noun phrases are names for newly introduced concepts:

While thousands of weary commuters scurry for homeward-bound transportation, a mountain of *letter mail* generated by the day's *business activities* is just beginning its own trip through culling, facing, canceling, enriching, and *sorting equipment* at Manhattan's main *post office*. . . .

And the freshest innovations—computer-controlled optical *character recognition* for sorting letter mail and the recently completed *bulk-mail system* for handling packages exclusively—are prime targets for critics grumbling about missent letters, *parcel damage*, and ever-increasing postal rates. Yet, like modern society's many other afflictions, the cure probably lies in pursuing even more imaginative technology. Jacob Rabinow, inventor of the *12–operator letter-sorting machine* now widely deployed throughout the *U.S. Postal Service* said as much in testimony before the Postal *Rate Commission*.

Chartered as a replacement for the original *U.S. Post Office Department* in 1971, the U.S. Postal Service (USPS) was launched with the expectation that *mail delivery* eventually could become a self-sufficient operation rather than continue forever as an outdated, heavily subsidized *government bureaucracy* . . . . During the early 1970s, several "state of the art" computer-controlled optical *reading/sorting systems* were contracted for, tested, and then installed at *USPS letter-mail facilities* in New York City, Boston, and Cincinnati.[1]

This passage has several interesting uses of noun compounds as names. Sometimes the noun compound appears only once, as a name, as with *bulk-mail system* in paragraph 2. Sometimes a noun compound appears once as a name and then again in another, larger noun compound. This occurs with *letter mail* and *USPS letter-mail facilities*. Finally, sometimes a multiword noun compound or name is shortened to an acronym, as with *USPS* from *U.S. Postal Service*. This new name (*USPS*) can then be used by itself or used to form other names: *USPS letter-mail facilities*. Acronyms can be used quite frequently if you are writing for an audience who knows what they mean.

Noun Compounds Used to Ease Shifting of Noun Phrases

Chapters 24–25 outline a method for establishing coherence in a text. This method involves ordering noun phrases so that given information comes before new information, topical information is placed in the subject position, and "lighter" units are placed before "heavier" ones.

It is sometimes easier to achieve this ordering if noun compounds are used, since they can be more easily shifted into different positions than can longer phrases and clauses. Noun compounds are especially useful for presenting the given information at the beginning of the sentence since there is a premium on keeping things short and light there. The flexibility offered by the noun phrase is illustrated below:

> The smooth flow of requests through the hospital's Correspondence Unit is marred by a number of operational problems. First, the Correspondence Unit has an average worker productivity of 59.1%, well below the suggested 65% level. Second, this low worker productivity has caused the *backlog of active and "waiting" requests* to rise above 2000 requests. Simply having this large a *backlog size* causes excessive processing delays. Third, these processing delays are often further increased by our method of assigning processing priority on the basis of patient registration number.

Notice how much easier it is to use the noun compound *backlog size* than it is to use the longer reference:

> Simply having this large a backlog of active and "waiting" requests causes excessive processing delays.

This sentence is too heavy before the verb and requires further cutting to make it acceptable.

REFERENCES

1   D. Mennie, "A Try at Automated 'Zip'," *IEEE Spectrum*, October 1976, p. 41.

ADDITIONAL READING

L. Bartolic, "Nominal Compounds in Technical English," in *English for Specific Purposes: Science and Technology*, edited by M. Todd-Trimble, L. Trimble, and K. Drobnic (Corvallis: Oregon State University English Language Institute, 1978), pp. 257–277.

J. M. Carroll, "The Role of Context in Creating Names," *Discourse Processes 3:*1–24 (1980).

J. M. Carroll and M. K. Tanenhaus, "Prolegomena to a Functional Theory of Word Formation," in *Papers from the Parasession on Functionalism*, edited by R. E. Grossman, L. J. San, and T. J. Vance (Chicago: Chicago Linguistics Society, Chicago, 1975), pp. 47–62.

P. Downing, "On the Creation and Use of English Compound Nouns," *Language 53*(4):810–842 (1977).

L. Gleitman and H. Gleitman, *Phrase and Paraphrase* (New York: Norton, 1970).

D. Kaufer and E. Steinberg, "On Revising Noun Compounds," *Journal of Advanced Composition 4*:51–64 (1984).

R. B. Lees, "The Grammar of English Nominalizations," *International Journal of Applied Linguistics*, 26:3 (1960).

J. Levi, *The Syntax and Semantics of Complex Nominals* (New York: Academic Press, 1978).

D. McNeill, "Speaking of Space," *Science 152*:875–880 (1966).

M. C. Potter and B. A. Faulconer, "Understanding Noun Phrases," *Journal of Verbal Learning and Verbal Behavior 18*:509–521 (1979).

K. Zimmer, "Some General Observations about Nominal Compounds," in *Stanford Working Papers on Language Universals* (Stanford, CA: Stanford University, 1971), pp. C1–C21.

K. Zimmer, "Appropriateness Conditions for Nominal Compounds," in *Stanford Working Papers on Language Universals* (Stanford, CA: Stanford University, 1972), pp. 3–20.

36

Vocabulary Building

Anyone who wishes to be an effective professional communicator must have a good vocabulary. Without a good vocabulary, you simply cannot put into practice all the principles we have been discussing. For reading, you need a large "recognition vocabulary" of at least 30,000 words; for writing, you need a fairly large "production vocabulary" of at least 15,000 words.

The best way to increase one's vocabulary, of course, is to read and hear the language as much as possible. The more you read and listen, the more exposure you'll have to new words. This chapter is designed to help you build up your vocabulary as efficiently as possible. It describes a strategy that allows you to expand your vocabulary at the same time as you pursue your studies or do your job, without memorizing word lists or constantly looking up words in a dictionary. Instead, you'll be using contextual clues to infer the meaning of new words.

36.1 USING CONTEXTUAL DEFINITIONS

Psychological research has demonstrated that people learn best when the words being learned are presented in a meaningful context. Apparently, the context allows the mind to form associations between the word being learned and other, related words. It gives us a more vivid image of the word's meaning to store in our memories. This suggests that the best way to expand one's vocabulary is through reading and listening, not by studying words in isolation. No doubt you already do this. You read English texts (like this one) and, when you encounter an unfamiliar word, you either look it up in a dictionary or try to guess its meaning from the context. Consulting a dictionary can be helpful, but it is also enormously time-consuming. It slows down your reading and makes it hard to follow the flow of logic in the text.

We recommend that you use context as much as possible to help you guess the meanings of words. Sometimes, for example, you can find "definitions" embedded in the text itself. Noticing these definitions can save you the time of looking them up in a dictionary. Consider this passage, for example:

> ## The Natural Sciences
>
> The development of modern science during the last three centuries has been at once so broad and so deep that a great deal of learning and thought is required for even the best of human minds to encompass a portion of it. This is in large part because progress in science is cumulative—science builds on and extends what has been observed and understood earlier. It grows by the interplay of experimental observations, imaginative reasoning, predictions based on this reasoning, and experiments to test the predictions. As the body of systematized observations and generalizations about the natural world has increased, it has been artificially subdivided into "different" scientific fields, and these have in turn been partitioned, all because human life is too short and the mind too limited to be able to learn all that has been observed and postulated.[1]

Suppose that when you began reading this passage you did not know the meaning of the word *cumulative*. Would you interrupt your reading and pick up your dictionary to find the definition for this word? No. You can see that there is a "definition" right there in the text: Progress in science is "cumulative" because it "builds on and extends what has been observed and understood earlier. . . ." This is not a formal definition of the word, but it is a vivid and memorable one, especially since it is followed by further description and elaboration ("It grows by the interplay of experimental observations, imaginative reasoning. . . .").

You can find many definitions like this in your reading, and you should always look for them before consulting your dictionary.

> **Principle 1** Whenever you come across a new word, first try to find an informal definition of it somewhere in the immediate context.

36.2 USING CONTEXTUAL CLUES

Of course, you are likely to encounter in your reading many unfamiliar words that are not defined for you. In such cases, it is even more tempting to reach immediately for the dictionary. Before you do, however, we would like you to consider the following three facts:

1 Every time you interrupt your reading to consult a dictionary, you lose valuable time and concentration.

2   It is not always necessary to know the exact meaning of every word in order to understand the meaning of a passage as a whole.

3   The context surrounding an unfamiliar word often contains enough clues for you to be able to guess the meaning of the word—if not the precise meaning, then at least a general meaning.

Given these facts, we urge you to use the context whenever you come across a new word, before you do anything else. If the context doesn't contain an informal definition of the word, it probably contains enough clues for you to guess the meaning.

What sorts of clues? Word-formation clues, syntactic clues, semantic clues, and rhetorical clues. The following is a step-by-step procedure that allows you to systematically inspect the context and find the clues you need.

Grammatical Function of the Word

Determine the grammatical function of the word. Is it being used as a noun? As a verb? An adverb? An adjective? To build your vocabulary, you must be able to properly identify the grammatical functions that words perform in sentences and phrases. Do not rely on a dictionary to do this for you! A dictionary will tell you what function a word *normally* plays, not what it plays in any particular context. In science and technology especially, words often perform a grammatical function different from that in ordinary English.

You can begin to diagnose a word's grammatical function by looking at its form, specifically at whatever ending, or suffix, it may have. For example, the *-tion* ending attached to a word normally indicates that the word is a noun (e.g., *optimization*) and an *-ize* ending normally indicates that a word is a verb (e.g., *optimize*). Also, of course, the vast majority of adverbs in English are marked by the *-ly* suffix (e.g., *optimally*). Below are several lists of suffixes commonly associated with grammatical functions; you should know them all by heart. A knowledge of these suffixes and others like them is a great help in determining the grammatical function of some word you've never seen before. You cannot rely entirely on such knowledge, however. Notice in particular that there are several cases where the same suffix can be used for either of two grammatical functions: *-al* can indicate either a noun or an adjective, *-ate* can indicate either an adjective or a verb, and so on. In addition, there are many words of every grammatical category that do not have any function-marking suffix at all, such as *comet, platform, shuttle, quick, stop, dark,* and *flood,* to mention but a few.

Some Common Noun-forming Suffixes

Suffix	Examples
-age	leverage, mileage, tonnage
-al	dismissal, approval, refusal

Some Common Noun-forming Suffixes

Suffix	Examples
-an (-ian)	technician, mathematician, historian
-ance (-ence)	conductance, acceptance, equivalence
-ation	filtration, refrigeration, determination
-ency (-ancy)	efficiency, transparency, buoyancy
-ent (-ant)	emollient, solvent, lubricant
-er (-or)	worker, manager, supervisor
-icide	pesticide, insecticide, fungicide
-ing	engineering, manufacturing, painting
-ist	physicist, biologist, typist
-itude	magnitude, amplitude, exactitude
-ment	equipment, shipment, easement
-ness	hardness, brittleness, dampness
-s	mathematics, physics, economics
-sion (-tion)	precision, corrosion, construction
-sis	osmosis, electrolysis, pyrolysis
-th	length, width, strength
-ture	mixture, furniture, expenditure
-ty (-ity)	certainty, activity, plasticity
-us	corpus, locus, terminus
-y	accuracy, recovery, synchrony

Some Common Verb-forming Suffixes

Suffix	Examples
-ate	activate, detonate, enumerate
-en	strengthen, weaken, worsen
-ify (-fy)	fortify, solidify, liquefy
-ize	sterilize, optimize, minimize

Some Common Adjective-forming Suffixes

Suffix	Examples
-able (-ible)	variable, malleable, flexible
-al	original, hexagonal, optical

Some Common Adjective-forming Suffixes (*Continued*)

Suffix	Examples
-ant (-ent)	resultant, insistent, dependent
-ary	secondary, sanitary, temporary
-ate	indeterminate, duplicate, intermediate
-ful	careful, powerful, useful
-ic	probabilistic, deterministic, magnetic
-ile	mobile, ductile, tensile
-ive	quantitative, negative, indicative
-less	careless, odorless, useless
-ous	porous, dangerous, hazardous
-y	oily, greasy, rubbery

A second and even more important way of diagnosing the grammatical function of a word in a particular sentence is to look at the other words in the sentence and see how they all fit together. For example, consider this sentence from an article on planetary exploration:

The probe will jettison the chute and continue its descent into oblivion.[2]

Even if you had never seen the words *probe, jettison, chute,* and *oblivion* before, you could accurately determine their grammatical function by observing how they fit with other words in the sentence. *Probe* must be a noun because it occupies the subject noun position ahead of the main verb and because it is preceded by an article (*the*). *Jettison* must be a verb because it follows the modal auxiliary *will*; modal auxiliaries are always followed by main verbs in full sentences. *Chute* must be a noun because it occupies the direct object noun position in the first part of the sentence, immediately after the main verb, and because it is preceded, as is *probe*, by a definite article. *Oblivion* must be a noun because it occupies the object-of-a-preposition position immediately following *into*. Notice that you could not have made these determinations by looking at these words in isolation: they do not carry any of the characteristic suffixes described earlier (in fact, most native speakers mistake *jettison* for a noun when they first encounter it).

❏ EXERCISE 36-1

Without referring to the lists of suffixes given above, see if you can guess the likely grammatical function of each of the following words:

bionic   seepage

duplicity   darken

exonerate	perusal
tertiary	optimist
fortitude	miscible
chalky	victimize
rapidity	aqueous
insolvency	prehensile
deliberately	verify
disruptive	fibrosis

❏ EXERCISE 36-2

Using whatever clues you can, try to determine the grammatical function of each of the italicized words in the following sentences:

A There is no *impediment* to bringing spacecraft and comet into line and allowing the craft to drift in and *bump* down.[3]

B The same gas molecules and atoms that absorb ultraviolet light may *emit* ultraviolet airglow, again at characteristic wavelengths that can be detected *spectroscopically*.[4]

C Much *Appalachian* coal appears to be low in organic sulfur and hence apt for *beneficiation*. Nonetheless, the *inapplicability* of beneficiation to all types of coal is an important *drawback* to the process.[5]

D During periods between meals, when the supply of nutrients in the *lumen* is minimal, the weaker *pumps* on the plasma side *revert* to their original direction and *pump* nutrients from the *plasma* into the cell, protecting the cell's nutritional level.[6]

Meaningful Parts of the Word

See if you can recognize any meaningful parts of the word. Many words are made up of two or more parts: a basic part, or root, and one or more suffixes or prefixes. For example, the word *inaudible* is made up of three parts: a prefix *in-* (meaning "not"), a root *-aud-* (meaning "hear"), and a suffix *-ible* (meaning "able to be"). Putting these meanings together, we arrive at a meaning for the word as a whole: "not able to be heard." Thus, even if you had never seen the word *inaudible* before, you could make an educated guess about its meaning if you knew the meaning of its parts.

Most English words, unfortunately, are not as readily analyzable as *inaudible* is. The English language is the product of thousands of years of development, having borrowed numerous words from other languages, which in turn may have borrowed them from still other languages, and significant changes of form and/or meaning may have occurred along the way. For example, the word *manufacture* is derived from the Latin word *manu* ("by hand") and *facere* ("to make"), and so it means literally "to make by hand." Centuries ago, that's what the word meant. Today, however, in the wake of the Industrial Revolution, the meaning of the word has been extended to include the making of something not only by hand but also by machines, by chemical processes, or by any other agency. Thus, in using roots and prefixes to help you figure out the meaning of an unfamiliar word, do not assume that the meaning of the parts will always give you the meaning of the whole. Often they will only give you a partial meaning, and you must use other devices to get the whole meaning.

The following lists contain a number of roots and prefixes that commonly occur in scientific and technical English. You should memorize as many of them as you can.

Some Common Word Stems

Stem	Original Meaning	Examples
aer (air)	air	aerial, aerodynamic, airport
ann (enn)	year	annual, anniversary, centennial
aqua	water	aqueous, aqueduct, aquifer
aud	hear	audience, audible, auditorium
bio	life	biology, biography, antibiotics
chron	time	chronology, chronic, synchronized
clos (clud)	shut	close, disclose, enclosure
corp	body	corporation, incorporated, corpuscle
cred	believe	credit, credo, incredible
dict	say	dictate, predict, contradict
duc	lead	product, reduce, transducer
fac (fic)	make, do	manufacture, satisfaction, efficient
flu	flow	fluid, effluence, influence
gen	birth	generate, genus, hydrogen
geo	earth	geology, geography, geometry
hydro	water	hydraulic, dehydrate, hydrofoil
ject	throw	eject, reject, injection
labor	work	laboratory, collaborate, elaborate
log	reason, science	logic, technology, analog
manu	hand	manual, manufacture, manuscript

Some Common Word Stems (*Continued*)

Stem	Original Meaning	Examples
mit (miss)	send	emit, omit, missile
nov (neo)	new	novel, innovate, neoplasm
ped (pod)	foot	pedal, podium, expedite
photo	light	photograph, photoelectric, photon
plic (ply)	fold	complicate, implicate, multiply
port	carry	portable, export, report
pos	put	deposit, composition, propose
puls	drive, push	impulse, propulsion, pulse
rota	wheel	rotation, rotary, rotate
scend	climb	ascend, descend, transcend
scop	look	microscope, telescopic, scope
sequ (secu)	follow	sequence, consequence, consecutive
solv	release	solve, solvent, resolve
spect	look	inspect, aspect, perspective
stat (stit)	state	state, static, constitute
strain	bind	restrain, constraint, stringent
tact	touch	tactile, contact, intact
tele	distant	television, telephone, telecommunication
tempor	time	temporal, contemporary, extemporaneous
therm	heat	thermodynamics, thermometer, thermal
tract	draw	traction, attract, contract
val	strength	value, valence, evaluation
vert	turn	invert, divert, conversion

Some Common Prefixes

Prefix	Meaning	Examples
a-	not	atypical, asymmetric, abnormal
ad- (at-, ap-, etc.)	to, toward	adhere, ad infinitum, attraction
anti-	opposite	antibiotic, anticatalyst, antimatter

Some Common Prefixes (*Continued*)

Prefix	Meaning	Examples
back-	behind	background, backlog, backfire
bi-	two	biweekly, bilateral, bivalent
centi-	1/100	centimeter, centigram, centiliter
co-(col-, con-, etc.)	together with	cooperate, collaborate, consolidate
counter- (contra-)	opposite	counterbalance, counterweight, contradict
de-	reverse, undo	deflate, decrease, defrost
di-	two	dimethyl, divide, dilemma
dia-	across, through	diameter, diagonal, diagnose
dis-	not	discontinuous, disproportionate, dishonest
dis-	reverse, undo	discharge, dislodge, dismantle
down-	down	downturn, downslope, downtime
e- (ex-)	out of	evaporate, eject, expel
en-	in, within	enclose, encapsulate, engage
extra-	beyond	extrapolate, extraordinary, extracurricular
in- (im-, ir-)	in	insert, immerse, irrigate
in- (im-, ir-, il-)	not	independent, inelastic, immobile
inter-	between	interpolate, intercontinental, interfuse
iso-	similar, same	isotope, isomorphic, isosceles
mal-	bad, badly	malfunction, malformed, malpractice
mis-	bad, badly	miscalculate, misfire, mispronounce
mono-	one	monomer, monorail, monotone
multi-	many	multiple, multiplex, multitude
non-	not	noncollapsible, nonmalleable, nonflammable
on-	on	ongoing, onlooker, on-line
out-	surpass	outperform, outdo, outclass
out-	out, outside	outline, outbuilding, outskirts
out-	proceeding from	outgrowth, outlook, output
over-	excessive	overwork, overproduce, oversell
over-	beyond	overlook, overlap, overtime
poly-	many	polygon, polymer, polytechnic

Some Common Prefixes (*Continued*)

Prefix	Meaning	Examples
post-	after	postpone, postmortem, postgraduate
pre-	before	preview, prevent, preformed
quadr-	four	quadrangle, quadrilateral, quadrant
re-	back	revert, reject, retract
re-	again	reapply, refasten, reiterate
retro-	back, back-wards	retrospect, retroactive, retrorocket
self-	by itself, by oneself	self-lubricating, self-employed, self-loading
self-	in itself, in one-self	self-evident, self-explanatory, self-assured
semi-	half	semicircular, semiconductor, semiannual
side-	to or on the side	sidetrack, sidelong, sidecar
sub-	under, below	submarine, subsoil, substrate
sub-	a lesser part of	subclass, subdivision, subroutine
super-	beyond, above	superstructure, supersonic, superheated
trans-	across	transmit, transaction, transect
tri-	three	triangle, tripod, trioxide
ultra-	extreme	ultrasound, ultraviolet, ultramicroscopic
un-	not	uncertain, unavoidable, unclear
un-	reverse	unfold, undo, unbend
under-	beneath, below	underground, underlie, undersurface
under-	inadequate	underdeveloped, underestimate, understaffed
uni-	one	uniform, unify, unique
up-	up	upgrade, upturn, upkeep
well-	well	well-timed, well-advised, well-defined

❑ EXERCISE 36-3

Without consulting a dictionary, see if you can guess the meaning of the following words. Use the root and prefix lists to help you. After you have finished, you may consult a dictionary to check your answers.

detoxify	implicate
reflux	centipede
circumspect	isotherm
interpose	disclosure
subversion	collaborate

Other Words Nearby

Look at other words for hints as to the meaning of the unfamiliar word. Often the immediate context surrounding an unfamiliar word contains clues that enable you to guess the word's meaning. Let's look at an example. Do you know what the word *blowout* means? Probably not. If you come across it in a context like the following, however, you can probably make a good guess as to its meaning just by noting some of the other words near it:

> Everyone is familiar with newspaper accounts of the shoreline damage that oil causes. The damage occurs year by year on different coasts and sometimes results from *blowouts* at offshore drilling platforms as well as from ships.[7]

Even if we knew nothing about offshore oil drilling, we would suppose that the word *damage* in this context refers to accidents of some sort. Furthermore, these seem to be specific kinds of accidents: they involve discharges of oil, which cause "shoreline damage." From such clues you can infer that the word *blowout*, in this context, means an explosion, a spill, a leak, or some other type of accident that releases oil into the sea. (*Explosion* comes closest to the exact meaning, as you might be able to guess if you know the word *blowup*; a *blowout* is an explosive release of oil out of the top of an oil well.)

Other Words in Parallel

Look for other words used in parallel with the unfamiliar word and see what common meanings they share. Often a new word may represent some item belonging to a list of items; if you know what all the other items are and what they all have in common, you may be able to guess the meaning—or at least narrow down the possible meanings—of the unidentified item. Or, you may come across a new word in some context that also contains a synonym or antonym of that word (two words are antonyms of each other if they are defined according to the same criteria but have opposite meanings). In such cases, writers typically put the related words (list items, synonyms, or antonyms) into parallel grammatical form—all nouns or all verbs or all adjectives, etc.—as discussed in Chapter 23. Thus, by looking for parallel grammatical form, you can often spot words related to the unfamiliar word you're trying to analyze, and these related words will provide clues for you.

Here is an example, from a description of measuring instruments on a spacecraft:

> A solar flux radiometer measured the energy reaching each level of the atmosphere from the sun. Two other devices also tallied heat radiation, one of them possessing a window made of diamond.[8]

Do you know what the word *tallied* means? Even if you've never seen it before, you can guess what it means by noticing that it is in parallel to the word *measured*:

> A solar flux radiometer measured the energy. . . .
> Two other devices also tallied heat radiation. . . .

Both words are main verbs, both are in the past tense, and both have measuring instruments for subjects and measurable quantities for objects. Indeed, these two words mean essentially the same thing, that is, they are synonyms. The connective word *also* supports this interpretation. So if you know what the word *measured* means, you know what *tallied* means.

Here is another example of parallel forms, this one being a two-item list from a discussion on plastic products:

> The insulated plastic coffee cup and the plastic egg carton are two products that often contain fluorocarbon gas.[9]

Do you know what a carton is? Even if you don't, you can guess what it is by noticing how the word *egg carton* parallels the word *coffee cup*. They are both noun compounds and the fact that they are both preceded by the same adjective (*plastic*) makes it especially easy to spot the parallelism. Just as a cup is a standard container for coffee, a carton is a standard container for eggs. So the words *cup* and *carton* have much in common, though they are not synonyms.

❑ EXERCISE 36-4

In the passages below, try to guess the meaning of the italicized words by looking at related words in the immediate context. Be prepared to explain your reasoning.

A Researchers at SPRL are currently collecting bottles of air at selected places to learn the actual amount of nitrous oxide entering the atmosphere. Such places include especially those *sites* that may be emanating N_2O at levels higher than used to occur in nature—fertilized fields at various seasons, *compost heaps*, animal feed lots, lawns, golf courses, and eutrophic lakes and ponds.[10]

B   The region around the Great Lakes contains some of the great population centers of the United States and Canada. To the *teeming* millions in this region the lakes hold promise of an inexhaustible source of water for municipal, industrial, agricultural, and recreational uses. But, as is well known, rapid *urban* and industrial development has *prematurely* exhausted the usefulness of some parts of the lakes, with the result that in those parts the waters are no longer *potable*, "fishable," or "swimable."[11]

Rhetorical Patterns and Visual Aids

Look at other aspects of the larger context, such as rhetorical patterns and visual aids. As is discussed in Chapters 22 and 23, properly written prose is always structured according to certain patterns of development, or rhetorical patterns. These patterns are most noticeable in units of prose larger than the single sentence. Paragraph units, for example, are often structured in a general-to-particular pattern: the first sentence of the paragraph contains some sort of generalization, and the subsequent sentences contain details supporting this generalization. Some other rhetorical patterns common to scientific and technical English are classification, comparison-contrast, problem-solution, process description, and logical analysis. Some rhetorical devices commonly used to fill out these patterns are examples, analogies, details, conditional statements, definitions, lists, and questions.

Often you can guess the meaning of an unfamiliar word if you can see how it fits into a rhetorical pattern. For example, do you know what the word *drawback* means? If not, see if you can't make a good guess at its meaning in the following context:

> The *drawbacks* of the bubble chamber are mainly two. First, the chamber is dumb and slow. It cannot be programmed except to a limited degree to distinguish interesting events from uninteresting ones; the piston must be withdrawn and a picture of every event taken before traces of the event vanish. This can result in millions of films that must be culled—the second *drawback*—by hand. Technicians, or scanners, must look at every film, and upon finding worthy images, they must measure the angles and curves by hand. Once derived, of course, the data from the bubble chamber films are precise, but thousands of technician-hours must first be invested.[12]

If you recognize the general-to-particular rhetorical pattern of this paragraph, you'll notice how all of the sentences following the first one (that is, all of the particulars) seem to be pointing out problems associated with the bubble chamber. If the first sentence is a proper generalization, then, it must be the case that the word *drawback* means, more or less, "problems." And this is roughly what it does mean.

Connective words can be very useful in analyzing rhetorical patterns. In the above case, the connectives *first* and *second* are used effectively to bring out the particularly problematic features of the bubble chamber and thus tie the discussion to the opening generalization. (Other connectives and their uses are discussed in Chapter 34.)

Visual aids are very important rhetorical devices, and you should use them whenever you can to help guess the meanings of unfamiliar words. For example, even though you may know something about the problems of cleaning up oil spills in the ocean, chances are you don't know what the terms *fiber bundle, honeycomb screen, vortex, boom, squeegee,* and *hydrofoil* mean. Do you? If not, you'll still be able to read the following excerpt and guess what they mean, simply by referring to the visual aids where indicated:

The recent test series dealt with devices that pick up spilled oil directly, without fences or corrals. They are a bit like carpet sweepers in that they draw oil off the surface of the ocean as it moves beneath or above them. Almost all the devices were contrived to slow down the oil, make the slick thicken, and then suck it off the surface. One means of thickening the oil layer was *fiber bundles* of polypropylene arrayed in the surface layer of oil and water (Fig. 1). The idea here is to provide some viscous resistance in the top layer and make it move more slowly than the rest of the current; an oil slick would then be made thick enough, after passing through the fibers, to be cleanly pumped away. In another scheme, a succession of *honeycomb screens* of diminishing gauge was assembled for the same purpose (Fig. 2). A third system sought to concentrate oil at the center of a *vortex*, where a suction pipe could then be applied to draw off the oil (Fig. 3). Another device used *booms* to extend a curtain of blowing air on the flowing oil slick (Fig. 4). Two systems used endless belts of oleophilic (oil-attracting) material. The belts of one of these systems would move at the same speed as the current, thereby reducing the slick's relative velocity to zero, and carry oil to a *squeegee* that would strip it from the belt for pumping away (Fig. 5). The other belt system used *hydrofoils* to slow and concentrate the slick while the belt served to elevate the oil from the water surface to a collector (Fig. 6).[13]

The visual aids accompanying this passage are virtually indispensable for understanding it. The author and illustrator seem to have known this, for they set up a form of parallelism between the italicized nouns and the visual aids, one series correlated with the other. Unfortunately, they were not completely consistent in carrying out this design. To be consistent, they should have used the same key terms in the visual aids as they used in the text. In four of the six figures, they did (Figures 1, 2, 4, and 6), but in the other two they didn't. In these latter two, you simply have to infer from the visual and written descriptions that *centrifuge* is a synonym for *vortex* and *driver/wringer* a synonym for *squeegee*.

Principle 2 If you encounter an unfamiliar word that is not defined in the text, try to guess or at least narrow down its meaning by using contextual clues.

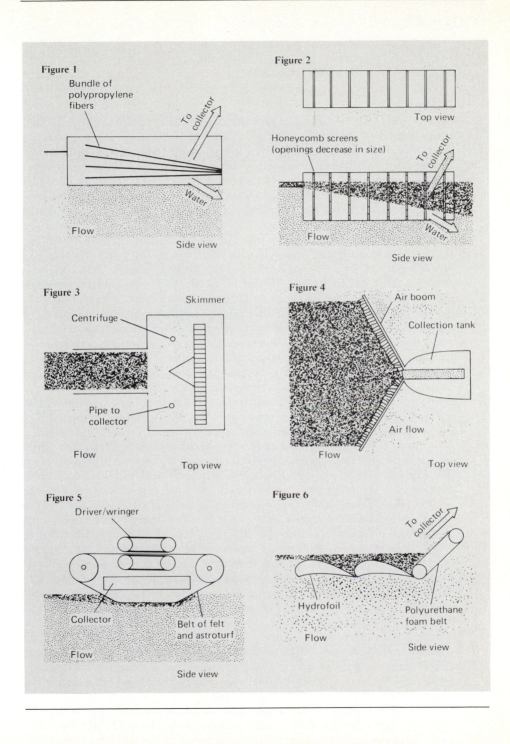

Figure 1
Bundle of polypropylene fibers
To collector
Water
Flow
Side view

Figure 2
Top view
Honeycomb screens (openings decrease in size)
To collector
Flow
Water
Side view

Figure 3
Skimmer
Centrifuge
Pipe to collector
Flow
Top view

Figure 4
Air boom
Collection tank
Air flow
Flow
Top view

Figure 5
Driver/wringer
Collector
Belt of felt and astroturf
Flow
Side view

Figure 6
To collector
Hydrofoil
Polyurethane foam belt
Flow
Side view

❑ EXERCISE 36-5

In the passage below, try to guess the meaning of the words in italics by using word analysis and contextual clues. Be prepared to explain your reasoning.

OSHA: Costly Crusaders

On April 28, 1971, Congress passed the Occupational Safety and Health Act, which created an organization whose purpose was to document and cut down the number of industrial accidents. OSHA (Occupational Safety and Health Administration) was to be the *watchdog* and enforcer of safety laws. OSHA is authorized to set safety and health standards for nearly all non-government employers. This included (at the time of its formation) 5 million employers and 60 million workers; only employers covered by other federal safety programs or family-operated enterprises were *exempt*.

OSHA, therefore, is a very powerful and influential organization. It has the power to investigate a company on a worker complaint basis or on its own *incentive*. If OSHA finds something which it feels is a safety hazard, it can impose a fine of up to $1000 a day for the violation or it can seek court orders to close the facility down. In the event of any accidental death or disaster injuring five or more workers, OSHA is automatically entitled to investigate the scene and to impose fines or shut the facility down (with a court order).

Many Americans tend to think that occupational hazards are a thing of the past, like *sweatshops* and child labor. This is a very dangerous misconception. Workplaces now are about twice as dangerous as our nation's streets and highways. OSHA officials felt this misconception was due to a lack of industrial accident statistics. Until OSHA was formed, a relatively small percentage of industrial accidents were being reported. Early OSHA officials estimated that once accidents were reported and investigated the number would double.

Much of OSHA's early dealings were with four target occupations which were abnormally unsafe and had a high rate of accidents. They were: *longshoremen* (the workers were in constant danger of falling crates and unstable loads), meat packers (packers often cut themselves on the band saws and cutting tools), loggers (workers are easily crushed by logs and are often in situations where one slip means death or serious injury), and metal stamping plants (presses and stampers generate a continuous 115 dB, the threshold of pain being 120 dB). These industries were required to protect their workers from such *adverse* conditions. The employers felt that protective devices such as steel-toe boots, hard hats, *goggles* and ear protectors would provide adequate personal protection along with guards and safety devices on the machines. OSHA, however, felt that protective wear was not the answer and began to inquire about engineering the safety into the machines and into the workplaces. This line of thought threw OSHA and businesses into deeper conflict. Businesses had already started criticism toward the way OSHA went about its business. Businessmen felt that OSHA was investigating poorly and that they were getting *bogged* down by trying to enforce many small and relatively insignificant things while letting the major issues slide.

Take, for instance, OSHA's latest main objective: noise reduction in factories and in products. OSHA at first had set noise levels at 90 dB. Workers were given ear protectors which met those standards, but OSHA felt that the machines should operate quietly enough to allow workers to take off the ear *muffs*.

However, the engineering involved to change the machines would cost billions of dollars. To make matters more *touchy*, OSHA is trying to decrease the noise level to 85 dB.[14]

36.3 CONSULTING AN AUTHORITY

Sometimes, no matter how many word-formation clues and contextual clues you use, you can't quite figure out what an unfamiliar word means. In such cases, you may have to consult an outside authority, that is, a knowledgeable person or a dictionary of some sort. Before taking the time required to do this, however, make sure you really need to know what the word means. Is it so important that you can't understand the passage without knowing the exact meaning of this word? Is the word likely to reappear later on? Can't you get by with just a vague idea of what it means or what it probably means? Remember, the time required to look up words in a dictionary or talk to an expert is time taken away from the reading process itself. There is a cost in both time and concentration whenever you consult an outside authority, and you must weigh the benefits against these costs.

When to Consult an Authority

Generally, you should consult an authority only for words that seem especially important in the particular passage you're reading. For example, if a word is a topical one—that is, if it represents the subject being discussed—we can assume it to be important. Topical words are often called key words in scientific and technical publications, and indeed topical words are key to both understanding and remembering the subject matter of discourse. For this reason, topical words are often used repeatedly, as are most important words in written prose. Thus, if a word is *used repeatedly*, it is probably an important one.

Sometimes, however, a word may be used only once and yet still be important enough to require exact knowledge. For instance, if it is part of an important *procedural* description, a description that you must know in step-by-step detail, then it is probably a word you should know the exact meaning of. Yet another circumstance signaling an important word is the use of examples or analogies to explain the meaning of important abstract concepts. Often an author will rely on the reader's knowledge of common words in trying to explain a difficult concept. These *explanatory* words, though appearing only once, briefly take on the importance associated with the concept being explained.

Whom to Consult

What sort of authority should you consult? In our view, the best authority is a knowledgeable person with whom you can talk directly, be it a supervisor, professor, fellow student, or someone else. Since the meaning of a word varies with the context in which it's used, there is a distinct advantage in being able to show your consultant the word in its actual context. That way, even if the consultant already knows the general meaning of the word, he or she can check it against the context and determine or verify its exact meaning. Also, when you deal with a consultant face to face, you can follow up his or her explanations with more questions if you like.

Dictionaries

A dictionary is also a useful authority, of course, provided it's one designed with your needs in mind. As a nonnative speaker of English, you probably need more information about words than is normally given in dictionaries designed for native speakers. For example, most nouns are normally used either as countable nouns or as uncountable nouns, and it is important to know the difference (Chapter 29). Native speakers learn very quickly whether a noun is normally countable or uncountable, but nonnative speakers often don't. Therefore a dictionary designed for nonnative speakers should indicate for each noun what its status is in terms of countability. Most dictionaries do not do this. Another feature you should look for in a dictionary is the use of sample sentences to supplement the definitions, as in the following example:

> envision (ĕn·vĭzh'ŭn), v.t. To have a mental picture of something beyond one's scope of vision. *It is difficult for most of us to envision the distant future.*

Sample sentences serve at least two important purposes. First, they make the meaning of a word more comprehensible. Definitions alone are often very abstract and thus harder to understand than concrete examples. Second, sample sentences illustrate how a word is typically used, not simply how it can be used. For instance, *envision* is typically used with a human subject (*us* in the above example) and a very inaccessible, possibly very abstract direct object. Although a single sample sentence cannot encompass all of the typical uses of a word, of course, it can at least be suggestive of them. Sample sentences are especially valuable in illustrating the use of idiomatic phrases and phrasal verbs. Among the dictionaries that satisfy these requirements are the *Longman Dictionary of Contemporary English*, the *Oxford Student's Dictionary of Current English*, and the *Oxford Advanced Learner's Dictionary of Current English*.

Even a well-designed dictionary, however, is likely to be deficient in at least one important respect: it will be out of date and incomplete in its treatment of scientific and technical words. The language of science and technology simply develops new terminology too quickly for dictionaries to keep up with it, and much of it is too specialized to be of interest to the general public. Thus, if you're interested in expanding your scientific and technical vocabulary, you might consider using a specialized dictionary if you can find one. Even a specialized dictionary, though, may be woefully incomplete and out of date. In particular, it may overlook many words from general everyday English which have highly specialized meanings when they are used in science and technology.

> Principle 3   If an unfamiliar word seems to be particularly important and you can't determine its meaning from contextual clues, consult an authority.

36.4 GETTING TO KNOW A WORD WELL

If you decide that a word is important enough to merit learning, try to learn it as well as you can. Don't be satisfied with simply learning what the word means; try to learn everything you can about it—its normal uses, its pronunciation, its related forms—so that you can use it productively in speaking and writing.

Usage

Note how the word is used. If it's a noun, is it countable or uncountable? What kinds of verbs or adjectives does it normally co-occur with? Does it occur in noun compounds and, if so, with what kinds of nouns? Is it restricted in the kinds of complements it can take?

If the word is a verb, is it a one-word verb or a two-word (phrasal) verb? Is it transitive or intransitive? Can it be followed by more than one object noun? Is it a state verb or an activity verb? What kinds of adverbs or nouns does it tend to co-occur with? What kinds of complements can it take? Does it occur in the passive form? If so, how often?

If the word is an adjective, what kinds of nouns or adverbs is it likely to co-occur with? Is it restricted in the kinds of complements it can take?

If the word is an adverb, what kinds of adjectives or verbs does it normally co-occur with?

Related Forms

Learn related forms. Most English words have one or more related forms, having the same root but differing in prefix or suffix and perhaps in grammatical function as well. For example, the word *adverse* has at least three closely related forms—*adversity, adversary,* and *adversely*—and a number of more distantly related ones—*converse, convert, inversion, subversive, perversity, animadversion,* and so on. The word *exempt* also has three closely related forms—*exempted, exemption, exemptible*—but only a few distantly related ones—*preempt, preemptive, peremptory,* and *peremptorily* are the only ones that come to mind. Notice how even with these two adjectives, *adverse* and *exempt,* there is considerable variation in the suffixes and prefixes one finds in their respective related words. Notice also that their closely related forms differ in grammatical function: *adverse* has two closely related noun forms (*adversity* and *adversary*) and a closely related adverbial form (*adversely*) but no closely related verb form at all, whereas *exempt* has a closely related noun form (*exemption*), verb form (*exempt, exempted*), and adjective form (*exemptible*) but no closely related adverb form.

Obviously, to learn the related forms of a word you cannot depend on a single pattern of relations. Each word has its own network of related forms, and you simply have to learn them one by one. The lists of roots, prefixes, and suffixes presented earlier in this chapter should help you at least to know what to look for in the way of possible related forms.

Pronunciation

Learn how the word and its related forms are pronounced. Although there are certain general principles of pronunciation in English, there are also many exceptions to these principles. Therefore, whenever you learn a new word, pay special attention to how it's pronounced. If you're consulting a knowledgeable person, ask him or her to pronounce the word for you, several times if necessary. If you're consulting a dictionary, make sure you know the phonetic symbols used to indicate proper pronunciation; a guide to these symbols can usually be found somewhere at the beginning or end of the dictionary. (For detailed instruction and practice in pronunciation, *see* Chapter 38.)

> Principle 4   Try to get to know words well, especially important words. This means learning what they mean, how they are used, how they are pronounced, and what their related forms are.

❑ EXERCISE 36-6

These final exercises are designed to test your ability to apply the principles described above. For each italicized word, first see if there isn't an informal definition of it somewhere in the passage. If not, try to guess the meaning of the word by using word-formation and contextual clues. For words that seem particularly important, make sure you know the exact meaning, even if you have to consult an outside authority. In general, try to learn as much as you can about each word within the time allotted.

A   Water Pollution: From Raw Sewage to Detergent Bubbles
to Carcinogenic Chemicals
Since the early 1960's the American public has become increasingly *uneasy* about the quality of its water. This uneasiness has been directed toward both *raw water* (the water in waterways and in the ground) and processed drinking water.

Public concern initially focused on massive pollution of our waterways by *untreated sewage* and other solid and liquid wastes from municipalities and industries. In the early 1960's the problem was dramatized by the *deterioration* of two great waterways, the Hudson River and Lake Erie. Oil slicks, dead fish, detergent bubbles, beer cans, and other *debris* had turned one of the most beautiful rivers in the world into "an open sewer." As for Lake Erie, about a fourth of the lake, encompassing 2600 square miles in the central basin, was all but *"dead."* The area was covered with a layer of *scum* and contained almost no oxygen and no fish. Across the nation, countless other once clear, cool, and lovely lakes and rivers also were rapidly deteriorating.

Subsequently *public complacence* was unsettled by a succession of scares about the presence of *insidious* chemicals in water. Detergent foam had emerged from household taps in some areas. Then various chemicals— DDT, mercury, and PCBs, among others—were discovered in unusually high concentrations in the *aquatic* food chain. These discoveries led to the issuance of fishing *bans* in Lake Erie, Lake St. Clair, the Hudson River, and other waterways. Mercury and other heavy metals were also detected in *trace quantities* in drinking water. Meanwhile various cancers were linked to drinking water in some localities. In 1974 asbestos-like fibers believed to be *carcinogenic* were found in the drinking water of Duluth, Minnesota. Suspected carcinogens were identified in New Orleans drinking water and in many water supplies throughout the country. Then the presence in drinking water of certain organic compounds believed to be carcinogenic *was linked* to the chlorination process.[15]

B Hit and Miss: Elastic Interactions

How big is a proton? Are the proton and the neutron, which weigh about the same, identical except for the neutron's lack of electrical charge? Are these particles hard, with sharp boundaries, like ball bearings, or soft, with *fuzzy* edges, like a powderpuff? And do they look the same at different energies? This last question is like asking whether one's hand looks the same to radar rays, under visible light, and under x-rays.

The proton and the neutron are the only strongly interacting particles found in natural, stable matter, and they form the nuclei of all atoms, together giving an atom almost its entire weight. Although physicists have now identified almost three hundred different elementary particle states that interact through the strong interaction, of these only the proton is permanently stable.

The proton and the neutron can be *mutually* studied by causing them to interact with one another and observing the results. Roughly speaking, when a neutron and a proton collide, either of two things can happen. If the collision is a *glancing* one, it is likely that the two particles will remain *intact* and scatter much like billiard balls. If, on the other hand, there is a *head-on* collision, the result can be the spectacular creation of as many as thirty new particles, mesons. These arise from a conversion of kinetic energy in the colliding particle into mass.

The central position of protons and neutrons in the subatomic world makes it *crucial* to understand in detail the structure, shape, and size of these particles. Many other experiments, if they are to work, depend upon good knowledge about protons and neutrons. And detailed knowledge about these important particles provides a *touchstone* for theories explaining the subatomic world; a *credible* theory must accord with the finely understood details of a few particles as well as provide *plausible* explanations of little-understood phenomena.

Jones and Longo are studying the glancing interactions. These are called elastic collisions, bounces, rather than crashes, or inelastic collisions, in which new particles result. Instruments are set up to determine how the beam and target particles *ricochet*. The experiment will learn about the surface of protons by observing the distribution pattern of neutrons and protons that have bounced off one another. The pattern will allow inferences about what happened when the neutron *grazed* the proton and hence about the proton's size and shape.

Jones *likens* the experiment to an exercise in analyzing the shape of an elephant by bouncing basketballs off it. With a large enough number of basketballs and a method of measuring where each basketball bounces away, a group of *blindfolded* scientists might eventually deduce the shape and size of the creature before them. If they could perform the same analysis again with tennis balls, they might discover details smaller than they had seen with basketballs. The benefits from a tennis-ball analysis, which might ensure that the blindfolded scientists did not mistake the elephant for a rhinoceros, correspond to the benefits of particle interactions at high energy-levels.[16]

REFERENCES

1 J. Wager, K. N. Trueblood, and C. M. Knobler, *Chem One* (New York: McGraw-Hill, 1976), p. 1.

2 B. Hiatt, "Planetary Exploration," *The Research News* 29(10–12):23 (1978).

3 Hiatt, "Planetary Exploration."

4 Hiatt, "Planetary Exploration," p. 26.

5 B. Hiatt, "Coal Technology," *The Research News* 24(12):13 (1974).

6 J. Wei, "Membranes as Mediators in Amino Acid Transport," *The Research News* 31(11/12):20 (1980).

7 B. Hiatt, "Oil on Troubled Waters," *The Research News* 27(2):3 (1975).

8 Hiatt, "Planetary Exploration," p. 15.

9 B. Hiatt, "Ozone Zone," *The Research News* 28(4/5):15 (1977).

10 Hiatt, "Ozone Zone," p. 19.

11 J. Wei, "Towards Cleaner Water," *The Research News* 29(8/9):14 (1978).

12 B. Hiatt, "What's the Matter: Particle Physics at the University of Michigan," *The Research News* 26(3/4):17 (1975).

13 Hiatt, "Oil on Troubled Water," pp. 12–13.

14   S. Hannah, "OSHA: Costly Crusaders," *Michigan Technic* 96(4):16 (1978).

15   J. Wei, "Trace Contaminants and Advanced Technology in Wastewater and Water Treatment," *The Research News* 29(8/9):5 (1978).

16   Hiatt, "What's the Matter," pp. 12–13.

ADDITIONAL READING

Practical

Ruth Gairns and Stuart Redman, *Working with Words: A Guide to Teaching and Learning Vocabulary* (Cambridge, Eng.: Cambridge University Press, 1986).

John Morgan and Mario Rinvolucri, *Vocabulary* (Oxford, Eng.: Oxford University Press, 1986).

Theoretical

Ronald Carter and Michael McCarthy, *Vocabulary and Language Teaching* (London: Longman, 1988).

Margaret G. McKeown and Mary E. Curtis, *The Nature of Vocabulary Acquisition* (Hillsdale, NJ: Erlbaum, 1987).

Thomas Huckin, Margot Haynes, and James Coady, *Second Language Reading and Vocabularly Acquisition* (Norwood, NJ: Ablex, 1991).

37

Informal Conversational Expressions

A great deal of scientific and technical communication is carried out by means of ordinary conversation—in the office, in the classroom, in the laboratory, in hallways, at lunch, over the telephone, at conferences, etc. In fact, it's probably safe to say that *most* scientific and technical communication takes place in this way. And this doesn't even take into consideration the amount of ordinary social communication that occurs through conversation—social communication that helps establish good relations between coworkers and thus helps make technical communication easier. For these reasons, it's well worth anyone's time to develop effective conversational skills.

One prominent feature of informal conversational language is the use of idiomatic expressions. Sometimes the meaning of such expressions cannot be determined by putting together the meanings of their component parts. *To kick the bucket*, for example, does not have anything to do with kicking or buckets (at least not in the idiomatic sense); rather, it means "to die." Because conversational expressions often have such restricted and unpredictable meanings, they have to be learned one by one, which makes it difficult to learn a large number of them. Nonetheless, you should try to learn as many as you can.

The following list is designed to help you get started. It contains 230 of the most commonly used conversational expressions in technical and scientific settings in the United States (though they are characteristic of American English, most are also used in British English). Each entry contains (1) the expression itself, (2) one or more synonyms for that expression, and (3) an example of how the expression is used.

610

INFORMAL CONVERSATIONAL EXPRESSIONS

Expression	Synonym(s)	Example
about-face	reversal	After first rejecting our offer, she did an about-face and accepted it.
about to	ready to, all set to	We were about to start the meeting when Jack called to say he'd be late.
above all	especially, most importantly	Above all, be sure to test all the samples using the same procedure.
account for	explain	We are unable to account for the extra chlorine in the water supply.
all along	the whole time, from the beginning	I felt all along that we were on the wrong track.
all in all	all things considered, in general	All in all, despite problems with the weather, we had a successful trip.
all set	ready	Everything is all set for the visitors' arrival.
allow for	take into consideration	Unfortunately, our calculations failed to allow for possible errors.
amount to	total, add up to	Our expenses will probably amount to more than $100,000.
appeal to	be attractive to	This new model should appeal to potential buyers.
arrive at	reach, make	The committee expects to arrive at a decision sometime next week.
ask about	inquire	I asked the project manager about hiring a new technician, but she said there were no funds for it in the budget.
ask for	request	We'll have to ask the manager for more supplies.
(not) at all	of any kind, in any way	They don't have any software specialists at all.
(be) at fault	wrong, deserving of blame	I'm afraid I'm at fault: I forgot to reset the stack switch.
(be) aware	know about	Are you aware of David's medical problems?
back to square one	at the starting point again	We're not making any progress with this approach. Let's go back to square one and try again.

INFORMAL CONVERSATIONAL EXPRESSIONS (*Continued*)

Expression	Synonym(s)	Example
back and forth	first in one direction, then in the opposite direction	A pendulum swings back and forth.
back up	(1) move backward	Back the truck up to the loading dock.
	(2) support, reinforce	Be sure to back up your recommendations with good arguments.
beat around the bush	avoid stating the main point	He didn't want to embarrass them, so instead of saying what was on his mind he kept beating around the bush.
blow up	detonate, explode	If too much pressure builds up in the sterilizer, it might blow up.
bottom line	main point	The bottom line is that they're not going to support us after our current grant expires.
(be) bound to	certain to	If we send out enough grant proposals to other agencies, we're bound to get some support somewhere.
brand-new	absolutely new	Barbara traded in her old car for a brand-new one.
break down	(1) stop functioning	The new machine has already broken down.
	(2) analyze, take apart	We shall begin by breaking the process down into 15 steps.
breakthrough	a major development or advance	IBM is about to make a new breakthrough in computer technology.
bring up	introduce a topic of discussion	I hate to bring up this topic at this time, but I think we really need to talk about it.
brush up on	renew one's knowledge of	I'll have to brush up on my French before going to Paris next year.
build on/upon	to use as a foundation	I'd like to get a job where I can build on my university training.
burn the midnight oil	work late at night, work overtime	To get my report finished on time, I'll have to burn the midnight oil.
burn up	destroy by fire or heat	Unfortunately those files were all burned up in that fire we had last month.

INFORMAL CONVERSATIONAL EXPRESSIONS (*Continued*)

Expression	Synonym(s)	Example
by and large	for the most part, in general	We've had some problems on this project, but by and large it's been quite a success.
by the way	incidentally, while we're on the subject	By the way—since you've raised the subject—why did you decide to use the 9200?
by word of mouth	by unofficial communication	Many job openings are advertised by word of mouth, not by posted advertisements.
call on	visit	May I call on you in your office sometime next week?
call (up)	telephone	I'll call you (up) as soon as I arrive. By the way, what's your number?
(be) capable of	able to	Ms. Gray's latest report shows that she is capable of handling tough assignments.
(be) careful to	take measures to, be sure to	You must be careful to protect our trade secrets from our competitors.
carry out	accomplish, complete	You will be expected to carry out your duties without extra help.
carry the ball	be in charge, have primary responsibility	We've done our share; let's let Jones carry the ball from now on.
catch up on	do what was postponed earlier	I hope to catch up on my reading during my vacation.
catch up with	overtake, draw even with a competitor after being behind	Although our competitors have been working on this process for a long time, I think we can catch up with them soon.
change one's mind	revise one's opinion about something	At first I thought their plan was good, but now I've changed my mind.
change over to	convert to	We shall change all our operations over to the SI system in January.
check over	inspect, examine	Check the equipment over one more time before you start.
check up on	investigate	We'd better check up on that company's credit rating before signing any contract with them.

INFORMAL CONVERSATIONAL EXPRESSIONS (*Continued*)

Expression	Synonym(s)	Example
check with	ask for advice or approval	It sounds like a good idea to me, but you should check with Dr. Lee before you go ahead with it.
come across	discover, find	I thought I had gotten all the bugs out of the program, but then I came across some new ones.
come along	progress, develop	How are your hydroponic tomatoes coming along?
come in handy	be useful	This new tool should come in handy.
come out	be published, be issued	The new OSHA regulations should come out sometime next week.
come to terms	reach an agreement	After three days of negotiations, the company and the union finally came to terms.
come to the point	state the main point	I wish you'd come to the point instead of beating around the bush so much.
come up with	develop, find	We should be able to come up with a solution to this problem by tomorrow.
consist of	be composed of	Brass consists of copper and zinc.
count on	depend on	You can count on that company to deliver the software on time.
cut down on	reduce	We can make a profit only if we cut down on expenses.
cut off	stop, terminate	Our supplier has threatened to cut off further shipments unless we pay our bill.
cut out	remove	The administration intends to cut all unnecessary items out of the budget.
day in and day out	regularly	Sue comes to work day in and day out, even when she's not feeling well.
day off	holiday	I'd like to take a day off next week so that I can visit my sick aunt.
deal with	confront, face	I'm too tired to deal with that problem right now; let's take it up tomorrow.
a dime a dozen	plentiful, commonplace	Skilled carpenters used to be a dime a dozen; now it's hard to find one anywhere.

INFORMAL CONVERSATIONAL EXPRESSIONS (*Continued*)

Expression	Synonym(s)	Example
depend on	rely on, count on	Linda's a good researcher; I think we can depend on her to design the experiment right.
(be) devoted to	be concerned with	Most of the staff meeting will be devoted to discussing the annual report.
do over	do again, redo	Your homework has too many mistakes in it; you'll just have to do it over.
down-to-earth	practical, realistic	We're spending too much time philosophizing about this; let's be a little bit more down-to-earth, OK?
drop off	fall off, decline	We expect their production of 360s to drop off as soon as they start manufacturing the 420x.
dwell on	be overly concerned with	Why dwell on the problems of the past? We've got enough things to worry about right now!
end up with	have as a result	If you hire that technician, you'll end up with serious trouble in the lab.
enlarge upon	explain in more detail	I see that this idea is new to you, so let me enlarge upon it for a minute.
faced with	confronted with	If this product doesn't sell, our firm will soon be faced with bankruptcy.
fall behind	proceed too slowly, lag behind	Production will fall behind if the supplies don't arrive soon.
fall off	decrease, drop off	The present rate of inflation should fall off by several percentage points in the next year.
feed into	flow into	The paper feeds into the machine, where it is printed on both sides.
few and far between	rare	Skilled carpenters used to be found all over the place; now they're few and far between.
figure on	plan on, expect to	Sam knew that going into business for himself would be difficult, but he didn't figure on going bankrupt.
figure out	solve	No matter how hard I try, I can't seem to figure out this homework problem.

INFORMAL CONVERSATIONAL EXPRESSIONS (*Continued*)

Expression	Synonym(s)	Example
fill in	insert material in	To complete this report form, all you have to do is fill in the blanks.
fill out	complete	Please fill out this application form before you leave.
find out	discover	We shall find out the truth one way or another.
finish up	complete	We can't finish up this job without more concrete.
focus (one's attention) on	concentrate on	In this report I shall focus on the steps we took to improve our analytical procedure.
follow up on	pursue, develop	Perez has good ideas, but he seldom follows up on them.
for good	forever, permanently	I think Shirley has left the academic world for good; she's got everything going for her in industry.
for the time being	for the present	The lab routine seems to be working fine for the time being. If problems come up later, we can always change it.
from A to Z	completely, from start to finish	Just to make sure everybody knows the procedure, let's go through it from A to Z.
from time to time	occasionally, once in a while	From time to time my husband has to go to the clinic for therapy.
get down to business	talk business after polite formalities	They spent 10 minutes talking about mutual acquaintances before getting down to business.
get going	start	I'm tired of sitting around waiting for everybody to show up; I wish we could get going.
get in touch with	contact by phone or by mail	Is there any way we can get in touch with you while you're on your trip?
get off	mail	I'll ask my secretary to get this material off to you as soon as possible.
get rid of	dispose of	We don't need this stuff any more. Will you please get rid of it.
give someone a buzz/ring	telephone	I'll give you a buzz at home tonight, OK?

INFORMAL CONVERSATIONAL EXPRESSIONS (*Continued*)

Expression	Synonym(s)	Example
gloss over	discuss superficially and thus hide certain weaknesses	Your report would be okay, Johnson, except that it glosses over some potentially very serious problems.
go ahead	proceed	When I heard it was on sale for 25% off, I decided to go ahead and buy it.
go by	be called by, use as a name	Your last name is hard for me to pronounce. Is there anything else you go by?
go-getter	an energetic, ambitious person	John's a real go-getter, isn't he? That's his third promotion in less than a year!
go halves/50–50	split the costs	I know you invited me out, Tim, but let's go halves on the bill, okay? It wouldn't be fair for you to pay it all yourself.
go over	review, re-examine	Let's go over this procedure one more time to be sure that everyone understands it.
go over someone's head	bring a complaint or proposal directly to your boss's supervisor without first getting your boss's permission	Sometimes the best way to get something done is by going over your boss's head. In some companies, though, it's the fastest way to get fired!
hand in	turn in, submit	Do you think you can hand in your report by 5:00?
hands-on experience	direct, personal experience	Kay should be just the person we need: she has a lot of hands-on experience with x-ray machines.
hang up (the phone)	put the receiver back on the hook	Excuse me, Sue, but I've got to hang up; my supervisor wants to talk to me.
have a lot on the ball	have many good qualities	Carol has a lot on the ball; I think she'll go far in her career.
have a word with	talk with	Max, could I have a word with you when you get a minute?
hinge on	depend on	The success of this new surgical technique hinges on how skilled the surgeon is.

INFORMAL CONVERSATIONAL EXPRESSIONS (*Continued*)

Expression	Synonym(s)	Example
hit the ceiling	become angry	Mr. Diaz will really hit the ceiling when he hears about this!
hold off (on)	postpone	Mr. Davis says he wants to hold off on the decision until the monthly review is completed.
hold up	delay	These repairs shouldn't hold up production for more than a few hours.
hold water	have any validity	I'm sorry, but your argument just doesn't hold water.
how come	why	How come you haven't complained about this before?
impose on/upon	inconvenience, be a burden to	I don't want to impose on you, but I need a letter of recommendation. Would you mind writing one for me?
improve on something	make something even better	I know it's a good product, Owens, but isn't there some way we can improve on it?
in a nutshell	in summary, in short	I'll just skip the details for the time being and tell you basically what happened. To put it in a nutshell, all our amoeba samples were contaminated.
in case (of)	in the event of	In case of an accident, call Dr. Wilson.
in charge of	responsible for	I'm putting you in charge of this part of this experiment.
in-house	within the company	Rather than using an outside agency, wouldn't it be more economical to develop our own in-house training program?
in the same boat	faced with the same problem	The models we're competing against all failed the collision test, too. In fact, it looks like we're all in the same boat on this.
(be) in the way	be an obstruction, a hindrance	I wish we could move this table somewhere else; it's just in the way here.
in time	within the required time	In order to finish my data-processing in time, I'll have to work night and day.

INFORMAL CONVERSATIONAL EXPRESSIONS (*Continued*)

Expression	Synonym(s)	Example
inquire about	ask about	The project manager just called to inquire about our progress.
interfere with	hinder	Professor Kim doesn't believe in interfering with the creative work of research scientists.
jot down	quickly write down	Let me just jot down your phone number so that I can get in touch with you.
jump to the conclusion that	make a quick judgment without adequate justification	Mr. Young looked at my first two printouts and jumped to the conclusion that I was arguing for Method A. In reality, though, I think Method B is better.
keep abreast of	keep up with, stay current with	We should be keeping abreast of the latest developments in our field.
keep down	maintain at a low level, suppress	I'd like to keep our expenses down, if at all possible.
keep in mind	bear in mind, not forget or overlook	Keep in mind that all the samples must be assayed before we can move on to the next stage of the experiment.
keep in touch	stay in communication (with)	We've enjoyed doing business with you. Please keep in touch.
keep on	continue	The one thing that drives me crazy is that Joe keeps on smoking in the lab even though we asked him not to.
keep out	prevent from entering	Be sure to keep all strangers out of the building.
keep pace with	be competitive with, grow at the same rate with	Without our own computer we cannot keep pace with our competitors.
keep track of	know the location or status of	It's hard for me to keep track of all my papers; maybe I need a filing system.
know-how	practical knowledge	Susan has the know-how it takes to get the job done right.
know the ropes	know the customary procedures, be experienced	It'll take a while before you know the ropes around here.

INFORMAL CONVERSATIONAL EXPRESSIONS (*Continued*)

Expression	Synonym(s)	Example
lag behind	move more slowly than, fall behind	The design group is lagging behind schedule.
lay off	temporarily dismiss from employment	If the national economy keeps getting worse, many companies will have to lay off some of their employees.
lean away from	avoid	We try to lean away from dependence on any one source of funding.
lean toward	favor	Although I lean toward option A, the department head seems to favor option B.
leave out	omit	Unfortunately, we left out some important details in our writeup of the experiment.
leave someone holding the bag	avoid responsibilities, thus forcing someone else to fulfill them	I think the construction firm wants to withdraw from its commitment and leave us holding the bag.
look for	seek	We aren't looking for perfection, but we do expect you to turn in reports on time.
look forward to	await with eagerness	I look forward to hearing from you at your earliest convenience.
look over	inspect, examine	I'd appreciate it if you would look these blueprints over before we start laying the foundation.
look up	search for and find	You can look her phone number up in the directory.
(it) looks like	(it) appears that, (it) seems to be the case that	It looks like the company is going to hire some new people.
make a mountain out of a molehill	exaggerate the real importance of something	Are you sure those cracks are really that dangerous, Mike? Are you sure you're not just making a mountain out of a molehill?
make certain	be sure	Make certain to turn out all the lights when you leave.
make do	accomplish one's purpose	We don't have new equipment, but we can make do with what we've got.
make sure	be certain	Please make sure that all the test tubes are thoroughly cleaned.

INFORMAL CONVERSATIONAL EXPRESSIONS (*Continued*)

Expression	Synonym(s)	Example
narrow down	reduce the number of	Given these data, I think we can narrow down the alternatives to Plan A and Plan D, right?
now and then	occasionally	Now and then I get the urge to take a long walk by myself.
old hand	someone with long experience in doing something	Fortunately my boss is an old hand at solving personnel problems like this.
on the whole	in general, for the most part	We had some bad weather, and the car needed repairs. Still, on the whole, it was a good trip.
on time	at the appointed time	Ms. Valdez always likes us to be on time for meetings.
once in a while	occasionally	We should have a day off once in a while, don't you think?
out-of-date	old-fashioned, obsolete	Too many of our plants are out-of-date; we should be building more new ones.
out of order	not functioning	Don't bother trying to get the TV to work; it's out of order.
out of the question	impossible, not even to be considered	Your proposed solution, I'm afraid, is out of the question; it's just far too expensive.
overstaffed	have too many employees	We're so overstaffed I'm afraid we'll have to lay some people off.
pay attention to	listen to, look at	Harry complains that his kids never pay attention to him when he's talking.
phase out	eliminate by stages	Management has decided to phase out our current accounting system and eventually replace it with a better one.
pick up	acquire, get	This project has been a nerve-wracking experience—I'll bet I've picked up a few gray hairs working on it.
plan on	expect to	You should plan on taking at least three weeks to complete this part of the project.

INFORMAL CONVERSATIONAL EXPRESSIONS (*Continued*)

Expression	Synonym(s)	Example
point of view	perspective, opinion	According to Bob's point of view, the rabbits make better test animals than the rhesus monkeys.
point out	direct someone's attention to, emphasize	When you show our new equipment to the inspection team, be sure to point out its safety features.
pull together	cooperate	Our company is going through tough times right now, but if we all pull together we can get by.
put back	return	After you've finished using the tape recorder, please put it back in the closet.
put off	postpone	I'd like to put off my decision until tomorrow.
put our heads together	consult, plan together	If we put our heads together, I'm sure we can find a solution to this problem.
put out	extinguish	Would you please put out your cigarette?
put two and two together	observe what's obvious, make an obvious inference	All you have to do is put two and two together and you can see that Robertson and Williams don't like each other.
qualify for	fulfill necessary conditions for	Because of your weak background in microbiology, you do not qualify for this position.
quite a few	a fairly large number of	I thought we'd only have one or two rejects in the entire sample, but actually we ended up with quite a few.
red tape	bureaucratic forms, official regulations	There's so much red tape involved, it may take us months to get our request approved.
reinvent the wheel	do outdated research	We want to do state-of-the-art research, not reinvent the wheel.
resort to	have recourse to (for lack of any other options)	Since we can't afford to buy a dialysis machine, I guess we'll have to resort to renting one.
right away/now	immediately	Let's buy it right away before prices go up.

INFORMAL CONVERSATIONAL EXPRESSIONS (*Continued*)

Expression	Synonym(s)	Example
run across/into	encounter, meet accidentally	While I was in Singapore on a business trip, I ran across an old friend of mine.
run around in circles	get nowhere, make no progress	I've been working on this design problem all day with nothing to show for it. I just seem to be running around in circles.
run low on	have only a small quantity of	We seem to be running low on supplies; let's put in an order for more.
run out of	use up, exhaust	Let's get the project finished before we run out of funds.
second-rate	of mediocre quality	Ellen got her degree at a second-rate school but has been a first-rate employee for us.
see eye to eye	agree	We just don't see eye to eye on this issue: she wants to keep using the same method and I want to try a new one.
set up	establish, create	Our company has just set up a new training program in intercultural communication for employees going overseas.
shoot for	aim for, have as a goal	Lou's been putting in a lot of overtime lately; he must be shooting for a promotion.
shoot the breeze/bull	talk casually	During lunch hour I like to shoot the breeze with my colleagues.
shorthanded	lacking sufficient help, understaffed	As shorthanded as we are, it will probably take us three weeks or more to fill your order.
shut down	stop, close	We don't want to shut down the assembly line if we can help it.
six of one, a half dozen of the other	there's no real difference between them	Why bother voting for either candidate? They're just six of one, a half dozen of the other.
sleep on it	think about it for a day or so	Your proposal sounds good, but I'd like to sleep on it a while before making a decision.
slow down/up	go slower, reduce the speed of	These machines are running too fast. We should slow them down.

INFORMAL CONVERSATIONAL EXPRESSIONS (*Continued*)

Expression	Synonym(s)	Example
so far	up to now	So far we haven't had any problems with cost overruns.
sooner or later	eventually	This stirrer may still be working okay, but don't forget, it's 20 years old. Sooner or later it's going to break down.
speak up	speak louder	We can't hear you in the back row. Could you speak up, please?
spin one's wheels	make no progress	I can't solve this homework problem; no matter what I try, I just keep spinning my wheels.
split hairs	make a big issue over minor details	There's no sense splitting hairs over such a petty matter.
stand a chance	have a chance	Do you think we stand a chance of having our paper accepted?
stand for	represent	In this equation S_l stands for the longitudinal tensile strength and S_r for the radial tensile strength.
stand out	be prominent	The color will stand out more if you put it against a white background.
start out with	begin with	Ms. Chen started out with three employees in 1974; today she has 75.
stick with	stay with, keep	I'd like to buy a new car, but for now I think I'll stick with the one I've got.
straighten out	put in order	We'll have to call in the supervisor to straighten this mess out.
switch on/off	activate/deactivate	Would you mind switching on that light over there?
take apart	disassemble	In order to see what's wrong with this motor, we'll have to take it completely apart.
take into account	consider, be aware of	In designing this machine tool, we must be sure to take into account all the design criteria.
take turns	alternate with	Alan, why don't you and Betty take turns checking the pressure readings?
talk shop	talk about business matters	My wife doesn't like it if I talk shop at parties.

INFORMAL CONVERSATIONAL EXPRESSIONS (*Continued*)

Expression	Synonym(s)	Example
team up with	work with	I'd like you to team up with Ed Kurolski on this project since he's been working in a related area.
tear down	demolish	The company intends to tear the old plant down to make room for a new one.
think of	remember, consider	Have you thought of all the possible consequences of going ahead with your plan?
think over	reflect on, meditate over	Mr. Peters, I'd like to think over your offer for a few days before making a decision.
through the grapevine	by rumor, by hearsay	We've heard through the grapevine that Frank's going to be our next section chief.
time off	free time, time away from the job	My daughter's in the hospital—I need some time off so I can go visit her.
turn down	deny, reject	I'm sorry to say your proposal has been turned down.
turn in	submit, hand in	Ms. Miller wants us to turn in our trip reports as soon as possible.
turn into	be converted into	Wine will turn into vinegar if you leave it uncorked for long.
turn on/off	activate/deactivate, switch on/off	Do you mind if I turn the air conditioner off?
turn out	happen, end up	We thought our experimental design was flawless, but it turns out that we failed to control an important variable.
understaffed	not have enough employees	Since we're so understaffed at the moment, most of us have been working overtime.
under the weather	sick, not very well	I'm feeling a little under the weather, Ms. Haddad. Could I take the afternoon off?
up-to-date	current	I wish I were more up-to-date about current developments, but I don't have much time for reading these days.

INFORMAL CONVERSATIONAL EXPRESSIONS (*Continued*)

Expression	Synonym(s)	Example
use up	exhaust, use completely	We've used up all of our travel funds for the year; any trips from now on will have to be paid for out of our own pockets.
used to	accustomed to	I'm not used to eating with chopsticks, but I'm willing to give them a try.
wear out	become useless	The fan belt in my VW is supposed to last 30,000 miles before it wears out.
wind up	complete, conclude	If we can wind up the first series of trial runs by next week, we'll have enough time to finish the other two on schedule.
wipe out	destroy, eliminate	If this new software doesn't sell, we'll be wiped out.
work out	(1) resolve	She and I argue a lot, but if we could just sit down and talk things over I think we could work out our differences.
	(2) develop	How are things working out at your new job?
would rather	prefer to	I would rather read *The New York Times* once a week than get the local paper every day.
wrestle with	struggle with, ponder, worry about	We've been wrestling with this problem for a long time.

ADDITIONAL READING

(American idioms unless noted otherwise)

L. A. Berman and L. Kirstein, *Idiom Workbook* (Silver Spring, MD.: IML, 1979).

R. J. Dixson, *Essential Idioms in English* (New York: Regents, 1971).

R. E. Feare, *Practice with Idioms* (New York: Oxford University Press, 1980).

L. Goldman, *Getting Along with Idioms: Basic English Expressions and Two-Word Verbs* (New York: Minerva Books, 1981).

Longman Dictionary of English Idioms (New York: Longmans, 1979) (mainly British English, advanced).

J. Seidl and W. McMordie, *English Idioms and How to Use Them*, 4th ed. (New York: Oxford University Press, 1978) (features British idioms).

38

Pronunciation

If you plan to use English for spoken communication, whether for oral presentations or for conversation, it is important that you pronounce the sounds correctly. This includes not only the individual sound segments (*p*, *b*, *s*, etc.) but also the stress patterns of words and the intonation patterns of phrases and sentences.

The best way to work on your pronunciation is to practice with a native speaker of English. If you can't find such a person, a good tape-recording of a native speaker will suffice. (A cassette tape keyed to this chapter is available; contact Professor Thomas Huckin, Department of English, University of Utah, Salt Lake City, UT, 84112, USA.) Using a live native speaker as a model is best, however, because he or she can both serve as a good model and tell you how well you are imitating.

In choosing a model, of course, you should consider the fact that American English and British English differ significantly in their pronunciation. Indeed, the differences can give rise to serious misunderstanding. Whether you should learn American English or British English is a choice that only you can make, based on such factors as your previous training in English, your current training, and your career plans.

38.1 ON USING THIS CHAPTER

The contents of this chapter are based on "standard American English" pronunciation (i.e., the pronunciation used by national television newscasters in America). This fact should be kept in mind, especially with regard to the material on vowels, word stress, and sentence intonation, since the pronunciation of these sounds varies noticeably from one English dialect to another, even within the United States. (The consonants and consonant clusters, by contrast, are pronounced pretty uniformly across dialects.) Thus, if you want to use the material to develop a British English pronunciation, you should first have it checked over and appropriately modified by a native speaker of British English.

Diagnosis

The best way to use the material in this chapter is to first have your pronunciation checked by a native speaker of English, to see what sounds are giving you trouble. This can best be done by having the native speaker sit down with you and listen while you read aloud a passage such as the one printed below. (If you provide your listener with the passage, it will be easier for him or her to make notes of the sounds that give you trouble.) Read over the passage ahead of time, so that you are familiar with it. Then, while you are reading it out loud, have your listener pay attention to your vowels and consonants, your word stress, and your sentence intonation. (The following checklist is designed to help your listener do this.)

Checklist for Listener of Reading Passage

While the student reads the passage aloud to you, listen to how he or she pronounces the various sound segments (vowels, consonants, consonant clusters, etc.). Make a note of those that are not pronounced correctly. On a second reading, try to note whether or not the student is putting stress on syllables that require stress and whether or not the student is applying proper intonation to the sentence as a whole. The following checklist gives more detailed suggestions of things to listen for.

1   *Sound segments.* Does the student pronounce the vowels correctly? The consonants? The consonant clusters (like the *skw* sound in *squeeze*)?

2   *Word stress.* Words containing more than one syllable typically have one syllable stressed or emphasized more than the others, as, for example, the first syllable of *criticized*. The stress should not only be located on the right syllable but should also be prominent enough to contrast sharply with the other syllables in the word, i.e., *CRITicized*. In general, does the student apply stress correctly?

3   *Sentence intonation.* Sentences should be uttered with fluctuating pitch, not in a monotone. In particular, the most important words of a sentence should be pronounced with relatively high pitch and the least important words with relatively low pitch. Questions requiring a "yes" or "no" answer should be pronounced with distinctly rising intonation at the end, whereas other questions as well as all statements should be pronounced with falling

intonation at the end. In general, does the student apply correct intonation to sentences?

Diagnostic Reading Passage

Engineers are sometimes criticized for designing devices that do not use the full value of the heat they consume. Do engineers deserve to be so criticized? If not, what can they say in their defense? One response that engineers can make to the charge is that the Carnot principle puts a theoretical limit on any attempt to squeeze more than a certain maximum of work out of a unit of heat.

Actually, engineers deserve to be congratulated in having created, in the modern steam-cycle electrical generating station, an engine that approaches the Carnot cycle in the effectiveness of its use of heat. The efficiency of the steam cycle, which is what powers the generators, is achieved when hot, high-pressure steam that enters the turbine chamber is directed ultimately to a condenser whose temperature is low (about 27°C, or 80°F) and whose pressure is reduced to about one tenth that of the atmosphere. Great care has been taken to maximize the temperature and pressure differences between the boiler, or internal reservoir of heat, and the condenser, or external reservoir to which heat is rejected.[1]

Practice

Once your listener has identified your pronunciation problems, you can begin working to solve them. This requires steady practice. Spend at least several minutes every day practicing each of the sounds that give you trouble. Then, from time to time, try to have your pronunciation checked again by your native-speaker listener. Do this until the listener is satisfied that your pronunciation, though perhaps not perfect, will at least not cause communication problems for other listeners.

For practice material, we suggest using the drill material included in the remainder of this chapter. It is divided into three sections: (1) sound segments, (2) word stress, and (3) sentence intonation.

38.2 SOUND SEGMENTS

This section provides practice material for the vowels and consonants of the language, including consonants that often occur together as "consonant clusters." We have not attempted to account for all of the sound segments of English, but only for those that traditionally prove troublesome for nonnative speakers.

Each practice list is headed by a phonetic symbol drawn from common dictionary usage. This is purely a convenience, necessitated by the fact that any one particular English sound can often be spelled in two or more different ways.

Vowels

There are 14 vowel sounds in English that form the core of the vowel system and so should be mastered. Most of these are not "pure" vowels; rather, they are spoken with the tongue moving slightly, causing the vowel quality to change somewhat as the vowel is being articulated. In some cases, a so-called single vowel is actually more of a diphthong, or two vowel sounds spoken in quick succession.

The best way to master these vowels is to try to imitate a native speaker or someone whose pronunciation is pretty much like that of a native speaker. The following lists can be used for this purpose. If you can find a good speaker to serve as a model, ask her or him to pronounce the words listed under each vowel sound that gives you trouble. Repeat each word in turn, trying to imitate the speaker exactly. If you can't do it immediately, don't give up! Often, if you keep working at it, you can perfect a sound through gradual approximation.

VOWELS

/a/	/ae/	/ā/
rod	add	rate
módel	ángle	decáy
operátion	reáction	operátion
hot	graph	equátion
próblem	stándard	scale
beyónd	mathemátics	paint
fáther	mechánical	weigh

/ɔ/	/e/	/ē/
lawn	set	speed
call	véctor	wheel
haul	accélerate	équal
ought	énergy	yield
saw	éffort	mean
áwkward	prevént	adiabátic
more	compléx	free

/i/	/ī/	/o/
línear	size	cold
ímpulse	light	coat
equilíbrium	inclíned	mótion

VOWELS (*Continued*)

/i/	/ī/	/o/
scientífic	idéal	Carnót
índex	gýroscope	flow
withín	pólaríze	télescope
bit	height	compónent

/au/	/oy/	/U/
now	noise	pull
abóut	destróy	push
down	emplóyer	book
loud	oil	refér
pówer	soil	úrgent
aróund	avóid	could
south	toy	good

/ū/	/ʌ/
rule	rúbber
redúce	ímpulse
joule	up
tool	indúction
Bernóulli	cóuple
humídity	nóthing
screw	númber

Sometimes the difference between two vowel sounds is quite small, and yet this difference can be the distinguishing factor between two entirely different words. For example, the words *age* and *edge* differ in their pronunciation only in how the vowel is pronounced: in *age* the vowel is pronounced with the tongue elevated to midheight in front and the jaw and tongue muscles somewhat tensed, whereas in *edge* the tongue is slightly lower in front and the muscles are more relaxed. Even though these movements differ slightly, it is important that you make them properly if you want to be understood.

The following exercises are designed to help you articulate the slight differences between vowel sounds that can distinguish different words. These are all common words. The words in each pair differ only in how the vowel is pronounced. Practice one pair at a time, going down the list until you can feel a clear difference when making the two vowel sounds. Try to have a native speaker listen to you while you recite.

SOME VOWEL CONTRASTS

/ē/	/i/	/ā/	/e/
feel	fill	age	edge
seeks	six	sale	cell
each	itch	weight	wet
decéased	desíst	sprayed	spread
leap	lip	láter	létter
wheel	will	attáined	atténd
reach	rich	rake	wreck

/o/	/ɔ/	/Ur/	/ɔr/
so	saw	first	forced
flow	flaw	were	wore
cold	called	burn	born
coat	caught	shirt	short
hole	haul	confírm	confórm
coke	caulk	stirred	stored
		fur	4
		expert	export

Consonants

English consonants in general are produced by momentarily restricting the stream of sound as it flows through the mouth. Each consonant is distinguished basically by three factors: (1) where the restriction occurs, (2) whether the restriction is complete or not, and (3) whether or not the vocal cords are vibrating. For example, the p sound is produced by completely restricting the sound flow at the lips and stopping the vocal cords from vibrating. The b sound is produced in the same way, except that the vocal cords continue to vibrate. We can characterize this one difference as a difference of "voicing": the p sound in English is "unvoiced" whereas the b sound is "voiced."

The 17 most troublesome English consonants for nonnative speakers are *b, ch, d, f, h, j, l, p, r, s, sh,* "soft" *th,* "hard" *th, v, w, y,* and *z.* They can be roughly described as follows:

b	Lips completely together, vocal cords vibrating. Example: *base*
ch	Body of tongue touching roof of mouth, then released; vocal cords not vibrating. Example: *check*
d	Tip of tongue touching roof of mouth just behind teeth, then released; vocal cords vibrating. Example: *detail*
f	Upper teeth touching lower lip; vocal cords not vibrating. Example: *fan*
h	Mouth slightly open, air flowing forcefully past vocal cords but vocal cords not vibrating. Example: *heat*
j	Like *ch* (body of tongue touching roof of mouth, then released) but vocal cords are vibrating. Example: *jet*
l	Tip of tongue touching roof of mouth behind teeth, sound stream flowing around sides of tongue; vocal cords vibrating. Example: *law*
p	Like *b* (lips completely together, then released) but vocal cords are not vibrating. Example: *push*
r	Lips curled out, extended forward, and held closely together, then quickly retracted; vocal cords vibrating. Example: *ratio*
s	Body of tongue almost touching roof of mouth behind teeth, allowing air to pass with hissing sound; vocal cords not vibrating. Example: *simple*
sh	Tongue almost touching roof of mouth but allowing air to pass continuously; vocal cords not vibrating. Example: *shut*
"soft" *th*	Tip of tongue held lightly between teeth, allowing air to pass; vocal cords not vibrating. Example: *thermal*
"hard" *th*	Like soft *th* but with vocal cords vibrating. Example: *this*
v	Like *f* (upper teeth against lower lip) but vocal cords are vibrating. Example: *value*
w	Lips briefly extended forward in the form of an *O*, then quickly retracted; vocal cords vibrating. Example: *wave*
y	Body of tongue tensed, held close to roof of mouth but not touching it, thus allowing air to pass; vocal cords vibrating. Example: *yield*
z	Like *s*, but with vocal cords vibrating. Example: *zero*

These are not complete or exact descriptions, of course, but only rough sketches; better descriptions, including diagrams of the articulatory movements involved, can be found in the references listed at the end of this chapter.

No description, however, no matter how exact it may be, will by itself enable you to achieve perfect pronunciation. The best way to work on your English pronunciation, as we mentioned earlier, is to try to imitate the pronunciation of a native speaker or a nonnative speaker whose pronunciation is very good. The following lists of words are best used this way. If you can find a suitable speaker of English to serve as a model, ask her or him to pronounce the words listed under each consonant sound that gives you trouble. Repeat each word in turn, trying to imitate the speaker exactly.

CONSONANTS

/b/	/ch/	/d/
base	check	detail
bus	reach	addítion
bend	touch	redúce
rúbber	charge	évidence
lab	catch	dock
absórb	arch	door
cúbic	chápter	áttitude

/f/	/h/	/j/
fan	heat	jet
fact	hot	join
fail	humídity	lógic
relíef	dehýdrate	gýroscope
laugh	unhóok	edge
síphon	héavy	énergy
44	hope	judge

/l/	/p/	/r/
law	push	rátio
línear	pówer	resístance
rélative	tip	rócket
eléctron	pump	rígid
parallél	mópping	súmmary
lével	prepáre	prefér
lówer	point	wire

/s/	/sh/	soft /th/
símple	shut	thérmal
sécond	sheet	théorem
scíence	percússion	length
force	coefficíent	width
subtráct	shéllfish	math

CONSONANTS (*Continued*)

/s/	/sh/	soft /th/
66	flash	thin
cénter of mass	relátion	earth

hard /th/	/v/	/w/
this	válue	wave
that	vápor	work
there	éven	awárd
ráther	évery	rewíre
lathe	solve	pówer
those	conservátion	dwélling
óther	vívid	twist

/y/	/z/
yield	zéro
yard	zone
únit	váporize
útilize	ózone
únion	phase
cálculate	éasy
ángular	zígzag

As with some of the vowels, the differences between certain consonants can be quite small. For example, as we noted before, the sounds *b* and *p* are articulated in exactly the same manner but for one difference: *b* is voiced and *p* isn't. The two words *bull* and *pull* are pronounced differently only insofar as one begins with the vocal cords vibrating and the other doesn't.

The following lists represent some of the most troublesome consonant contrasts for nonnative speakers. In all of the pairs listed (except for hard *th* and soft *th*), the two words are pronounced exactly alike except for the particular consonant being contrasted. Practice with one pair at a time; try to make the difference in consonants as clear as you can.

SOME CONSONANT CONTRASTS

/l/	/r/	/f/	/v/
lock	rock	fan	van

SOME CONSONANT CONTRASTS (*Continued*)

/l/	/r/	/f/	/v/
lamp	ramp	few	view
law	raw	life	live
flame	frame	fee	V
glow	grow	proof	prove
colléct	corréct	face	vase
tool	tour	first	versed
lével	léver	half	have
false	force		
rúbble	rúbber		

soft /th/	hard /th/	/s/	/sh/	/z/
thésis	these	sip	ship	zip
north	northern	sue	shoe	zoo
bath	bathe	see	she	Z (zee)
thin	this	said	shed	Z (zed)
éther	éither	face	fácial	phase
width	withering	class	clash	
wrath	rather	loose		lose
thérapy	their	hats	hatch	
		fúrnace	fúrnish	

/p/	/b/	/j/	/y/
pan	ban	jet	yet
P	B	joke	yoke, yolk
push	bush	jarred	yard
pump	bump	jot	yacht
rip	rib	jeer	year
cámper	cámber		
plaque	black		
pole	bowl		

Consonant Clusters

Many English words contain strings of consonants, or consonant clusters. For example, the word *scrapes* begins with three consecutive consonant sounds (*s, k, r*) and ends with two more (*p, s*); if we spelled the word according to the way it's pronounced, it would look like this: *skraps*.

There are more than 40 different kinds of consonant clusters in English, of which 30 commonly cause trouble for nonnative speakers: *bl, br, byu, dr, fl, fr, fyu, gl, gr, kl, kr, ks, kt, kw, kyu, pl, pr, ps, sk, skr, sl, sp, spl, spr, st, str, sw, thr, tr,* and *ts*. The following lists are provided to give you practice with them. If you have trouble pronouncing a particular consonant cluster, here's a trick you can try. Pretend there's a space between the consonants and then work on them separately, gradually reducing the space so that it disappears. For example, let's say you're having trouble with the *bl* cluster at the beginning of a word like *blue*. Divide the word into two parts with a space between the *b* and the *l*, like this: *b lue*. Pronounce each part separately, pausing briefly in between. When you feel you have command of each consonant separately, gradually begin shortening the pause between them until, in effect, there is no pause:

b   lue

b   lue

b lue

b lue

b lue

b lue

blue

There, you've got it!

CONSONANT CLUSTERS

/bl/	/br/	/byu/
blue	bridge	beáuty
bleach	break	bureáucracy
bláckboard	cálibrate	bútane
blíster	bróken	abúse
blast	álgebra	debut
block	broad	imbúe
blówtorch	bright	bútyl

CONSONANT CLUSTERS (*Continued*)

/dr/	/fl/	/fr/
drive	flow	free
draw	flúid	frame
draft	flat	fríction
address	éffluent	refrígerate
drill	flux	refráction
drum	defláte	frózen
drý cell	flight	infraréd

/fyu/	/gl/	/gr/
few	glass	grow
fúsion	glue	grádual
fúture	gleam	ingrédient
refúel	glow	mílligram
sulfúric	glaze	grádient
refúse	conglomerátion	grew
diffúsion	Éngland	green

/kl/	/kr/	/ks/
clear	crack	lacks
clamp	cross	makes
clutch	mícroscope	áxiom
núclear	incréase	óptics
decláre	crítical	x́-ray
cýclotron	crúde	ínflux
conclúde	crane	compléx

/kt/	/kw/	/kyu/
efféct	quálity	cube
cóntact	quart	perpendícular
reáctor	equátion	cúrious
véctor	báckward	mércury

CONSONANT CLUSTERS (*Continued*)

/kt/	/kw/	/kyu/
duct	quótient	Curie
predíct	equípment	cúcumber
liked	requést	excúse

/pl/	/pr/	/ps/
plan	príntout	ellípse
plane	proof	keeps
plot	prime	hopes
unplúg	propórtions	pumps
supplý	compréssion	sýnapse
expláin	recíprocal	tapes
súrplus	prevént	loops

/sk/	/skr/	/sl/
skill	screw	slant
scálar	scratch	slow
schédule	descríbe	slip
disk	discréte	slot
task	súbscript	slope
Pascál	scrape	slate
skýscraper	scrawl	slice

/sp/	/spl/	/spr/
spécial	splash	spring
space	splice	spray
spárk	split	spread
spíral	expláin	sprócket
respónd	explóre	dispróve
speak	explóde	exprópriate
expéct	displáy	expréss

CONSONANT CLUSTERS (*Continued*)

/st/	/str/	/sw/
steel	strong	swing
stándards	strúcture	sweep
státic	strain	switch
elástic	stróntium	persúade
cónstant	extréme	unswépt
cost	restríct	Swéden
resísts	abstract	swell

/thr/	/tr/	/ts/
three	try	únits
through	true	convérts
throw	tráffic	its
thróttle	traíning	creátes
thrust	trigonómetry	sátellites
unthréaded	Detróit	debts
thréatened	contról	prétzel

38.3 WORD STRESS

In spoken English the different vowel sounds of a word are often pronounced with different degrees of emphasis, or "stress." The word *vector*, for example, has two vowel sounds, the first of which is given much greater stress than the second: in pitch, in length, and in loudness. That is, a stressed vowel is normally pronounced at a higher pitch and is normally longer and louder than an unstressed vowel. Conversely, an unstressed vowel is pronounced with lower pitch and is shorter and weaker; indeed, an unstressed vowel often even loses its identity as a particular type of vowel. For example, the *o* vowel in *vector*, being unstressed, is pronounced not as an *o* but rather as a neutral, unidentifiable, "weak" vowel. (We shall hereafter use the symbol ə for this type of vowel.) Putting all these facts together, the word *vector* should be pronounced like this:

$$V E C$$
$$t ə r$$

Try pronouncing it yourself. Make sure you produce a sharp difference between the two vowels, even if you have to exaggerate. Nonnative speakers are often reluctant to make the difference as sharp and clear as it should be, so force yourself if necessary.

There are actually not two but three degrees of stress in English. The example given above involved only two vowels, and so two degrees of stress were sufficient to describe it. Longer words, however, often invoke a third, intermediate degree, which we shall call "partial stress." Thus, the three degrees of word stress in English are full stress, partial stress, and no stress. Let us take the word *vectorial* to illustrate the difference. This word is pronounced with four vowel sounds, the second of which receives full stress: *vecTORial*. The first and third vowels receive partial stress, and the fourth vowel receives no stress at all. Putting these facts together, we can depict the correct pronunciation of *vectorial* roughly as follows:

$$\text{vec} \quad \overset{\textstyle\text{T O R}}{} \quad \overset{\textstyle i}{\partial l}$$

Try it yourself.

These two examples illustrate several important points about the English stress system. First of all, there is normally only one fully stressed vowel in a word. Secondly, the location of this fully stressed vowel can vary from one word to the next, even when two words are related (as are *vector* and *vectorial*, for example). Thirdly, the spelling system used for written English is not a reliable guide for pronunciation. Written vowels are not always pronounced (e.g., the underscored *e* in *refine̲d*). Those vowels which are pronounced but which are unstressed often are not given the pronunciation suggested by the spelling (e.g., the *o* in *vector* is not pronounced like an *o*). Two words with the same spelling may have different stress patterns (e.g., the word *contrast*: when used as a noun it is pronounced *CONtrast*; when used as a verb, however, it is pronounced *conTRAST*).

❑ EXERCISE 38-1

The following is a list of words commonly used in science and technology. For each one, indicate by means of an accent mark (´) which vowel you think should be fully stressed. Do not use the dictionary while doing the exercise, though you may use it later to check your answers. The first two have already been done for you.

párticle	arithmetic	dynamic
mátter	resistance	automobile
energy	calculate	minimum
cyclotron	machine	potential
biology	technical	pressure
reverse	process	variable
physics	turbine	

We mentioned above that even when two words are obviously related, they may not have full stress on the same vowel. This is particularly striking in the case of certain two-vowel words that are spelled exactly alike, such as the example we gave above regarding the words *conTRAST* (a verb) and *CONtrast* (a noun). In such cases, the verb form has full stress on the second vowel and the nonverb form (usually a noun) has it on the first vowel. Some other examples of this type are as follows:

Verb	Nonverb
conflíct	cónflict
decréase	décrease
extráct	éxtract
impórt	ímport
inclíne	íncline
incréase	íncrease
objéct	óbject
perféct	pérfect
permít	pérmit
presént	présent
progréss	prógress
recórd	récord
rejéct	réject

A more common occurrence concerns two or more words having the same stem but different suffixes, so that they are related but of different length. In such cases, the longer word often has its fully stressed vowel to the right of where it is in the shorter word. For example, consider our old friends *vector* and *vectorial*. These words are obviously related, both having the same stem. However, the shorter form has full stress on the first vowel (*véctor*) whereas the longer form has it on the second vowel (*vectórial*). In other words, the location of the full stress has shifted one step to the right for the longer word.

Although this kind of stress shift is a common phenomenon in English, there are many exceptions to the rule. That is, there are many cases where one member of a related word pair is longer than the other and yet has the same stressed vowel, e.g., *resíst* and *resístance*. In fact, there are even a few cases where the longer word has its stressed vowel to the *left* of where the stressed vowel is in the shorter word.

❑ EXERCISE 38-2

Below are some sets of related words. Examine each and see if you can correctly mark the location of the fully stressed vowel. (The first six words are already marked for you.) As in the previous exercise, do not use your dictionary until after you've completed the exercise.

périod	dénse
periódic	dénsity
periodícity	densitómeter
instinct	supplement
instinctive	supplementary
argument	social
argumentative	socialist
argumentation	socialistic
distribute	electron
distribution	electronic
electric	plastic
electricity	plasticity
sequence	molecule
sequential	molecular
develop	alternate
development	alternative
developmental	alternation
insulate	component
insulation	componential
microscope	probable
microscopic	probabilistic
separate	economy
separable	economic
separation	economize
separability	economically
category	concept
categoric	conceptual
categorize	conceptually
categorically	conceptualize
categorization	conceptualization

Although all English words have a standard stress pattern (as indicated in any good dictionary), there are cases where this pattern should be violated for purposes of pointing out a contrast to some other word nearby. Take, for example, the word *internal*. Normally, this word is pronounced with the second vowel receiving full stress: *intérnal*. However, if you recall the last sentence of the diagnostic reading passage, you'll remember that the writer had deliberately set up a contrast between the words *internal* and *external*.

> Great care has been taken to maximize the temperature and pressure differences between the boiler, or *internal* reservoir of heat, and the condenser, or *external* reservoir to which heat is rejected.[1]

In such a case, since the difference between these two words lies in the first vowel sound of each (*in-* versus *ex-*), it is appropriate, indeed required, to stress these two vowels instead of the normal second vowels. In other words, in order to point out the contrast between these two words in a context such as this one, you should pronounce them: *ínternal* and *éxternal*.

38.4 NOUN COMPOUNDS

When two nouns are strung together and used as a single word, or noun compound, full stress is usually applied only to some vowel in the first noun, not the second. For example, consider the nouns *gravity* and *meter*. When these words are pronounced independently, they are stressed like this: *grávity* and *méter*. When they are joined together, however, only the first word is stressed: *grávity meter*. Some other examples are *círcuit breaker*, *vácuum tube*, *spárk plug*, *aír conditioner*, *Bernóulli effect*, *pháse reaction*, *ignítion system*, *héat loss*, *óil pump*, and *fíre extinguisher*. An example from the diagnostic reading passage is *génerating station*.

    This stress pattern is usually found only when the two words are in fact <u>nouns</u>. Do not expect to find it for adjective plus noun combinations, even when such a combination may be commonly used as a "single" word. For example, the term *chemical reaction* is commonly used as a single word, just as *phase reaction* is. Since the word *chemical* is an adjective, not a noun (note the adjectival suffix *-ical*), the combination is accordingly pronounced with stress on <u>both</u> words: *chémical reáction*. Some other adjective plus noun examples are *hydraúlic préss*, *inclíned pláne*, *mechánical shóvel*, *kinétic énergy*, and *Bóyle's láw*.

❑ EXERCISE 38-3

The following are two-word combinations used as single words. Some are noun compounds, others are adjective plus noun combinations. Indicate by means of one or more accent marks what the stress pattern should be. (Be sure to put the accent on the correct vowel when there is more than one vowel in the word!) Try pronouncing each combination or compound while you are doing the exercise.

water softener	electric motor
molecular weight	fracture mechanics
edge dislocation	network molecules
energy gap	linear density
photographic film	heat pump
thermal conductivity	vacuum cleaner
radiation detector	nuclear reactor
shock absorber	particle accelerator
polynomial equation	ionic bond
valence electrons	x-ray diffraction
periodic table	Bragg's law
civil engineering	computer specialist
crystal-field theory	plane-polarized light

38.5 SENTENCE INTONATION

Correct pronunciation of English includes more than just sound segments and word stress; it also includes the intonation of sentences. English sentences are typically spoken with up-and-down pitch variations. These variations in pitch have important communicative functions, and so listeners are trained to listen for them. In particular, the most important words of a sentence are normally pronounced with relatively high pitch and the least important words are pronounced with relatively low pitch. Also, the difference between types of sentences is marked by a distinctive pitch pattern at the end: questions requiring a "yes" or "no" answer are pronounced with rising intonation, whereas all statements (and all other types of questions) are pronounced with abruptly falling intonation at the end.

To simplify the following discussion, we shall assume that English sentence intonation is constructed from two basic tones: high and low. These terms should be interpreted in a relative sense, not in any absolute one. That is, every native speaker of English makes a distinction between these two tones, though the exact pitch of each tone differs from speaker to speaker. In fact, even for the same speaker either of these tones can be pronounced at somewhat varying levels of pitch.

Signaling Important Words

One of the principal functions of sentence intonation is to draw the listener's attention to the most important words of the sentence. What do we mean by "important words"? These are usually any of the following:

1   Nouns, verbs, adjectives, or adverbs that represent *new information*, that is, information that has not been mentioned before or even suggested

2   Any word or word prefix that represents an *opposition or contrast* of some sort

3   Any word that the speaker feels deserves *special emphasis*

We can illustrate the first two of these categories with examples drawn from the diagnostic reading passage. The passage begins like this:

> Engineers are sometimes criticized for designing devices that do not use the full value of the heat they consume. Do engineers deserve to be so criticized? If not, what can they say in their defense?

In the first sentence, all of the nouns, verbs, adjectives, and adverbs, of course, represent new information: this is where the concepts conveyed by these words are first introduced. The speaker can signal the fact that this is new information—and thus important information—by pronouncing each of these words (i.e., the italicized ones) with high pitch:

> *Engineers* are *sometimes criticized* for *designing devices* that do *not use* the *full value* of the *heat* they *consume*.

(Conversely, each of the other words in the sentence should be pronounced with low pitch.)

In the second sentence, only one word (*deserve*) represents new information:

> Do engineers *deserve* to be so criticized?

The remaining words of this sentence either represent given information (*engineers, so criticized*) or are auxiliaries (*do, to be*).

In the third sentence, the word *not* is an important word because it represents an opposition or contrast:

> If *not*, . . .

Therefore, it should be pronounced with high pitch. Other words that should also be given high pitch in this sentence (because they represent new information) are *what, say,* and *defense*.

Signaling the Sentence Type

The other principal function of sentence intonation is to alert the listener to what type of sentence it is. In particular, sentence intonation is used to distinguish yes or no questions (i.e., those questions requesting a "yes" or "no" answer) from all other sentence types: yes or no questions are pronounced with rising intonation (low → high), and all other sentence types (statements, commands, exclamations, other types of questions, etc.) are pronounced with abruptly falling intonation (high → low).

Unlike the signaling of important words, which can occur anywhere in the sentence, the signaling of sentence type is normally done at or near the end of the sentence. Specifically, *the signaling of sentence type begins with the last important word in the sentence.* For example, consider the second sentence in the reading passage:

Do engineers *deserve* to be so criticized?

This is a yes or no question, and so it requires rising intonation. The last important word in the sentence is *deserve* (which happens to be the *only* important word in this case), and so the rising intonation pattern should begin on this word (to be more exact, on the *fully stressed* vowel of this word):

Do engineers de ^{s}erve to be so critized?

In other words, the sentence is pronounced with low pitch until the second vowel of the word *deserve*, where it rises quickly to high pitch. This high pitch is then maintained to the end of the sentence.

The third sentence, however, is not a yes or no question; accordingly, it requires falling intonation. The last important word in this sentence is *defense*, which, like *deserve*, is normally stressed on the second vowel. Thus, the falling intonation pattern should begin on this second vowel. That is, the speaker will begin to pronounce this vowel sound with high pitch but will quickly let the pitch drop to a low level at the end of the vowel:

. . . de$^{f^{e}}$↘nse.

If we combine this end-of-sentence signaling with the signaling associated with important words, we end up with the following intonation contour for this sentence:

If ^{n}o$_{t,}$ what can they ^{say} in their de$^{f^{e}}$↘nse?

In other words, the speaker will pronounce all four of the important words with high pitch but will make sure that the pitch on the last one drops abruptly from high to low. (The pitch on *not* should also drop somewhat because it marks the end of a clause.)

Signaling the Items of a List

A third important use of intonation occurs when a speaker is reciting a list. In such cases, each item of the list (except the last) should be pronounced with a *sustained* high pitch. For example, if we have a sentence like this:

> The three states of matter are solids, liquids, and gases.

we note that it contains an unformatted list of three items: *solids, liquids, gases*. The first two items should each be pronounced with sustained high pitch (i.e., high pitch not only on the first, stressed vowel but also on the second, unstressed vowel); this makes it immediately clear to the listener that a list is being presented. The third item, on the other hand, receives high pitch only on the first vowel, not on the second. The resulting intonation pattern will be roughly as follows:

$$\ldots \quad \text{solids, liquids,} \quad \underset{\text{and}}{} \quad \text{gas}^{e}{}_{s}$$

Notice how the words *solids* and *liquids*, in contrast to *gases*, are pronounced at a sustained high pitch level even through their unstressed second vowels.

A special kind of list is one that asks the listener to make a choice from among two or more items. Here, too, each item of the list except the last should be pronounced with sustained high pitch. Consider, for example, this sentence:

> Is the test *destructive* or *nondestructive?*

Normally the word destructive is pronounced with high pitch only on the stressed second vowel:

$$\text{de}^{\text{struc}}{}_{\text{tive}}$$

In the sentence above, however, it is part of a two-item list and so should be pronounced differently, with sustained high pitch carrying through the last, unstressed vowel:

$$\ldots \text{de}^{\text{structive}} \quad \ldots$$

This way, the listener is alerted right away to the fact that he or she is being asked to make a choice between two possible answers. To make matters complete, we can describe the intonation contour for the full sentence as being approximately like this:

$$\text{Is the test de}^{\text{structive}} \quad \text{or} \quad {}^{\text{non}}\text{destructive?}$$

NOTE: If you listen to a native speaker's sentence intonation, you will find that it actually contains considerably more variation of pitch than we are depicting here. Intonation can be used in many different ways to express many different—often subtle—shades of meaning. Unfortunately, it takes years of constant interaction with native speakers to master such subtleties. What we are emphasizing here are the more basic aspects of English intonation, those which form the foundation of the intonation system and which, happily, can be readily mastered by applying the principles described above.

❑ EXERCISE 38-4

Convert each of the following sentences into a form that indicates the intonation pattern it should have. You may assume that the important words of the sentence are those appearing in italics. (The first sentence has already been done for you.)

A   *No silicon analog* of *graphite* is *known*.

Answer: No sil i con an a log of graph i te is kno w n.

B   The *range* of *speeds observed* for *chemical reactions* is *enormous*.

C   Is there *good evidence* that *forces exist* between *nonpolar molecules*?

D   *What* is the *weight* of an *aluminum* atom?

E   *Every applied scientist* and *engineer—mechanical, civil, electrical,* or *other*—is *vitally concerned* with the *materials available* to *him* or *her*.

F   *Which* is a *simpler concept: speed? or velocity?*[2]

❑ EXERCISE 38-5

Read over the following passage and underline what you think are the most important words. Then, using these underlined words as aids, convert each sentence into a form indicating the intonation pattern it should have. When you have finished, check your result by reading it aloud to your teacher or native-speaker listener.

How big is a proton? Are the proton and the neutron, which weigh about the same, identical except for the neutron's lack of electrical charge? Are these particles hard, with sharp boundaries, or soft, with fuzzy edges? The central position of protons and neutrons in the subatomic world makes it crucial to understand in detail the structure, shape, and size of these particles. Detailed knowledge about these important particles provides a testing ground for theories explaining the subatomic world.[3]

REFERENCES

1   B. Hiatt, "Heat into Work: The Second Law of Thermodynamics," *The Research News* 28(11/12):5 (1977).
2   L. H. Van Vlack, *Elements of Materials Science and Engineering*, 3d ed. (Reading, MA: Addison-Wesley, 1977), p. 3.
3   B. Hiatt, "What's the Matter: Particle Physics at the University of Michigan," *The Research News* 26(3/4):12 (1975).

ADDITIONAL READING

J. Gilbert, *Clear Speech: Pronunciation and Listening Comprehension in American English* (Cambridge, Eng.: Cambridge University Press, 1982). (Tapes available.)

J. Morley, *Improving Spoken English* University of Michigan Press, (Ann Arbor, MI: 1979). (Tapes available).

D. L. F. Nilsen and A. P. Nilsen, *Pronunciation Contrasts in English* (New York: Regents, 1973).

E. C. Trager, *PDs in Depth* (Portland, OR: ELS, 1982). (Tapes available).

APPENDIXES

A

Punctuation, Grammar, and Style

A.1 SENTENCE PUNCTUATION

Punctuation marks serve as useful "road signs" to a reader. They help the reader see the grammatical structure of a sentence and thus more easily grasp its meaning. Punctuation can also help the reader see the rhetorical relationships between sentence elements.

The most commonly used—and *mis*used—punctuation marks in scientific and technical writing are:

The comma	,
The semicolon	;
The colon	:
Parentheses	( )
Dashes	— —

These punctuation marks should be studied and used effectively by every technical writer. Unfortunately, they often are not. Many inexperienced writers believe—mistakenly—that punctuation marks are supposed to indicate the sound pattern of a sentence, that is, the pauses one would make if reading the sentence aloud. It is true that vocal pauses and punctuation marks often coincide, but this is not always the case, and even where they do coincide, a vocal pause by itself is hardly a reliable guide as to which punctuation mark to use.

In short, punctuation marks do not indicate sound patterning (at least not directly); instead, they indicate *grammatical structure*, *meaning*, and *rhetorical relationships*.

Independent Clauses

Three marks of punctuation are used to show a close relationship between independent clauses:

1 Use a *comma* if the two independent clauses are coordinate to each other, and the second clause begins with a coordinating conjunction (*and, but, or, so,* etc.):

> The pressure in a gas or in a liquid is different at points that differ in height, and properties that depend on the pressure are also different.

> Computer-linked scanning devices have been developed for bubble chamber films, but they cannot function without human helpers.

2 Use a *semicolon* if the two independent clauses are coordinate to each other, but there is no coordinating conjunction:

> The structure of a simple liquid such as argon or methane is determined primarily by the repulsive forces between molecules; the attractive forces act only as a "glue" that holds the fluid together.

> There are many geometries that are different from Euclidean; however, only the elliptic and hyperbolic geometries are called *non*-Euclidean.

3 Use a *colon* if the second clause elaborates on or explains the first:

> All chemical equations should be balanced: they should show the same number of atoms of a given kind and the same net charge on each side.

> The concept of randomness can best be explained by the use of examples: the sequence of heads and tails in tossing a perfect coin is said to be random.

Example

Technology never exists in a social vacuum: it is embodied in products, processes, and people. By the same token, technology does not "transfer," or circulate, in a social void; it circulates throughout diverse institutional channels and mechanisms. . . . Thus social values cannot be separated either from technology itself or from its mode of transfer. When these values, which are usually determined by the "parent" culture, conflict with those of developing countries, the technology transferred can deter—rather than promote—those countries' social objectives

STATEMENT FOLLOWED BY EXPLANATION

STATEMENT FOLLOWED BY CONTRASTING STATEMENT

> The most significant sources of technology transfer are consulting firms and the transnational corporations (TNCs) for manufacturing, extraction, and service.

TNCs disseminate product and process technologies. But "decisional technologies" exist as well, and these are vital: they comprise the know-how required to diagnose complex problems and formulate solutions. As Argentine physicist Jorge Sabato puts it: "The ability to conduct a feasibility study with its own means is the touchstone revealing when a country has conquered technological autonomy."[1]

STATEMENT FOLLOWED BY ELABORATION

Introductory Subordinate Elements

Use a comma to separate introductory subordinate elements from the rest of the sentence (and thus make it easier for the reader to locate the sentence subject):

> Since most current plantings are transplants from the greenhouse and are several months old, they can tolerate stronger herbicides.

> However, the supply of rubber has so far been able to keep up with the slow increase in demand.

Parenthetical Information

Three different marks can be used to indicate parenthetical (amplifying or digressive) information:

1 Use *commas* in most cases:

> The so-called British system of weights and measures, which is now on its way out in Britain, has its roots in ancient history.

> Galileo originally intended to be a doctor, on his father's advice that there was much more money in healing ailments than in proving theorems.

2 Use *parentheses* for references, abbreviations, definitions, qualifications, and other subsidiary information that may be of interest only to certain readers:

> It has long been known that the vertebrate electroretinogram (ERG) in response to flashes of light consists of various components that have been ascribed to electrical sources located mainly in the receptor layer and in the inner nuclear layer of the retina.

> The nucleus of each atom is composed of protons and neutrons (except for the isotope of hydrogen of atomic weight 1, whose nucleus consists of a single proton).

3 Use *dashes* to give more prominence than is allowed by other options within the sentence (unless writing for conservative readers):

Mixtures can be categorized as either homogeneous—uniform in properties throughout the sample—or heterogeneous.

Many of the experimental methods of modern chemistry were developed originally by men and women regarded as physicists—for example, the methods of spectroscopy and those of structure determination by the diffraction of x-rays.

Example

At the small end of the physical world, there is a realm in which it is difficult, *if not impossible*, to measure or manufacture anything. There, a stray cosmic ray can destroy a wire only a few atoms thick. There, the Heisenberg uncertainty principle prevents simultaneous measurement of both the velocity and position of a particle. There, the very act of examining a structure—*with the beam of an electron microscope, for example*—will degrade or destroy it.

    It is a realm too fascinating—*and promising*—to resist. When engineers explore it, they find a wonderland of faster switches, more efficient circuits, higher reliability, and lower costs. As scientists work on scales of atomic diameters and angstroms (*an angstrom equals a ten-billionth, 10^{-10}, of a meter*), they are discovering new and unexplored phenomena that increase their understanding of the physical world.[2] [Italics added.]

> PARENTHETICAL INFORMATION
>
> PARENTHETICAL INFORMATION EMPHASIZED
>
> PARENTHETICAL SUBSIDIARY INFORMATION

4 If you must write sentences with highly interrupting material, use dashes or parentheses to set off the interrupting material. Highly interrupting material includes parenthetical units that are long or grammatically complex or that appear in the middle of highly connected units such as lists.

> And the freshest innovations—computer-controlled optical character recognition for sorting letter mail and the recently completed bulk-mail system for handling packages exclusively—are prime targets for critics grumbling about missent letters, parcel damage, and ever-increasing postal rates.

Lists

Unformatted lists—those not separated from the text—are set off either by a colon or by dashes or parentheses:

1 Use a *colon* if the list ends the sentence:

> The new technology is most widely known for its microelectronics: pocket calculators, microcomputers, and other products that have revolutionized information processing and communication.

To produce 100,000 Btu's, one needs the following quantities of various fuels: 2½ lb of natural gas, or 5 lb of petroleum, or 8 lb of coal, or 14 lb of wood.

2 Use *dashes* or *parentheses* if the list does not end the sentence:

Every aspect of the peptide neurotransmitters—how they are made, what they do, where and how they act, how they are disposed of, and even whether they are true transmitters or serve some other function— is being studied with great intensity.

The charge of an electron has been shown to occur in integral multiples (1, 2, 3, 4, etc.) of its basic value; therefore magnetic charges should also occur in integral multiples of some basic value.

To separate the items of a list:

1 Use *commas* in most cases:

The three parallel lines are as follows: $y = y_1$, $y = \frac{1}{2}(y_1 + y_2)$, and $y = y_2$.

There are many words in our language that indicate sets: *school* of fish, *swarm* of bees, *herd* of cattle, *squadron* of planes, etc.

2 Use *semicolons* if the items are long or are punctuated internally:

We have assembly plants in the following locations: Cedar Rapids, Iowa; Totowa, New Jersey; Edmonds, Washington; and Castle Rock, Colorado.

An efficient project cost system must accomplish three important objectives: (1) check actual and predicted costs of ongoing projects against the estimated cost; (2) obtain production rates for use in estimating of new work and create historical files; and (3) forecast the project final cost.

In formatted, displayed lists, that is, those set off from the body of the text, punctuation is optional:

The positions available are

Recovery plant supervisor

Management analyst

Administrative aide

Clerk-typist

The positions available are the following:

Recovery plant supervisor,

Management analyst,

Administrative aide, and

Clerk-typist.

A.2 WORD AND PHRASE PUNCTUATION

At the word and phrase level, the most frequently misused punctuation marks are:

The apostrophe	’
The hyphen	-
Quotation marks	" "
Underlining	_____

These are most often misused with regard to the functions discussed below.

Possession

1   Add an *apostrophe* and an *s* to all nouns except plurals ending in *s*:

Roberta's watch	the company's policy
my assistant's desk	the people's concern
the children's toys	the men's room

EXCEPTION: Personal pronouns (*it, they, you,* etc.) never take the apostrophe to show possession. The correct possessive forms of these words are *its, their, your, our, my, her,* and *his* when they are used as adjectives and *its, theirs, yours, ours, mine, hers,* and *his* when they are used as pronouns.

In cases of joint possession, attach *'s* only to the last noun: *Dun and Bradstreet's projections, John and Mary's sailboat.* When referring to combined cases of individual possession, on the other hand, attach *'s* to each of the nouns: *John's and Mary's birthdates.*

2   Add just an *apostrophe* (no *s*) to plural nouns ending in *s*:

the workers' pension plan	our parents' friends
the Joneses' house	the neighbors' garden

Occasionally you may want to show possession with a singular noun which ends in *s* or with an *s* sound. In such cases, follow rule 1 above unless the result would be hard to pronounce:

Bass's new hiking boots   but   Sears' new hiking boots

Alice's restaurant   but   Moses' restaurant

Noun Compounds

Nonspecialist readers may have difficulty interpreting long noun compounds. There are two simple ways to reduce such problems:

1   Use an occasional *preposition*. Scientific and technical English abounds with terms made up of long strings of nouns: terms like *diffusion convection mass transfer resistance* and *4PbO·PbSO₄ lead acid battery paste formation*. Although such noun compounds may cause no trouble for specialist readers, they can be very confusing to nonspecialists, especially if there are not enough clues in the immediate context to serve as a guide to interpretation. Therefore, avoid using such terms out of context (for example, in titles or abstracts); instead of *3B pump maintenance recall rate*, try *rate of recalls for maintenance of the 3B pump* or *maintenance recall rate for the 3B pump*. It may take a little more space, but it enables the nonspecialist reader to grasp the meaning of the term much more easily and quickly.

2   Use an occasional *hyphen*. Another way of carving up long compound nouns into more digestible pieces is to pair off closely related words by means of hyphens:

> Diffusion-convection mass-transfer resistance
>
> High-frequency semiconductor layers
>
> Automatic score-keeping feature
>
> Dow coal-liquefaction process
>
> High-energy laser-beam weapons
>
> Manual materials-handling tasks

This grouping of related words helps prevent momentary misinterpretations on the part of a reader who is not familiar with such terms, especially when adjectives are involved. Notice, for instance, how *manual materials handling tasks* without the hyphen could momentarily be misconstrued as tasks involving the handling of manual materials. By inserting a hyphen between *materials* and *handling* (*manual materials-handling tasks*), the writer alerts the reader to the fact that the adjective *manual* modifies *materials-handling* or *tasks*, not *materials*.

End-of-Line Hyphenation

To divide a word at the end of a line, use a *hyphen*. If you have started to write
a word at the end of a line but find you don't have room to complete it, you can
hyphenate it and carry a part over to the beginning of the next line. Bear in
mind, though, that this device exacts a slight toll on the reader: it briefly
interrupts the smooth flow of words and forces the reader to guess at the meaning
of the hyphenated word before seeing the second part of it. So, do not overuse
this technique and, when you do use it, try to make it as easy as possible for
the reader to guess the meaning of the full word after seeing only part of it. The
following rules are designed for that purpose.

1 *Never divide a syllable.* Syllables are the building blocks of words. You can
   often guess the meaning of an unfamiliar word just by putting together the
   "meanings" of the syllables it is made up of. If you see the syllable *chem-*
   leading off a word, you can pretty well guess what the rest of the word might
   be, but if you see only half the syllable (*ch-*), you have no idea what follows:
   chart, chapter, chiropractic, chlorine, choices, chest, to mention but a few.
   So be sure to maintain the integrity of syllables when you divide a word. If
   you don't have time to check the dictionary (all good dictionaries show the
   syllable divisions of words), you can use these two rules of thumb: (1) all
   syllables have at least one vowel sound; (2) all syllables can be pronounced
   easily and, if necessary, can be shouted easily to someone at a distance. *Chem*
   satisfies these conditions, but *ch-* doesn't.

INCORRECT	CORRECT
ma-rket	mar-ket
repo-rt	re-port
thr-ough	through
proje-ct	pro-ject
dupl-ic-ate	du-pli-cate
persp-e-ctive	per-spec-tive
phl-ebi-ti-s	phle-bi-tis

2 *If there is a natural break in a word, try to divide it there.* Many words
   divide naturally into two parts: *work-shop, self-disciplined, turn-key, breakwater, anti-toxin, electro-magnetic, thermo-meter,* etc. By dividing the word at the end of the line according to its natural parts,
   you make it easier for the reader to recognize the two parts. Notice
   how funny it looks, for example, to see a word divided like this: *sol-
   id-state.* It's much easier and more natural to read it divided as *solid-
   state.*

Titles of Written Works

<u>Underline</u> (or *italicize* if you have access to a computer or typewriter with an italics font) the names of books, journals, magazines, monographs, and other publications that are bound separately:

> J. Galbraith, <u>Designing Complex Organizations</u> (Menlo Park, CA: Addison-Wesley, 1973).

> D. O'Brien et al., "Responsibility for Inspection," *Journal of the American Concrete Institute, 69*(6): 320–333 (June 1972).

Put *quotation marks* around the names of articles, reports, memos, and other pieces of writing that are either not published or not bound separately:

> J. D. Borcherding, "An Exploratory Study of Attitudes That Affect Human Resources in Building and Industrial Construction," Department of Civil Engineering, Stanford University, Stanford, CA, 1972.

> W. Larkin and D. Burns, "Sentence Comprehension and Memory for Embedded Structures," *Memory and Cognition, 5* : 17–22 (1977).

> M. J. Zakkak, "Cost Control Simulation." Thesis presented to the University of Texas, Austin, TX, 1976, in partial fulfillment of the requirements for the degree of Doctor of Philosophy.

NOTE: Long formal technical reports are often cited with the descriptive name in quotation marks but the report number underlined or in italics:

> "PCS, Project Control System/360," <u>GH 20-0376-3</u>, International Business Machines Corp., 1971.

> S. H. Wearne, "Contractual Responsibilities for the Design of Engineering Plants," *Report No. TMR5*, School of Technological Management, University of Bradford, England, 1975.

A.3 CAPITALIZATION

When referring to something by its official name, *capitalize* all words in the name other than articles, short prepositions, and conjunctions:

Personal names:	Phyllis Gomez, Lawrence Van Vlack
Personal titles:	Project Officer, Professor of Mechanical Engineering, M.D.
Places:	115 South Cache Street Jackson, Wyoming United States of America
Organizations:	Du Pont Chemical Company, Arizona Girls Leadership Club, Brooklyn College

Days, months, eras:	Monday, Memorial Day, Christmas, October, the Twentieth Century, the Eighties
Programs:	the Professionals-in-Training Program, the Stanford University Degree Program in Physics
Natural languages:	French, Spanish, Chinese
Computer languages:	BASIC or Basic, FORTRAN or Fortran, ALGOL, Pascal
Model names:	the IBM Selectric, the Ford Mustang

A.4 DOCUMENTING SOURCES

When using outside sources of information for a formal report, article, or thesis, you should explicitly cite those sources. By doing so, you give proper credit to the creator of that information (and thus show yourself to be an ethical person) and you help "situate" your writing for a knowledgeable reader. The documenting of sources is normally done through both (a) internal referencing at the point of use and (b) full bibliographic referencing at the end of the text. There are many systems for documenting sources, and you should select the one that is customary in the type of document you are writing. For example, if you are writing a company report and your company has its own style of documentation, then obviously you should use that style. If you are writing an article for *Physical Review*, you should use the style it uses (which is based on the American Institute of Physics' Style Manual).

The two most commonly used systems for documenting sources in technical and scientific communication are the Number style and the American Psychological Association (APA) style. Each has its own way of handling both internal referencing and end referencing.

Number Style

The Number style is preferred in most engineering and natural science documents. In this format, each internal reference is indicated by a single number keyed to the full list of references at the end. The number may be enclosed in parentheses or brackets, or may be written in superscript— whichever is expected in that particular publication or field. For example, here are the first two sentences of an article from *The Journal of Algebra:* [3]

> The study of certain operations on rational languages leads to the consideration of pseudovarieties generated by power semigroups or power monoids (cf. [11, 16, 12]). The case of monoids has been treated successfully by Margolis and Pin [9] (see also [13]).

The full references are given in list form at the end of the article. In this particular article, there are 16 end references, and they are arranged in alphabetical order and numbered in ascending order. Those cited in the excerpt above are written as follows:

9. S. M. MARGOLIS AND J. E. PIN, Power monoids and finite J-trivial monoids, *Semigroup Forum* **29** (1984), 125–135.
. . .
11. J. F. PERROT, Variétés de langages et opérations, *Theor. Comput. Sci.* **17** (1978), 197–210.
12. J. E. PIN, Variétés de langages et monoïde des parties, *Semigroup Forum* **20** (1980), 11–47.
13. J. E. PIN, "Variétés de langages et variétés de semigroupes," Thèse d'Etat, Université de Paris 7, 1981.
. . .
. . .
16. H. STRAUBING, Recognizable sets and power sets of finite semigroups, *Semigroup Forum* **18** (1979), 331–340.

A common alternative to the alphabetical arrangement is the order-of-presentation arrangement. The internal references are numbered in ascending order, according to their order of presentation in the text, and the end references are numbered and ordered in the same way. Under this format, since the Perrot article is cited first, it would be given the reference number 1; the Straubing article is cited next and so it would be given the number 2; and so on. Thus the excerpt given above would look like this:

The study of certain operations on rational languages leads to the consideration of pseudovarieties generated by power semigroups or power monoids (cf. [1, 2, 3]). The case of monoids has been treated successfully by Margolis and Pin [4] (see also [5]).

and the list of references at the end would start off like this:

1. J. F. PERROT, Variétés de langages et opérations, *Theor. Comput. Sci.* **17** (1978), 197–210.
2. H. STRAUBING, Recognizable sets and power sets of finite semigroups, *Semigroup Forum* **18** (1979), 331–340.
3. J. E. PIN, Variétés de langages et monoïde des parties, *Semigroup Forum* **20** (1980), 11–47.
4. S. M. MARGOLIS AND J. E. PIN, Power monoids and finite J-trivial monoids, *Semigroup Forum* **29** (1984), 125–135.
5. J. E. PIN, "Variétés de langages et variétés de semigroupes," Thèse d'Etat, Université de Paris 7, 1981.
. . .

The full citation should give the author's name; the title of the document (article, thesis, book, etc.); the title of the book or journal in which it appears, including, if appropriate, the volume number; the date of publication; and, if the document is part of a larger volume, the appropriate page numbers.

The exact form of the full citation may differ somewhat from one field to another and even within fields. Sometimes the author's last name is given first and the first name or initial is given second. Sometimes the title of the document is omitted. Sometimes the date of publication is given at the end, after the page numbers. And sometimes quotation marks are put around the title of articles and chapters but not around titles of separate publications like books and theses. You should acquaint yourself with the system being used by the particular journal or the particular organization you are writing for, and you should adhere consistently to it.

APA Style

The APA style is the most widely preferred style in the social sciences, certain natural sciences, and business. In this format, each internal reference is indicated by the author's last name and the date of publication. The end references are given in alphabetical order. If the *Journal of Algebra* were using this system, the above excerpt would appear as follows:

> The study of certain operations on rational languages leads to the consideration of pseudovarieties generated by power semigroups or power monoids (cf. [Perrot, 1978; Straubing, 1979; Pin, 1980]). The case of monoids has been treated successfully by Margolis and Pin (1984) (see also Pin, 1981).

The end references are listed in alphabetical order. The author's name appears first, followed usually by the year of publication (in parentheses), the title of the document (only the first word is capitalized), the title of the journal in which it appears (italicized or underlined), the volume number (in bold print), and the relevant pages. Sometimes the year of publication appears instead after the journal title. For separately bound documents (i.e., books, dissertations, or reports), the author's name is followed by the year of publication (in parentheses), the title (italicized or underlined, with only the first word capitalized), the address of the publisher (city only, if large; otherwise, city and state), colon, and the name of the publisher. For unpublished dissertations, the name of the university is given instead of a publisher. Thus, the list of references for the citations given above would look like this:

> Margolis, S. M. and Pin, J. E. (1984). Power monoids and finite J-trivial monoids, *Semigroup Forum* **29**, 125–135.
> . . .
> Perrot, J. F. (1978). Variétés de langages et opérations, *Theor. Comput. Sci.* **17**, 197–210.
> Pin, J. E. (1980). Variétés de langages et monoïde des parties, *Semigroup Forum* **20**, 11–47.
> Pin, J. E. (1981). *Variétés de langages et variétés de semigroupes*, Thèse d'Etat, Université de Paris 7.
> . . .
> . . .
> Straubing, H. (1979). Recognizable sets and power sets of finite semigroups, *Semigroup Forum* **18**, 331–340.

The APA system is more fully described in the *Publication Manual of the American Psychological Association*, 3d ed. (Washington: American Psychological Association, 1983).

Some discipline-specific variants to the Number system and the APA system can be found in the style manuals listed under Additional Reading at the end of this appendix.

A.5 COMMON GRAMMATICAL AND STYLISTIC ERRORS

Technical/professional writing is expected to be done according to the conventions of Standard English (the kind of English described as such in dictionaries and grammar books and taught in school). If you violate these conventions, your readers may draw negative inferences about your background or level of education, and this, of course, could do great harm to your professional image. Therefore, if you are not completely familiar with these conventions, you should buy one of the reference books listed under Additional Reading at the end of this appendix and study it.

In our experience, the most common grammatical and stylistic errors in technical/professional writing are as follows.

The "Comma Splice"

If you run two independent clauses together with only a comma separating them, you have created what is called a "comma splice."

> Faulty
> Technology never exists in a social vacuum, it is embodied in products, processes, and people.

Although this is considered perfectly correct in many other languages, it is not correct in Standard English. For ways to avoid comma splices when you are trying to connect two independent clauses, use a semicolon, colon, or coordinating conjunction. There are subtle differences in which one you choose (see Section A.1). In the example above, a colon would probably be the best choice:

> Correct
> Technology never exists in a social vacuum: it is embodied in products, processes, and people.

The Sentence Fragment

Subordinate clauses cannot stand alone but must be attached to a main clause. Subordinate clauses begin with a subordinating conjunction (*since, although, if*, etc.) and have their main verb in the *to* or *-ing* form. If you punctuate a subordinate clause as if it were an independent sentence, you are creating a "sentence fragment."

Faulty
Since specimens from each batch are fully tested and qualified. It is expected that any generic defects will be detected during testing.

Correct
Since specimens from each batch are fully tested and qualified, it is expected that any generic defects will be detected during testing.

Faulty
The Government DECAS inspectors have since accepted my new inventory list as the official list. *Thus, making WRC responsible for only the parts that our revised inventory records show.*

Correct
The Government DECAS inspectors have since accepted my new inventory list as the official list, *thus making WRC responsible for only the parts that our revised inventory records show.*

Faulty Subject-Verb Agreement

The subject and main verb of a sentence should agree in number; that is, they should either both be singular or both be plural. With complex subjects that have two or more nouns in them, the *head noun* (not necessarily the nearest noun) dictates whether the verb should be singular or plural.

Faulty
Each of the runners *were* going at top speed.

Correct
Each of the runners *was* going at top speed.

Pronoun-Antecedent Agreement

Pronouns should agree in number and gender with their antecedents: *Mary* went to *her* desk. Sometimes writers wanting to avoid the sexism of the generic *he* use the *they* form instead:

Faulty
If *a student* turns in an assignment late, *they* will be marked down for it.

Although this usage is certainly nonsexist, it violates the grammatical rule of pronoun- antecedent agreement. A better way of getting around this problem would be to use *he or she* :

Correct
If *a student* turns in an assignment late, *he or she* will be marked down for it.

Alternatively, you could pluralize both the antecedent and the pronoun:

Correct
If *students* turn in an assignment late, *they* will be marked down for it.

For further discussion of this problem, see Chapter 27.

Parallelism

All items of a list should be in parallel grammatical form. (See Chapter 23.)

Choppy Style

If you write too many short sentences, your writing can develop a choppy rhythm that might irritate some of your readers. To avoid this, look for opportunities to make an occasional longer sentence by combining two shorter ones. Don't just combine *any two* sentences, though; make sure they're closely related in meaning. (See Chapter 25.)

Excessive Use of the Passive Voice

If you have a sentence where some agent is acting on some object (e.g., *Dr. Ruth examined the laboratory results*), you can turn it into a "passive voice" version by switching the object and subject and changing the form of the verb: *The laboratory results were examined by Dr. Ruth.* The passive voice serves many good purposes and has been unfairly maligned in many style books. For example, it can be used to put given information and/or topical information into subject position. (See Chapters 24 and 25.) It allows the agent to be deleted (e.g., *The laboratory results were examined.*) and thus lets you create more concise sentences (when the agent is unimportant) and more tactful sentences (when you don't want to explicitly cite the agent).

Do not, however, get into the habit of using the passive voice without thinking. If you use the passive voice excessively, your writing will become depersonalized and "heavy"; it may even begin to sound like that ponderous bureacratic language that everybody hates to read. Good technical and scientific writers use passive sentences only when the occasion calls for them, which is about one-fourth of the time. You should try to do likewise.

Ambiguous Reference

Technical and professional writing should be clear and precise, without ambiguity. When you use pronouns or modifiers, make sure it's clear what those pronouns are referring to and what those modifiers are modifying.

Unclear
Ground loops associated with coaxial cables are totally eliminated with fiber optic cables. Because *they* are dielectric, *they* do not attact lightning. Nor do *they* act as antennas.

Do the *they's* refer to *ground loops* or to *fiber optic cables?* To most readers, it's not immediately clear. This is a better version:

> Clear
>
> Ground loops associated with coaxial cables are totally eliminated with fiber optic cables. Because they are dielectric, fiber optic cables do not attact lightning. Nor do they act as antennas.

Wordiness

As is discussed in Chapter 26, you should always try to prune your writing of unnecessary words. Consider, for example, this sentence:

> Wordy
>
> It is my purpose in this report to present and justify the modifications that I recommend in order to increase the performance of your Corsair.

With a little editing, the writer could have said approximately the same thing in eight fewer words:

> Better
>
> My purpose in this report is to recommend certain modifications for improving the performance of your Corsair.

Inappropriate Jargon

Although technical or professional jargon helps promote communication among specialists, it has just the opposite effect when used with nonspecialists. If you are writing for nonspecialists, try to minimize your use of jargon. And when you do use jargon terms, be sure to define them. (See Chapter 27 for further discussion.)

REFERENCES

1 Denis Goulet, "The Dynamics of International Technology Flow," *Technology Review*, May 1978, p. 32.
2 William J. Cromie, "Room at the Bottom," *Mosaic*, May-June 1981, p. 25.
3 J. Almeida, "On Power Varieties of Semigroups," *Journal of Algebra 120*:1 (1989).

ADDITIONAL READING

American Chemical Society, *Handbook for Authors* (Washington, DC: American Chemical Society, 1978).
American Mathematical Society, *A Manual for Authors*, rev. ed.(Providence, RI: American Mathematical Society, 1980).

CBE Style Manual Committee, *Council of Biology Editors Style Manual: A Guide for Authors, Editors, and Publishers in the Biological Sciences*, 4th ed. (Washington, DC: Council of Biology Editors, 1978).

The Chicago Manual of Style, 13th ed. rev. (Chicago: University of Chicago Press, 1982).

John C. Hodges and Mary E. Whitten, *Harbrace College Handbook*, 10th ed. (San Diego, CA: Harcourt Brace Jovanovich, 1986).

Randolph Quirk, Sidney Greenbaum, Geoffrey Leech, and Jan Svartvik, *A Grammar of Contemporary English* (London: Longman, 1972).

US Government Printing Office Style Manual, rev. ed. (Washington, DC: Government Printing Office, 1984).

Joseph M. Williams, *Style: Ten Lessons in Clarity and Grace*, 3rd ed. (Glenview, IL: Scott Foresman, 1989).

B

Summary Checklist

The following is a checklist you can use for producing any type of writing. It contains general guidelines that have been discussed at length in the chapters noted. For more specific advice about particular kinds of writing (letters, memos, proposals, etc.), you may want to consult chapters 10–18.

Generating Ideas (Ch. 2)

1 Have you generated good ideas about your topic?   Yes [ ]   No [ ]

2 If not, have you tried brainstorming?   Yes [ ]   No [ ]

3 If not, have you tried systematic prompts?   Yes [ ]   No [ ]

4 Have you used social and ethical considerations?   Yes [ ]   No [ ]

5 Have you formed a potential thesis or point?   Yes [ ]   No [ ]

Identifying Audiences and Purposes (Ch. 3)

6 Have you identified the communication's uses and routes?   Yes [ ]   No [ ]

7 Have you identified all possible audiences?   Yes [ ]   No [ ]

		Yes	No
8	Have you identified the concerns, goals, values, and needs of each audience?	[ ]	[ ]
9	Have you identified each audience's preferences for and objections to different kinds of arguments?	[ ]	[ ]
10	Do you have your ostensible purpose clearly in mind?	[ ]	[ ]
11	Do you have your other purposes clearly in mind?		

Constructing Arguments (Ch. 4)

		Yes	No
12	Have you constructed some good arguments?	[ ]	[ ]
13	Have you based your arguments on criteria that are appropriate for the intended audience(s)?	[ ]	[ ]
14	Have you built an appropriate case?	[ ]	[ ]

Stating Problems (Ch. 5)

		Yes	No
15	Have you clearly stated the problem you're addressing? That is, are you sure the audience knows the A-but-B terms?	[ ]	[ ]
16	Have you indicated your strategy for dealing with it?	[ ]	[ ]
17	Have you clearly stated the purpose of the communication?	[ ]	[ ]

Drafting and Word Processing (Ch. 6)

		Yes	No
18	Have you written a rough draft?	[ ]	[ ]
19	Have you written a second draft?	[ ]	[ ]

Testing and Revising (Ch. 7)

		Yes	No
20	Have you tested out your document on readers who resemble your target readers?	[ ]	[ ]
21	Have you revised the document accordingly?	[ ]	[ ]

Selecting Visual Elements (Ch. 8)

22   Have you selected appropriate places to use visual aids?   Yes [ ]   No [ ]

23   Have you selected the best types of visual aids to use?   Yes [ ]   No [ ]

Creating Visual Elements (Ch. 9)

24   Have you made your visual aids relevant, clear, and truthful?   Yes [ ]   No [ ]

25   Have you made them appropriately *visual*?   Yes [ ]   No [ ]

26   Have you integrated them into the text?   Yes [ ]   No [ ]

27   Have you formatted your document properly?   Yes [ ]   No [ ]

Editing for Readability (Chs. 21–28)

28   Have you made the important concepts in the document clear to all your readers?   Yes [ ]   No [ ]

29   Do your paragraphs have good topic statements and clear patterns of organization?   Yes [ ]   No [ ]

30   Have you used keywords prominently?   Yes [ ]   No [ ]

31   Have you used parallelism for all items in all lists?   Yes [ ]   No [ ]

32   In your sentences, have you consistently put "given" information before "new"?   Yes [ ]   No [ ]

33   Have you put "light" noun phrases before "heavy" ones?   Yes [ ]   No [ ]

34   Have you placed topical information in subject position?   Yes [ ]   No [ ]

35   Have you combined closely related sentences?   Yes [ ]   No [ ]

36   Have you used signal words to indicate your flow of thought?   Yes [ ]   No [ ]

37   Have you cut out unnecessary words?   Yes [ ]   No [ ]

38 Have you checked your words for accuracy?

Yes   No
[ ]   [ ]

39 Have you checked for comprehensibility?

Yes   No
[ ]   [ ]

40 Have you checked for tone?

Yes   No
[ ]   [ ]

41 Have you proofread the entire document?

Yes   No
[ ]   [ ]

Sample Letters and Reports

Letters

McNamee, Porter and Seely

3131 South State Street / Ann Arbor, Michigan 48104 / (313) 665-6000

Consulting Engineers

January 4, 1990

Mr. Donald P. Ziemke
City Manager
City Hall
7800 Shaver Road
Portage, MI 49081

Re: Portage-Kalamazoo Rate Negotiations

Dear Mr. Ziemke:

As requested by Mr. Williams we have attempted to develop a rate that would be a fair compromise between the $531 per million gallons ($0.14/M^3) based on the 1987 agreement and the $946/mg ($0.25/M^3) requested by the City of Kalamazoo.

The 1987 agreement does not reference interest costs, and the City of Kalamazoo would be justified in asking that the interest costs associated with borrowing money be passed on to the system's customers. The data needed to develop an accurate interest cost for Portage are not readily available; however, assigning all of Kalamazoo's interest cost for 1991 to the treatment plant and calculating Portage's share based on design capacity at the plant would give a 1991 cost of $101,393 and an interim cost of $75,031. This equates to an additional commodity charge of $75,031/494 mg (1,871,100 M^3) = $152/mg, or $0.04/M^3. This rate would be expected to increase because of increases in actual interest rates over the rates originally estimated by Kalamazoo. A reallocation of interest costs to both the sewer system and the WWTP may also cause a change in the rate. As mentioned previously, the information to calculate an exact allocation is not currently available.

Based on the numbers as presented, the commodity rate for Portage should be $531 + $152, or $683, per million gallons. The existing compromise rate of $700 reflects this cost plus a small adjustment for increased interest costs or differences resulting from the revised allocation figures.

In the interest of the City, this should be changed to an OM&R rate of $239 per mg plus a fixed monthly capital charge of approximately $18,000 and a demand charge of $760 per month.

Mr. Donald P. Ziemke
Page Two
January 4, 1990

If you have any questions, or comments, please advise.

Very truly yours,
McNAMEE, PORTER AND SEELEY

BY _____

Raymond J. Smit

RJS:dr
Enclosure

McNamee, Porter and Seely

3131 South State Street / Ann Arbor, Michigan 48104 / (313) 665-6000

Consulting Engineers

January 18, 1990

Mayor and City Council
City of Portage
7800 Shaver Road
Portage, MI 49081

Attention: Mr. Donald P. Ziemke, City Manager

Dear Mayor and Members of Council:

In accordance with your request on January 12, 1990, bids were received for constructing new production wells at Portage. A total of five bids were received and are summarized on the attached tabulation.

The lowest and best bid is that of Leo Riegler Corporation of Muskegon, Michigan, at an estimated cost for two wells of $48,090. We recommend that a contract be awarded to Riegler at the unit prices shown in their bid proposal.

Please advise us of your wishes regarding this recommendation, and we will prepare the necessary contract documents for signatures of the contractor.

In addition to the work of Leo Riegler, we would propose to retain the services of L. M. Miller and Associates to review the well-drilling operations and pumping tests. L. M. Miller and Associates would be working as a subcontractor to us. Their charges will range from $5,000 to $6,000, and in addition we may expect to incur an additional $3,000 expense in our work in reviewing the progress of construction. Accordingly, we would appreciate receiving your purchase order for services at our hourly rate basis at a cost not to exceed $9,000.

Should you have any questions regarding this project, please call on us.

Very truly yours,
McNAMEE, PORTER AND SEELEY

By _____

Raymond J. Smit

RJS:bh
Enclosure
cc: Lynn Miller

McNamee, Porter and Seely

3131 South State Street / Ann Arbor, Michigan 48104 / (313) 665-6000

Consulting Engineers

February 5, 1990

Mr. G. Dallas Williams, P.E.
City of Portage
7800 Shaver Road
Portage, Michigan 49081

      Re: Master Stormwater Management Report

Dear Mr. Williams:

      Pursuant to our phone conversation of February 4, 1990, we would like to request a 60–day time extension for the completion date for the Master Stormwater Management Report. Therefore, we are requesting April 29, 1990, as the completion date for this project. We understand you will contact Marcy Brooks of the Southcentral Michigan Planning Council with this request and obtain her agency's approval for the time extension. Please contact us when our signature is required on any forms due to the time extension request.

      We will be submitting a Final Phase 1 report at the end of February 1990 and are currently proceeding with the Phase 2 portion of the report.

      If you have any questions, or if there is any problem with our time extension request, please contact us.

                Very truly yours,
                McNAMEE, PORTER AND SEELEY

By _____
                Raymond J. Smit
                Partner

By _____
                John S. Wise
                Project Engineer

RJS:JSW:dr
cc: Lynn Miller

McNamee, Porter and Seely

3131 South State Street / Ann Arbor, Michigan 48104 / (313) 665-6000

Consulting Engineers

February 1, 1990

Newkirk Electric Associates, Inc.
1875 Roberts Street
Muskegon, MI 49442

Attn: Mr. Robert Mudget

     RE: City of Portage
          Improvement to Pump Controls

Dear Mr. Mudget:

Mr. Robert Kimmer has asked that MP&S come to the City of Portage job site and help solve the problems remaining with our Portage contract.

We have agreed that David Matthews from our office will be there at 9:00 a.m. Tuesday, February 9. Mr. Matthews called your office on January 29, 1990, and left word of this meeting with Joe Breckring. Mr. Matthews offered to help you solve the remaining problems and wanted to know if the above date was mutually acceptable. Mr. Matthews also suggested that certain parts and test equipment be brought to analyze and solve the problems in a methodical manner.

You called back at 4:30 stating you would not go back to the job site without a purchase order. May we point out in the specifications page 46–2 under "Responsibility and Coordination"—"It shall be the responsibility of the contractor to furnish a complete and fully operable system. The contractor shall be responsible for all details which may be necessary to properly install, adjust and place in operation the complete installation."

Also, under "Service and Start-up" page 46–5—"The supplier's representative shall revisit the job site for 8 hours per day as often as necessary after installation until all trouble is corrected and the equipment has passed the acceptance test and is operating satisfactorily to the engineer."

The acceptance test as defined on page 46–5 has not yet taken place. The engineer to this date has not been requested to witness the operation for acceptability.

We are offering in good faith to help you solve the problems that remain and request that your technical people come prepared to solve these problems on February 9.

Newkirk Electric Associates, Inc.
February 1, 1990
Page Two

If you do not cooperate to complete this work, we will be forced to invoke the liquidated damages clause and such other measures at our disposal. May we receive your favorable response to completing this project beginning February 9.

Very truly yours,
McNAMEE, PORTER AND SEELEY

By _____
Raymond J. Smit
Partner

By _____
David R. Matthews
Associate

DRM:RJS:bh

cc: City of Portage
Dallas Williams
Robert Kimmer

McNamee, Porter and Seely

3131 South State Street / Ann Arbor, Michigan 48104 / (313) 665-6000

Consulting Engineers

To:	Leo Riegler Corporation	Date:	2/11/90
	2121 Glade Street	Project:	City of Portage
	Muskegon, MI 49444		Water Works Improvements
Attn:	David Riegler		Production Wells
		Contract No.:	81-W-1

Gentlemen: Agency No.: _____

☒ We are enclosing   ☒ We are sending under separate cover

The following:

☐ Drawings ☐ Specifications ☐ Change Orders

☐ Pay Estimate ☒ Contract Documents ☐ _____

5 copies of contract documents on the above referenced.

For:

☐ Your approval ☐ Your corrections ☐ Your comments

☒ Your execution ☐ Your file ☐ Your signature

Remarks:

PLEASE SIGN BUT DO NOT FILL IN THE DATE on the following as indicated: Agreement page No. 22; execute page 23 if it pertains to you. Performance Bond Page No. 24. Payment Bond Page No. 26. DATES WILL BE FILLED IN BY THE OWNER. After signing all copies, please return all copies to this office for further processing. Also return with the signed documents the necessary insurance papers as called for on page 7 of INSTRUCTIONS TO BIDDERS AND ARTICLE 8, GENERAL STIPULATIONS, pages 17, 18, and 19.

CC: Transmittal only/Owner   By: Richard G. Walterhouse/ml
     File(2)

Informal Reports

ACE MANUFACTURING COMPANY
23200 Dearborn Avenue
Detroit, Michigan 48227

October 9, 1989

M Consulting Services
G. G. Brown Laboratories
Ann Arbor, Michigan 48105

Gentlemen:

We are planning to install several high-speed punch presses in each of our new plant locations. Both groups of presses will be supported by block foundations which rest directly upon a thick layer of sand. At one of the sites the sand is white, and at the other it is black. Samples of each type of sand have been sent to you under separate cover.

Our engineers have obtained some data on the in situ characteristics of each sand deposit. The natural field void ratio of the white sand is 0.65, and the dry unit weight of the black sand in situ is 118 lb/ft³.

Our concern is the possibility that these machines will settle because of the vibratory loadings transferred from the machines to the underlying sand. Please send us your expert opinion as to whether these foundations will or will not settle when located on the white or black sands.

Very truly yours,

J. C. Starr
Chief Engineer

JCS/dd

M. Consulting Services
G. G Brown Laboratories
Ann Arbor, Michigan 48105

October 16, 1989

ACE Manufacturing Company
23200 Dearborn Avenue
Detroit, Michigan 48227

Attention: Mr. J. C. Starr, Chief Engineer

Re: Soil Settlement Tests at New Plant Locations

Dear Mr. Starr:

In your letter of October 9, 1989, you were concerned with the possibility of soil settlement, through vibratory loadings, if high-speed punch presses are installed at your new plant locations; you sent sand samples from these locations for us to inspect. Accordingly, we have run tests on these samples to determine if indeed settlement is possible. The purpose of this report is to inform you of the test results, which lead us to recommend that your company forgo the installation of the presses at either of the new plant locations.

Summary

We performed three tests on the sands: a specific gravity test, a minimum void ratio test, and a maximum void ratio test. The relative density, a measure of looseness or compactness of soil, was calculated for each sand using these test results along with the information your engineers supplied. The results indicate that soil settlement is very possible, and therefore we would not recommend the installation of the punch presses.

Test Descriptions

We determined the specific gravity of the sands by taking the ratio of two weights: the weight, in air, of a given volume of soil versus the weight, in air, of an equal volume of distilled water. We determined the minimum and maximum void ratios using local standardized tests: the "air pliviation" method and the "elephant trunk" method, respectively. The minimum void ratio test allows the soil particles to take up a minimum amount of volume. The maximum void ratio test does just the opposite. Detailed test descriptions and calculations of specific gravity, minimum and maximum void ratio, and relative density can be found in the attached appendix. The results are presented here:

	Ottawa sand	Turkish emery
Specific gravity	2.65	3.99
Minimum void ratio	.53	.78
Maximum void ratio	.71	1.15
Relative density	33.3%	10.8%

Conclusion/Recommendation

Soils having a relative density greater than 70% are quite dense or compact and would settle very little. Our tests indicate that the relative densities of the sands tested are considerably below this and therefore settlement is quite possible. After carefully studying the data and taking into consideration the limitations of the tests, we would not recommend the installation of the presses at either of the new plant locations.

Very truly yours,

Timothy J. Burns

Timothy J. Burns, PE

Attachments:

Data calculations
Test descriptions

FORD RESEARCH CENTER

Kennedy Avenue
Dearborn, Michigan 48500

Date: November 1, 1989

To: Dr. Daniel Smith
             Director of the Electronic Circuit Design Group

From: John Leong
             Numerical Analysis Group Director

Subject: Proposal for the implementation of the Sparse Matrix Algorithm as an efficient
             subroutine for solving large systems of linear equations

Ref: Sparse Matrices and Their Applications, by Rose (Plenum)
             Numerical Methods, by Dahlquist (Prentice-Hall)

Dist: Nick Mente, Research and Development Director
             David Hill, Computer System Director
             Cathy Hope, Budget Director

You stated in your memo of October 25, 1989, that your designs often involve large systems of linear equations (SLE), but our present subroutine for solving them is very inefficient, especially if there are many zero entries in the SLE. In order to increase the efficiency, we have found a mathematical algorithm which is especially suitable for solving large SLE efficiently; we call it the Sparse Matrix Algorithm (SMA). In this report, we introduce the new algorithm and propose that it be implemented as a subroutine in our computer system.

Summary

The Sparse Matrix Algorithm is highly recommended to be implemented as a system subroutine for solving large SLE. The SMA is efficient because it operates on entries of the SLE matrix selectively. It requires less computer storage area and less computer processing time. Tests were performed which showed that the SMA is more efficient than the present Gaussian Elimination Algorithm. It can be implemented in the computer system easily, and considerable computer processing expenditure can be saved after its implementation.

Introduction

The present subroutine for solving large SLE uses the Gaussian Elimination Algorithm. A brief description of the subroutine is shown in Appendix A, and a detailed description is found in Computer System Memo 451. This algorithm is inefficient in solving large SLE because it operates on zero entries and results nonselectively. However, the designs of the Electronic Circuit Design Group often involve large SLE which have many zero entries. The inefficiency increases exponentially with larger SLE. Consequently, a new algorithm is needed, and the SMA is found to be efficient in the application.

Sparse Matrix Algorithm

The algorithm is efficient in solving large SLE which have many zero entries because it operates on entries selectively. It does not operate on zero entries and does not store zero results. These features lead to efficient usage of computer storage area and shorter duration of computer processing time:

1. Efficient usage of computer storage area—
    The algorithm neglects all zero entries and transforms the original matrix A of the large SLE into two smaller matrices, the compacted matrix Ā and the pointer matrix P. The Ā matrix stores all non-zero entries while the P matrix saves their original positions in the matrix A. The two smaller matrices require less storage area exponentially than the original matrix as the system increases in size. Moreover, the SMA does not store any zero results, which also saves storage area. (Appendix B)

2. Shorter duration of computer processing time—
    Since the SMA operates selectively, it has fewer operating steps to be performed in solving the large SLE. This algorithm requires considerably less computer processing time.

A brief description of the SMA is shown in Appendix C, and a detailed description is found in the first suggested reference (pp. 1–40).

Testing

Simple computer simulated tests were done to compare the Gaussian Elimination Algorithm and the Sparse Matrix Algorithm. Besides needing less computer processing time, the SMA also needs less computer storage area in solving large SLE (Appendix D). A detailed testing report is found in Numerical Analysis Memo 815.

Implementation

The SMA can be implemented as a subroutine in our computer system by developing one of our own with the assistance of our Computer System Group. Moreover, the subroutine can be bought from HP, IBM, or Honeywell for a reasonable price (Appendix E). In any case, we can have higher efficiency and considerably less computer processing expenditure because more than fifty percent of the designs of the Electronic Circuit Design Group involve large systems of linear equations. We expend twenty dollars for each computer processing minute (Computer System Memo 15).

Conclusion

The Sparse Matrix Algorithm is efficient in solving large systems of linear equations. It can be implemented as a subroutine in our computer system without much difficulty. Its implementation is highly recommended. At the same time, the present Gaussian Elimination Subroutine should be retained as it is still useful in solving small systems of linear equations.

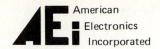

American
Electronics
Incorporated

To: Mr. John Nicol
           Chief Optoelectronics Engineer
           Optoelectronics Department

From: Khalil Najafi
           Assistant Optoelectronics Engineer
           Optoelectronics Department

Date: October 30, 1989

Subject: Liquid Crystal Displays: Analysis of Failure and Recommendations for Solution

Dist: Mr. Edward Jones, President
           American Electronics Incorporated

Foreword

Although becoming fairly well established as displays in small instruments, Liquid Crystal
Displays (LCDs) are still regarded with some skepticism by many designers. Since American
Electronics Incorporated (AEI) plans to use LCDs instead of Light Emitting Diodes (LEDs)
for its future products, I was assigned by Mr. John Nicol to investigate the LCD's modes
of failure, to provide solutions to prevent these failures, and to propose different kinds of
tests to detect these modes of failure.

The following report describes the results of my investigation: it explains the shortcomings
of LCDs and proposes means for overcoming and detecting these.

Summary

Investigations show that LCDs can be used as good display devices if the problems stated
below are eliminated through the adoption of the solutions suggested. LCDs were found to
have problems in two major areas:

1   problems arising as a result of the presence of moisture, oxygen, and a high temperature
    environment, which can cause very severe damage to the LCDs;

2   problems due to the loss of contact between the connector and the indium oxide leads
    on the LCD.

These problems can be avoided by following recommended procedures:

1   use of good insulating materials and of protective seals against moisture and temper-
    ature;

2   replacement of the current method of creating a contact between the connector and
    the indium oxide leads on the LCD by a display that can be soldered directly to the
    circuit board.

Finally, any occurrence of these problems can be detected by the following recommended
tests:

a   test for tolerance against temperature,

b test for moisture content, and

c test for high electric fields.

All of these tests should be carried out by the quality control or inspection department.

Discussion

The discussion provides details on the following topics:

1 Failure due to temperature and moisture, and proposed prevention measures.

2 Failure due to loss of contact, and proposed prevention measures.

3 Recommended quality control procedures.

1. FAILURE DUE TO TEMPERATURE AND MOISTURE and Proposed Prevention Measures

One of the main problems facing LCDs is their strong sensitivity to temperature and moisture. These factors can change the behavior of LCDs rather drastically and result in gradual degradation of the device. Two basic categories can be recognized in this case: damages due to combination of high temperature and high humidity and damages due to high temperature only.

1.1 Damages Due to a Combination of Temperature and Moisture

The most damaging environment that an LCD is likely to experience is produced by the combination of high temperature and high humidity. Depending on how the display is constructed, three failure modes can occur under this condition: failure of the polarizers, failure of the adhesive on the polarizer and reflector, and, finally, failure of the liquid crystal mixture.

1.1.1 Failure of the Polarizers

The most common failure occurs with the polarizers. Regular polarizing material will begin to degrade after a short period of time (4 days) when subjected to a temperature of 50°C and 95% relative humidity. Once moisture penetrates the plastic backing of the polarizing material, it affects the polarizing properties. The digit area of the display will start to lose contrast, gradually turning from black to brown until it finally disappears completely.

1.1.1.1 Preventing Failure of the Polarizers

To provide protection against the effects of high temperature and humidity, the use of a polarizing material known as "K-sheet," introduced by Polaroid Corporation, is recommended. This material has polarization properties which are relatively unaffected by moisture.

1.1.2 Failure of the Adhesive

Moisture can also attack the adhesive on the polarizer and reflector and cause it to begin peeling away from the glass. If the alignment is lost, dark patches will appear in the viewing area like those as already described.

1.1.2.1 Preventing Failure of the Adhesive

To prevent this kind of failure, the use of a truly hermetic seal is recommended to protect the adhesive.

1.1.3 Failure of the Liquid Crystal Mixture

If any of the components in the liquid crystal mixture (such as Schiff-base compounds) are attacked by moisture, the liquid crystal molecules will start to break down, thereby lowering the clearing point of the mixture and the effective upper operating temperature of the display. Due to the ionic by-products of the chemical reactions, the electrical currents of the display will also show a very significant increase.

1.1.3.1 Preventing Failure of the Liquid Crystal Mixture

The procedure described in part 1.1.2.1 should be adopted.

1.2 Failure Due to High Temperature Only

Failure can occur when an LCD is exposed to high temperatures only. The twisted nematic display requires the glass plates to be specially treated to impress a uniform alignment on the liquid crystal molecules next to the glass surface. The quality of this alignment can be degraded by high temperature because of decomposition of the interface layer.

1.2.1 Preventing Failure Due to High Temperature

The procedure used in part 1.1.2.1 should be adopted.

2. FAILURE DUE TO LOSS OF CONTACT

Problems can arise due to loss of contact between the connector and the indium oxide leads on the LCD. The user will often blame the LCD for an electrical failure such as an open segment when in fact it is the connector which is responsible.

2.1 Preventing Failure Due to Loss of Contact

This problem was solved by the introduction of the DIL displays with substrate chips by Liquid Crystal Displays in 1984. They offer a display that can be soldered directly in a circuit board to reduce the possibility of any disconnections.

3. RECOMMENDED QUALITY CONTROL PROCEDURES

It is recommended that AEI provide its quality control or inspection department with the equipment to measure the following characteristics or to make at least the following tests:

1. The electrical current through the cell at room temperature, AC square wave at approximately 32 Hz variable in range 5 to 12 volts to match the range cells. (Normal operating temperature range -10°C to +50°C.)

2. The clearing point of the fluid, measured by raising the temperature until the viewing area turns black. This requires a dry-environment oven.

3. Electrical operation of the display at the specified upper operating temperature for 30 days, followed by tests (1) and (2) above to detect any changes.

4. Excursions of temperature outside the operating range. Both low and high temperatures less than -40°C and more than +50°C.

5. Humidity testing 95% at 50°C.

6. Response time.

OCEAN UTILIZATION INC.

2714 Johnson
Corvallis, OR 97330
(903) 555-8527

Date: 9 November 1989

To: Mr. Edward Blye, Fleet Co-ordinator
            Del Mar Fish Company

From: Jeffrey D. Paduan, Consultant   *J.D.P.*

Subject: Breakwater Design Recommendations for the New Inlet Beach Harbour Facility

Dist.:   Jack Doner, Construction Supervisor

FOREWORD

On August 12, 1989, Ocean Utilization Inc. was contacted by the Del Mar Fish Company regarding the relocation of the Del Mar Fishing Fleet at Inlet Beach, Florida. This relocation may require the construction of some form of breakwater to protect the fleet from heavy wave action in the nearshore area. I was assigned to investigate the conditions at Inlet Beach to determine if a breakwater was actually needed and, if so, to recommend the most economical design. The purpose of this report is to provide the results of my investigation.

SUMMARY

Because the direction of the most intense waves and the predominant waves are not the same at Inlet Beach, I am recommending that two parallel breakwaters be constructed. This is based on the determination that some form of breakwater construction is clearly necessary to protect the Del Mar fleet given the wave conditions that exist. The breakwaters should be about 30 feet wide in order to sufficiently damp out the incoming wave energy and should extend about 300 feet from shore in order to sufficiently clear the dangerous nearshore area. The breakwaters can be constructed of readily available stones, and the cost of one breakwater appears to be about $500,000, for a total of $1,000,000 for t._ two breakwaters.

1.) BACKGROUND

1.1) Nearshore Hazards

The area that is most hazardous for ships to navigate is the area near the shore. There are two reasons for this:

1) Increased chance of the ship running aground in shallow water

2) More severe and treacherous wave conditions near the shore

This report is concerned with the second of these reasons. Although it seems paradoxical that waves would be more treacherous for ships near the shore than for ships far from the shore, it can be explained using a simplified theory of water waves. It can be shown that waves in shallow water have speeds that are proportional to the depth of the water. Because waves approaching the shore find themselves in water that is becoming progressively less deep, the speed of the waves decreases and the height is forced to increase in order to maintain the same energy level.

1.2)   What Is the Surf Zone?

For the purpose of this report, we shall define the "surf zone" to be the area extending from the shore out to a depth where the waves are no longer affected by the bottom. This depth is defined by wave theory as L/2, or one half the wavelength of the incoming wave. Our investigation will yield an average wavelength, L, and therefore a characteristic surf zone width, $\underline{d}$, depending on the depth at our location.

1.3)   What Is a Breakwater?

A breakwater is a physical device that blocks or damps wave action. Breakwaters are used to provide an alley of safe passage for ships and boats moving, across the surf zone, from the harbor to the open water. The definition of a breakwater is general. It includes everything from a simple pile of stones to an ordered pile of concrete blocks to a solid wall of cement. Logically, these different breakwater designs, utilizing different materials, have varying degrees of effectiveness at shunting wave action. The particular design chosen for any one location is a function of wave conditions at that location and of the degree of protection required by the ships that will utilize the breakwater.

1.3a)   Breakwater Width

The width of a breakwater depends on two things: the materials of which it is constructed and the size of the waves attacking it. If the breakwater is solid, it need only be wide enough to be stable. Cement breakwaters usually are 10 to 20 feet wide at the bottom and converge at the top to only a few feet. These relatively narrow breakwaters are 100% efficient at damping out wave energy because they are solid and let no water pass through. Stone breakwaters, on the other hand, or other designs using loose-fitting, porous materials allow some of the water and wave energy to pass through, and so in order to be effective they must be wide enough to damp out the required amount of wave energy. It has been shown that for most applications porous breakwaters must be at least as wide as one third of the wavelength. This width can range, typically, from 10 to 50 feet.

1.3b)   Breakwater Length

Breakwaters should be long enough to extend through the surf zone. This length will depend on the wavelength and the water depth at a particular location.

1.4)   Critical Parameters for Describing Nearshore Waves

1.4a)   The Predominant Wave Height

The predominant wave height is defined as the average height of waves approaching from the predominant wave direction.

1.4b)   The Predominant Wave Direction

The predominant wave direction is defined as the direction from which waves are most likely to approach the beach. Far offshore this direction is usually the same for large stretches of coastline, but because the topography or depths of a particular beach turn waves due to refraction (see section 1.5), the predominant wave direction will be much different in the surf zone than offshore and slightly different for adjacent stretches of beach.

1.4c)   The Direction of Most Intense Waves

The direction of most intense waves is defined as the direction from which the largest waves approach the beach. The direction will depend on which direction the strongest storms typically come from and on the topography of the beach. Every beach has a unique topography. Although many flat, gently sloping beaches may be nearly identical, irregularities such as an underwater trough or mound can focus wave energy and create very large waves.

1.5)   Wave Refraction

The dependence of wave speeds in shallow water on depth leads to a phenomenon known as refraction. Refraction means the bending of waves. For example, if a wave approaches a beach at an angle, the part of the wave closest to the shore will reach the shallow water first and be slowed down while the offshore portion of the wave is still traveling relatively fast. The effect of this is to line up waves more nearly parallel to the beach and to decrease the angle at which they approach the beach. If deep holes or tall mounds are present in the beach topography, refraction can cause wave energy to be intensified on a mound and dispersed over a hole or trough. This phenomenon is illustrated, graphically, in the Appendix. Oceanographers use lines to represent wave fronts. The gaps between successive lines contain equal amounts of energy; therefore the lines that are farther apart represent less intense, smaller waves and the lines that are close together represent more intense, larger waves.

1.5a)   The Wave Refraction Program

Oceanographers at O.U.I. have developed a computer simulation program that uses wave theory and water depths to model refraction phenomena.[1] The user must fill in the depth grid and specify the spacing; then he must indicate the direction from which the waves are approaching far offshore and the wavelength of the waves. The program then provides a graphic output depicting the waves and how they are refracted as they approach the beach.

2.)   DATA

2.1)   Offshore Wave History

In order to determine the relative frequencies of the directions from which waves approach the Inlet Beach site, the Coast Guard records from ships traveling across the Gulf of Mexico were consulted.[2] We also examined the records of wind and wave data from the National Weather Service's Tampa, Florida, station.[3]

2.2)   Surf Zone Topography

In order to determine the topography in the surf zone, we took depth measurements with a small boat and a commercial depth gauge made by Shakespeare (Model 1021). The measurements were taken every 50 feet out to a distance of 1000 feet from shore, and then the process was repeated every 50 feet along the shore for a one-half-mile stretch of beach centered at the Inlet Beach site. This gave us a depth grid with a 50-foot resolution and over 1000 data points for use in the wave refraction program.

3.)   ANALYSIS AND RESULTS

3.1)   The Predominant Wave Height Indicates a Need for a Breakwater

The predominant wave height at the Inlet Beach facility is of sufficient height to warrant the construction of a breakwater. This was determined by using the refraction program and the depth data described above. As waves are bent by the decreasing depths in the nearshore zone, they are crowded together. This crowding shows up in the refraction program and allows us to calculate how much the wave height increases from offshore to nearshore. Using the offshore wave data, the predominant offshore wave height was determined to be about 4 feet. When this height was corrected for refraction, the predominant wave height was calculated to be about 6 feet. Six feet is 30% higher than the maximum average wave height allowable for the Del Mar fishing fleet according to the fleet design criteria you sent us.[4]

3.2)   The Predominant Wave Direction Indicates a Need for an Eastern Breakwater

The predominant wave direction is from the southeast. This direction suggests that the east side of the ship tract through the surf zone should be protected by a breakwater. The predominant wave direction was determined by first examining the offshore wave history and determining that the predominant offshore wave direction is southeast at precisely 45 degrees to the shore west of the site (see Appendix). Waves approach from this direction 91% of the time. In order to compensate for wave refraction in the surf zone, the surf zone depth grid, the predominant offshore wave direction, and the average wavelength (see section 4.2) were fed into the refraction program. The refraction program predicted the predominant wave direction to be from the southeast at 20 degrees to the shore west of the site.

3.3)   Direction of Most Intense Waves Indicates a Possible Need for an Eastern Breakwater

The most intense waves approach the Inlet Beach site from the southwest. This suggests that it may be necessary to construct a breakwater on the west side of the ship tract through the surf zone. The direction of the most intense waves was determined in two stages. First, the wave history was consulted to find out what direction the most intense offshore waves come from and how often they approach from this direction. It was found that the most intense offshore waves approach from the south and that waves approach from this direction only 1% of the time. Because any breakwater is going to be ineffective on waves approaching straight on as these waves from the south will and because waves approach from this direction only 1% of the time, we decided to ignore the contribution to the direction of most intense waves by these waves approaching from the south. Second, we used the wave refraction program to test all possible offshore wave directions to see what direction produces the most intense waves in the surf zone. It turned out that offshore waves approaching from the southwest at 45 degrees to the shore east of the site produce the largest waves in the surf zone (see Appendix). These waves in the surf zone average 12 feet high and approach the site at an angle of about 55 degrees to the shore east of the site. Waves approaching the site from an offshore direction of 45 degrees to the shore east of the site were taken as the most intense waves and their direction as the direction of the most intense waves. It happens that waves approach from this direction 6% of the time.

3.4)   Surf Zone Profile Indicates an Optimum Breakwater Angle

It turns out that a breakwater built at the Inlet Beach site would be more effective if it were inclined at an angle toward the west shore. The optimum angle is about 85 degrees to the shore west of the site (see Appendix). This angle was determined by first realizing that inclining the breakwater to the west would more directly confront the waves approaching from the predominant wave direction and that if the angle was not less than 70 degrees the breakwater would still effectively combat the waves from the direction of the most intense waves. Next we see that the determining factor becomes the surf zone width. We don't want to make the breakwater any longer than we have to to get across the surf zone. We looked at the depth data in the surf zone to determine the angle that would give the best protection but not increase the cost too much. The surf zone width is pretty regular straight offshore from the site, and so it became a somewhat subjective choice as to the best angle. We chose an angle of about 85 degrees because we felt that any smaller angle would increase the length, and therefore the cost, of the breakwater by an amount that is not justified by the increased effectiveness.

4.)   CONSTRUCTION SPECIFICATIONS

4.1)   Breakwater Materials

A breakwater constructed of loosely packed stones would be an effective yet relatively inexpensive breakwater for the Inlet Beach site. Stones were determined to be a sufficient building material based on a report published by the Army Corps of Engineers' Coastal Engineering Research Center.[5] At the Center they did extensive testing on different materials for breakwaters. The report compares three different materials: stones, rhombohedrally shaped concrete blocks, and solid concrete walls. The strength and endurance of the three materials were given in terms of the maximum predominant wave height that each could routinely withstand, the maximum wave height that each could withstand for short periods of time (less than 5% of the time), and the percent of initial cost that would be required each ten years to maintain the structure. The figures showed the stones to be least effective, followed by the blocks and then the concrete wall with the highest effectiveness. However, the predominant wave conditions at Inlet Beach are some 26% lower than the maximum given for a stone breakwater. Also, the stone breakwater is much cheaper than either of the other two, averaging about 60% cheaper to construct, and the maintenance cost is comparable at 10% per 10 years, or 1% of the initial cost per year.

4.2)   Breakwater Width

Because a stone breakwater is a porous breakwater, the width of the breakwater will depend on the average wavelength, L, of the incoming waves. Wave theory suggests that the breakwater be at least L/3 feet wide to insure that the breakwater is not narrow enough to be a nodal point in the wave field and therefore ineffective. The average wavelength at Inlet Beach is about 60 feet; therefore a breakwater that is 30 feet wide would adequately damp out the wave energy and provide a safety margin of some 50% over the theoretically suggested width.

4.3)   Breakwater Length

The length is determined by the distance across the surf zone. With an average wavelength of 60 feet, the edge of the surf zone is defined at a depth of 30 feet. From the depth data in the surf zone, it is evident that it would require a breakwater

300 feet long inclined 85 degrees to the shore west of the site to reach a depth of 30 feet and thereby clear the surf zone.

4.4) Breakwater Separation

If two breakwaters are used, the question of their separation distance becomes critical. Clearly, they must be at least twice as far apart as the ships that will travel between them are wide, but it is also critical that they are not too far apart. Model studies have shown that it is better if the insert distance, ID, is not greater than one wavelength, L.[6] This suggests a separation distance, SD, of about 140 feet (see Appendix).

5.) PRELIMINARY COST ESTIMATES

5.1) Cost per Breakwater

We have obtained estimates of production cost from the Brown Construction Co. of Miami and the Watertech Co. of Corpus Christi, totalling $500,000[7] and $600,000[8], respectively. Both of these estimates are for stone breakwaters 30 feet wide and include material costs. The Watertech Co. estimate comes from a formula based on the average water depth, which is about 20 feet. For an average depth of 20 feet they quote a price of $2000 per running foot, totaling $600,000. The Brown Co. estimate is based on a more complicated formula: ($1000)(Q)(dX), where X is the distance offshore and Q = 1 for a depth that is less than or equal to 10 feet and Q = .1(y − 10) + 1 for y = depth greater than 10 feet. This formula requires an integration of the beach slope across the surf zone. If one assumes a smooth, straight-line slope of 1/10, which is indicated by the depth data, the formula yields a total cost of $500,000 per breakwater.

5.2) Cost Analysis of Two Breakwaters

In 11 years the cost of the second breakwater would be recovered from revenues on fishing days that would otherwise be lost. Using the gross income figures you supplied, I determined that a loss of 6% of your fishing days due to severe waves offshore of the site would result in a revenue loss of $51,000 per year. The additional cost of some $555,000 ($500,000 initial plus $55,000 maintenance) for the second breakwater would therefore be regained in 11 years at these revenue rates.

6.) RECOMMENDATIONS

I am recommending that you construct twin stone breakwaters, 30 feet wide, 300 feet long, 140 feet apart, and inclined 85 degrees to the shore west of your facility at Inlet Beach, Florida (see Appendix). This breakwater configuration should adequately protect your vessels from both the average waves and the most intense waves approaching the site. The cost for such a breakwater system appears to be about $1,000,000 according to an estimate given by the Brown Construction Co. of Miami. I recommend, however, that you investigate other marine construction firms before making any final decision, and I can suggest several others if you would like me to.

REFERENCES

1 O.U.I. Wave Refraction Program, based on Airy linear wave theory (1957), adapted at O.U.I. 1976

2 United States Coast Guard, Reports of Ships of Opportunity, Gulf of Mexico, 1946–1979

3   National Weather Service, Climatological Weather Bulletin 1950–1959, 1960–1969, 1970–1979, Tampa, Fla., Station

4   Del Mar Fleet Specifications, personal communication, Edward Blye, Aug. 12, 1981

5   Coastal Engineering Research Center, Technical Report #78–1333465–2, 1978

6   "Optimum Breakwater Spacing, A Model Study," Univ. of Michigan Technical Report, #79–004356–1, 1979

7   Brown Construction Company, 2600 Rickenbacker Causeway, Miami, Fla., production bid for Del Mar Fish Company—Inlet Beach Facility Ship Protection System, via Jason Clark, Sept. 18, 1981

8   Watertech Inc., 18000 Gorman, Corpus Christi, Tex., Del Mar Construction Bid, via George Leilin, Sept. 23, 1981

Bell, Swartzel, and Thomas
Consulting Engineers
Ann Arbor, Michigan

Date: April 20, 1989

To: G. B. Williams, President
            D. E. Hammer, Project Manager
            Technical Specialties Company
            Ann Arbor, Michigan

From: Douglas J. Thomas   *DJT*

Subject: Design Proposal for Scrubbing Section of Dual-Alkali Flue Gas Desulfurization
            System

Foreword

In a letter from D. E. Hammer and G. B. Williams dated January 27, 1988, our company was retained by Technical Specialties to design the scrubbing and flue gas disposal sections of a dual-alkali flue gas desulfurization (FGD) system being built for a client in West Virginia. The design has been completed. This report includes:

1. A description of the process

2. A description of the process control scheme

3. Specifications for all required equipment

4. A discussion of design judgments

5. Summaries of capital and operating costs

Summary

The design was started by determining the rate of mass transfer in the system. Using the data, the scrubber was sized and the final material balance was closed. The flue gas disposal system was then designed, considering the desire to eliminate a condensing vapor plume. The scrubbing liquor recirculation system was sized using the liquid rate dictated by mass transfer and assuming 30 minutes of residence time.

The best scrubber configuration was found to be a single vertical spray tower equipped with an internal demister. The spray tower is a vertical cylinder 30 feet in diameter and 85 feet high. It is equipped with 230 spray nozzles, 46 in each of 5 levels, delivering a total of 2300 gallons per minute of scrubbing liquor. The liquor recirculation system consists of one large recirculation tank, a pump, and a soda ash/water reaction tank. In addition to the existing stack, a fan and a fired heater are required for flue gas disposal.

Minimum sulfur dioxide removal in this system should be 73 percent, a safe margin above the required 62.4 percent. Capital cost of the installed system is $3.5 million, and annual utility costs are $360,000.

Introduction

Technical Specialties required a scrubber design for a demonstration dual-alkali FGD system the company had contracted to build. There were no commercial-scale scrubber designs available for this chemistry, and so it was necessary to design a scrubber from first principles using pilot plant scrubbing liquor concentration data. The system was designed to clean the flue gas resulting from combustion of coal of known composition. The flue gas cleaned by this process must comply with a sulfur emission regulation.

The scrubber works by absorbing sulfur dioxide from the flue gas into the scrubbing liquor, an aqueous solution of sodium bisulfite, sodium sulfite, and sodium sulfate. The sulfur dioxide thus absorbed subsequently reacts with sodium sulfite and water to form sodium bisulfite. Sulfur is eliminated from the system in the form of calcium sulfite and calcium sulfate by the scrubbing liquor regeneration system, the design of which was left to Technical Specialties.

The main technical problems encountered in this design were selecting a scrubber type, quantifying mass transfer of sulfur dioxide from the gas to the scrubbing liquor, and devising a method for flue gas disposal.

1. Synopsis of Process

The incoming flue gas, stream F1 on the Process and Instrumentation Flow Diagram, Figure 1, contains 0.146 mole percent sulfur dioxide when the coal is burned with 15 percent excess air. At least 62.4 percent of this sulfur dioxide must be removed before the flue gas is discharged to the atmosphere in stream F3. This is accomplished by mass transfer of sulfur dioxide from the gas to the liquid phase in the countercurrent spray column. The sulfur dioxide, after dissolving in the scrubbing liquor, undergoes an essentially instantaneous irreversible reaction with sodium sulfite and water to form sodium bisulfite. The scrubbing liquor returns to the recirculation tank, where it is mixed with regenerated liquor, stream F2, and the sodium hydroxide from the soda ash/water reactor. A portion of the contents of the recirculation tank is withdrawn to the regeneration loop, stream F4, where sulfur is removed in the form of calcium sulfite and calcium sulfate. The material balance for this process appears in the appendix.

A sketch of the spray tower appears in the appendix [Figure A-1]. Particulate-free flue gas enters through a duct near the base of the tower. The gas rises through a distributor intended to flatten out the gas velocity profile in the column in order to reduce entrainment of scrubbing liquor droplets. Flue gas rises at approximately 5.4 feet/sec. through 50 feet of gas-liquid contacting area. At 10-foot intervals in the contacting section, banks of 46 spray nozzles spray 10 gpm each in a pattern designed to cover the entire cross section of the tower. The liquor flow to three of these spray banks may be shut off when it is desired to run the system at less than design capacity. Immediately above the last level of spray nozzles in the tower, a wire mesh demister is installed. The demister was designed within the column in order to save the cost of a separate vessel and to keep gas velocity through the demister low enough that re-entrainment would not be significant. A portion of the make-up water to the system is used as wash water for the demister.

The scrubbed flue gas exits the tower through ducts to a fan, where the pressure is increased from -5.5 inches of water to atmospheric pressure. The gas enters the desulfurization system at -0.5 inches of water and undergoes a total pressure drop of 5 inches of water in the scrubber, demister, and ductwork.

In order to eliminate a condensing vapor plume in the stack gas, the flue gas is

Figure C-1 ❏ PROCESS AND INSTRUMENTATION FLOW DIAGRAM (FLUE GAS SCRUBBING)

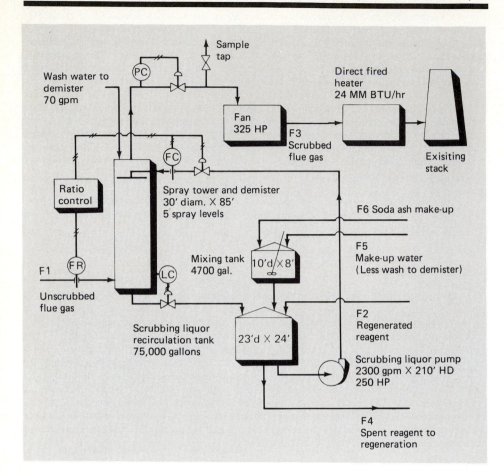

reheated in a direct-fired heater. The flue gas is then drawn out of the system by the existing stack.

Scrubbing liquor is circulated by the pump downstream of the recirculation tank. Liquor returns to the tank from the tower by gravity flow.

2. Process Control Scheme

Scrubbing liquor flow to the tower is maintained in constant proportion to the flue gas flow rate. A ratio control system is used to compare flow rates of gas and liquor and to adjust the flow of liquor by means of a pneumatically operated valve on the liquor feed line. All controls in this system are pneumatic for reliability in continuous service in a corrosive environment exposed to the elements.

Pressure in the tower is kept within specifications by a pressure control loop on the tower outlet duct. It was deemed preferable to control pressure at this location rather than to restrict flow of flue gas from the boiler to the scrubber.

A liquid-level control loop is specified for the liquor drain in the tower base. It is intended to maintain five feet of liquid to prevent flue gas bypass of the scrubbing section or intake of atmospheric air through this drain.

A sample tap is provided in the exiting flue gas duct so that samples for evaluating the system performance may be withdrawn. Flue gas analysis facilities are to be supplied by the client.

3. Equipment Specifications

Specifications for the scrubbing liquor pump, flue gas fan, scrubbing liquor recirculation tank, soda ash/water reaction tank, and spray tower are in the appendix.

Material selection was often dictated by consideration of corrosion resistance. The materials of construction of all major pieces of equipment are listed in the following table.

Materials of Construction

Equipment Item	Material
Scrubber	
Shell	316L Stainless Steel
Piping	"
Demister	"
Distributor	"
Nozzles	Brass
Pump	
Casing	Natural Rubber-Lined Carbon Steel
Impeller	Hard Nickel Alloy
Fan	
Casing	Natural Rubber-Lined Carbon Steel
Impeller	Inconel 625
Recirculation Tank	A36 Carbon Steel with Fiberglass Reinforced Polyester Lining
Fired Heater	Carbon Steel
Soda Ash/Water Reaction Tank	316L Stainless Steel

4. Discussion of Design

Several types of scrubber were considered for this system, but the vertical spray tower was decided to be the best option. Packed towers and tray towers were eliminated from consideration because of their tendencies for fouling and corrosion with the particulate-loaded, corrosive liquor used by this process. Venturi scrubbers were rejected because of difficulty in modeling mass transfer in scaled-up units. It was also difficult to quantify mass transfer in fluidized bed scrubbers, removing them from consideration. Horizontal spray scrubbers were seen to offer no advantage over vertical towers, but caused concern that flue gas could channel through an unsprayed portion of the cross section.

The rate of mass transfer depends upon the solubility of sulfur dioxide in the scrubbing liquor. However, no better value for this solubility than the Henry's Law constant for sulfur dioxide in water was found. The salts in the liquor would be expected to reduce sulfur dioxide solubility. The sensitivity of the mass transfer rate to variation in solubility was tested by decreasing the solubility value by a factor of 1000. This caused a mere 3 percent change in the required liquid rate, and so an exact value for solubility was not considered necessary.

The partial pressure of oxygen in the scrubber was too low and the concentration of sulfite ions in the liquor too high for oxygen absorption to be significant. The effect of dissolution and reaction of oxygen was not considered.

The minimum liquid rate required for mass transfer was multiplied by a factor to account for entrainment, drop size distribution, drop evaporation, and drop coalescence to a film on the tower walls. This factor was arbitrarily set equal to 5 after a study of the data presented by Pigford and Pyle (ref. 6)

The flue gas exiting the spray tower was assumed to approach saturation at a temperature approaching the temperature of the scrubbing liquor. This fixed the make-up water flow rate. In order to eliminate the condensing vapor plume which this gas would produce on contact with the atmosphere, the design calls for heating it 100°F above its dew point. Flexibility should be allowed for in the heat duty of the heater should 100-degree heating prove to be insufficient or excessive.

More information is needed regarding the quantity of particulates in the flue gas. If solid matter accumulates in the recirculation tank, it may be necessary to provide agitation for this vessel.

5. Costs

Capital costs were calculated using the method of Guthrie (ref. 2) and adjusting for inflation using Marshall and Swift's index. The capital cost breakdown for the sections of the scrubbing system that we designed are given in the table that appears on the following page.

Capital Cost

Item	Installed Cost (1989)
Scrubber	$1,950,000
Pump	62,693
Fan	395,861
Tank	129,555
Heater	361,819
Reactor Tank	76,981
	2,976,909
Contingency and Contractor Fee, 18%	535,844
Total	$3,512,753

Utility costs are given in the following table. Cost data are from the Chemical Engineers' Handbook (ref. 5), corrected for inflation using the Marshall and Swift Index.

Utility Costs

Electricity	$108,031/year
Water	11,043
Natural Gas	244,045
Total	$363,119/year

Conclusions

The sulfur emission regulations for this plant can be met by using the dual-alkali scrubbing system detailed in this report. A spray tower 30 feet in diameter with 50 feet of gas-liquid contacting should provide more than adequate surface for mass transfer. A liquor rate of 2300 gpm is required for assurance of adequate absorption. A fan, a direct-fired heater, and the existing stack are required for disposal of the scrubbed flue gas. The instrumentation and control system is designed to allow turndown of the system capacity. Liquor feed rate, spray tower pressure, and liquid level in the tower are controlled. Provision is made for flue gas sampling, but facilities for testing are left to the client. The system's present cost is $3.5 million, installed. It will cost $360,000 per year for utilities to run.

Details of the particulate removal system and the scrubbing liquor regeneration loop are left to the principal contractor, Technical Specialties.

References

1 HS Fogler, The Elements of Chemical Kinetics and Reactor Calculations, Prentice-Hall Inc., Englewood Cliffs, New Jersey, 1974.

2 KM Guthrie, "Capital Cost Estimating," Chemical Engineering, March 24, 1969.

3 L Kroop, Master's Thesis, Chemical Engineering Department, University of Michigan, 1981.

4 JJ Matley, "Cost File," Chemical Engineering, January 26, 1981.

5 RH Perry and CH Chilton, editors, Chemical Engineers' Handbook, 5th Edition, McGraw-Hill, New York, 1973.

6 RL Pigford and C Pyle, "Performance Characteristics of Spray Type Absorption Equipment," Industrial and Engineering Chemistry, July 1951.

7 MR Tek and JO Wilkes, "Fluid Flow and Heat Transfer," Department of Chemical Engineering, University of Michigan, 1974.

8 RC Weast, editor, CRC Handbook of Chemistry and Physics, 59th Edition, CRC Press, Boca Raton, Florida, 1978.

9 CJ West, editor, The International Critical Tables, Vol. 3, McGraw-Hill, New York, 1933.

10 RN Wimpress, "Rating Fired Heaters," Hydrocarbon Processing and Petroleum Refiner, October 1963.

Appendix

MASS BALANCE

Flow Rate lb moles/hr

Stream	Total	CO_2	H_2O	SO_2	N_2	O_2	$NaHSO_3$	Na_2SO_3	NA_2SO_4	$NACl$	$NaNO_3$	Na_2CO_3	Inserts
F1	24,994.3	3645.4	1539.8	36.5	19,113.5	659.1	—	—	—	—	—	—	4,258 lb/hr
F2	2,450.0	—	2388.7	—	—	—	—	22.8	35.5	2.6	0.36	—	—
F3	31,397.9	3645.4	7970.0	9.92	19,113.5	659.1	—	—	—	—	—	—	—
F4	2,299.5	—	2227.4	—	—	—	12.9	17.3	36.9	4.6	0.36	—	1,281 lb/hr
F5	6,268.7	—	—	—	—	—	—	—	—	—	—	—	—
F6	*	—	—	—	—	—	—	—	—	—	—	*	—

*Soda ash flow rate was not determined because of inconsistencies in the pilot plant data.

Figure A-1 ❏ SPRAY TOWER DESIGN

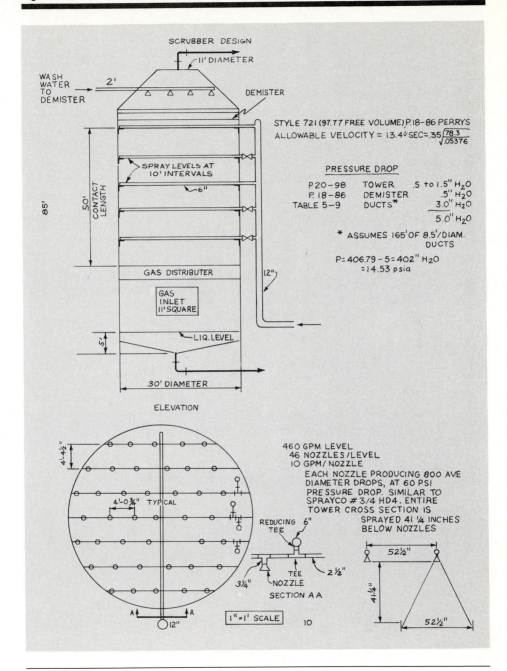

SCRUBBER DESIGN

11' DIAMETER

WASH WATER TO DEMISTER   2'

DEMISTER

STYLE 721 (97.77 FREE VOLUME) P.18-86 PERRYS
ALLOWABLE VELOCITY = 13.4↓/SEC=.35$\sqrt{\frac{78.3}{.05376}}$

SPRAY LEVELS AT 10' INTERVALS

6"

50' CONTACT LENGTH

85'

PRESSURE DROP

P.20-98	TOWER	.5 to 1.5" H₂O
P.18-86	DEMISTER	.5" H₂O
TABLE 5-9	DUCTS*	3.0" H₂O
		5.0" H₂O

* ASSUMES 165' OF 8.5'/DIAM. DUCTS

P= 406.79 - 5 = 402" H₂O
= 14.53 psia

GAS DISTRIBUTER

12"

GAS INLET 11" SQUARE

LIQ. LEVEL

5'

30' DIAMETER

ELEVATION

4'-4½"

4'-0¾" TYPICAL

460 GPM LEVEL
46 NOZZLES/LEVEL
10 GPM/NOZZLE
EACH NOZZLE PRODUCING 800 AVE
DIAMETER DROPS, AT 60 PSI
PRESSURE DROP. SIMILAR TO
SPRAYCO #3/4 HD4. ENTIRE
TOWER CROSS SECTION IS
SPRAYED 41 ¼ INCHES
BELOW NOZZLES

REDUCING TEE   6"

3/4"   TEE NOZZLE   2½"

SECTION A A

52½"

41¼"

52½"

A   A

12"

1"=1' SCALE   10

<u>PUMPS</u>

Page 1 of 1
By   D.J. Thomas
Date: 2-16-88   .

Item No.   P-1
Name: Scrubbing Liquor Pump
No. Req'd   1
<u>Function</u>:

<u>Type</u>:
  Pump <u>CENTRIFUGAL</u>
  Drive _____

<u>Materials Handled</u>:
  Name of Material: <u>SCRUBBING LIQUOR</u>
  Pumping temp., °F <u>135</u>
  Specific gravity at t°F <u>1.26</u>
  Viscosity at t°F <u>1.35×10^{-3} lb/ft^2</u>
  Normal Flow: lb/hr <u>1,444,000</u>   GPM <u>2280</u>
  Specified capacity, GPM <u>2300</u>
  Suction vessel (origin) <u>RECIRCULATION TANK</u>
  Discharge vessel (destination) <u>SCRUBBER TOWER</u>

<u>Head Data</u>:

	Suction Side		Discharge Side		
	<u>psia</u>	<u>ft.</u>	<u>psia</u>	<u>ft.</u>	
Press. over liquid	15		75	110.34	
Static head				95.00	
Friction drop				2414	combined
Velocity head				1.32	
Total head			114	209.1	

Driver Power, HP:
  Normal calcd.   20L
  Maximum calcd.   250   @ 20% excess capacity
  Specified   _____

COMPRESSORS

Page 1 of 1
By: D.J. Thomas
Date: 2-16-88

Item No.:   C-1
Name: FLUE GAS FAN
Number Required: 1

Function:

Type:
  Compressor:
  Driver:

Process Requirements:

  Material Compressed:

Component	Wt.%	Mol%	MW
CO_2			44
H_2O			18
SO_2			64
N_2			28
O_2			32

Flow rate:
  lb/hr
  Mol/hr   25,000
  SCFM (60°F, 14.7 psia)   158,152
  Mean spec. ht. ratio, c_p/c_v
  Mean MW   ~30

	Suction	Discharge
Pressure, psia.	14.5	14.7
Temperature, °F	~150	~150
ACFM	185,083	185,525

Compressor Design:

No. of Cylinders		Volumetric efficiency, %	
		CFM at suction	
Compression ratio		conditions	185,083
Stroke, in.		Ideal gas power, HP	148
Cylinder diam., in.		Compression eff., %	55
Speed, RPM		Shaft power, HP	265
Displacement, CFM		Calc. Driver Power, HP	
Clearance, %		Spec. Driver Power, HP	325 @ 20%

excess capacity

TANKS

Page 1 of 1
By: D.J. Thomas
Date: 2-16-88

Item No.:   T-1
Name: SCRUBBING LIQUOR RECLAMATION TANK
Number Required: 1

Function:

Type:

Process Data:

Material handled: SCRUBBING LIQUOR
Composition:

Component	Wt. %	Mol %
H_2O	79.67	
$NaHSO_3$	7.67	
Na_2SO_3	14.32	
Na_2SO_4	10.42	
NaCl	.32	
$NaNO_3$	.06	
Inerts	2.54	

Temperature, °F: Operating pressure, psig.:
  Normal   135   Normal   0
  Maximum   140   Maximum   0
  Sp. G. at t °F 1.26

Flow rate:
  GPM   2386
  lb./hr.   1,505,000

Size Data:
  Residence time   ~30 min
  Volumetric capacity   74,592 gal

Design Specifications:
  Material of construction   A36 CS, FIBERGLASS REINFORCED POLYESTER
                              LINING
  Position   VERTICAL
  Design temp., °F.   135
  Design, pressure, psig.   0
  Diameter, ft.   23
  Length, ft.   24
  Wall thickness, in.   ⅛″

Corrosion allowance, in.   NONE _____

Insulation req'd (Yes or No) No _____

   For what   _____

SODA ASH/WATER REACTION TANK

Material Handled:
   Soda ash
   Water   155gpm

Temperature
   Normal   70°F
   Maximum   100

Size Data
   Residence time   30 minutes
   Volumetric capacity   4700 gallons

Design Specifications
   Material of construction   316L stainless steel
   Position   Vertical
   Design temp.   70°F
   Design pressure   0 psig
   Diameter   10 feet
   Height   8 feet
   Wall thickness   .125 inch
   Corrosion allowance   none
   Insulation   none

Long Formal Report

FUNCTIONAL
SPECIFICATION

OLDSMOBILE DIVISION
GENERAL MOTORS CORPORATION

FASCIA SYSTEM

Prepared by
DIGITAL EQUIPMENT CORPORATION
Software Services

Great Lakes District
21333 Haggerty Road
Novi, Michigan 48050

November 15, 1981
Version 1.0

Summary Table of Contents

APPENDIXES

A   TERMS AND CONDITIONS

B   SOFTWARE PRODUCT DESCRIPTIONS

C   CHANGE ORDERS

D   EXAMPLE DOCUMENTATION

E   SAMPLE ACCEPTANCE PLAN

F   INDEX

FIGURES

SECTION ONE   INTRODUCTION AND OVERVIEW

CHAPTER 1   INTRODUCTION AND OVERVIEW

CHAPTER 4   SYSTEM REQUIRMENTS

CHAPTER 5   SYSTEM FEATURES

CHAPTER 6   SYSTEM CONVENTIONS

SECTION FOUR   KNOWLEDGE

CHAPTER 7   SYSTEM INPUTS

CHAPTER 8   SYSTEM OUTPUTS

CHAPTER 12 DOCUMENTATION

CHAPTER 13 TRAINING

SECTION SIX CONDITIONS

CHAPTER 14 SYSTEM ACCEPTANCE

CHAPTER 15 COMMUNICATIONS

CHAPTER 16 SPECIFICATION CHANGES

CHAPTER 17 RESPONSIBILITIES

CHAPTER 18 PROPOSAL BIDDING INFORMATION

APPENDIXES

APPENDIX A   TERMS AND CONDITIONS

APPENDIX B   SOFTWARE PRODUCT DESCRIPTIONS

APPENDIX C   CHANGE ORDERS

APPENDIX D   EXAMPLE DOCUMENTATION

APPENDIX E   SAMPLE ACCEPTANCE PLAN

APPENDIX F   INDEX

Partial Answer Key

❑ EXERCISE 3-2

(one version; other versions are possible)

Dear Mr. Poirot:

I'm writing to follow up on our recent meeting in Lyons and to request more information about your new MAX synthesizer. It may be just what my company needs. Before putting in an order, though, I need to have more details. For example, how does it handle in-between pitch levels? How does it handle off-key problems? Sometimes we have many off-key problems, with the F_1 signal fluttering wildly and the F_2 signal being even worse. If MAX can solve these kinds of problems, my company will definitely be interested.

Sincerely,

❑ EXERCISE 4-1

A How has a 40 Wh/bl energy density been achieved? How easy is it to achieve such a density? How has a $50 per kWh cost been achieved? Does this cost include development and setup? What are the comparable figures for the latest batteries? What were the "major problems with bearings and energy conversion"? Are there any other problems involved in the use of flywheels? Specifically, what kind of research is being conducted by each of the people listed in the Summary?

C   What "problems of competition" are there? What are the "social advantages of larger membership"? What is "walking distance"? How slightly do rent prices vary? What will the annex cost? How much does living in the fraternity house cost? Where does that "excess revenue" come from, exactly? How do you figure that "at least six members" would have to live in the annex? How does the management at Plaza Apartments feel about having a fraternity annex there? If they approve of the idea, would they be willing to lease space at long-term reduced rates? How would this move allow Alpha Epsilon Zeta to compete financially with larger fraternities, dormitories, and co-ops?

❑ EXERCISE 4-2

A   Character and credentials.

C   Mainly an appeal to logic and reason, but the phrase "stockpiled like scarce materials against the day when it per chance may be needed" is an appeal to emotion.

❑ EXERCISE 4-3

Sample 2: Criteria case. The main criteria are

1   Effectiveness: information must be centralized so as to allow quick job status reports, optimal inventory levels, efficient job shop scheduling, etc.

2   Affordability: it should be cost-efficient.

3   Desirability: the system must be able to be phased in without disrupting company operations unduly.

4   Technical feasibility: the system should be technically feasible without having to invest in extremely expensive equipment or personnel.

❑ EXERCISE 5-1

A   "Aluminum is a desirable material . . . cooling systems. However, the exact degree of corrosion, especially in environments where chloride ion solutions are present, is presently unknown. Aluminum itself is known to undergo localized corrosion, or pitting, when exposed to chloride ion solutions. But aluminum surfaces are always covered with a passive oxide or hydroxide film, and it is not known how chloride ions interact with this film. This interaction is currently being studied . . . tests."

C   "The ceaseless health warnings issued by the media—regarding DDT, PCB, polyvinyl chloride, nitrous oxides, fallout, x-rays, etc.—suggest that none of us has a chance to live past the age of 25. But most of us do. Thus, there is something spurious about these warnings. It is because of the uncertainties and complexities of the concept of safety that crises of this type often arise."

❑ EXERCISE 5-3

1 "an invasion of privacy": this would probably be of primary concern to students, perhaps also to administrators sensitive about possible legal ramifications.

2 "poorly designed, yielding poor results": this would probably be of primary concern to counselors, perhaps also to instructors of the chemistry and math courses involved, and to students.

3 "wastes valuable student time during orientation week": this would probably be of primary concern to administrators, especially those who supervise orientation week; it might also be of concern to students wanting to make the best use of their time during this week.

❑ EXERCISE 5-4

A Question/task: "This interaction is currently being studied . . . pitting corrosion."
Purpose: "The purpose of this report . . . tests."

C Question/task: [Given all this dangerous technology,] "how do we ever manage to survive past the age of 25?"
Purpose: Neither stated nor implied.

❑ EXERCISE 5-5

A A term: We don't want to have slowdowns and repairs in the incineration area of the Sewage Treatment Plant.
but (B term:) "Slowdowns and necessary repairs have occurred . . . ."

B A term: At present we know of no life on Mars.
but (B term:) "Of all the extraterrestrial bodies . . . Mars is the most plausible habitat for extraterrestrial life in the solar system."

❑ EXERCISE 9-1

E Figure D-1 was designed by a student, Naoyesu Seki, for Table 9-3. It emphasizes the fact that Yougan Dairy's growth is occurring mainly through growth in diversified products.

❑ EXERCISE 10-1

The opening sentence sounds too stuffy and is also too vague about what kind of position she's applying for. The first subparagraph is "not directly related

Figure D-1 ❑ GROSS SALES FOR YOUGAN DAIRY PRODUCTS, 1985–89

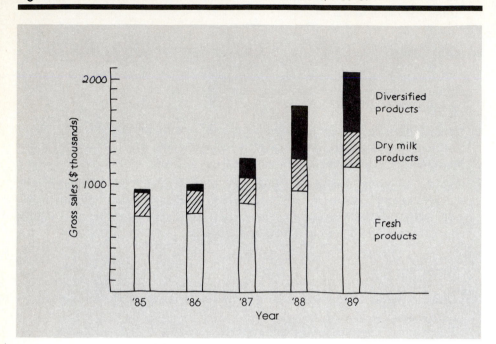

to (her) field of study and interest" and therefore should be relegated to a less prominent position. The second subparagraph should be promoted to initial position, and it should say more about the writer's accomplishments as a student. The third subparagraph could be deleted.

(Résumé) The second sentence under Career Plans should be more specific. Her educational background and accomplishments need to be expanded on and given more prominence. The Work Experience section should be demoted. Reverse chronological ordering should be used for listing work experience and education.

❑ EXERCISE 12-1

A *Summary*: We subjected each of the two samples to two different neuron activation analyses: short half-life and long half-life. The combined results for the six major elements in the samples are as follows:

(give table from Results section)

❏ EXERCISE 22-1

A i

B i

❏ EXERCISE 22-2

B   At the time of its explosion, Mount St. Helens was probably the most closely watched volcano in the world. Scientists from the Geological Survey's Hawaiian Volcano Observatory, the University of Washington Geophysics Program, and many other government and university research groups had come to study it. They flew over it in helicopters and landed cautiously on its flanks to install instruments and collect data. They monitored its behavior with a network of seismographs, tiltmeters, magnetometers, and other devices. They studied the deformation of the north flank with photogrammetric techniques and analyzed changes in the chemical composition of the gases vented from the crater at the summit. Yet in spite of this impressive array of scientific expertise and technological resources, both the suddenness of the explosive eruption and its devastating force took the scientists by surprise. (H. Lansford, "Vulcan's Chimneys: Subduction-Zone Volcanism," *Mosaic 12*(2), National Science Foundation, March-April 1981, p. 46)

D   Until 1922, no one knew how a signal crosses the junction between one nerve cell and another. That year Otto Loewi, a German physiologist, made the far-reaching discovery that nerve signals are transmitted across these synapses by chemicals. Loewi called his chemical neurotransmitter *Vagusstoff* ("vagus stuff"), for the cranial neuron with which he worked; it has since been identified as acetylcholine. (E. Edelson, "The Neuropeptide Explosion," *Mosaic 12*(3), National Science Foundation, May-June 1981, p. 15)

❏ EXERCISE 22-3

A   Pattern: Listing
    Features: 1   parallelism
                2   descending order of importance
                3   connectives (*and, also, indeed*)
C   Pattern: General-to-particular, extended definition
    Features: 1   decreasing frame of reference
                2   elements of a definition (term, class, differential)

❏ EXERCISE 22-4

The original paragraph divisions were at line 7 (. . . *Helens./Few* . . .) line 17 (. . . */Paricutin* . . .), line 20 (. . . *America,/A second* . . .), line 25 (. . . *bases./ Shield* . . .), line 32 (. . . *meters./Lava* . . .), and line 34 (. . . *vent./The* . . .).

❑ EXERCISE 23-1

(suggested answers; others are possible)

A   The building is 140 feet long, 78 feet wide, and four stories high.

C   The widest employment of DDT is in the control of insects of public importance (as a mosquito larvicide, as an antimalaria spray, and as an antityphus dust).

E   In this particular case the most important variables are the following:
1   pressure and temperature of the boiler
2   type of fuel
3   amount of oxygen
4   temperature of fuel

G   We recommend the purchase of the New Orleans heat exchanger because (i) it can be obtained for $25,000, a savings of 80% over the new cost, (ii) it could increase production by 28% if needed, 8% higher than requested, and (iii) it will help recoup losses incurred in our benzene plant.

I   This buoy terminal has three components that rotate as a single unit:
1   A Rotating Mooring Bollard which allows floating mooring lines to lie on the lee side when weather conditions become bad.
2   A Rotating Cargo Manifold which carries products from the terminal-to-tanker hoses to the multiproduct distribution unit in the center of the buoy.
3   A Rotating Balance Arm which not only maintains the buoys on an even keel but also provides an accommodation ladder.

❑ EXERCISE 23-2

(suggested answers; others are possible)

A   My present occupation is repairing printing machines, duplicating machines, and typewriters.

C   This technology consists of three methods of scrambling: coupling the light source and optical fiber in low-order mode; splicing, bending, and tightening the fiber near its connectors; and installing the scrambler into the existing fiber optic in intervals of 1 km along the route.

❑ EXERCISE 23-3

(suggested answers; others are possible)

A   In order to meet the job requirements it became clear that a microcomputer would be needed. Microcomputers are very compact and portable. They are easy to program and easy to operate. They can be used for a wide variety of data processing operations and are capable of storing large amounts of data both inside and outside the computer itself. Most importantly, they are very inexpensive compared to other large-scale systems.

C   In order to reduce neutron leakage and at the same time get better and sharper pictures in this experiment, we suggest decreasing the diameter of the collimator. At present, considerable neutron leakage is occurring because the collimator is not very close to the x-ray film. If we minimize this distance, we can not only reduce the leakage but also improve our picture resolution, since it will make the neutron beams maximally parallel.

❑ EXERCISE 24-1

A   The main conveyor line in Building 9 is used to transport three of Construction Equipment's products. This line has three stations, which are located at various positions along it. At each of these stations, a particular worker is assigned to one of Construction Equipment's three products. Each time a worker's assigned product passes by, the worker is responsible for lifting it off the main conveyor line and placing it in the appropriate chute, which transfers it to another conveyor line.

❑ EXERCISE 25-1

(suggested answers; others are possible)

A   *Stage 1*: Labor unions are permanent and powerful institutions in America. Unions are based in a labor movement which has existed in the United States for about 180 years. In the past 40 years, labor unions have become an integral part of the workplace. Unions represent more than 19 million workers. Labor unions influence and determine the wage rates, working hours, and working conditions of both unionized and nonunionized industries.

Stage 2 [after considering the material from Chapter 26]: Labor unions are permanent and powerful institutions in America. They are based in a labor movement which has existed in the United States for about 180 years. In the past 40 years, they have become an integral part of the workplace. They now represent more than 19 million workers and influence and determine the wage rates, working hours, and working conditions of both unionized and non-unionized industries.

C   Several recent developments have encouraged a reevaluation of wind-powered vessels for the American merchant fleet. The most significant of these developments is the sharp rise in the cost of energy required to drive a fuel-powered ship. A second development is the improvement of technology associated with wind-powered ships.

D   General Appraisal of the EPA Plan

Version 1: The plan prepared by the EPA lacks several characteristics. First, it does not clearly outline program priorities, nor does it relate priorities to overall program goals. Second, it does not provide a clear planning process or guidelines for future updates of the plan. Finally, it does not provide a clear rationale for the budgetary directives. For example, the plan offers no basis for the dominant expenditure on developing control technology over the 5–year period covered by the plan.

Version 2: Several essential characteristics are lacking in the EPA plan. First, program priorities are not clearly outlined nor are they related to overall program goals. Second, the planning process is vague, and no guidelines are offered for future updates of the plan. Finally, the budgetary directives have no clear rationale. For example, the plan offers no basis for the dominant expenditure on developing control technology over the 5–year period covered by the plan.

❏ EXERCISE 26-1

(suggested answers; others are possible)

A   In 1992 the Cleveland Engine Plant will begin making diesel engines along with the gas engines already in production. Since there is only one area for machining pistons, both gas pistons and diesel pistons will have to be machined there.

C   Electric cars have a much lower maximum speed than gas cars, and are barely fast enough to maintain freeway speeds. This causes problems with utilization of electric cars on freeways, almost limiting their use to residential streets. If electric cars are to be used on freeways, their maximum speed must be increased.

❏ EXERCISE 26-5

(suggested answers; others are possible)
Minimalist approach, eliminating all words which duplicate the meaning of another word.

A   We are reevaluating the budget.

C   We estimate system cost at $1.2 million.

E   We recommend the continued use of the double-alkali reactant.

G   At our meeting of 9/4/89, we discussed the numerous complaints about the Hudson elevator.

Rhetorically sensitive approach which eliminates wordiness but retains the emphasis of the original.

A   We are presently reevaluating the budget.

C   We estimate that the system would cost about $1.2 million.

E   We recommend the continued use of the double-alkali reactant.

G   At our meeting of 9/4/89, we discussed the fact that there have been numerous complaints about excessive wait times for the Hudson elevator.

❏ EXERCISE 27-1

policeman → police officer
man-hours → person-hours
old wives' tale → traditional lore, superstition
errand boy → messenger, helper, "gofer"
male nurse → nurse

❏ EXERCISE 27-2

1   If we work together, . . .

3   We would be negligent if we didn't send at least a letter expressing our sorrow.

5   It looks like we'll have to start all over.

7   Do you think we can deal with this?

9   There are so many bureaucratic regulations involved, it may take us months to get this project going.

❏ EXERCISE 27-3

1   discard, reject, abandon, throw away, get rid of

3   succor, assist, aid, help, lend a hand

5   confidant, acquaintance, friend, pal, sidekick

❏ EXERCISE 28-2

B   There are at least 24 errors of spelling, punctuation, or grammar in this letter.

❏ EXERCISE 29-1

rocket—C	program—C
inflation—U	satellite—C
carbon dioxide—U	ecology—U
friction—U	radar—U
atom—C	motor—C
aluminum—U	book—C
machine—C	sodium nitrite—U

❏ EXERCISE 29-2

gravity—U; acceleration—U; body—C; weight—U

❏ EXERCISE 29-3

A   In space the sun is always shining, so there is no problem with storing electricity. On earth, however, the sun shines only half the time in good weather and not at all in bad. Thus, electricity must be stored for sunless periods. Currently this is done with lead-acid storage batteries, similar to those used in automobiles. A day's electricity for an average single-family house can be stored in batteries occupying the space of a closet; a row of such "closets" in the basement stores power for sunless periods.

C   Vectors are quantities that have both a magnitude and a direction. Examples of physical quantities that are vectors are force, velocity, and acceleration. Thus, when one states that a coordinate system attached to the earth, one is specifying a vector quantity, velocity, with a magnitude (100 kilometers per hour) and a direction (north).

Scalars are quantities that have a magnitude only. Examples of physical quantities that are scalars are mass, distance, speed, and density. Thus, when one states only the fact that a car is moving at 100 kilometers per hour one has specified a scalar, speed, since only a magnitude (100 kilometers per hour) is given (that is, no direction is specified).

❏ EXERCISE 29-4

(sample answers)
amperes of current; joules of work; measurements of temperature; causes of pollution; experiments in electrolysis; forms of combustion; volts of electric potential.

❑ EXERCISE 29-5

A 1 Water (U), container (C)
   3 communication (U), lifeblood (U), organization (C)
   5 changes (C), finance (U)
B 1 people (C), fitness (U), strength (U), looks (U)
   3 decisions (C), earnings (U), effort (U)
   5 year (C), physicist (C)
C 1 Mechanics (U), forced (U)
   3 Mass (U), mechanics (U), property (C), language (U)
   5 Pressure (U), dynes (C)

❑ EXERCISE 30-1

the 1950s, the ground, the same radar zone, the sky, the atmosphere.

❑ EXERCISE 30-2

"One of the most challenging engineering problems of the century" . . .
"Dwarfing anything constructed in the past for the study of the universe" . . .
"The largest movable land-based structure ever constructed in the world."
      These were only three of the descriptors enthusiastically applied to the
U.S. Navy's attempt in the late 1950s to build a huge radio telescope with a
600-foot-diameter antenna that would be fully steerable. The structure, which
would have been twice as large as any fully steerable telescope built since,
would have stood more than 750 feet above the ground on a 7-acre foundation
near Sugar Grove, West Virginia. Unfortunately for astronomers, however,
in 1961–62 original cost estimates of $52 million were reevaluated and
reestimated to be between $200 and $300 million for the future. As a result,
the so-called Big Dish was abandoned. Now all that remains of that dream to
see more of the sky and further into space than ever before is the concrete
foundation for the telescope tracks and the pintle bearing.

❑ EXERCISE 30-3

the days (FM), the frail, canvas-covered aircraft (G-FM), the human body
(G), the now relatively frail human occupant (G-PM), these important forces
(PM), the fundamental principles (FM), the physiology (FM), the history (FM),
the physiological effects (G-FM implied), the clinical response (G-FM), the
flight surgeon (G), the conventional terminology (FM), these forces (PM).

❑ EXERCISE 30-4

A It is easy to be aware of the revolution brought by the internal combustion
   engine. The effects of the revolution are part of our world. Most of us
   own a car and know something of a piston, a cylinder, and horsepower.

It is also easy to be aware of the revolution brought about by the electronic tube. Most of us own radios and television sets and know something about waves and electrical interference. Unfortunately, it is harder to be aware of the chemical revolution because its products are hidden. Yet this revolution is important, because it is the basis for the other revolutions. An automobile could not run without fuel, which is a chemical. A car or an electronic tube could not be built without metal, fabric, adhesive, glass, paint, and plastics which are chemicals or the results of chemical processes. If we are to understand our world, we must be aware of the chemical revolution.

B   The classic treatment of rotational physics usually found in an engineering textbook and in an advanced text on gyrodynamics is ordinarily quite rigorous in its mathematical treatment of the subject. This paper is intended to present the subject in such manner, first, as to relieve physicians and medical officers of the labor required to gain rigorous insight into the subject and, second, to eliminate the nonessentials associated with classic developments which are not pertinent to the work at hand.

      To this end, the treatment used in this paper will be a largely intuitive, extensively graphical one, and the use of mathematics beyond algebra will be studiously avoided.

❑ EXERCISE 30-5

B   By far the most complete wind data available in the United States reside in the records of the U.S. Weather Bureau (ESSA) stations, which number more than 1000. Many of the stations have been in operation for one or more decades, and ESSA has provided an excellent record-keeping center at Asheville, North Carolina, so that data can be retrieved easily.

      A typical "surface" wind observation is usually obtained from a slow-moving cup anemometer and vane assembly, mounted in a well-exposed location about 30 feet above the ground or a building structure. A vast majority of the sites are at airports, and urban observations are comparatively rare. The standard observation procedure is to note the indicated wind speed and direction for a brief period each hour, and record them to 10 degrees (formerly 16 compass points) and the nearest mile per hour. Normally a contacting device is associated with the anemometer, from which one can obtain the time required for one mile of wind flow to pass the instrument. Observations called PIBALS or RABALS are taken at many of the stations, using either visual or electronic tracking of a rising balloon several times per day to determine the variation of wind with height. The data are processed to reflect the wind at 1000-foot intervals above sea level.

C   On December 17, 1903, Orville and Wilbur Wright successfully achieved sustained flight in a power-driven aircraft. The first flight that day lasted only 12 seconds over a distance of 37 meters (120 feet). This is about the length of the *Space Shuttle Orbiter*. The fourth flight of the day (and the longest flight) traveled 260 meters (852 feet) in 59 seconds. The initial notification of the event to the world was a telegram to the Wrights' father.

Sixty-six years later, a human being first stepped on the lunar surface. An estimated 500 million people throughout the world saw the event on television or listened to it on radio as it happened. This was surely a/an historic event.

Historic events are spectacular. The space program, however, has always been much more than a television spectacular. Today, space transportation is working in many ways for us all, and we have come to expect this.

A whole new era of transportation has come into being in the 1980s with the advent of the *Space Shuttle*. As a transportation system to Earth orbit, it offers the workhorse capabilities of such earthbound carriers as trucks, ships, and airlines. It is as vital to the nation's future in space as the more conventional carrier of today is to the country's economic life and well-being.

E   The provisions of the Clean Air Act Amendments of 1970 and the Federal Water Pollution Control Act of 1972 include requirements to limit the emission of pollutants from various points of discharge such as the automobile tailpipe and wastewater effluent. It is fairly obvious that such measures are required if we are to maintain our air and water in a sufficiently clean state for protection of the public health and welfare. What is not obvious is the exact extent to which it is necessary to limit these discharges in order to reach sufficient purity. As long as costs rise exponentially with degree of purification, we may expect only enough public pressure to control discharges for adequate protection and no more. This economic fact underlines the critical necessity for the development of a valid environment measurement system.

There are two fundamental elements to such a system. First, we must establish as accurately as possible the dose-response relationships so that the most proper ambient air and water quality standards can be set. Secondly, we must establish the relationship between pollutant concentrations at the point of discharge and at the point of human contact. Basic to this/the measurement system are requirements that:

1   the health effect can be measured with sufficient accuracy.
2   an accurate model is available.
3   pollutant measurements at ambient and source concentrations are internally consistent.

The part of the measurement system with which we at the National Bureau of Standards have been primarily concerned is the third of these. In the discussion that follows, it is presumed that the other two require-ments are met.

❏ EXERCISE 31-1

B had, did, was, is, are, provided

C was, has been, has complained, is or has been, believes, is, is, seems, is

D have been, has been, have been, are, differentiated, demonstrated, is

❏ EXERCISE 32-1

(answers chosen by the original authors; certain others are possible)

A must, may, should, should

B may, should, should, should, must, may, may, should

❏ EXERCISE 32-2

(best answers, in the sense of being the most accurate; others are possible)

A will probably not

B must

C will probably

❏ EXERCISE 32-4

(answers chosen by the original authors; certain others are possible)

A should, may, should, can, might

B can, may, must, might, can

❏ EXERCISE 33-1

A Any celestial body that emits light waves may be expected to emit radio waves because both wave forms are part of the same electromagnetic spectrum.

B The concepts which underlie these methods were known previously but could not be economically utilized.

C OK as is.

❏ EXERCISE 33-2

A OK as is.

B The flow of current through a wire conductor is equal to the number of electrons which (or that) pass through its cross-sectional area in a particular unit of time.

C   Thermodynamics is a science <u>which</u> (or <u>that</u>) deals with heat and work and those properties of substances <u>which</u> (or <u>that</u>) bear a relation to heat and work.

D   A property can be defined as any quantity <u>which</u> (or <u>that</u>) depends . . .

E   OK as is.

❏ EXERCISE 33-3

A   A pump is a device that induces fluid flow through pipes . . . .

B   Pulley systems provide a mechanical advantage which is equal to . . . .

C   . . . into the development of fuels and oxidizers which will give . . . .

D   . . . that people are now accustomed to in air-conditioned cars.

❏ EXERCISE 33-4

A   . . . studies <u>that</u> . . .

B   . . . studies <u>that</u> . . .

C   Cannot be changed.

D   Cannot be changed.

❏ EXERCISE 33-5

A   . . . something smaller than the atom.

B   "Water treatment" is a term traditionally applied to . . . .

C   In many power plants the air used for combustion . . . .

❏ EXERCISE 33-6

(More context is needed to make an accurate determination. The following are the choices made by the original authors.)

A   . . . is a giant engine attached to an electrical generator.

B   . . . through a throttle valve so designed and so located that . . . .

C   . . . The momentum change associated with this increase . . . .

D   Leave as is.

E   . . . for making the computations involved.

❏ EXERCISE 33-7

A   The model allows those <u>using</u> it to compress real time . . . .

B   Some of the liquid refrigerant <u>emerging</u> from the condenser . . . .

C   Cannot be changed.

D   Tankers <u>plying</u> traditional shipping lanes in ocean waters . . . .

E   . . . the potential dangers <u>attending</u> the disposal . . . .

❏ EXERCISE 33-8

A   . . . to a vocabulary <u>consisting of</u> an infinite sequence of natural numbers . . . .

B   Leave as is.

C   . . . A person <u>wishing</u> information on a topic chooses a set of identifiers <u>which best</u> <u>describes</u> the topic . . . and to select papers <u>having</u> sets of identifiers <u>containing</u> one or more identifiers common to the topic identifiers . . . .

❏ EXERCISE 33-9

A   . . . the same value they had at the beginning.

B   . . . the research attention they are getting.

C   . . . of the air pollution problem we now face.

❏ EXERCISE 34-1

(answers)

A   Indeed, In others words, For example, Thus.

B   For example, However, Thus, first of all, for example, Since, On the other hand, Since.

❏ EXERCISE 35-1

B   polymer semiconductor junctions

C   a biomechanics research program

D   building project cost control

❑ EXERCISE 35-2

(suggested answers, all taken from the actual technical reports).

A At the Administrator's request in December 1979, an exhaust emission test program was initiated to gather data on approximately 20 high-mileage diesel engine passenger cars.

B The 360X diesel engine was selected for the purpose of determining its emission deterioration over the in-use life of the vehicle.

C In an attempt to predict generator overheating more accurately, I investigated alternative monitoring methods based on electrical insulation degradation at elevated temperatures.

❑ EXERCISE 35-3

(possible answers; correct answers can vary according to context)

B surveys of (the results of) tests of the breath of drivers stopped by police alongside the road for suspicion of drunk driving

C a center where entertainment products for automobile travel (car radio, tape player, etc.) can be purchased

D the removal of insoluble particles from toluene

E a system for disposing of waste materials on land

❑ EXERCISE 36-1

bionic—ADJECTIVE
duplicity—NOUN
exonerate—VERB
tertiary—ADJECTIVE
fortitude—NOUN
chalky—ADJECTIVE
rapidity—NOUN

seepage—NOUN
darken—VERB
perusal—NOUN
optimist—NOUN
miscible—ADJECTIVE
victimize—VERB
aqueous—ADJECTIVE

❑ EXERCISE 36-2

1 impediment (N), bump (V)

2 emit (V), spectroscopically (ADV)

3 Appalachian (ADJ), beneficiation (N), inapplicability (N), drawback (N)

❏ EXERCISE 36-3

detoxify: to remove the poison (toxin) from
reflux: a flowing back
circumspect: cautious, prudent (taking care to "look around")
interpose: to put between, to intervene
subversion: the act of undermining ("turning under") or overthrowing an
    established system from underneath

❏ EXERCISE 36-4

A   *sites:* places (used as a synonym of the sentence subject)
    compost heaps: piles or heaps of vegetable matter (leaves, grass clippings,
       garbage, etc.) left to decompose and turn into fertilizer

B   *teeming:* many, overflowing ("teeming millions" is a synonym for "great
      population" in this context)
    urban: municipal (see preceding sentence)
    prematurely: before the appropriate time has arrived, contrary to "prom-
      ises" (see preceding sentence)
    potable: drinkable (see other words in parallel and guess at meaning)

❏ EXERCISE 36-5

watchdog: guardian
exempt: excluded
incentive: initiative, desire
sweatshops: factories or workplaces with poor working conditions
longshoremen: dockworkers (men who work "along the shore")
adverse: dangerous, undesirable
goggles: eye protectors
bogged down: slowed down, sidetracked
(*letting* the major issues) *slide:* (letting the major issues) go unchecked
earmuffs: ear protectors
touchy: difficult, delicate

❏ EXERCISE 36-6

A   *uneasy:* concerned, worried about
    raw water: unprocessed water in waterways and in the ground
    untreated sewage: raw waste material carried in sewers
    deterioration: worsening
    debris: pieces or remains of things
    "dead": unable to support life, esp. large life forms like fish
    scum: a thin layer of garbage floating on liquid
    public complacence: people's confidence

insidious: secretly harmful
aquatic: pertaining to waterways
bans: prohibitions
trace quantities: very small but measurable quantities
carcinogenic: cancer-causing
was linked to: was found to be related to or caused by

B *fuzzy:* undefined, unclear
mutually: together, in a reciprocal manner
glancing: indirect
intact: in one piece, unbroken
head-on: direct
crucial: vitally important
touchstone: basis, foundation
credible: believable
plausible: reasonable
ricochet: bounce, change direction
grazed: barely touched while passing
likens: compares
blindfolded: having one's eyes covered with a (folded) handkerchief

Author Index

738

Whitley, M. Stanley, 495
Whitten, Mary E., 669
Wiegand, V. K., 427
Wikstrom, T., 353
Williams, Joseph M., 485, 669
Wilson, Albert G., 577
Winkler, Victoria M., 33, 53
Winsor, Dorothy, 39, 54

Witte, Stephen P., 468
Wolff, M. F., 529
Wright, Patricia, 161, 184, 331, 353
Wylie, Brian D., 194, 195

Yankelovich, Nicole, 119, 123
Young, J. G., 454

Young, Richard E., 53, 71, 84, 95, 96, 108

Zappen, James P., 55
Zemke, Ron, 392
Zimmer, K., 586
Zimmerman, Guy, 577

Subject Index

Two-way nouns, 510–511

Unblocked format, 208, 212–214
Uncountable nouns, 503–504, 508–509
Unity, 408
Usage, 605
User testing, 405

Ventura, 120
Verb complements, 539
Verb tense, 414, 531–535
Verbs, 531–541
  irregular, 538–541
  modal, 542–551
  passive, 538
  stative, 536
Visual aids, 137–183, 340, 399–402, 599–600
  bar graphs, 152–154, 163–165
  clarity, 168–170
  computer experimentation, 156

conventional presentation, 145–146
creation of, 162–184
description and clarification, 142–144
design of, 162–172
highlighting points, 144–145
integration into text, 172–173
line drawings, 155–157
line graphs, 151, 152, 163, 165–169
making truly visual, 139–141
photographs, 153, 155, 157
pie diagrams, 152, 153, 155
relevance, 162–167
selecting type, 147–158
tables, 153, 156
truth of, 168–172
types and uses, 151
when to use, 141–146

Visual elements, 137–183
  creation of, 162–184
  formatting, 176–183
Visual perception, conventions of, 147–150
Vocabulary building, 494, 587–609
Vowels, 629–632

Warnings, 343–346
Whistleblowing, 91–94
White space, 398
whiz-deletion, 559
Word prefixes, 594–596
Word processing, 110–123, 129, 130
Word stems, 593–594
Word stress, 640–644
Wordiness, 405, 470, 478–479, 481–483, 668
Words, 486–495, 587–609
Working memory, 396–397, 442–443, 457
Writer's block, 111